UNIT CONVERSION FACTORS   (continued)

Power

$$1\ \text{W} = 0.239\ \text{cal/s} \qquad 1\ \text{cal/s} = 4.184\ \text{W}$$
$$1\ \text{W} = 3.414\ \text{Btu/h} \qquad 1\ \text{Btu/h} = 0.293\ \text{W}$$
$$1\ \text{cal/s} = \text{:} \qquad\qquad\qquad\qquad \text{cal/s}$$

Viscosity

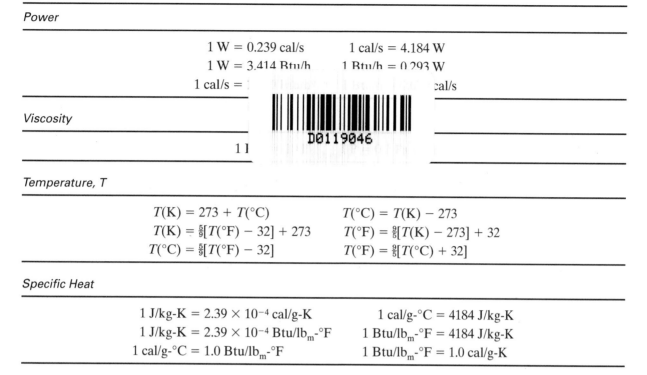

Temperature, T

$$T(\text{K}) = 273 + T(^\circ\text{C}) \qquad\qquad T(^\circ\text{C}) = T(\text{K}) - 273$$
$$T(\text{K}) = \tfrac{5}{9}[T(^\circ\text{F}) - 32] + 273 \qquad T(^\circ\text{F}) = \tfrac{9}{5}[T(\text{K}) - 273] + 32$$
$$T(^\circ\text{C}) = \tfrac{5}{9}[T(^\circ\text{F}) - 32] \qquad\qquad T(^\circ\text{F}) = \tfrac{9}{5}[T(^\circ\text{C}) + 32]$$

Specific Heat

$$1\ \text{J/kg-K} = 2.39 \times 10^{-4}\ \text{cal/g-K} \qquad 1\ \text{cal/g-}^\circ\text{C} = 4184\ \text{J/kg-K}$$
$$1\ \text{J/kg-K} = 2.39 \times 10^{-4}\ \text{Btu/lb}_\text{m}\text{-}^\circ\text{F} \qquad 1\ \text{Btu/lb}_\text{m}\text{-}^\circ\text{F} = 4184\ \text{J/kg-K}$$
$$1\ \text{cal/g-}^\circ\text{C} = 1.0\ \text{Btu/lb}_\text{m}\text{-}^\circ\text{F} \qquad 1\ \text{Btu/lb}_\text{m}\text{-}^\circ\text{F} = 1.0\ \text{cal/g-K}$$

STANDARD PREFIXES, SYMBOLS, AND MULTIPLICATION FACTORS

Prefix	Symbol	Factor by Which Unit Has to Be Multiplied
Tera	T	10^{12}
Giga	G	10^{9}
Mega	M	10^{6}
Kilo	k	10^{3}
Hecto	h	10^{2}
Deca	da	10^{1}
Deci	d	10^{-1}
Centi	c	10^{-2}
Milli	m	10^{-3}
Micro	μ	10^{-6}
Nano	n	10^{-9}
Pico	p	10^{-12}
Femto	f	10^{-15}
Atto	a	10^{-18}

MECHANICAL BEHAVIOR OF MATERIALS

Marc André Meyers

University of California, San Diego

Krishan Kumar Chawla

University of Alabama at Birmingham

Prentice Hall
Upper Saddle River, New Jersey 07458

Library of Congress Cataloging-in-Publication Data

Meyers, Marc A.
    Mechanical Behavior of Materials/Marc A. Meyers,
  Krishan K. Chawla
        p.   cm.
    Includes bibliographical references and index.
    ISBN: 0-13-262817-1
    1. Materials.   2. Mechanical engineering.
    I. Chawla, Krishan K.   II. Title.
    TA403.M554   1998
    620.1′12—dc21   98–22454
                        CIP

Acquisitions editor: **WILLIAM STENQUIST**
Editor-in-Chief: **MARCIA HORTON**
Production editor: **IRWIN ZUCKER**
Managing editor: **BAYANI MENDOZA DE LEON**
Director of production and manufacturing: **DAVID W. RICCARDI**
Copy editor: **BRIAN BAKER**
Cover director: **JAYNE CONTE**
Manufacturing buyer: **PAT BROWN**
Editorial assistant: **MARGARET WEIST**

 © 1999 by Prentice-Hall, Inc.
Simon & Schuster/A Viacom Company
Upper Saddle River, New Jersey 07458

The author and publisher of this book have used their best efforts in preparing this book. These efforts include the development, research, and testing of the theories and programs to determine their effectiveness. The author and publisher make no warranty of any kind, expressed or implied, with regard to these programs or the documentation contained in this book. The author and publisher shall not be liable in any event for incidental or consequential damages in connection with, or arising out of, the furnishing, performance, or use of these programs.

Printed in the United States of America

10 9 8 7 6 5 4 3 2 1

ISBN 0-13-262817-1

Prentice-Hall International (UK) Limited, London
Prentice-Hall of Australia Pty. Limited, Sydney
Prentice-Hall Canada Inc., Toronto
Prentice-Hall Hispanoamericana, S.A., Mexico
Prentice-Hall of India Private Limited, New Delhi
Prentice-Hall of Japan, Inc., Tokyo
Simon & Schuster Asia Pte. Ltd., Singapore
Editora Prentice-Hall do Brasil, Ltda., Rio de Janeiro

Lovingly dedicated to the memory of my mother, Marie-Anne.

Marc André Meyers

Lovingly dedicated to my parents, Manohar L. and the late Sumitra Chawla.

Krishan Kumar Chawla

We dance round in a ring and suppose.
But the secret sits in the middle and knows.

Robert Frost

Contents

3   PLASTICITY   112

Preface

Courses in the mechanical behavior of materials are standard in both mechanical engineering and materials science/engineering curricula. These courses are taught, usually, at the junior or senior level. This book provides an introductory treatment of the mechanical behavior of materials with a balanced mechanics–materials approach, which makes it suitable for both mechanical and materials engineering students. The book covers metals, polymers, ceramics, and composites and contains more than sufficient information for a one-semester course. It therefore enables the instructor to choose the path most appropriate to the class level (junior- or senior-level undergraduate) and background (mechanical or materials engineering). The book is organized into 15 chapters, each corresponding, approximately, to one week of lectures. It is often the case that several theories have been developed to explain specific effects; this book presents only the principal ideas. At the undergraduate level the simple aspects should be emphasized, whereas graduate courses should introduce the different viewpoints to the students. Thus, we have often ignored active and important areas of research. Chapter 1 contains introductory information on materials that students with a previous course in the properties of materials should be familiar with. In addition, it enables those students unfamiliar with materials to "get up to speed." The section on the theoretical strength of a crystal should be covered by all students. Chapter 2, on elasticity and viscoelasticity, contains an elementary treatment, tailored to the needs of undergraduate students. Most metals and ceramics are linearly elastic, whereas polymers often exhibit nonlinear elasticity with a strong viscous component. In Chapter 3, a broad treatment of plastic deformation and flow and fracture criteria is presented. Whereas mechanical engineering students should be fairly familiar with these concepts, (Section 3.2 can therefore be skipped), materials engineering students should be exposed to them. Two very common tests applied to materials, the uniaxial tension and compression tests, are also described. Chapters 4 through 9, on imperfections, fracture, and fracture toughness, are essential to the understanding of the mechanical behavior of materials and therefore constitute the core of the course. Point, line (Chapter 4), interfacial, and volumetric (Chapter 5) defects are discussed. The treatment is introductory and primarily descriptive. The mathematical treatment of defects is very complex and is not really essential to the understanding of the mechanical behavior of materials at an engineering level. In Chapter 6, we use the concept of dislocations to explain work hardening; our understanding of this phenomenon, which dates from the 1930s, followed by contemporary developments, is presented. Chapters 7 and 8 deal with fracture from a macroscopic (primarily mechanical) and a microstructural viewpoint, respectively. In brittle materials, the fracture strength under tension and compression can differ by a factor of 10, and this difference is discussed. The variation in strength from specimen to specimen is also significant and is analyzed in terms of Weibull statistics. In Chapter 9, the different ways in which the fracture resistance of materials can be tested is described. In Chapter 10, solid solution, precipitation, and dispersion strengthening, three very important mechanisms for strengthening metals, are presented. Martensitic transformation and toughening (Chapter 11) are very effective in metals and ceramics, respectively. Although this effect has been exploited for over 4,000 years, it is only in the second half of the 20th century that a true scientific understanding has been gained; as a result, numerous new applications have appeared, ranging from shape-memory alloys to maraging steels, that exhibit strengths higher than 2 GPa. Among novel materials with unique properties that have been developed for advanced applications are intermetallics, which often contain ordered structures. These are presented in Chapter 12. In Chapters 13 and 14, a detailed treatment of the fundamental mechanisms responsible for creep and fatigue, respectively, is presented. This is supplemented by a description of the principal testing and data analysis methods for these two phenomena. The last chapter of the book deals with composite materials. This important topic is, in some schools, the subject of a separate course. If this is the case, the chapter can be omitted.

This book is a spinoff of a volume titled *Mechanical Metallurgy* written by these authors and published in 1984 by Prentice-Hall. That book had considerable success in the United States and overseas, and was translated into Chinese. For the current volume, major changes and additions were made, in line with the rapid development of the field of materials in the 1980s and 1990s. Ceramics, polymers, composites, and intermetallics are nowadays important structural materials for advanced applications and are comprehensively covered in this book. Each chapter contains, at the end, a list of suggested readings; readers should consult these sources if they need to expand a specific point or if they want to broaden their knowledge in an area. Full acknowledgment is given in the text to all sources of tables and illustrations. We might have inadvertently forgotten to cite some of the sources in the final text; we sincerely apologize if we have failed to do so. All chapters contain solved examples and extensive lists of homework problems. These should be valuable tools in helping the student to grasp the concepts presented.

By their intelligent questions and valuable criticisms, our students provided the most important input to the book; we are very grateful for their contributions. We would like to thank our colleagues and fellow scientists who have, through painstaking effort and unselfish devotion, proposed the concepts, performed the critical experiments, and developed the theories that form the framework of an emerging quantitative understanding of the mechanical behavior of materials. In order to make the book easier to read, we have opted to minimize the use of references. In a few places, we have placed them in the text. The patient and competent typing of the manuscript by Jennifer Natelli, drafting by Jessica McKinnis, and editorial help with text and problems by H. C. (Bryan) Chen and Elizabeth Kristofetz are gratefully acknowledged. Krishan Chawla would like to acknowledge research support, over the years, from the U.S. Office of Naval Research, Oak Ridge National Laboratory, Los Alamos National Laboratory, and Sandia National Laboratories. He is also very thankful to his wife, Nivedita; son, Nikhilesh; and daughter, Kanika, for making it all worthwhile! Kanika's help in word processing is gratefully acknowledged. Marc Meyers acknowledges the continued support of the National Science Foundation (especially R. J. Reynik and B. MacDonald), the U.S. Army Research Office (especially G. Mayer, A. Crowson, K. Iyer, and E. Chen), and the Office of Naval Research. The inspiration provided by his grandfather, Jean-Pierre Meyers, and father, Henri Meyers, both metallurgists who devoted their lives to the profession, has inspired Marc Meyers. The Institute for Mechanics and Materials of the University of California at San Diego generously supported the writing of the book during the 1993–96 period. The help provided by Professor R. Skalak, director of the institute, is greatly appreciated. The Institute for Mechanics and Materials is supported by the National Science Foundation. The authors are grateful for the hospitality of Professor B. Ilschner at the École Polytechnique Fédérale de Lausanne, Switzerland during the last part of the preparation of the book.

Marc André Meyers
La Jolla, California

Krishan Kumar Chawla
Birmingham, Alabama

1

Materials: Structure, Properties, and Performance

1.1 INTRODUCTION

Everything that surrounds us is matter. The origin of the word matter is *mater* (Latin) or *matri* (Sanskrit), for *mother*. In this sense, human beings anthropomorphized that which made them possible—that which gave them nourishment. Every scientific discipline concerns itself with matter. Of all matter surrounding us, a portion comprises materials. What are materials? They have been variously defined. One acceptable definition is "matter that human beings use and/or process." Another definition is "all matter used to produce manufactured or consumer goods." In this sense, a rock is not a material, intrinsically; however, if it is used in aggregate (concrete) by humans, it becomes a material. The same applies to all matter found on earth: A tree becomes a material when it is processed and used by people, and a skin becomes a material once it is removed from its host and shaped into an artifact.

The successful utilization of materials requires that they satisfy a set of properties. These properties can be classified into thermal, optical, mechanical, physical, chemical, and nuclear, and they are intimately connected to the structure of materials. The structure, in its turn, is the result of synthesis and processing. A schematic framework that explains the complex relationships in the field of the mechanical behavior of materials, shown in Figure 1.1, is Thomas's iterative tetrahedron, which contains four principal elements: mechanical properties, characterization, theory, and processing. These elements are related, and changes in one are inseparably linked to changes in the others. For example, changes may be introduced by the synthesis and processing of, for instance, steel. The most common metal, steel has a wide range of strengths and ductilities (*mechanical properties*), which makes it the material of choice for numerous applications. While low-carbon steel is used as reinforcing bars in concrete and in the body of automobiles, quenched and tempered high-carbon steel is used in more critical applications such as axles and gears. Cast iron, much more brittle, is used in a variety of applications, including automobile engine blocks.

1

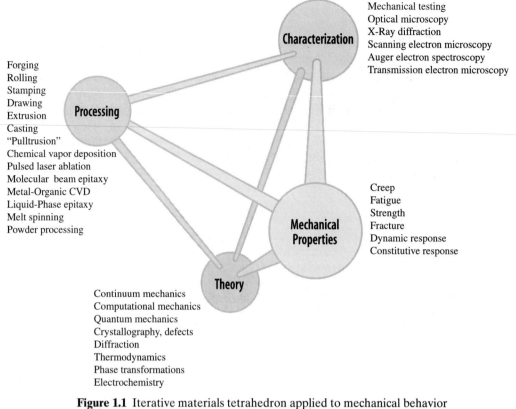

Forging
Rolling
Stamping
Drawing
Extrusion
Casting
"Pulltrusion"
Chemical vapor deposition
Pulsed laser ablation
Molecular beam epitaxy
Metal-Organic CVD
Liquid-Phase epitaxy
Melt spinning
Powder processing

Mechanical testing
Optical microscopy
X-Ray diffraction
Scanning electron microscopy
Auger electron spectroscopy
Transmission electron microscopy

Creep
Fatigue
Strength
Fracture
Dynamic response
Constitutive response

Continuum mechanics
Computational mechanics
Quantum mechanics
Crystallography, defects
Diffraction
Thermodynamics
Phase transformations
Electrochemistry

Figure 1.1 Iterative materials tetrahedron applied to mechanical behavior of materials. (Courtesy of G. Thomas)

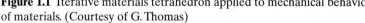

These different applications require, obviously, different mechanical properties of the material. The different properties of the three materials, resulting in differences in performance, are due to differences in the internal structure of the materials. The understanding of the structure comes from *theory*. The determination of the many aspects of the micro-, meso-, and macrostructure of materials is obtained by *characterization*. Low-carbon steel has a primarily ferritic structure (body-centered cubic; see section 1.3.1), with some interspersed pearlite (a ferrite–cementite mixture). The high hardness of the quenched and tempered high-carbon steel is due to its martensitic structure (body-centered tetragonal). The relatively brittle cast iron has a structure resulting directly from solidification, without subsequent mechanical working such as hot rolling. How does one obtain low-carbon steel, quenched and tempered high-carbon steel, and cast iron? By different *synthesis* and *processing* routes. The low-carbon steel is processed from the melt by a sequence of mechanical working operations. The high-carbon steel is synthesized with a greater concentration of carbon (>0.5%) than the low-carbon steel is (0.1%). Additionally, after mechanical processing, the high-carbon steel is rapidly cooled from a temperature of approximately 1,000°C by throwing it into water or oil; it is then reheated to an intermediate temperature (tempering). The cast iron is synthesized with even higher carbon contents (~2%). It is poured directly into the molds and allowed to solidify in them. Thus, no mechanical working, except for some minor machining, is needed. These interrelationships among structure, properties, and performance, and their modification by synthesis and processing, constitute the central theme of materials science and engineering. The tetrahedron of Figure 1.1 lists the principal processing methods, the most important theoretical approaches, and the most used characterization techniques in materials science today.

   The selection, processing, and utilization of materials have been part of human culture since its beginnings. Anthropologists refer to humans as "the toolmakers," and this is indeed a very realistic description of a key aspect of human beings responsible for their ascent and domination over other animals. It is the ability of humans to manufacture and use tools, and the ability to produce manufactured goods, that has allowed technological, cultural, and artistic progress and that has led to civilization and its development. Materials were as important to a Neolithic tribe in the year 10,000 B.C. as they are to us today. The only difference is that today more complex synthetic materials are available in our society, while Neolithics had only natural materials at their disposal: wood, minerals, bones, hides, and fibers from plants and animals. Although these naturally occurring materials are still used today, they are vastly inferior in properties to synthetic materials.

1.2 *MONOLITHIC, COMPOSITE, AND HIERARCHICAL MATERIALS*

The early materials used by humans were natural, and their structure varied widely. Rocks are crystalline, pottery is a mixture of glassy and crystalline components, wood is a fibrous organic material with a cellular structure, and leather is a complex organic material. Human beings started to synthesize their own materials in the neolithic ceramics first, then metals, and later, polymers. In the 20th century, simple monolithic structures were used first. The term *monolithic* comes from the Greek *mono* (one) and *lithos* (stone). It means that the material has essentially uniform properties throughout. Microstructurally, monolithic materials can have two or more phases. Nevertheless, they have properties (electrical, mechanical, optical, and chemical) that are constant throughout. Table 1.1 presents some of the important properties of metals, ceramics, and polymers. Their detailed structures will be described in Section 1.3. The differences in their structure are responsible for differences in properties. Metals have densities ranging from 3 to 19 g cm^{-3}; iron, nickel, chromium, and niobium have densities ranging from to 7 to 9 g cm^{-3}; aluminum has a density of 2.7 g cm^{-3}; and titanium has a density of 4.5 g cm^{-3}. Ceramics tend to have lower densities, ranging from 5 g cm^{-3} (titanium carbide; TiC = 4.9) to 3 g cm^{-3} (alumina; Al$_2$O$_3$ = 3.95; Silicon carbide; SiC = 3.2). Polymers have the lowest densities, fluctuating around 1 g cm^{-3}. Another marked difference among these three classes of materials is their ductility (ability to undergo plastic deformation). At room temperature, metals can undergo significant plastic deformation. Thus, metals tend to be ductile, although there are a number of exceptions. Ceramics, on the other hand, are very brittle, and the most ductile ceramics will be more brittle than most metals. Polymers have a behavior ranging from brittle (at temperatures below their glass transition temperature) to very deformable (in a nonlinear elastic material, such as rubber). The fracture toughness is a good measure of the resistance of a material to failure and is generally quite high for metals and low for ceramics and polymers. Ceramics far outperform metals and polymers in high-temperature applications, since many ceramics do not oxidize even at very high temperatures (the oxide ceramics are already oxidized) and retain their strength to such temperatures. One can compare the mechanical, thermal, optical, electrical, and electronic properties of the different classes of materials and see that there is a very wide range of properties. Thus, monolithic structures built from primarily one class of material cannot provide all desired properties.

   In the field of biomaterials (materials used in implants and life-support systems), developments also have had far-reaching effects. The mechanical performance of implants is critical in many applications, including hipbone implants, which are subjected to high stresses, and endosseous implants in the jaw designed to serve as the base for teeth.

TABLE 1.1 Summary of Properties of Main Classes of Materials

Property	Metals	Ceramics	Polymers
Density (g/cm³)	from 2 to 20	from 1 to 14	from 1 to 2.5
Electrical conductivity	high	low	low
Thermal conductivity	high	low	low
Ductility or strain-to-fracture ratio (%)	4–40	<1	2–4
Tensile strength (MPa)	100–1,500	100–400	—
Compressive strength (MPa)	100–1,500	1,000–5000	—
Fracture toughness (MNm$^{-3/2}$)	10–30	1–10	2–8
Maximum service temperature (°C)	1,000	1,800	250
Corrosion resistance	low to medium	superior	medium
Bonding	metallic (free-electron cloud)	ionic or covalent	covalent
Structure	mostly crystalline (Face-centered cubic-FCC Body-centered cubic-BCC; Hexagonal closed packed-HCP)	complex crystalline structure	amorphous or semicrystalline polymer

Figure 1.2 shows the most successful design for endosseous implants in the jawbone. With this design, the tooth is fixed to the post and is effective. A titanium post is first screwed into jawbone and allowed to heal. The tooth is then fixed to the post, and is effectively rooted into the jaw. *Biocompatibility* is a major concern for all implants, and ceramics are especially attractive because of their (relative) chemical inertness. Metallic alloys such as Vitallium® (a cobalt-based alloy) and titanium alloys also have proved to be successful, as have polymers such as polyethylene. A titanium alloy with a solid core surrounded by a porous periphery (produced by sintering of powders) has shown considerable potential. The porous periphery allows bone to grow and affords very effective fixation. Two new classes of materials that appear to present the best biocompatibility with bones are the Bioglass® and calcium phosphate ceramics. Bones contain calcium and phosphorus, and Bioglass is a glass in which the silicon has been replaced by those two elements. Thus, the bone "perceives" these materials as being another bone and actually bonds with it. Biomechanical properties are of great importance in bone implants, as are the elastic properties of materials. If the stiffness of a material is too high, then when implanted, the material will carry most of the load placed on it and the adjacent bone. This could in turn lead to a weakening of the bone, since bone growth and strength are dependent on the stresses that the bone is subjected to. Thus, the elastic properties of bone and implant should be similar. Polymers reinforced with strong carbon fibers are also candidates for such applications. Metals, on the other hand, are stiffer than bones and tend to carry most of the load. With metals, the bones would be shielded from stress, which could lead to bone resorption and loosening of the implant.

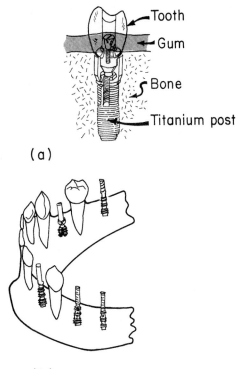

Figure 1.2 (a) Endosseous implants for anchoring artificial teeth and bridges. (b) Details of implant. (R. Skalak, private communication.)

Although new materials are being developed continuously, monolithic materials, with their uniform properties, cannot deliver the range of performance needed in many critical applications. *Composites* are a mixture of two classes of materials (metal–ceramic, metal–polymer, or polymer–ceramic). They have unique mechanical properties that are dependent on the amount and manner in which their constituents are arranged. Figure 1.3 shows schematically how different composites can be formed. Composites consist of a matrix and a reinforcing material. In making them, the modern materials engineer has at his or her disposal a very wide range of possibilities. However, the technological problems involved in producing some of them are immense, although there is a great deal of research addressing those problems. Figure 1.4 shows three principal kinds of reinforcement in composites: particles, continuous fibers, and discontinuous (short) fibers. The reinforcement usually has a higher strength than the matrix, which provides the ductility of the material. In ceramic-based composites, however, the matrix is brittle, and the fibers provide barriers to the propagating cracks, increasing the toughness of the material.

The alignment of the fibers is critical in determining the strength of a composite. The strength is highest along a direction parallel to the fibers and lowest along directions perpendicular to it. For the three kinds of composite shown in Figure 1.4, the polymer matrix plus (aramid, carbon, or glass) fiber is the most common combination if no high-temperature capability is needed.

Composites are becoming a major material in the aircraft industry. Carbon/epoxy and aramid/epoxy composites are being introduced in a large number of aircraft parts. These composite parts reduce the weight of the aircraft, increasing its economy and payload. The major mechanical property advantages of advanced composites over metals are better stiffness-to-density and strength-to-density ratios and greater resistance to fatigue. The values

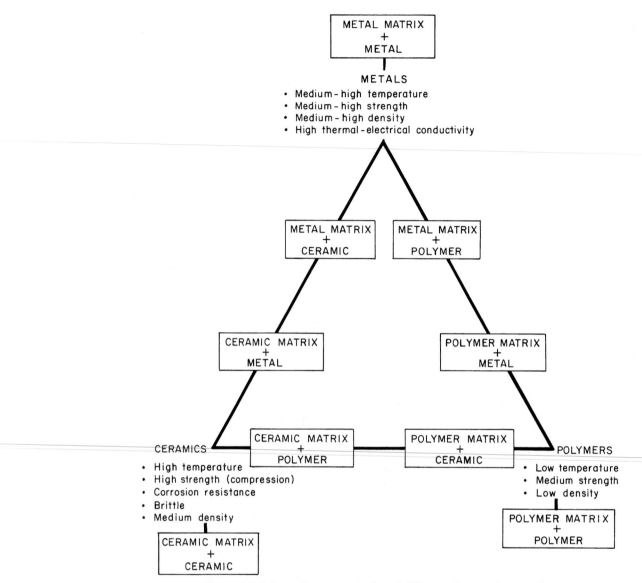

Figure 1.3 Schematic representation of different classes of composites.

given in Table 1.2 apply to a unidirectional composite along the fiber reinforcement orientation. The values along other directions are much lower, and therefore, the design of a composite has to incorporate the anisotropy of the materials. It is clear from the table that composites have advantages over monolithic materials. In most applications, the fibers are arranged along different orientations in different layers. For the central composite of Figure 1.4, these orientations are 0, 45°, 90°, and 135° to the tensile axis.

Can we look beyond composites in order to obtain even higher mechanical performance? Indeed, we can: Nature is infinitely imaginative.

Our body is a complex arrangement of parts, designed, as a whole, to perform all the tasks needed to keep us alive. Scientists are looking into the make up of soft tissue (skin, tendon, intestine, etc.), which is a very complex structure with different units active at different levels complementing each other. The structure of soft tissue has been called a *hierarchical* structure, because there seems to be a relationship between the ways in which it operates at

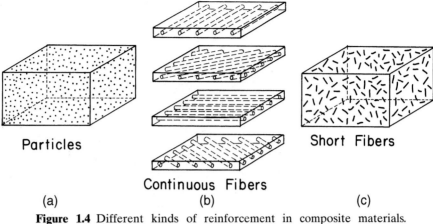

Particles   Continuous Fibers   Short Fibers

(a)   (b)   (c)

Figure 1.4 Different kinds of reinforcement in composite materials. (a) Composite with particle reinforcement. (b) Composite with continuous fibers with four different orientations (shown separately for clarity). (c) Composite reinforced with short, discontinuous fibers.

different levels. Figure 1.5 shows the structure of a tendon. This structure begins with the tropocollagen molecule, a triple helix of polymeric protein chains. The tropocollagen molecule has a diameter of approximately 1.5 mm. The tropocollagen organizes itself into microfibrils, subfibrils, and fibrils. The fibrils, a critical component of the structure, are crimped when there is no stress on them. When stressed, they stretch out and then transfer their load to the fascicles, which compose the tendon. The fascicles have a diameter of approximately 150–300 μm and constitute the basic unit of the tendon. The hierarchical organization of the tendon is responsible for its toughness. Separate structural units can fail independently and thus absorb energy locally, without causing the failure of the entire tendon. Both experimental and analytical studies have been done, modeling the tendon as a composite of elastic, wavy fibers in a viscoelastic matrix. Local failures, absorbing energy, will prevent catastrophic failure of the entire tendon until enormous damage is produced.

Materials engineers are beginning to look beyond simple two-component composites, imitating nature in organizing different levels of materials in a hierarchical manner. E. Baer

TABLE 1.2 Specific Modulus and Strength of Materials Used in Aircraft

Material	Elastic Modulus Density (GPa/g cm^{-3})	Tensile Strength Density (MPa/g cm^{-3})
Steel (AISI 4340)	25	230
Al (7075-T6)	25	180
Titanium (Ti-6Al-4V)	25	250
E Glass/Epoxy composite	21	490
S Glass/Epoxy composite	47	790
*Kevlar®/Epoxy composite	55	890
HS (High Tensile–Strength) Carbon/Epoxy composite	92	780
HM (high modulus) Carbon/Epoxy composite	134	460

*Kevlar is an aramid (aromatic polyamide) fiber produced by Dupont; There are two types: Kevlar 29 (for ballistic protection, ropes, cables) and Kevlar 49 (aerospace, marine, automotive applications).

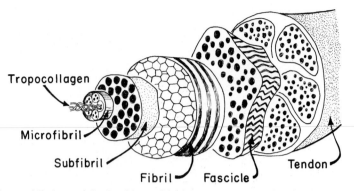

Figure 1.5 A model of a hierarchical structure occurring in the human body. (Adapted from E. Baer, *Sci. Am.* 254, No. 10 (1986) 179)

(*Sci. Am.* 254, No. 10 (1986) 179) suggests that the study of biological materials could lead to new hierarchical designs for composites. One such example is shown in Figure 1.6, a layered structure of liquid-crystalline polymers consisting of alternating core and skin layers. Each layer is composed of sublayers which, in their turn, are composed of microlayers. The molecules are arranged in different arrays in different layers. The lesson that can be learned from this arrangement is that we appear to be moving toward composites of increasing complexity.

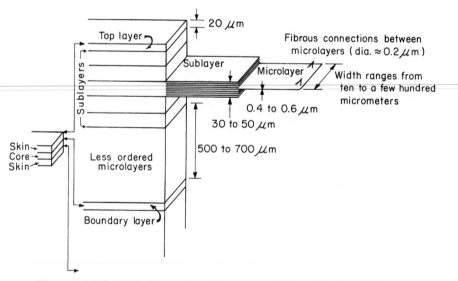

Figure 1.6 Schematic illustration of a proposed hierarchical model for a composite (not drawn to scale). (Courtesy of E. Baer)

EXAMPLE 1.1

Discuss advanced materials used in bicycle frames.

This is a good case study, and the instructor can "pop" similar questions on an exam, using different products. For our specific example here, we recommend the insightful article by M. F. Ashby, *Met. and Mat. Trans.*, A 26A (1995) 3057. Ashby states that "*Materials and processes underpin all engineering design.*"

Figure E1.1.1 shows a bicycle, with forces F_1 and F_2 applied to the frame by the pedals. These forces produce bending moments and torsions in the frame tubes. In bicycle frames, *weight* and *stiffness* are the two primary requirements. Stiffness

Figure E1.1 Bending moments (M_1 and M_2) and torsional torques (T_1 and T_2) generated in bicycle frame by forces F_1 and F_2 applied to pedals.

is important because excessive flexing of the bicycle upon pedaling absorbs energy that should be used to propel the bicycle forward. This requires the definition of new properties, because just the strength or endurance limit (the stress below which no failure due to fatigue occurs) and Young's modulus (defined in Chapter 2) are not sufficient. In conversations, we always say that aluminum bicycles are "stiffer" than steel bicycles, whereas steel provides a more "cushioned" ride. An aluminum bicycle can indeed be stiffer than a steel bicycle, although E_{st} ($= 210$ GPa) $\approx 3\,E_{Al}$ ($= 70$ GPa). We will see shortly how this can happen and what is necessary for it to occur. The forces F_1 and F_2 cause bending moments (M_1 and M_2), respectively. The bending stresses in a hollow tube of radius r and thickness t are[1]

$$\sigma = \frac{Mr}{I},$$

where I is the moment of inertia, M the bending moment, and r the radius of the tube. Setting $\sigma = \sigma_e$, the endurance limit, and substituting the expression for the moment of inertia $I = \pi r^3 t$, we obtain the thickness of the tube, t, from:

$$M = \frac{\sigma_e \pi r^3 t}{r}.$$

From strength considerations, the mass per unit length of the bicycle frame is

$$\frac{m}{L} = 2\pi r t \rho = \frac{2M}{r}\left(\frac{\rho}{\sigma_e}\right) \tag{E 1.1.1}$$

[1]Students should consult their notes on the mechanics of materials or examine a book such as *Engineering Mechanics of Solids,* by E. P. Popov (Englewood Cliffs, NJ: Prentice Hall, 1990).

where ρ is the density of the frame. Now, the radius of curvature ρ' of a circular beam under bending is given by the Bernoulli–Euler equation,

$$\frac{1}{\rho'} = \frac{d^2\nu}{dx^2} = \frac{M}{EI},$$

where ν is the deflection of the beam. Substituting for I, we obtain

$$\frac{1}{\rho'} = \frac{M}{E\pi r^3 t}, \quad \text{or} \quad \pi r t = \frac{M\rho'}{r^2 E}.$$

From bending considerations, the mass per unit length is

$$\frac{m}{L} = 2\pi r t \rho = \frac{2M\rho'}{r^2}\left(\frac{\rho}{E}\right) \tag{E 1.1.2}$$

A similar expression can be developed for the torsion, which is important in pedaling. The torsion is shown in Figure E1.1.1 as T_1 and T_2. Since M, the applied moment, is given by the weight of cyclist, it is constant for each frame. Likewise, the maximum curvature $1/\rho'$ can be fixed. The quantity m/L has to be minimized for both strength and stiffness considerations. Ashby accomplished this by plotting (σ_e/ρ) and (E/ρ), whose reciprocals appear in Equations (E1.1.1) and (E1.1.2), respectively. (See Figure E1.1.2.) The computations assume a con-

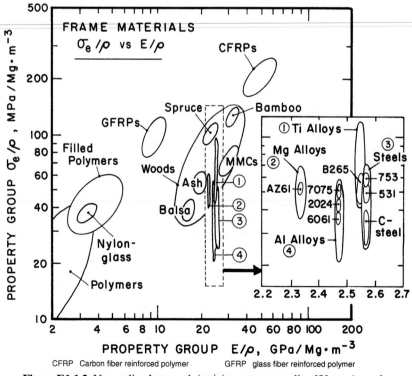

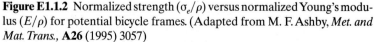

Figure E1.1.2 Normalized strength (σ_e/ρ) versus normalized Young's modulus (E/ρ) for potential bicycle frames. (Adapted from M. F. Ashby, *Met. and Mat. Trans.*, **A26** (1995) 3057)

stant r, but varying tube thickness t. The most common candidate metals (steels, titanium, and aluminum alloys) are closely situated in the figure. The expanded window in this region shows a clearer separation of the various alloys. Continuous carbon/fiber reinforced composites (CFRPs) are the best materials, and polymers and glass/fiber reinforced polymer composites (GFRPs) have insufficient stiffness. By relaxing the requirement of constant r and allowing different tube radii, the results are changed considerably. This example illustrates how material properties enter into the design of a product and how compound properties $(E/\rho, \sigma/\rho)$ need to be defined for a specific application. It can be seen from Equations (E1.1.1) and (E1.1.2) that strength scales with r and stiffness with r^2. By varying r, it is possible to obtain aluminum bicycle frames that are stiffer than steel. Now the student is prepared to go on a bike ride!

EXAMPLE 1.2

Suppose you are a design engineer for the ISAACS bicycle company. This company traditionally manufactures chromium–molybdenum (Cr–Mo) steel frames. The racing team is complaining that the bicycles are too "soft" and that stiffer bicycles would give them a competitive edge. Additionally, the team claims that competing teams have aluminum bikes which are considerably lighter. You are asked to redesign the bikes, using a precipitation hardenable aluminum alloy (7075 H4).

a) Calculate the ratio of the stiffness of the two bikes if the tube diameters are the same.
b) What would you do to increase the stiffness of the two bikes?
c) If the steel frame weighs 4 kg, what would the aluminum frame weigh? State your assumptions.

Given:

	σ_e (MPa)	Density (kg/m³)	E (GPa)	G (GPa)
7075 Al	500	2700	70	27
4340 Steel	1350	7800	210	83

Steel tube diameter, $2r = 25$ mm
Wall thickness, $t = 1.25$ mm

Solution: The mass per unit length, from strength considerations, is

$$\frac{m}{L} = 2\pi r t \rho = \frac{2M}{r}\left(\frac{\rho}{\sigma_e}\right).$$

The mass per unit length, from bending considerations, is

$$\frac{m}{L} = 2\pi r t \rho = \frac{2M\rho'}{r^2}\left(\frac{\rho}{E}\right).$$

where ρ' is the radius of curvature and M is the bending moment applied by cyclist.

The radius of curvature ρ' is a good measure of the stiffness; the larger ρ', the higher is the stiffness, for a fixed M.

a) $r_{Al} = r_{St} = 12.5$ mm.

For the two metals, we have:

	Steel	Aluminum
ρ/σ_e	5.77	5.4
ρ/E	37.14	38.57

The mass-to-length ratios are

$$\frac{\left(\dfrac{m}{L}\right)_{st}}{\left(\dfrac{m}{L}\right)_{Al}} = \frac{\dfrac{2M}{r}\left(\dfrac{\rho}{\sigma_e}\right)\Big|_{st}}{\dfrac{2M}{r}\left(\dfrac{\rho}{\sigma_e}\right)\Big|_{Al}} = 1.06.$$

For the same weight, we calculate the ratio of the radii of curvature from bending:

$$1.06\,\frac{\rho'_{Al}}{\rho'_{st}} = \frac{\left(\dfrac{\rho}{E}\right)_{st}}{\left(\dfrac{\rho}{E}\right)_{Al}} = 0.96,$$

$$\frac{\rho'_{Al}}{\rho'_{st}} = \frac{0.96}{1.06} = 0.91.$$

Thus, the stiffness is approximately the same for each metal.

b) we increase diameter of the tubes. This is possible because the wall thickness of aluminum bikes is approximately three times the wall thickness of steel bikes.[2] For instance, we can increase the diameter to 50 mm![3]

c) Let us assume that, for aluminum, $2r_{Al} = 50$ mm. Then

$$\frac{\left(\dfrac{m}{L}\right)_{st}}{\left(\dfrac{m}{L}\right)_{Al}} = x = \frac{\dfrac{2M\rho'_{st}}{r_{st}^2}\left(\dfrac{\rho}{E}\right)_{st}}{\dfrac{2M\rho'_{Al}}{r_{Al}^2}\left(\dfrac{\rho}{E}\right)_{Al}},$$

$$x\,\frac{\rho'_{Al}r_{st}^2}{\rho'_{st}r_{Al}^2} = \frac{\left(\dfrac{\rho}{E}\right)_{st}}{\left(\dfrac{\rho}{E}\right)_{Al}}.$$

[2]Since the wall thickness is larger, we can produce larger tube diameters without danger of collapse by buckling.
[3]A 50-mm steel tube would have walls that would be exceedingly thin; indeed, it could be dented by pressing it with the fingers.

Going back to the strength equation, we obtain

$$x = \frac{\left(\dfrac{m}{L}\right)_{st}}{\left(\dfrac{m}{L}\right)_{Al}} = \frac{\dfrac{2M}{r_{st}}\left(\dfrac{\rho}{\sigma_e}\right)_{st}}{\dfrac{2M}{r_{Al}}\left(\dfrac{\rho}{\sigma_e}\right)_{Al}} = 2\frac{5.77}{5.4} = 2.14.$$

If the total weight of the steel frame is 4 kg, then

$$\frac{w_{Al}}{w_{st}} = \frac{\left(\dfrac{m}{L}\right)_{Al}}{\left(\dfrac{m}{L}\right)_{st}} \cdot w_{st} = \frac{4}{2.14} = 1.86.$$

The stiffness ratio will be

$$\frac{\rho'_{Al}}{\rho'_{st}} = \frac{1}{x}\frac{r^2_{Al}}{r^2_{st}}\frac{\left(\dfrac{\rho}{E}\right)_{st}}{\left(\dfrac{\rho}{E}\right)_{Al}} = \frac{4}{2.14}\frac{37.14}{38.54} = 1.80,$$

or

$$\rho'_{Al} = 1.8\rho'_{st}.$$

The aluminum bike is almost twice as stiff!

1.3   STRUCTURE OF MATERIALS

The *crystallinity,* or periodicity, of a structure, does not exist in gases or liquids. Among solids, the metals, ceramics, and polymers may or may not exhibit it, depending on a series of processing and composition parameters. Metals are normally crystalline. However, a metal cooled at a superfast rate from its liquid state—called *splat cooled*—can have an amorphous structure. (This subject is treated in greater detail in Section 1.3.4.) Silicon dioxide (SiO_2) can exist as amorphous (fused silica) or as crystal (crystoballite or trydimite). Polymers consisting of molecular chains can exist in various degrees of crystallinity.

Readers not familiar with structures, lattices, crystal systems, and Miller indices should study these subjects before proceeding with the text. Most books on materials science, physical metallurgy, or X-rays treat the subjects completely. A brief introduction is presented next.

1.3.1   Crystal Structures

To date, seven crystal structures describe all the crystals that have been found. By translating the unit cell along the three crystallographic orientations, it is possible to construct a three-dimensional array. The translation of each unit cell along the three principal directions by distances that are multiples of the corresponding unit cell size produces the crystalline lattice.

Up to this point, we have not talked about atoms or molecules; we are just dealing with the mathematical operations of filling space with different shapes of blocks. We now introduce atoms and molecules, or "repeatable structural units." The unit cell is the smallest repetitive unit that will, by translation, produce the atomic or molecular arrangement. Bravais established that there are 14 space lattices. These lattices are based on the seven crystal structures. The points shown in Figure 1.7 correspond to atoms or groups of atoms. The 14 Bravais lattices can represent the unit cells for all crystals. Figure 1.8 shows the indices used for directions in the cubic system. The same symbols are employed for different structures. We simply use the vector passing through the origin and a point (m, n, o):

$$\mathbf{V} = m\mathbf{i} + n\mathbf{j} + o\mathbf{k}.$$

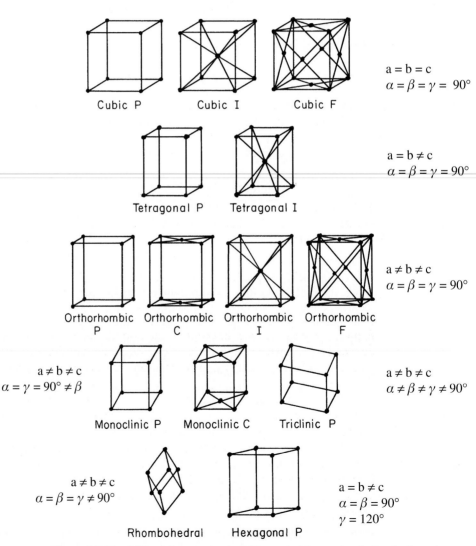

Figure 1.7 The 14 Bravais space lattices (P = primitive or simple; I = body-centered cubic; F = face-center cubic; C = base-centered cubic).

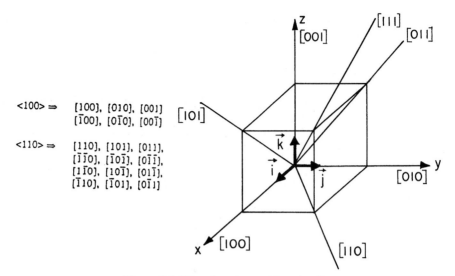

Figure 1.8 Directions in a cubic unit cell.

The notation used for a direction is

$$[m\, n\, o].$$

When we deal with a family of directions, we use the symbol $<m\, n\, o>$.
The following family encompasses all equivalent directions:

$$<m\, n\, o> \Rightarrow [m\, n\, o], [m\, o\, n], [o\, m\, n], [o\, n\, m], [n\, m\, o],$$

$$[m\, \bar{n}\, o], [m\, o\, \bar{n}], [o\, m\, \bar{n}], [o\, \bar{n}\, m], [\bar{n}\, m\, o], \dots$$

Note that for the negative, we use a bar on top. For planes, we use the Miller indices, obtained from the intersection of a plane with the coordinate axes. Figure 1.9 shows a plane and its intercepts. We take the inverse of the intercepts and multiply them by their common denominator so that we end up with integers. In Figure 1.9 (a), we have

$$\frac{1}{1}, \frac{1}{1}, \frac{1}{1/2} \Rightarrow (112).$$

Figure 1.9 (b) shows an indeterminate situation. Thus, we have to translate the plane to the next cell, or else translate the origin. The indeterminate situation arises because the plane passes through the origin. After translation, we obtain intercepts $(-1, 1, \infty)$. By inverting them, we get $(\bar{1}\, 10)$.

For hexagonal structures, we have a slightly more complicated situation. We represent the hexagonal structure by the arrangement shown in Figure 1.10. The atomic arrangement in the basal plane is shown in the top portion of the figure. Often, we use four axes (x, y, k, z) with unit vectors $(\vec{i}, \vec{j}, \vec{k}, \vec{l})$ to represent the structure. This is mathematically unnecessary, because three indices are sufficient to represent a direction in space from a known origin. Still, the redundancy is found by some people to have its advantages and is described here. We use the intercepts to designate the planes. The hatched plane (prism plane) has indices

$$\frac{1}{1}, \frac{1}{-1}, \frac{1}{\infty}, \frac{1}{\infty}.$$

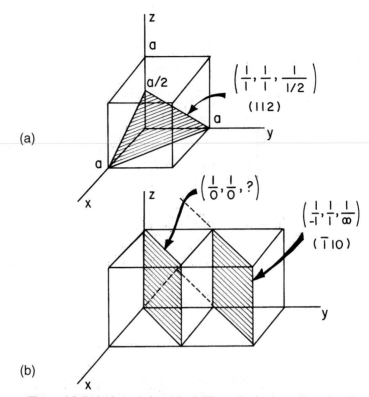

(a)

(b)

Figure 1.9 Indexing of planes by Miller rules in the cubic unit cell.

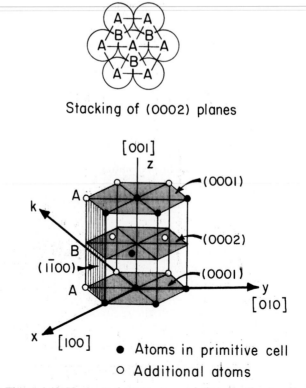

Stacking of (0002) planes

Figure 1.10 Hexagonal structure consisting of a three-unit cell.

After determining the indices of many planes, we learn that one always has

$$h + k = -i.$$

Thus, we do not have to determine the index for the third horizontal axis. if we use only three indices, we can use a dot to designate the fourth index, as follows:

$$(1\bar{1} \cdot 0).$$

For the directions, we can use either the three-index notation or a four-index notation. However, with four indices, the $h + k = -i$ rule will not apply in general, and one has to use special "tricks" to make the vector coordinates obey the rule.

EXAMPLE 1.3

Write the indices for the directions and planes marked in Figure E1.3.

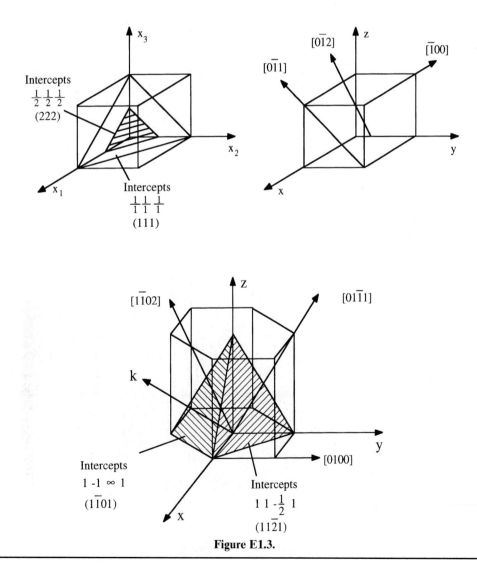

Figure E1.3.

1.3.2 Metals

The metallic bond can be visualized, in a very simplified way, as an array of positive ions held together by a "glue" consisting of electrons. These positive ions, which repel each other, are attracted to the "glue," which is known as an electron gas. Ionic and covalent bonding, on the other hand, can be visualized as direct attractions between atoms. Hence, these types of bonding—especially covalent bonding—are strongly directional and determine the number of neighbors that one atom will have, as well as their positions.

The bonding—and the sizes of the atoms in turn—determines the type of structure a metal has. Often, the structure is very complicated for ionic and covalent bonding. On the other hand, the directionality of bonding is not very important for metals, and atoms pack into the simplest and most compact forms; indeed, they can be visualized as spheres. The structures favored by metals are the face-centered cubic (FCC), body-centered cubic (BCC), and hexagonal close-packed (HCP) structures. In the periodic table, of the 81 elements to the right of the Zindl line, 53 have either the FCC or the HCP structure, and 21 have the BCC structure; the remaining 8 have other structures. The Zindl line defines the boundary of the elements with metallic character in the table. Some of them have several structures, depending on temperature. Perhaps the most complex of the metals is plutonium, which undergoes six polymorphic transformations.

Transmission election microscopy can reveal the positions of the individual atoms of a metal, as shown in Figure 1.11 for molybdenum. The regular atomic array along a [001] plane can be seen. Molybdenum has a BCC structure.

Figure 1.12 shows the three main metallic structures. The positions of the atoms are marked by small spheres and the atomic planes by dark sections. The small spheres do not correspond to the scaled-up size of the atoms, which would almost completely fill the available space, touching each other. For the FCC and HCP structures, the coordination number (the number of nearest neighbors of an atom) is 12. For the BCC structure, it is 8.

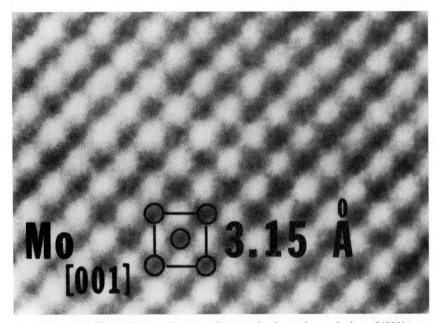

Figure 1.11 Transmission electron micrograph of atomic resolution of (001) plane in molybdenum showing body-centered cubic arrangement of atoms. (Courtesy of R. Gronsky)

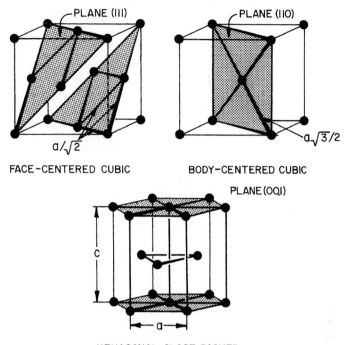

Figure 1.12 Most closely packed planes in (a) FCC; (b) BCC; (c) HCP.

The planes with the densest packing are indicated in the figure. They are $(1\bar{1}1)$, $(1\bar{1}0)$, and (00.1) for the FCC, BCC, and HCP structures, respectively. These planes have an important effect on the directionality of deformation of the metal, as will be seen in Chapters 4 and 6. The distances between the nearest neighbors are also indicated in the figure. The reader should try to calculate them as an exercise. These distances are $a/\sqrt{2}$, $(a\sqrt{3}/2)$, and a for the FCC, BCC, and HCP structures, respectively.

The similarity between the FCC and HCP structures is much greater than might be expected from looking at the unit cells. Planes (111) and (00.1) have the same packing, as can be seen in Figure 1.13. This packing, the densest possible of coplanar spheres, is shown in Figure 1.13(a). The packing of a second plane similar to, and on top of, the first one (called A) can be made in two different ways; Figure 1.13 (b) indicates these two planes by the letters B and C. Hence, either alternative can be used. A third plane, when placed on top of plane B, would have two options: A or C. If the second plane is C, the third plane can be either A or B. If only the first and second layers are considered, the FCC and HCP structures are identical. If the position of the third layer coincides with that of the first (the ABA or ACA sequence), we have the HCP structure. Since this packing has to be systematically maintained in the lattice, one would have $ABABAB \ldots$ or $ACACAC \ldots$ In case the third plane does not coincide with the first, we have one of the two alternatives ABC or ACB. Since this sequence has to be systematically maintained, we have $ABCABCABC \ldots$ or $ACBACBACB \ldots$ This stacking sequence corresponds to the FCC structure. We thus conclude that the only difference between the FCC and HCP structures (the latter with a theoretical c/a ratio of 1.633) is the stacking sequence of the most densely packed planes. The difference resides in the next neighbors and in the greater symmetry of the FCC structure.

Figures 1.13 (c) and (d) show photographs of ideal ball stackings. The $ABA \ldots$ sequence of layers, characteristic of HCP structure (Figure 1.13(c)) is compared with the $ABCA \ldots$ sequence for the HCP structure (Figure 1.13(d)).

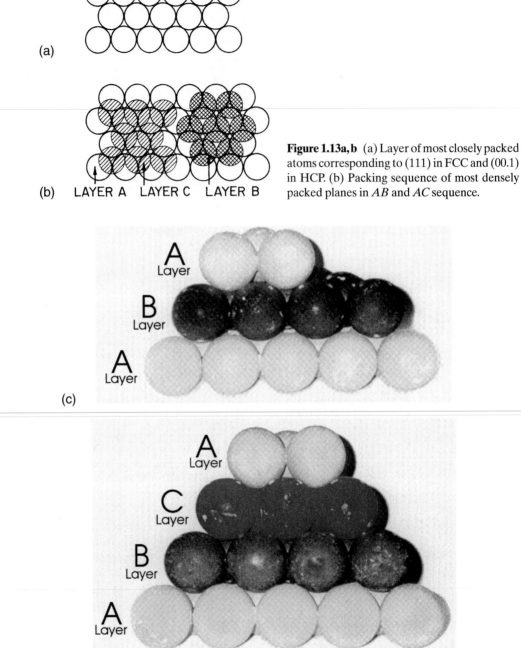

(a)

(b)   LAYER A   LAYER C   LAYER B

Figure 1.13a, b (a) Layer of most closely packed atoms corresponding to (111) in FCC and (00.1) in HCP. (b) Packing sequence of most densely packed planes in *AB* and *AC* sequence.

(c)

(d)

Figure 1.13c, d (c) Photograph of ball model showing the *ABAB* sequence of the HCP structure. (d) Photograph of ball model showing the *ABCABC* sequence of the FCC structure.

In addition to the metallic elements, intermediate phases and intermetallic compounds exist in great numbers, with a variety of structures. For instance, the beta phase in the copper–manganese–tin (Cu–Mn–Sn) system exhibits a special ordering for the composition Cu_2MnSn. The unit cell (BCC) is shown in Figure 1.14. However, the ordering of the Cu, Mn, and Sn atoms creates a superlattice composed of four BCC cells. This super-

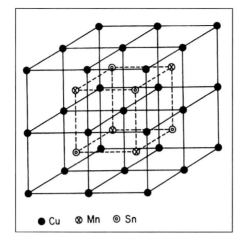

● Cu   ⊗ Mn   ◎ Sn

Figure 1.14 β-ordered phase in Heusler alloys (Cu_2MnSn). (Reprinted with permission from M. A. Meyers, C. O. Ruud, and C. S. Barrett, *J. Appl. Cryst,* 6 (1973) 39)

lattice is FCC; hence, the unit cell for the ordered phase is FCC, whereas that for the disordered phase has a BCC unit cell. This ordering has important effects on the mechanical properties and is discussed in Chapter 11.

Table 1.3 lists some of the most important intermetallic compounds and their structures. Intermetallic compounds have a bonding that is somewhat intermediate between metallic and ionic/covalent bonding, and have properties that are most desirable for high-temperature applications. Nickel and titanium aluminides are candidates for high-temperature applications in jet turbines and airplane applications.

TABLE 1.3 Some Important Intermetallic Compounds and Their Structure.

Compound	Melting Point (°C)	Type of Structure
Ni_3Al	1,390	LI_2 (ordered FCC)
Ti_3Al	1,600	DO_{19} (ordered hexagonal)
TiAl	1,460	LI_0 (ordered tetragonal)
Ni–Ti	1,310	CsCl
Cu_3Au	1,640	B_2 (ordered BCC)
FeAl	1,250–1,400	B_2 (ordered BCC)
NiAl	1,380–1,638	B_2 (ordered BCC)
$MoSi_2$	2,025	$C11_b$ (tetragonal)
Al_3Ti	1,300	DO_{22} (tetragonal)
Nb_3Sn	2,134	A15
Nb_5Si_3	2,500	(tetragonal)

EXAMPLE 1.4

Determine the ideal c/a ratio for the hexagonal structure.

Solution: The atoms in the basal A plane form a closely packed array, as do the atoms in the B plane going through the mid plane. If we take three atoms in the basal plane, with an atom in the B plane resting among them, we have constructed a tetrahedron. The sides of the tetrahedron are $2r = a$, where r is the atomic radius. The height of this tetrahedron is $c/2$, since the distance between planes is c. Hence, the problem is now reduced to finding the height, $c/2$, of a regular tetrahedron. In Figure E1.4, we have

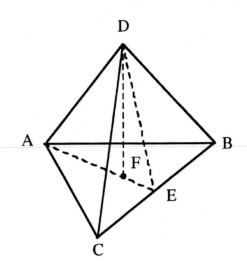

Figure E1.4

$$DF = \frac{C}{2},$$

$$AB = AC = BC = AD = DB = DC = a.$$

For triangle AEC,

$$AE^2 + EC^2 = AC^2,$$

$$AE = \sqrt{a^2 - \frac{a^2}{4}} = \frac{a}{2}\sqrt{3}.$$

For triangle DFE,

$$EF^2 + DF^2 = DE^2.$$

But

$$EF = \frac{1}{3}AE = \frac{a}{6}\sqrt{3},$$

$$DE = AE = \frac{a}{2}\sqrt{3},$$

$$DF = \left(\frac{3a^2}{4} - \frac{3a^2}{36}\right)^{1/2},$$

$$\frac{c}{2} = a\left(\frac{2}{3}\right)^{1/2},$$

$$\frac{c}{a} = 2\left(\frac{2}{3}\right)^{1/2}.$$

Thus,

$$\frac{c}{a} = 1.633.$$

EXAMPLE 1.5

If the copper atoms have a radius of 0.128 nm, determine the density in FCC and BCC structure.

(i) In FCC structure, $4r = \sqrt{2}a_0$

$$a_0 = \frac{4}{\sqrt{2}}r = \frac{4}{\sqrt{2}} \times 0.128 \text{ nm}$$

$$a_0 = 0.362 \text{ nm}$$

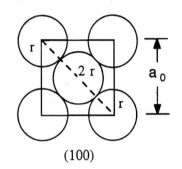

(100)

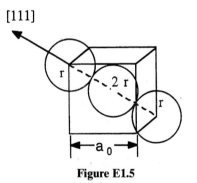

Figure E1.5

There are 4 atoms per unit cell in FCC. Atomic mass (or weight) of copper is 63.54 g/g.mole. So, density of copper (ρ) in FCC structure is

$$\rho = \frac{63.54 \times 4}{(0.362 \times 10^{-7})^3 \times (6.02 \times 10^{23})} = 8.89 \text{ g/cm}^3$$

$$\uparrow$$

Avogadro's number

(ii) In BCC structure, $4r = \sqrt{3}a_0$

$$a_0 = \frac{4}{\sqrt{3}}r = \frac{4}{\sqrt{3}} \times 0.128 \text{ nm}$$

$$a_0 = 0.296 \text{ nm}$$

There are 2 atoms per unit cell in BCC.

$$\rho = \frac{63.54 \times 2}{(0.296 \times 10^{-7})^3 \times (6.02 \times 10^{23})} = 8.14 \text{ g/cm}^3$$

The stable form of Cu is FCC. Only under unique conditions, such as Cu precipitates in iron, is the BCC form stable (because of the constraints of surrounding material).

EXAMPLE 1.6

Sketch the 12 members of the $\langle 110 \rangle$ family for a cubic crystal. Indicate the four $\{111\}$ planes. You may use several sketches.

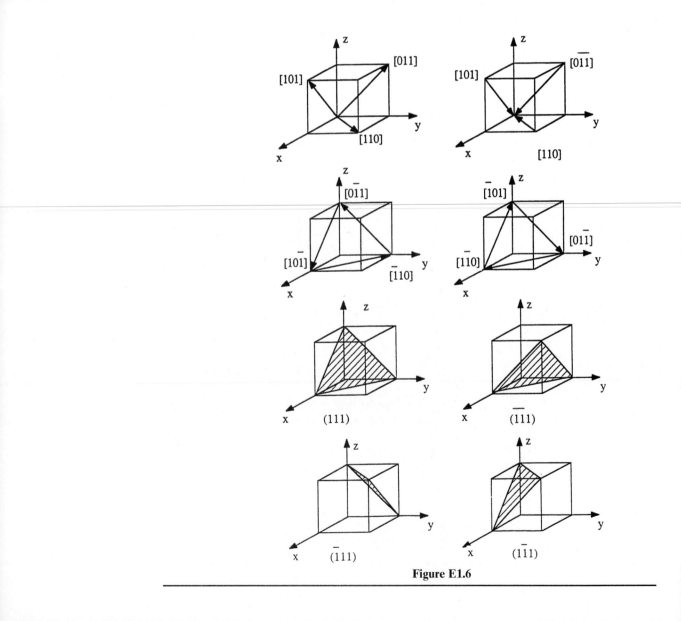

Figure E1.6

1.3.3 Ceramics

The name ceramic comes from the Greek KERAMOS (pottery). The production of pottery made of clay dates from 6500 BC. The production of silicate glass in Egypt dates from 1500 BC. The main ingredient of pottery is a hydrous aluminum silicate that becomes plastic when mixed, in fine powder form, with water. Thus, the early utilization of ceramics; included both crystalline and glassy materials. Portland cement is also a silicate ceramic; by far the largest tonnage production of ceramics today—glasses, clay products (brick, etc.), cement—are silicate based.

However, there have been dramatic changes since the 1970s and a wide range of new ceramics has been developed. These new ceramics are finding applications in computer memories (due to their unique magnetic applications), in nuclear power stations (UO_2 fuel rods), in rocket nose cones and throats, in submarine sonar units (piezoelectric barium titanate), in jet engines (as coatings to metal turbine blades) as electronic packaging components (Al_2O_3, SiC substrates), as electrooptical devices (lithium niobate, capable of transforming optical into electrical information and vice-versa), as optically transparent materials (ruby and yttrium garnet in lasers, optical fibers), as cutting tools (boron nitride, synthetic diamond, tungsten carbide), as refractories, as military armor (Al_2O_3, SiC, B_4C) and in a variety of structural applications.

The structure of ceramics is dependent on the character of the bond (ionic, covalent, or partly metallic), on the sizes of the atoms, and on the processing method. We will first discuss the crystalline ceramics. Transmission electron microscopy has reached the point of development where we can actually image individual atoms, and Figure 1.15 shows a beautiful picture of the zirconium atoms in ZrO_2. The much lighter oxygen atoms cannot be seen but their positions are marked in the electron micrograph. By measuring the atomic distances along two orthogonal directions, one can see that the structure is not

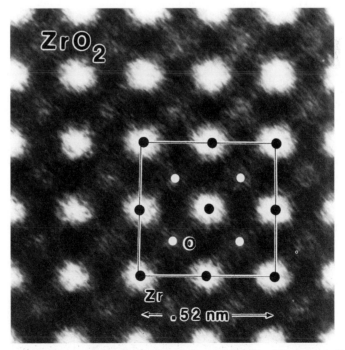

Figure 1.15 Transmission electron micrograph of ZrO_2 at high resolution, showing individual Zr atoms and oxygen sites. (Courtesy of R. Gronsky)

cubic, but tetragonal. The greater complexity of ceramics, in comparison to metallic structures, is evident from Figure 1.15. Atoms of different sizes have to be accommodated by a structure, and bonding (especially covalent) is highly directional. We will first establish the difference between ionic and covalent bonding.

The electronegativity value is a measure of an atom's ability to attract electrons. Compounds in which the atoms have a large difference in electronegativity are principally ionic, while compounds with the same electronegativity are covalent. In ionic bonding one atom loses electrons and is therefore positively charged (cation). The atom that receives the electrons becomes negatively charged (anion). The bonding is provided by the attraction between positive and negative charges, compensated by the repulsion between charges of equal signs. In covalent bonding the electrons are shared between the neighboring atoms. The quintessential example of covalent bonding is diamond. It has four electrons in the outer shell, which combine with four neighboring carbon atoms, forming a tridimensional regular diamond structure, which is a complex cubic structure. Figure 1.16 shows the diamond structure. The bond angles are fixed and equal to 70° 32′. The covalent bond is the strongest bond, and diamond has the highest hardness of all materials. Another material that has covalent bonding is SiC.

As the difference of electronegativity is increased, the bonding character changes from pure covalent to covalent-ionic, to purely ionic. Ionic crystals have a structure determined largely by opposite charge surrounding an ion. These structures are therefore established by the maximum packing density of ions. Compounds of metals with oxygen (MgO, Al_2O_3, ZrO_2, etc.) and with group VII elements ($NaCl$, LiF, etc.) are largely ionic. The most common structures of ionic crystals are presented in Figure 1.17. Evidently, one has more complex structures in ceramics than in metals because the combinations possible between the elements are so vast.

Ceramics also exist in the glassy state. Silica in this state has the unique optical property of being transparent to light, which is used technologically to great advantage. The building blocks of silica in crystalline and amorphous forms are the silica tetrahedra. Silicon bonds to four oxygen atoms, forming a tetrahedron. The oxygen atoms bond to just two silicon atoms. Numerous structures are possible, with different arrangements of the tetrahedra. Pure silica crystallizes into quartz, crystobalite, and trydimite. Because of these bonding requirements, the structure of silica is fairly open and, consequently, gives the mineral a low density. Quartz has a density of 2.65 g cm^{-3}, compared with 3.59 g cm^{-3} and 3.92 g cm^{-3}, for MgO and Al_2O_3, respectively. The structure of crystobalite (Figure 1.17(h)) shows clearly that each Si atom (open circle) is surrounded by four oxygen atoms (filled circles), while each oxygen atom binds two Si atoms. A complex cubic structure results. However, an amorphous structure in silica is more common when the mineral is cooled from the liquid state. Condensation of vapor on a cold substrate is another

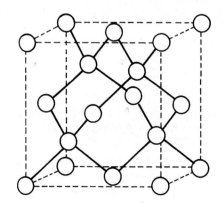

Figure 1.16 Crystal structure of diamond.

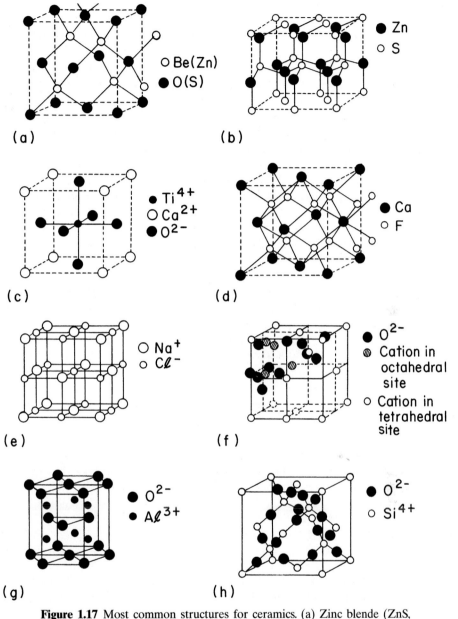

Figure 1.17 Most common structures for ceramics. (a) Zinc blende (ZnS, BeO, SiC). (b) Wurtzite (ZnS, ZnO, SiC, BN). (c) Perovskite (CoTiO$_3$, BaTiO$_3$, YCu$_2$Ba$_3$O$_{7-x}$). (d) Fluorite (ThO$_2$, UO$_2$, CeO$_2$, ZrO$_2$, PuO$_2$). (e) NaCl (KCl, LiF, KBr, MgO, CaO, VO, CsCl, MnO, NiO). (f) Spinel (FeAl$_2$O$_4$, ZnAl$_2$O$_4$, MoAl$_2$O$_4$). (g) Corundum (Al$_2$O$_3$, Fe$_2$O$_3$, Cr$_2$O$_3$, Ti$_2$O$_3$, V$_2$O$_3$). (h) Crystobalite (SiO$_2$—quartz).

method by means of which thin, glassy films are made. One can also obtain glassy materials by electro-deposition, as well as by chemical reaction. Chapter 3 describes glassy metals in greater detail. Figure 1.18 provides a schematic representation of silica in its crystalline and glassy forms in an idealized two-dimensional pattern. The glassy state lacks long-range ordering; the three-dimensional silica tetrahedra arrays lack both symmetry and periodicity.

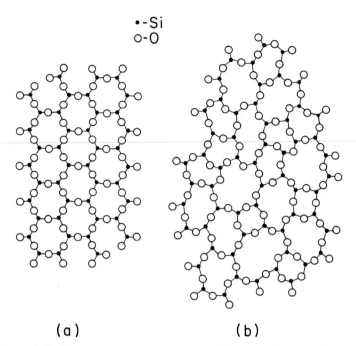

Figure 1.18 Schematic representation of (a) ordered crystalline and (b) random-network glassy form of silica.

EXAMPLE 1.7

Determine the C–C–C–bonding angle in polyethylene.

The easiest manner to visualize the bonding angle is to assume that one C atom is in the center of a cube and that it is connected to four other C atoms at the edges of the cube. (See Figure E1.7.) Suppose all angles are equal to α.

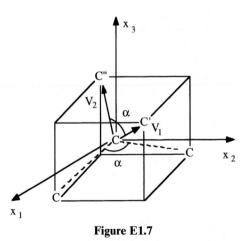

Figure E1.7

The problem is best solved vectorially. We set the origin of the axes at the center of the carbon atom and have

$$\vec{V}_1 = \frac{1}{2}\vec{i} + \frac{1}{2}\vec{j} + \frac{1}{2}\vec{k},$$

$$\vec{V}_2 = -\frac{1}{2}\vec{i} - \frac{1}{2}\vec{j} + \frac{1}{2}\vec{k}.$$

The angle between two vectors is (see Chapter 6 or any calculus text)

$$\cos \alpha = \frac{\frac{1}{2}\left(-\frac{1}{2}\right) + \frac{1}{2}\left(-\frac{1}{2}\right) + \frac{1}{2} \cdot \frac{1}{2}}{\sqrt{\frac{1}{4} + \frac{1}{4} + \frac{1}{4}} \cdot \sqrt{\frac{1}{4} + \frac{1}{4} + \frac{1}{4}}} = -\frac{1}{3}.$$

so

$$\alpha = 109.47°.$$

(*Note:* When we have double bonds, the angle is changed.)

1.3.4 Glasses

As described earlier, glasses are characterized by a structure in which no long-range ordering exists. There can be short-range ordering, as indicated in the individual tetrahedral arrays of SiO_4^{-4} in Figure 1.18, which shows both the crystalline and glassy forms of silica. Over distances of several atomic spacings, the ordering disappears, leading to the glassy state. It is possible to have glassy ceramics, glassy metals, and glassy polymers.

The structure of glass has been successfully described by the *Zachariasen* model. The *Bernal* model is also a successful one. It consists of drawing lines connecting the centers of adjacent atoms and forming polyhedra. These polyhedra represent the glassy structure of glass. Glassy structures represent a less efficient packing of atoms or molecules than the equivalent crystalline structures. This is very easily understood with the "suitcase" analog. We all know that by throwing clothes randomly into a suitcase, the end result is often a major job of sitting on the suitcase to close it. Neat packing of the same clothes occupies less volume. The same happens in glasses. If we plot the inverse of the density (called *specific volume*) versus temperature, we obtain the plot shown in Figure 1.19. Contraction occurs as the temperature is lowered. If the material crystallizes, there is a discontinuity in the specific volume at the melting temperature T_m. If insufficient time is allowed for crystallization, the material becomes a supercooled liquid, and contraction follows the liquid line. At a temperature T_g, called the *glass transition temperature,* the supercooled liquid is essentially solid, with very high viscosity. It is then called a glass. This difference in specific volume between the two forms is often referred to as *excess volume.*

In ceramics, reasonably low cooling rates can produce glassy structures. The regular arrangement of the silica tetrahedra of Figure 1.18(a) requires a significant amount of time. The same is true for polymeric chains, which need to arrange themselves into regular crystalline arrangements. For metals, this is more difficult. Only under extreme conditions it is possible to obtain solid metals in a noncrystalline structure. Figure 1.20 shows a crystalline and a glassy alloy with the same composition. The liquid state is frozen in, and the structure resembles that of glasses. It is possible to arrive at these special structures by

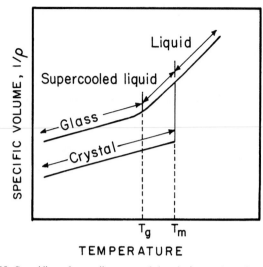

Figure 1.19 Specific volume (inverse of density) as a function of temperature for glassy and crystalline forms of *materials.*

cooling the alloy at such a rate that virtually no reorganization of the atoms into periodic arrays can take place. The required cooling rate is usually on the order of 10^6 to 10^8 s^{-1}. It is also possible to arrive at the glassy state by means of solid-state processing (very heavy deformation and reaction) and from the vapor.

The original technique for obtaining metallic glasses was called splat cooling and was pioneered by Duwez and students.[†] An alloy in which the atomic sizes are quite dissimilar, such as Fe–B, is ideal for retaining the "glassy" state upon cooling. This technique consisted of propelling a drop of liquid metal with a high velocity against a heat-conducting surface such as copper. The interest in these alloys was mainly academic at the time. However, the unusual magnetic properties and high strength exhibited by the alloys triggered worldwide interest, and subsequent research has resulted in thousands of papers. The splat-cooling technique has been refined to the point where 0.07- to 0.12-mm-thick wires can be ejected from an orifice. Production rates as high as 1,800 m/min can be obtained. Sheets and ribbons can be manufactured by the same technique. An alternative technique consists of

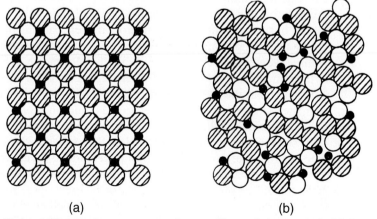

(a)   (b)

Figure 1.20 Atomic arrangements in crystalline and glassy metals. (a) Crystalline metal section. (b) Glassy metal section. (Courtesy of L. E. Murr)

[†] W. Klement, R. H. Willens, and P. Duwez, *Nature* 187 (1960) 869.

vapor deposition on a substrate (sputtering). This seems a most promising approach, and samples with a thickness of several millimeters have been successfully produced.

1.3.5 Polymers

From a microstructural point of view, polymers are much more complex than metals and ceramics. On the other hand, they are cheap and easily processed. Polymers have lower strengths and moduli and lower temperature-use limits than do metals or ceramics. Because of their predominantly covalent bonding, polymers are generally poor conductors of heat and electricity. Polymers are generally more resistant to chemicals than are metals, but prolonged exposure to ultraviolet light and some solvents can cause degradation of a polymer's properties.

Chemical Structure. Polymers are giant chainlike molecules (hence, the name *macromolecules*), with covalently bonded carbon atoms forming the backbone of the chain. Polymerization is the process of joining together many monomers, the basic building blocks of polymers, to form the chains. For example, the ethyl alcohol polymer has the chemical formula

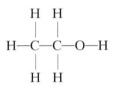

The monomer vinyl chloride has the chemical formula C_2H_3Cl, which, on polymerization, becomes polyvinyl chloride (PVC). The structural formula of polyvinyl chloride is represented by

where n is the degree of polymerization.

Types of Polymers. The difference in the behavior of polymers stems from their molecular structure and shape, molecular size and weight, and amount and type of bond (covalent or van der Waals). The different chain configurations are shown in Figure 1.21. A *linear polymer* consists of a long chain of atoms with attached side groups (Figure 1.21a). Examples include polyethylene, polyvinyl chloride, and polymethyl methacrylate. Note the coiling and bending of the chain. *Branched polymers* have branches attached to the main chain (Figure 1.21b). Branching can occur with linear, cross-linked, or any other types of polymers. A *crossed-linked* polymer has molecules of one chain bonded with those of another (Figure 1.21c). Cross-linking of molecular chains results in a three-dimensional network. It is easy to see that cross-linking makes sliding of molecules past one another difficult, resulting in strong and rigid polymers. *Ladder polymers* have two linear polymers linked in a regular manner (Figure 1.21d). Not unexpectedly, ladder polymers are more rigid than linear polymers.

Yet another classification of polymers is based on the type of the repeating unit. (See Figure 1.22) When we have one type of repeating unit—for example, *A*—forming the

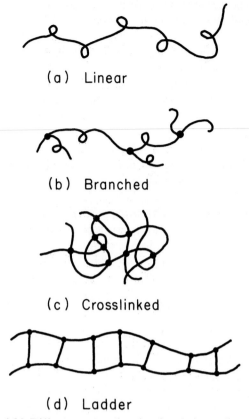

Figure 1.21 Different types of molecular chain configurations.

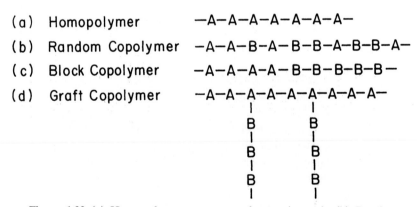

Figure 1.22 (a) Homopolymer: one type of repeating unit. (b) Random copolymer: two monomers, A and B, distributed randomly. (c) Block copolymer: a sequence of monomer A, followed by a sequence of monomer B. (d) Graft copolymer: Monomer A forms the main chain, while monomer B forms the branched chains.

polymer chain, we call it a *homopolymer. Copolymers,* on the other hand, are polymer chains having two different monomers. If the two different monomers, *A* and *B,* are distributed randomly along the chain, then we have a *regular,* or *random, copolymer.* If, however, a long sequence of one monomer *A* is followed by a long sequence of another monomer *B,* we have a *block copolymer.* If we have a chain of one type of monomer *A* and branches of another type *B,* then we have a *graft copolymer.*

Tacticity has to do with the order of placement of side groups on a main chain. It can provide variety in polymers. Consider a polymeric backbone chain having side groups. For example, a methyl group (CH_3) can be attached to every second carbon atom in the polypropylene chain. By means of certain catalysts, it is possible to place the methyl groups all on one side of the chain or alternately on the two sides, or to randomly distribute them in the chain. Figure 1.23 shows tacticity in polypropylene. When we have all the side groups on one side of the main chain, we have an *isotactic* polymer. If the side groups alternate from one side to another, we have a *syndiotactic* polymer. When the side groups are attached to the main chain in a random fashion, we get an *atactic* polymer.

Thermosetting Polymers and Thermoplastics. Based on their behavior upon heating, polymers can be divided into two broad categories:

(i)   thermosetting polymers
(ii) thermoplastics

When the molecules in a polymer are cross-linked in the form of a network, they do not soften on heating. We call these cross-linked polymers *thermosetting* polymers. Thermosetting polymers decompose upon heating. Cross-linking makes sliding of molecules past one

Isotactic polypropylene

Syndiotactic polypropylene

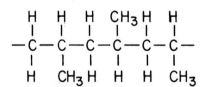

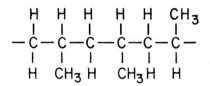

Atactic polypropylene

Figure 1.23 Tacticity, or the order of placement of side groups.

another difficult, which produces a strong and rigid polymer. A typical example is rubber cross-linked with sulfur, i.e., vulcanized rubber. Vulcanized rubber has 10 times the strength of natural rubber. Common examples of thermosetting polymers include phenolic, polyester, polyurethane, and silicone. Polymers that soften or melt upon heating are called *thermoplastics*. Suitable for liquid flow processing, they are mostly linear polymers—for example, low- and high-density polyethylene and polymethyl methacrylate (PMMA).

Polymers can have an amorphous or partially crystalline structure. When the structure is amorphous, the molecular chains are arranged randomly, i.e., without any apparent order. Thermosetting polymers, such as epoxy, phenolic, and unsaturated polyester, have an amorphous structure. Semicrystalline polymers can be obtained by using special processing conditions. For example, by precipitating a polymer from an appropriate dilute solution, we can obtain small, platelike crystalline lamellae, or crystallites. Such solution-grown polymer crystals are characteristically small. Figure 1.24 shows a transmission electron micrograph of a lamellar crystal of poly (ϵ-caprolactone). Note the formation of new layers of growth spirals around screw dislocations. The screw dislocations responsible for crystal growth are perpendicular to the plane of the micrograph. Polymeric crystals involve molecular chain packing, rather than the atomic packing characteristic of metals. Molecular chain packing requires a sufficiently stereographic regular chemical structure. Solution-grown polymeric crystals generally have a lamellar form, and the long molecular chains crystallize by folding back and forth in a regular manner. Lamellar polymeric crystals have straight segments of molecules oriented normal to the lamellar panes. Figure 1.25 depicts some important chain configurations in a schematic manner. The flexible, coiled structure is shown in Figure 1.25a, while the chain-folding configuration that results in crystalline polymers is shown in Figure 1.25b. Under certain circumstances, one can obtain an extended and aligned chain structure, shown in Figure 1.25c Such a structure, typically obtained in fibrous form, has very high strength and stiffness. A semicrystalline configuration called a fringed micelle structure is shown in Figure 1.25d. Almost all so-called semicrystalline polymers are, in reality, mixtures of crystalline and amorphous regions. Only by using very special techniques, such as solid-state polymerization, is it possible to prepare a 100% crystalline polymer. Polydiacetylene single crystals in the form of lozenges and fibers have been prepared by solid-state polymerization.

Partially crystallized, or semicrystalline, polymers can also be obtained from melts. Generally, because of molecular chain entanglement, the melt-formed crystals are more irregular than those obtained from dilute solutions. A characteristic feature of melt-formed

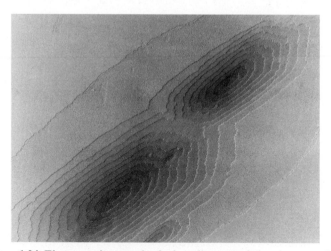

Figure 1.24 Electron micrograph of a lamellar crystal showing growth spirals around screw dislocations. (Courtesy of H. D. Keith)

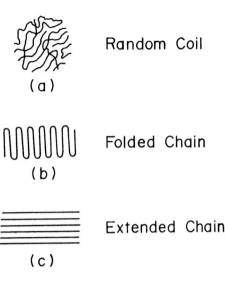

Figure 1.25 Some important chain configurations. (a) A flexible, coiled chain structure. (b) A folding chain structure. (c) An extended and aligned chain structure. (d) A fringed micelle chain structure.

polymers is the formation of *spherulites*. When seen under cross-polarized light in an optical microscope, the classical spherulitic structure shows a Maltese cross pattern. (See Figure 1.26a) Figure 1.26b presents a schematical representation of a spherulite whose diameter can vary between a few tens to a few hundreds of micrometers. Spherulites can nucleate at a variety of points, as, for example, with dust or catalyst particles, in a quiescent melt and then grow as spheres. Their growth stops when the neighboring spherulites impinge upon each other. Superficially, the spherulites look like grains in a metal. There are, however, differences between the two. Each grain in a metal is a single crystal, whereas each spherulite in a polymer is an assembly of radially arranged, narrow crystalline lamellae. The fine-scale structure of these lamellae, consisting of tightly packed chains folding back and forth, is shown in Figure 1.26c. Amorphous regions containing tangled masses of molecules fill the spaces between the crystalline lamellae.

Degree of Crystallinity. The *degree of crystallinity* of a material can be defined as the fraction of the material that is fully crystalline. This is an important parameter for semi-crystalline polymers. Depending on their degree of crystallinity, such polymers can show a range of densities, melting points, etc. It is worth repeating that a 100-percent crystalline polymer is very difficult to obtain in practice. The reason for the difficulty is the long chain structure of polymers: Some twisted and entangled segments of chains that get trapped between crystalline regions never undergo the conformational reorganization necessary to achieve a fully crystalline state. Molecular architecture also has an important bearing on a polymer's crystallization behavior. Linear molecules with small or no side groups crystallize easily. Branched chain molecules with bulky side groups do not crystallize as easily. For

(b)

(a)   (c)

Figure 1.26 Spherulitic structures. (a) A typical spherulitic structure in a melt-formed polymer film. (Courtesy of H. D. Keith) (b) Schematic of a spherulite. Each spherulite consists of an assembly of radially arranged narrow crystalline lamellae. (c) Each lamella has tightly packed polymer chains folding back and forth. Amorphous regions fill the spaces between the crystalline lamellae.

example, linear, high-density polyethylene can be crystallized to 90 percent, while branched polyethylene can be crystallized only to about 65 percent. Generally, the stiffness and strength of a polymer increase with the degree of crystallinity.

Like crystalline metals, crystalline polymers have imperfections. It is, however, not easy to analyze these defects, because the topological connectivity of polymer chains leads to large amounts and numerous types of disorder. Polymers are also very sensitive to damage by the electron beam in TEM, making it difficult to image them. Generally, polymer crystals are highly anisotropic. Because of covalent bonding along the backbone chain, polymeric crystals show low-symmetry structures, such as orthorhombic, monoclinic, or triclinic. Deformation processes such as slipping and twinning, as well as phase transformations that take place in monomeric crystalline solids, also can occur in polymeric crystals.

Molecular Weight and Distribution. Molecular weight is a very important attribute of polymers, especially because it is not so important in the treatment of nonpolymeric materials. Many mechanical properties increase with molecular weight. In particular, resistance to deformation does so. Of course, concomitant with increasing molecular weight, the processing of polymers becomes more difficult.

The molecular weight of a polymer is given by the product of the molecular weight of the repeat unit (the "mer") and the number of repeat units. The molecular weight of the ethylene repeat unit ($-CH_2-CH_2-$) is 28. We write the chemical formula as H $(-CH_2-CH_2-)_n$ H. If n, the number of repeat units, is 10,000, the high-density polyethylene will have a molecular weight of 280,002. In almost all polymers, the chain lengths are not equal, but rather, there is a distribution of chain lengths. In addition, there may be more than one species of chain in the polymer. This makes for different parameters describing the molecular weight.

The number-averaged molecular weight (M_n) of a polymer is the total weight of all of the polymer's chains divided by the total number of chains:

$$M_n = \sum N_i M_i / \sum N_i.$$

where N_i is the number of chains of molecular weight M_i.

The weight-averaged molecular weight (M_w) is the sum of the square of the total molecular weight divided by the total molecular weight. Thus,

$$M_w = \sum N_i M_i^2 / \sum M_i N_i.$$

Two other molecular weight parameters are

$$M_z = \sum N_i M_i^3 / \sum N_i M_i^2.$$

and

$$M_v = \left[\sum N_i M_i^{(1+a)} / \sum N_i M_i \right]^{1/a},$$

where a has a value between 0.5 and 0.8.

Typically, $M_n : M_w : M_z = 1:2:3$. Figure 1.27 shows a schematic molecular weight distribution curve with various molecular weight parameters indicated. Molecular weight distributions of the same polymer obtained from two different sources can be very different.

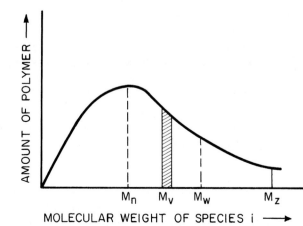

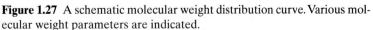

Figure 1.27 A schematic molecular weight distribution curve. Various molecular weight parameters are indicated.

Also, molecular weight distributions are not necessarily single peaked. For single-peaked distributions, M_n is generally near the peak—that is, the most probable molecular weight. The weight-averaged molecular weight, M_w, is always larger than M_n. The molecular weight characterization of a polymer is very important. The existence of a very high-molecular-weight tail can make processing very difficult because of the enormous contribution of the tail to the melt viscosity of a polymer. The low end of the molecular weight distribution, however, can be used as a plasticizer.

It is instructive to compare some monomers with low- and high-molecular-weight polymers. A very common monomer is a molecule of water, H_2O, with a molecular weight of 18. Benzene, on the other hand, is a low-molecular-weight organic solvent; its molecular weight is 78. By contrast, natural rubber has a molecular weight of about 10^4, and polyethylene, a common synthetic polymer, can have a molecular weights greater than this. Polymers having such large molecular weights are sometimes called *high polymers*. Their molecular size is also very great.

EXAMPLE 1.8

A polymer has three species of molecular weights: 3×10^6, 4×10^6, and 5×10^6. Compute its number-averaged molecular weight M_n and weight-averaged molecular weight M_w.

Solution: For the number-averaged molecular weight, we have

$$M_n = \frac{\sum N_i M_i}{\sum N_i}$$

$$= \frac{3 \times 10^6 + 4 \times 10^6 + 5 \times 10^6}{3} = 4 \times 10^6.$$

The weight-averaged molecular weight is

$$M_w = \frac{\sum N_i M_i^2}{\sum N_i M_i}$$

$$= \frac{(3 \times 10^6)^2 + (4 \times 10^6)^2 + (5 \times 10^6)^2}{3 \times 10^6 + 4 \times 10^6 + 5 \times 10^6}$$

$$= \frac{50 \times 10^{12}}{12 \times 10^6} = 4.17 \times 10^6$$

EXAMPLE 1.9

Estimate the molecular weight of polyvinyl chloride with degree of polymerization, n, equal to 800.

Solution: The molecular weight of each *mer* of polyvinyl chloride (C_2H_3Cl) is

$$2(12) + 3(1) + 35.5 = 62.5.$$

For $n = 800$, the molecular weight is $800 \times 62.5 = 50,000$ g/mole.

EXAMPLE 1.10

Discuss how a polymer's density changes as crystallization proceeds from the melt.

Answer: The density increases and the volume decreases as crystallization proceeds. This is because the molecular chains are more tightly packed in the crystal than in the molten or noncrystalline polymer. This phenomenon is, in fact, exploited in the so-called *density* method to determine the degree of crystallinity.

Quasi Crystals. Quasi crystals represent a new state of solid matter. In a crystal, the unit cells are identical, and a single unit cell is repeated in a periodic manner to form the crystalline structure. Thus, the atomic arrangement in crystals has positional and orientational order. Orientational order is characterized by a rotational symmetry; that is, certain rotations leave the orientations of the unit cell unchanged. The theory of crystallography holds that crystals can have twofold, threefold, fourfold, or sixfold axes of rotational symmetry; a fivefold rotational symmetry is not allowed. A two-dimensional analogy of this is that one can tile a bathroom wall using a single shape of tile *if and only if* the tiles are rectangles (or squares), triangles, or hexagons, but not if the tiles are pentagons. One may obtain a glassy structure by rapidly cooling a vapor or liquid well below its melting point, until the disordered atomic arrangement characteristic of the vapor or liquid state gets frozen in. The atomic packing in the glassy state is dense but random. This can be likened to a mosaic formed by taking an infinite number of different shapes of tile and randomly joining them together. Clearly, the concept of a unit cell will not be valid in such a case. The atomic structure in the glassy state will have neither positional nor orientational order.

Quasicrystals are not perfectly periodic, but they do follow the rigorous theorems of crystallography. They can have any rotational symmetry axes which are prohibited in crystals. It is worth reminding the reader that a glassy structure shows an electron diffraction pattern consisting of diffuse rings for all orientations. A crystalline structure has an electron diffraction pattern that depends on the crystal symmetry.

Schectman et al. discovered that a rapidly solidified (melt-spun) aluminum-manganese alloy showed fivefold symmetry axis.[4] They observed a metastable phase that showed a sharp electron diffraction pattern with a perfect icosahedral symmetry. [Remember that sharp electron diffraction patterns are associated with the orderly atomic arrangement in crystals and icosahedral symmetry is forbidden in crystals.] At first, this was thought to be a paradox. However, some very careful and sophisticated electron microscopy work showed conclusively that it was indeed an icosahedral (twenty-fold) symmetry. Al-Mn alloys containing 18 to 25.3 weight percent Mn examined by transmission electron microscopy showed the same anomalous diffraction. In particular, Al-25.3 wt % Mn alloy consisted almost entirely of one phase which has a composition close to Al_6Mn. The selected area diffraction pattern of Al_6Mn showed a fivefold symmetry. This new kind of structure is neither amorphous nor crystalline; rather the new phase in this alloy had a three-dimensional icosahedral symmetry.

Perhaps, it would be in order for us to digress a bit and explain this icosahedral symmetry. *Icosahedral* means twenty faces. An icosahedron has twenty triangular faces, thirty edges, and twelve vertices. Consider the two-dimensional case. As pointed out earlier, one can tile a bathroom wall without leaving an open space (a *crack*) by hexagons. Three hexagons can be tightly packed without leaving a crack. Three pentagons, however, cannot be tightly packed. The reader may try this out. In three dimensions, four spheres pack tightly to form a tetrahadron. Twenty tetrahedrons can, with small distortions, fit tightly into an icosahedron. Icosahedrons have fivefold symmetry (five triangular faces meet at

[4]D. Schectman, I. A. Blech, D. Gratias, and J. W. Cahn, Phys. Rev. Lett., 53 (1984) 1951.

each vertex) and they *cannot* fit together tightly, i.e., complete space filling is not possible with them. An icosahedron, therefore, cannot serve as a unit cell for a crystalline structure. Therefore, structures, are known as quasi crystals.

1.3.6   Liquid Crystals

A liquid crystal is a state of matter that shares some properties of liquids and crystals. Like all liquids, liquid crystals are fluids; however, unlike ordinary liquids, which are isotropic, liquid crystals can be anisotropic. Liquid crystals are also called *mesophases*. The liquid crystalline state exists in a specific temperature range, below which the solid crystalline state prevails and above which the isotropic liquid state prevails. That is, the liquid crystal has an order between that of a liquid and a crystalline solid. In a crystalline solid, the atoms, ions, or molecules are arranged in an orderly manner. This very regular three-dimensional order is best described in terms of a crystal lattice. Because of a different periodic arrangement in different directions, most crystals are anisotropic. Now consider a crystal lattice with rod-shaped molecules at the lattice points. In this case, we now have, in addition to a positional order, an orientational order. An analogy that is used to qualitatively describe the order in a liquid crystal is as follows: If a random pile of pencils is subjected to an external force, it will undergo an ordering process very much akin to that seen in liquid crystals. The pencils, long and rigid, tend to align themselves, with their long axes approximately parallel. By far the most important characteristic of liquid crystals is that their long molecules tend to organize according to certain patterns. The order of orientation is described by a directed line segment called the *director*. This order is the source of the rather large anisotropic effect in liquid crystals, a characteristic that is exploited in electrooptical displays or the so-called liquid-crystal displays. Another important application of liquid crystals is the production of strong and stiff organic fibers such as aramid fiber, in which a rigid, rodlike molecular arrangement is provided by an appropriate polymer solution in the liquid crystalline state.[5] When a polymer manifests the liquid crystalline order in a solution, we call it a *lyotropic* liquid crystal, and when the polymer shows the liquid crystalline state in the melt, it is called a *thermotropic* liquid crystal. The three types of order in the liquid crystalline state are nematic, smectic, and cholesteric, shown schematically in Figure 1.28. A nematic order is an

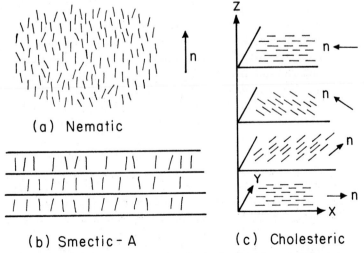

(a) Nematic

(b) Smectic - A   (c) Cholesteric

Figure 1.28 Different types of order in the liquid crystalline state.

[5]See K. K. Chawla, *Fibrous Materials* (Cambridge, U.K.: Cambridge University Press, 1998).

approximately parallel array of polymer chains that remains disordered with regard to end groups or chain units; that is, there is no positional order along the molecular axis. Figure 1.28a shows this type of order, with the director vector n as indicated. In smectic order, we have one-dimensional, long-range positional order. Figure 1.28b shows smectic-A order, which has a layered structure with long-range order in the direction perpendicular to the layers. In this case, the director is perpendicular to the layer. Other more complex smectics are B, C, D, F, and G. The director in these may not be perpendicular to the layer, or there may exist some positional order as well. Cholesteric-type liquid crystals, shown in Figure 1.28c, have nematic order with a superimposed spiral arrangement of nematic layers; that is, the director n, pointed along the molecular axis, has a helical twist.

1.3.7 Biomaterials

The mechanical properties of biological materials are, of course, of great importance, and the design of all living organisms is optimized for the use of these properties. Biological materials cover a very broad range of structures. The common feature is the hierarchical organization of the structure, such that failure at one level does not generate catastrophic fracture: The other levels in the hierarchy "take up" the load. Figure 1.29 demonstrates this fact. Figure 1.29a shows the response of the ureter of three animals: guinea pig, dog, and rabbit. This muscle is a thick-walled cylindrical tube that has the ability to contract until the closure of the inner hole is complete. With a nonlinear elastic mechanical response, the ureter is not unlike other soft tissues in that regard: Its stiffness increases with loading, and the muscle becomes very stiff after a certain strain is reached. The unloading and loading responses are different, as shown in the figure, and this causes a hysteresis. Increases in length of 50% can be produced. Bone, on the other hand, is a material with drastically different properties: Its strength and stiffness are much higher, and its maximum elongation is much lower. The structure of bones is quite complex, and they can be considered composite materials. Figure 1.29b illustrates the strength (in tension) of dry and wet bone. The maximum tensile strength is approximately 80 MPa, and Young's modulus is about 20 GPa.

The abalone shell and the shells of bivalve molluscs are often used as examples of a naturally occurring laminated composite material. These shells are composed of layers of calcium carbonate, glued together by a viscoplastic organic material. The calcium carbonate is hard and brittle. The effect of the viscoplastic glue is to provide a crack-deflection layer so that cracks have difficulty propagating through the composite. Figure 1.30 shows cracks that are deflected at each soft layer. The toughness of this laminated composite is vastly superior to that of a monolithic material, in which the crack would be able to propagate freely, without barriers. The effect is shown at two scales: the mesoscale and the microscale. At the mesoscale, layers of calcium carbonate have a thickness of approximately 500 μm. At the microscale, each calcium carbonate layer is made up of small brick-shaped units (about 0.5×7.5 μm longitudinal section), glued together with the organic matter. The formation of this laminated composite results in a fracture toughness and strength (about 4 MPa m$^{1/2}$ and approximately 150 MPa, respectively) that are much superior to those of the monolithic $CaCO_3$. The composite also exhibits a hierarchical structure; that is, the layers of $CaCO_3$ and organic glue exist at more than one level (at the micro- and mesolevels). This naturally occurring composite has served as inspiration for the synthesis of B_4C–Al laminate composites, which exhibit a superior fracture toughness.[6] In these synthetic composites, there is a 40% increase in both fracture toughness and strength over monolithic B_4C–Al cermets. *Biomimetics* is the field of

[6]M. Sarikaya, K. E. Gunnison, M. Yasrebi, and I. A. Aksay, *Mater. Soc. Symp. Proc.*, 174 (1990) 109.

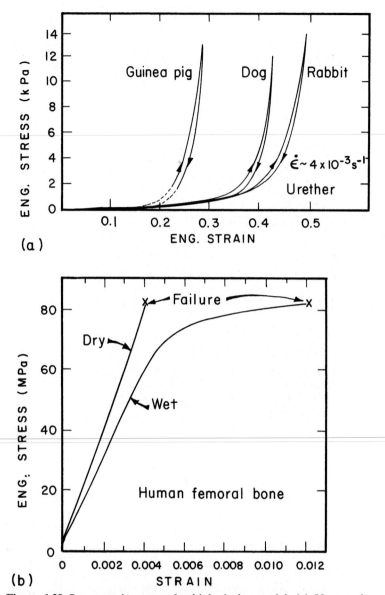

Figure 1.29 Stress–strain curves for biological material. (a) Ureter after F. C. P. Yin and Y. C. Fung, *Am. J. Physiol.* 221 (1971), 1484. (b) Human femur bone (after F. G. Evans, *Artificial Limbs,* 13 (1969) 37.

materials science in which inspiration is sought from biological systems for the design of novel materials.

Another area of biomaterials in which mechanical properties have great importance is bioimplants. Complex interactions between the musculoskeletal system and these implants occur in applications where metals and ceramics are used as replacements for hips, knees, teeth, tendons, and ligaments. The matching of material and bone stiffness is important, as are the mechanisms of bonding tissue to these materials. The number of scientific and technological issues is immense, and the field of bioengineering focuses on these.

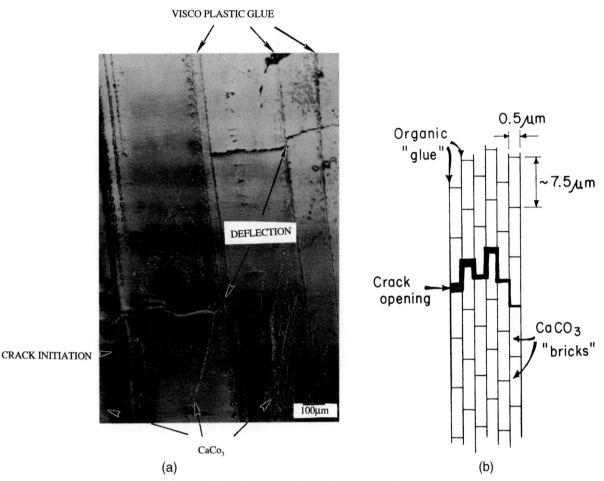

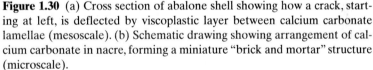

Figure 1.30 (a) Cross section of abalone shell showing how a crack, starting at left, is deflected by viscoplastic layer between calcium carbonate lamellae (mesoscale). (b) Schematic drawing showing arrangement of calcium carbonate in nacre, forming a miniature "brick and mortar" structure (microscale).

1.3.8 Porous and Cellular Materials

Wood, cancellous bone, styrofoam, cork, and insulating tiles of the Space Shuttle are examples of materials that are not compact; their structure has air as a major component. The great advantage of cellular structures is their low density. Techniques for making foam metals, ceramics, and polymers have been developed, and these cellular materials have found a wide range of applications, in insulation, in cushioning, as energy-absorbing elements, in sandwich panels for aircraft, as marine buoyancy components, in skis, and more.

The mechanical response of cellular materials is quite different from that of bulk materials. The elastic loading region is usually followed by a plateau that corresponds to the collapse of the pores, either by elastic, plastic buckling of the membranes or by their fracture. The third stage is an increase in the slope, corresponding to final densification. Figure 1.31a shows representative curves for polyethylene with different initial densities. The plateau occurs at different stress levels and extends to different strains for different initial

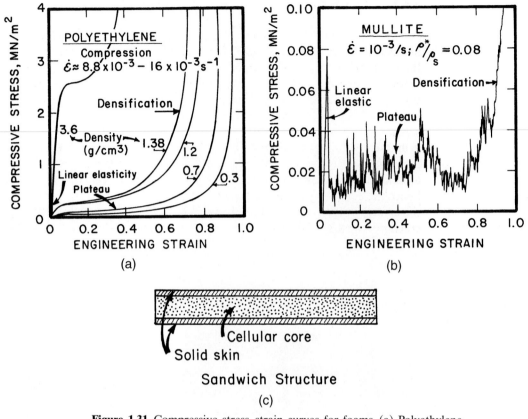

Figure 1.31 Compressive stress–strain curves for foams. (a) Polyethylene with different initial densities. (b) Mullite with relative density $P^x / Ps = 0.08$ (adapted from L. J. Gibson and M. F. Ashby, *Cellular Solids: Structure and Properties* (Oxford, UK: Pergamon Press 1988), pp. 124, 125). (c) Schematic of a sandwich structure.

densities. The bulk (fully dense) polyethylene is shown for comparison purposes. Cellular mullite, an alumina–silica solid solution, exhibits a plateau marked by numerous spikes, corresponding to the breakup of the individual cells (Figure 1.31b). Materials with initial densities as low as 5% of the bulk density are available as foams. Figure 1.31(c) shows a very important use of foams: Sandwich structures, composed of end sheets of solid material in which a foam forms the core region, have numerous applications in the aerospace industry. The foam between the two panels makes them more rigid; this is accomplished without a significant increase in weight.

1.4 THEORETICAL STRENGTH OF A CRYSTAL

The theoretical strength of a material with a perfect lattice is of great importance. It is very high, and experimental efforts have been made to reach it. Theoretical strength is determined by the nature of the interatomic forces, by the temperature (which causes atoms to vibrate), and by the stress state of the material. In this section, we shall make calculations for two states of stress: uniaxial normal stress and shear stress. These two states determine two different types of failure: cleavage and shear, respectively. The stresses required for failure under the two situations will be calculated, and the theoretical strength should be the lower of the two values.

1.4.1 Theoretical Tensile Strength by Orowan's Method

A material is said to *cleave* when it breaks under normal stress and the fracture path is perpendicular to the applied stress. The process involves the separation of the atoms along the direction of the applied stress. Orowan developed a simple method for obtaining the theoretical tensile strength of a crystal.[7] With his method, no stress concentrations at the tip of the crack are assumed; instead, it is assumed that all atoms separate simultaneously once their separation reaches a critical value. Figure 1.32 shows how the stress required to separate two planes will vary as a function of the distance between the planes. The distance is initially equal to a_0. Naturally, $\sigma = 0$ for $a = a_0$; σ will also be zero when the separation is infinite. The exact form of the curve of σ versus a depends on the nature of the interatomic forces. In Orowan's model, the curve is simply assumed to be a sine function—hence the generality of the model. The area under the curve is the work required to cleave the crystal. This work of deformation—and here there is a certain similarity with Griffith's crack propagation theory to be presented in Chapter 3—cannot be lower than the energy

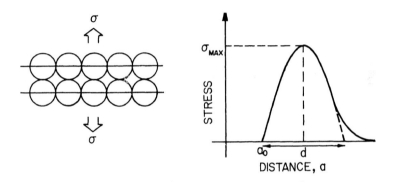

Figure 1.32 Stress required to separate two atomic layers.

of the two new surfaces created by the cleavage. If the surface energy per unit area is γ and the cross-sectional area of the specimen is A, the total energy is $2\gamma A$ (two surfaces formed). The stress dependence on plane separation is then given by the following equation, admitting a sine function and assuming a periodicity of $2d$:

$$\sigma = K \sin \frac{2\pi}{2d} (a - a_0). \tag{1.1}$$

K can be determined by the following artifice: When a is close to a_0, the material responds linearly to the applied loads (Hookean behavior). Assuming that the elastic deformation is restricted to the two planes shown in Figure 1.32a and that the material is isotropic, the fractional change in the distance between the planes, da/a_0, is defined as the incremental strain $d\varepsilon$.

$$\frac{da}{a_0} = d\varepsilon,$$

$$\frac{d\sigma}{d\varepsilon} = \frac{d\sigma}{da/a_0} = E, \tag{1.2}$$

where E is Young's modulus, which is defined as $d\sigma/d\varepsilon$ in the elastic region.

[7]E. Orowan, "Fracture and Strength of Solids," Rep. Prog. Phys. 12 (1949) 185.

Thus,

$$a_0 \frac{d\sigma}{da} = E.$$

Taking the derivative of Equation 1.1 and substituting it into Equation 1.2 for $a = a_0$,

$$a_0 \frac{d\sigma}{da} = K \frac{\pi}{d} a_0 \cos \frac{\pi}{d} (a - a_0) = E,$$

$$K = \frac{E}{\pi} \frac{d}{a_0}. \tag{1.3}$$

However, d is not known; to determine d, the area under the curve has to be equated to the energy of the two surfaces created:

$$\int_{a_0}^{a_0+d} \sigma \, da = 2\gamma. \tag{1.4}$$

Substituting Equation 1.1 into Equation 1.4, we get

$$\int_{a_0}^{a_0+d} K \sin \frac{2\pi}{2d} (a - a_0) \, da = 2\gamma. \tag{1.5}$$

From a standard mathematics text, the preceding integral can be evaluated:

$$\int \sin ax \, dx = -\frac{1}{a} \cos ax. \tag{1.6}$$

A substitution of variables is required to solve Equation 1.5; applying the standard Equation 1.6, we have $a - a_0 = y$; therefore, $da = dy$, and

$$K \int_0^d \sin \frac{\pi}{d} y \, dy = 2\gamma,$$

$$K \frac{d}{\pi} = \gamma,$$

and

$$d = \frac{\pi\gamma}{K}. \tag{1.7}$$

The maximum value of σ is equal to the theoretical cleavage stress. From Equation 1.1, and making the sine equal to 1, we have, from Equation 1.3,

$$\sigma_{\max} = K = \frac{E}{\pi} \frac{d}{a_0}. \tag{1.8}$$

Substituting Equation 1.7 into Equation 1.8 yields

$$K = \sigma_{max} = \frac{E\gamma}{a_0 K},$$

and

$$K^2 = (\sigma_{max})^2 = \frac{E\gamma}{a_0},$$

or

$$\sigma_{max} = \sqrt{\frac{E\gamma}{a_0}}. \qquad (1.9)$$

According to Orowan's model, the surface energy is given by

$$\gamma = \frac{Kd}{\pi} = \frac{E}{a_0}\left(\frac{d}{\pi}\right)^2 \qquad (1.10)$$

It has been experimentally determined that d is approximately equal to a_0. Hence,

$$\gamma \simeq \frac{Ea_0}{10} \quad \text{and} \quad \boxed{\sigma_{max} \simeq \frac{E}{\pi}.} \qquad (1.11)$$

We can conclude from Eq. 1.9 that, in order to have a high theoretical cleavage strength, a material must have a high Young's modulus and surface energy and a small distance a_0 between atomic planes. Table 1.4 presents the theoretical cleavage strengths for a

TABLE 1.4 Theoretical Cleavage Stresses According to Orowan's Theory*

Element	Direction	Young's Modulus (GPa)	Surface Energy (J/m²)	σ_{max} (GPa)	σ_{max}/E
α-Iron	<100>	132	2	30	0.23
	<111>	260	2	46	0.18
Silver	<111>	121	1.13	24	0.20
Gold	<111>	110	1.35	27	0.25
Copper	<111>	192	1.65	39	0.20
	<100>	67	1.65	25	0.38
Tungsten	<100>	390	3.00	86	0.22
Diamond	<111>	1,210	5.4	205	0.17

*Adapted with permission from A. Kelly, *Strong Solids,* 2d ed. (Oxford, U.K.: Clarendon Press, 1973), p. 73.

number of metals. The greatest source of error is γ: it is not easy to determine γ with great precision in solids, and the values used in the table come from different sources and were not necessarily determined at the same temperature.

1.4.2 Theoretical Shear Stress

Frenkel performed a simple calculation of the theoretical shear strength of crystals by considering two adjacent and parallel lines of atoms subjected to a shear stress;[8] this configuration is shown in Figure 1.33 where a is the separation between the adjacent planes and b is the interatomic distance. Under the action of the stress τ, the top line will move in relation to the bottom line; the atoms will pass through successive equilibrium positions A, B, C, for which τ is zero. When the applied shear stress is enough to overcome these barriers, plastic deformation will occur, and the atoms will move until a shear fracture is produced. The stress is also zero when the atoms are exactly superimposed; in that case, the equilibrium is metastable. Between these values the stress varies cyclically with a period b. Frenkel assumed a sine function, as we would expect:

$$\tau = k \sin \frac{2\pi x}{b}. \tag{1.12}$$

For small displacements,

$$\tau \simeq k \frac{2\pi x}{b}. \tag{1.13}$$

Since, for small displacements, one can consider the material to be deformed elastically, we have

$$\tau = G \frac{x}{a}, \tag{1.14}$$

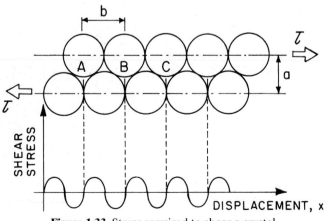

Figure 1.33 Stress required to shear a crystal.

where x/a is the shear strain and G is the shear modulus. Substituting Equation 1.14 into Equation 1.13, we have

$$k = \frac{Gb}{2\pi a}.$$ (1.15)

Substituting Equation 1.15 into Equation 1.12 yields

$$\tau = \frac{Gb}{2\pi a} \sin \frac{2\pi x}{b}.$$

The maximum of τ occurs for $x = b/4$:

$$\tau_{max} = \frac{Gb}{2\pi a}.$$ (1.16)

For FCC materials, the relationship between a_0 (the lattice parameter), a, and b can be calculated. Drawing a unit cell, the student will be able to show that $b = a_0/\sqrt{2}$; the spacing between adjacent planes, is given by (see crystallography textbooks):

$$d_{nkr} = \frac{a_0}{\sqrt{h^2 + k^2 + l^2}}$$

For (111):

$$d_{111} = a_0/\sqrt{3}$$

This is equal to a in Fig. 1.33.
Substituting b and a into Equation 1.16, we obtain

$$\tau_{max} \simeq \frac{G}{5.1}.$$ (1.17)

More complex models have been advanced in which the sine function is replaced by; more precise curves expressing the interaction energy. The method used by Kelly (Mackenzie's method) is an example. Kelly took into account the distortion of the planes. Table 1.5

TABLE 1.5 Theoretical Shear Strength[a]

Element	G (GPa)	τ_{max} (GPa)	τ_{max}/G
Iron	60.0	6.6	0.11
Silver	19.7	0.77	0.039
Gold	19.0	0.74	0.039
Copper	30.8	1.2	0.039
Tungsten	150.0	16.5	0.11
Diamond	505.0	121.0	0.24
NaCl	23.7	2.8	0.12

[a]From A. Kelly, *Strong Solids* (Oxford, U.K. Clarendon Press, 1973), p. 28.

shows the stresses calculated by Mackenzie's method. Note that the ratio τ_{max}/G varies between 0.039 and 0.24. Consequently, it is fairly close to Frenkel's ratio (0.18), obtained by the simpler method.

The theoretical strengths derived in previous sections are on the order of gigapascals; unfortunately, the actual strength of materials is orders of magnitude below that. There are two main reasons this is true.

Crack Growth: Real materials can have small internal cracks, at whose extremities high-stress concentrations are set up. Hence, the theoretical cleavage strength can be achieved at the tip of the crack at applied loads that are only a fraction of that stress. Griffith's theory (see Chapter 7) explains this situation very clearly. These stress concentrations are much lower in ductile materials, since plastic flow can take place at the tip of a crack, blunting the crack's tendency to grow.

Dislocations and Plastic Flow: Before the theoretical shear stress is reached, dislocations are generated and move in the material; if they are already present, they start moving and multiply. These dislocations are elementary carriers of plastic deformation and can move at stresses that are a small fraction of the theoretical shear stress. They will be discussed in detail in Chapter 4.

In sum, cracks prevent brittle materials from obtaining their theoretical cleavage stress, while dislocations prevent ductile materials from obtaining their theoretical shear stress.

EXAMPLE 1.11

Estimate the theoretical shear and cleavage strength for copper and iron. From Table 2.5 in Chapter 2, we have the following data:

$$
\begin{array}{lll}
\text{Iron} & E = 211.4 \text{ GPa} & G = 81.6 \text{ GPa} \\
\text{Copper} & E = 129.8 \text{ GPa} & G = 48.3 \text{ GPa}
\end{array}
$$

For the shear strength, we assume, to a first approximation, that $b = a$. Thus,

$$
\boxed{\tau_{max} = \frac{G}{2\pi}} \quad \text{and} \quad
\begin{array}{ll}
\text{Fe:} & \tau_{max} = 13 \text{ GPa} \\
\text{Cu:} & \tau_{max} = 7.7 \text{ GPa}
\end{array}
$$

For the cleavage strength,

$$
\sigma_{max} = \sqrt{\frac{E\gamma}{a_0}} \quad \text{and} \quad \gamma \sim \frac{Ea_0}{10}.
$$

so

$$
\sigma_{max} \approx \sqrt{\frac{E^2}{10}} \quad \boxed{\approx \frac{E}{3.16}.}
$$

Therefore, we have

$$\text{Fe:}\quad \sigma_{\max} = 66.9 \text{ GPa}$$

$$\text{Cu:}\quad \sigma_{\max} = 41.1 \text{ GPa}$$

The actual tensile strength of pure Fe and Cu is on the order of 0.1 GPa. Since these metals fail by shear, the actual shear strength is equal to 0.05 GPa.

To achieve the theoretical strength of a crystalline lattice, there are two possible methods: (1) eliminating all defects and (2) creating so many defects, that their interactions render them inoperative. The first approach has yielded some materials with extremely high strength. Unfortunately, this has been possible only in special configurations called "whiskers." The second approach is the one more commonly pursued, because of the obvious dimensional limitations of the first; the strength levels achieved in bulk metals have steadily increased by an ingenious combination of strengthening mechanisms, but are still much lower than the theoretical strength. Maraging steels with useful strengths up to 2 GPa have been produced, as have patented steel wires with strengths of up to 4.2 GPa; the latter are the highest strength steels.

Figure 1.34 compares the ambient-temperature strength of tridimensional, filamentary, and whisker materials. The whiskers have a cross-sectional diameter of only a few micrometers and are usually monocrystalline (although polycrystalline whiskers have also been developed). Whiskers are one of the strongest materials developed by human beings. The dramatic effect of the elimination of two dimensions is shown clearly in Figure 1.35. The strongest whiskers are ceramics. Table 1.6 provides some illustrative examples. Iron whiskers with a strength of 12.6 GPa have been produced, compared with 2 GPa for the strongest bulk steels. The value 12.6 GPa is essentially identical to the theoretical shear stress, because the normal stress is twice the shear stress. In general, FCC whiskers tend to be much weaker than BCC whiskers and ceramics. For instance, Cu whiskers have a strength of about 2 GPa. This is consistent with the much lower theoretical shear strength exhibited by copper whiskers. In Table 1.6, silver, gold, and copper have $\tau_{\max}/G$ ratios of 0.039. Hence, they are not good whisker materials. Figure 1.35 shows a stress–strain curve for a copper whisker. The specimen had a length between 2 and 3 mm and a cross-sectional

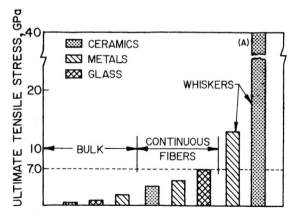

Figure 1.34 Theoretical strength of tridimensional materials, continuous fibers, and whiskers. The strength of the SiC whisker produced by the Philips Eindhoven Laboratory is indicated by (A).

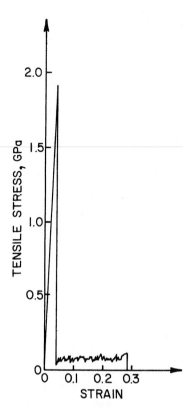

Figure 1.35 Stress–strain curve of a copper whisker with a fiber direction <100>. The whisker diameter is 6.8 μm. (Adapted with permission from K. Yoshida, Y. Goto, and M. Yamamoto, *J. Phys. Soc. Japan,* 21 (1966) 825.)

diameter of 6.8 μm. The stress drops vertically after the yield point, with a subsequent plateau corresponding to the propagation of a Lüders band.

In the elastic range, the curve deviates slightly from Hooke's law and exhibits some temporary inflections and drops (not shown in the figure). In many cases, for both metals and nonmetals, failure occurs at the elastic line, without appreciable plastic strain. When plastic deformation occurs, as, for example, in copper and zinc, a very large yield drop is observed. Although the strength of whiskers is not completely understood, it is connected to the absence of dislocations. It is impossible to produce a material virtually free of dislocations—in other words, perfect. However, for whiskers, dislocations can easily escape out of the material during elastic loading. Their density and mean free path are such that they will not interact and produce other sources of dislocation. Hence, the yield point is the stress re-

TABLE 1.6 Tensile Strength of Whiskers at Room Temperature*

Material	Maximum Tensile Strength (GPa)	Young's Modulus (GPa)
Graphite	19.6	686
Al_2O_2	15.4	532
Iron	12.6	196
SiC	20–40	700
Si	7	182
AlN	7	350
Cu	2	192

*Adapted with permission from A. Kelly, *Strong Solids* (Oxford, U.K.: Clarendon Press, 1973), p. 263.

quired to generate dislocations from surface sources. The irregularities observed in the elastic range indicate that existing dislocations move and escape out of the whisker. At a certain stress, the whisker becomes essentially free of dislocations. When the stress required to activate surface sources is reached, the material yields plastically, or fails.

EXAMPLE 1.12

Calculate the stresses generated in a turbine blade if its cross-sectional area is 10 cm^2 and the mass of each blade is 0.2 kg.

This is an example of a rather severe environment where the material properties must be predicted with considerable detail. For example, the blade may be in a jet engine. Figure E1.11 shows a section of the compressor stage of a jet. The individual blades are fixed by a dovetail arrangement to the turbine vanes. Assume a rotational velocity $\omega = 10,000$ rpm and a mean radius $R = 0.5$ m. The centripetal acceleration in the bottom of each turbine blade is

$$a_c = \omega^2 R = \left[10,000 \times \frac{1}{60} \times 2\pi \right]^2 \times 0.5 = 5.4 \times 10^5 \text{m/s}^2.$$

The stress that is generated is

$$\sigma = \frac{F}{A} = \frac{ma_c}{A} = \frac{0.2 \times 5.4 \times 10^5}{10 \times 10^{-4}} = 100 \text{ MPa},$$

where F is the centripetal force and A is the cross-sectional area. This stress of 100 MPa is significantly below the flow stress of nickel-based superalloys at room temperature, but can be quite significant at higher temperatures.

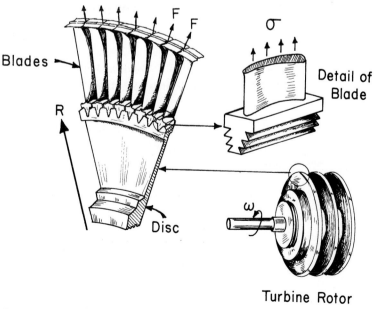

Figure E1.11 Turbine blade subjected to centripetal force during operation.

SUGGESTED READINGS

Materials in General

J. F. SHACKELFORD. *Introduction to Materials Science for Engineers,* 4th ed. Upper Saddle River, NJ: Prentice Hall, 1996.

W. F. SMITH. *Principles of Materials Science and Engineering,* 3rd ed. New York: McGraw Hill, 1996.

D. R. ASKELAND. *The Science and Engineering of Materials,* 3rd ed. Boston: PVUS, 1994.

Metals

C. S. BARRETT and T. B. MASSALSKI. *Structure of Metals,* 3rd rev. ed. Oxford: Pergamon, 1980.

M. A. MEYERS and K. K. CHAWLA. *Mechanical Metallurgy.* Englewood Cliffs, NJ: Prentice-Hall, 1984.

Ceramics

W. D. KINGERY, H. K. BOWEN, and D. R. UHLMANN. *Introduction to Ceramics,* 2nd ed. New York, J. Wiley, 1976.

Y.-M. CHIANG, D. BIRNIE III, and W. D. KINGERY, *Physical Ceramics,* New York, J. Wiley, 1997.

Polymers

D. C. BASSETT. *Principles of Polymer Morphology.* Cambridge, U.K.: Cambridge University Press, 1981.

A. HILTNER (ed.). *Structure–Property Relationships of Polymeric Solids*. New York: Plenum Press, 1983.

R. J. YOUNG. *Introduction to Polymers.* London: Chapman and Hall, 1986.

B. WUNDERLICH. *Macromolecular Physics, Vol. 1: Crystal Structure.* New York: Academic Press, 1973.

B. WUNDERLICH. *Macromolecular Physics, Vol. 2: Crystal Nucleation.* New York: Academic Press, 1976.

Composite Materials

K. K. CHAWLA. *Composite Materials: Science & Engineering.* 2nd ed. New York: Springer-Verlag, 1998.

Quasicrystals

D. GRATIAS. "Quasicrystals." *Contemp. Phys.* 28:3 (1987) 219.

D. R. NELSON. "Quasicrystals." *Sci. Amer.* 254:8 (Aug. 1986) 43.

Liquid Crystals

A. CIFERRI, W. R. KRIGBAUM, and R. B. MEYER (eds.). *Polymer Liquid Crystals.* New York: Academic Press, 1982.

Biomaterials

J. F. V. VINCENT. *Structural Biomaterials.* Princeton, NJ: Princeton University Press, 1991.

Y. C. FUNG. *Biomechanics: Mechanical Properties of Living Tissues.* New York: Springer, 1981.

Cellular Materials

L. J. GIBSON and M. F. ASHBY. *Cellular Solids: Structure and Properties.* Oxford, U.K.: Pergamon Press, 1988

EXERCISES

Section 1.3

1.1   A jet turbine rotates at a velocity of 15,000 rpm. Calculate the stress acting on the turbine blades if the turbine disc radius is 70 cm and the cross-sectional area is 15 cm².

1.2   The material of the jet turbine blade in Problem 1.1, Superalloy IN 718, has a room-temperature yield strength equal to 200 MPa; it decreases with temperature as

$$\sigma = \sigma_0\left(1 - \frac{T - T_0}{T_m - T}\right)$$

where T_0 is the room temperature and T_m is the melting temperature in K ($T_m = 1,700$ K). At what temperature will the turbine flow plastically under the influence of contripetal forces?

1.3   **(a)**   Describe the mechanical properties that are desired for a tennis racket, and recommend different materials for the different parts of the racket.

   (b)   Describe the mechanical properties that are desired in a golf club, and recommend different materials for the different parts of the club.

Section 1.4

1.4   On eight cubes that have one common vertex, corresponding to the origin of axes, draw the family of {111} planes. Show that they form an octahedron and indicate all <110> directions.

1.5   The frequency of loading is an important parameter in fatigue. Estimate the frequency of loading (in cycles per second, or Hz) of an automobile tire in the radial direction when the car speed is 100 km/h and the wheel diameter is 0.5 m.

1.6   Indicate, by their indices and in a drawing, six directions of the <112> family.

1.7   The density of Cu is 8.9 g/cm³ and its atomic weight (or mass) is 63.546. It has the FCC structure. Determine the lattice parameter and the radius of atoms.

1.8   The lattice parameter for W(BCC) is $a = 0.32$ nm. Calculate the density, knowing that the atomic weight (or mass) of W is 183.85.

1.9   The unit cell of the CsCl is shown in Figure 1.17 (NaCl structure). The radius of Cs^+ is 0.169 nm and that of Cl is 0.181 nm. (a) Determine the packing factor of the structure, assuming that Cs^+ and Cl^- ions touch each other along the diagonals of the cube. (b) Determine the density of CsCl if the atomic weight of Cs is 132.905 and of Cl is 35.453.

1.10 MgO has the same structure as NaCl. If the radii of O^{2-} and Mg^{2+} ions are 0.14 nm and 0.070 nm, respectively, determine (a) the packing factor and (b) the density of the material. The atomic weight of O_2 is 16 and that of Mg is 24.3.

1.11 Germanium has the diamond cubic structure with interatomic spacing of 0.245 nm. Calculate the packing factor and density. (The atomic weight of germanium is 72.6.)

1.12 The basic unit (or mer) of polytetrafluoroethylene (PTFE) or teflon is C_2F_4. If the mass of the PRFE molecule is 45,000 amu, what is the degree of polymerization?

1.13 Using the representation of the orthorhombic unit cell of polyethylene (see Figure Ex1.13), calculate the theoretical density. How does this value compare with the density values of polyethylene obtained in practice?

1.14 A pitch blend sample has five different molecular species with molecular masses of 0.5×10^6, $0.5 \times 10^7, 1 \times 10^7, 4 \times 10^7$, and 6×10^7. Compute the number-averaged molecular weight and weight-averaged molecular weight of the sample.

1.15 Determine the theoretical cleavage stress for tungsten pulled in tension along [100] if the surface energy is 1,650 mJ/m² and $E_{[100]} = 411$ GPa. The atomic radius for tungsten is 0.137 nm.

1.16 Determine the theoretical cleavage stress for tungsten using the approximate expression.

1.17 Calculate the theoretical shear strength for tungsten (G = 166 GPa).

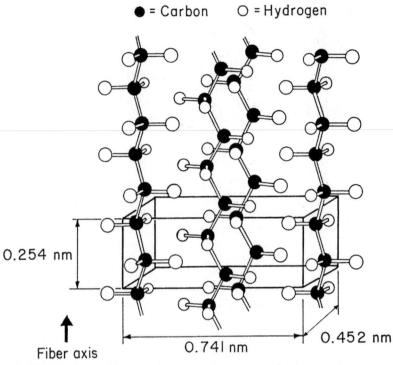

Figure E1.13 Crystalline form of polyethylene with orthorhombic unit cell.

1.18 Determine the theoretical cleavage strength for α Si_3N_4 if its surface energy is equal to $30\,J/m^2$ and Young's modulus is equal to 360 GPa; the unit cell dimensions are $a = 0.775$ nm and $c = 5.62$ nm.

1.19 For a cubic system, calculate the angle between
  (a) [100] and [111]
  (b) [111] and [11$\bar{2}$]
  (c) [11$\bar{2}$] and [221]

1.20 Recalculate the bicycle stiffness ratio for a titanium frame. (See Examples 1.1 and 1.2) Find the stiffness and weight of the bicycle if the radius of the tube is 25 mm. Assume the following information:

$$\text{Alloy: Ti—6\% Al—4\% V}$$
$$\sigma_e = 1{,}150\text{ MPa}$$
$$\text{Density} = 4.5\text{ g/cm}^3$$
$$E = 106\text{ GPa}$$
$$G = 40\text{ GPa}$$

1.21 Calculate the packing factor for NaCl, given that $r_{Na} = 0.186$ nm and $R_{cl} = 0.107$ nm.

1.22 Determine the density of BCC iron structure if the iron atom has a radius of 0.124 nm.

1.23 Draw the following direction vectors in a cubic unit cell:

a [100] and [110], b [112], c [$\bar{1}$10], d[$\bar{3}2\bar{1}$].

2

Elasticity and Viscoelasticity

2.1 INTRODUCTION

Elasticity deals with elastic stresses and strains, their relationship, and the external forces that cause them. An *elastic strain* is defined as a strain that disappears instantaneously once the forces that cause it are removed. The theory of elasticity for Hookean solids—in which stress is proportional to strain—is rather complex in its more rigorous treatment. However, it is essential to the understanding of micro- and macromechanical problems. Examples of the former are stress fields around dislocations, incompatibilities of stresses at the interface between grains, and dislocation interactions in work hardening; examples of the latter are the stresses developed in drawing, and rolling wire, and the analysis of specimen–machine interactions in testing for tensile strength. This chapter is structured in such a way as to satisfy the needs of both the undergraduate and the graduate student. A simplified treatment of elasticity is presented, in a manner so as to treat problems in an undergraduate course. Stresses and strains are calculated for a few simplified cases; the tridimensional treatment is kept at a minimum. A graphical method for the solution of two-dimensional stress problems (the Mohr circle) is described. On the other hand, the graduate student needs more powerful tools to handle problems that are somewhat more involved. In most cases, the stress and strain systems in tridimensional bodies can be better treated as tensors, with the indicial notation. Once this tensor approach is understood, the student will have acquired a very helpful visualization of stresses and strains as tridimensional entities. Important problems whose solutions require this kind of treatment involve stresses around dislocations, interactions between dislocations and solute atoms, fracture mechanics, plastic waves in solids, stress concentrations caused by precipitates, the anistropy of individual grains, and the stress state in a composite material.

2.2 LONGITUDINAL STRESS AND STRAIN

Figure 2.1 shows a cylindrical specimen being stressed in a machine that tests materials for tensile strength. The upper part of the specimen is screwed to the crosshead of the machine. The coupled rotation of the two lateral screws causes the crosshead to move. The load cell is a transducer that measures the load and sends it to a recorder; the increase in length of the specimen can be read by strain gages, extensometers, or, indirectly, from the velocity of motion of the crosshead. Another type of machine, called a servohydraulic machine, is also used. Assuming that at a certain moment the force applied on the specimen by the machine is F, there will be a tendency to "stretch" the specimen, breaking the internal bonds. This breaking tendency is opposed by internal reactions, called *stresses*. The best way of visualizing stresses is by means of the method of analysis used in the mechanics of materials: The specimen is "sectioned," and the missing part is replaced by the forces that it exerts on the other parts. This procedure is indicated in the figure. In the situation shown, the "resistance" is uniformly distributed over the normal section and is represented by three modest arrows at A. The normal stress σ is defined as this "resistance" per unit area. Applying the equilibrium-of-forces equation from the mechanics of materials to the lower portion of the specimen, we have

$$\sum F = 0$$

$$F - \sigma A = 0$$

$$\sigma = \frac{F}{A} \tag{2.1}$$

This is the internal resisting stress opposing the externally applied load and avoiding the breaking of the specimen. The following stress convention is used: Tensile stresses are positive and compressive stresses are negative. In geology and rock mechanics, on the other hand, the opposite sign convention is used because compressive stresses are much more common.

As the applied force F increases, so does the length of the specimen. For an increase dF, the length l increases by dl. The normalized (per unit length) increase in length is equal to

$$d\varepsilon = \frac{dl}{l},$$

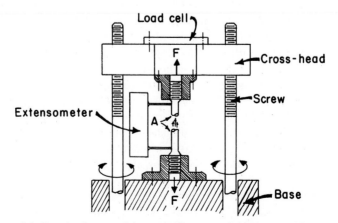

Figure 2.1 Sketch of screw-driven tensile-strength testing machine.

or, upon integration,

$$\varepsilon = \int_{l_0}^{l_1} \frac{dl}{l} = \ln \frac{l_1}{l_0} \tag{2.2}$$

where l_0 is the original length. This parameter is known as the *longitudinal true strain.*

In many applications, a simpler form of strain, commonly called *engineering* or *nominal strain,* is used. This type of strain is defined as

$$\varepsilon_n \equiv \varepsilon_e = \frac{\Delta \ell}{\ell_0} = \frac{\ell_1}{\ell_0} - 1. \tag{2.2a}$$

When the strains are reasonably small, the engineering (or nominal) and true strains are approximately the same. We will use subscripts t for true values and e for engineering values. It can be easily shown that

$$\varepsilon_t = \ln(1 + \varepsilon_e). \tag{2.2b}$$

The elastic deformation in metals and ceramics rarely exceeds 0.005, and for this value, the difference between ε_t and ε_e can be neglected.

In a likewise fashion, a *nominal* (or *engineering*) *stress* is defined as

$$\sigma_e = \frac{F}{A_0}. \tag{2.2c}$$

where A_0 is the original area of cross-section.

The relationship between the true stress and the engineering stress is

$$\frac{\sigma_t}{\sigma_e} = \frac{A_0}{A}.$$

During elastic deformation, the change in cross-sectional area is less than 1% for most metals and ceramics; thus $\sigma_e \cong \sigma_t$. However, during plastic deformation, the differences between the true and the engineering values become progressively larger. More details are provided in Chapter 3 (Section 3.1.2).

The sign convention for strains is the same as that for stresses: Tensile strains are positive, compressive strains are negative. In Figure 2.2, two stress–strain curves (in tension) are shown; both specimens exhibit elastic behavior. The full lines describe the loading trajectory and the dashed lines describe the unloading. For perfectly elastic solids, the two kinds of lines should coincide if thermal effects are neglected. The curve of Figure 2.2(a) is characteristic of metals and ceramics; the elastic regime can be satisfactorily described by a straight line. The curve of Figure 2.2(b) is charcteristic of rubber; σ and ε are not proportional. Nevertheless, the strain returns to zero once the stress is removed. The reader can verify this by stretching a rubber band. First, you will notice that the resistance to stretching increases slightly with extension. After considerable deformation, the rubber band "stiffens up," and further deformation will eventually lead to rupture. The whole process (except failure) is elastic. A conceptual error often made is to assume that elastic behavior is *always* linear; the latter example shows very clearly that there are notable exceptions. However, for metals, the stress and strain can be assumed to be proportional in the elastic regime; these materials are known as Hookean solids. For polymers, viscoelastic effects are very important. Viscoelasticity results in different trajectories for loading and unloading, with the formation of a hysteresis loop. The area of the hysteresis loop is

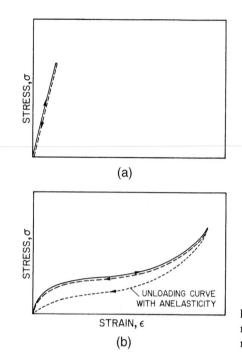

Figure 2.2 Stress–strain curves in an elastic regime. (a) Typical curve for metals and ceramics. (b) Typical curve for rubber.

the energy lost per unit volume in the entire deformation cycle. Metals also exhibit some viscoelasticity, but it is most often neglected. Viscoelasticity is due to time-dependent microscopic processes accompanying deformation. An analogy that applies well is the attachment of a spring and dashpot. The spring represents the elastic portion of the material, the dashpot the viscoelastic portion.

In 1678, Robert Hooke performed experiments that demonstrated the proportionality between stress and strain. He proposed his law as an anagram—"ceiiinosssttuv," which rearranged, forms the Latin *Ut tensio sic vis.* The meaning is "As the tension goes, so does the stretch." In its most simplified form, we express this law as

$$E = \frac{\sigma}{\varepsilon}, \tag{2.3}$$

where E is Young's modulus. For metals and ceramics, E has a very high value—for example, 210 GPa for iron. Chapter 4 devotes some effort to the derivation of E for materials from first principles. E depends mainly on the composition, crystallographic structure, and nature of the bonding of elements. Heat and mechanical treatments have little effect on E, as long as they do not affect the former parameters. Hence, annealed and cold-rolled steel should have the same Young's modulus; there are, of course, small differences due to the formation of the cold-rolling texture. E decreases slightly with increases in temperature.

In monocrystals, E shows different values for different crystallographic orientations. In polycrystalline aggregates that do not exhibit any texture, E is isotropic: It has the same value in all directions. The values of E given in tables (e.g., Tables 2.3–2.5) are usually obtained by dynamic methods involving the propagation of elastic waves, not from conventional stress–strain tests. An elastic wave is passed through a sample; the velocities of the longitudinal and shear waves, V_ℓ and V_s, respectively, are related to the elastic constants by means of the following mathematical expressions (ρ is the density, E is Young's modulus, and G is the shear modulus):[1]

$$V_\ell = \sqrt{\frac{E}{\rho}} \qquad V_s = \sqrt{\frac{G}{\rho}}.$$

EXAMPLE 2.1

Calculate the material properties E, G, and ν of SiC, given the graphs of the longitudinal and shear sound velocities obtained using ultrasonic equipment. (See Figure E2.1). Here, $\rho = 3.18 \times 10^3$ kg/m³ and the length of specimen is $L = 4$ mm.

Solution: We take equivalent peaks, marked by arrows, in sequential signal packets. We must remember that the pulse reflects at the free surface, and therefore, we have to take twice the length of the pulse. We have

$$V_\ell = \frac{2L}{t_2 - t_1} = \frac{2 \times 4 \times 10^{-3}}{(1.16 - 0.52) \times 10^{-6}} = 12.5 \times 10^3 \text{m/s},$$

$$V_\ell = \sqrt{\frac{E}{\rho}},$$

$$E = \rho V_\ell^2 = 3.18 \times 10^3 \times (12.5 \times 10^3)^2 = 496.9 \times 10^9 \text{ Pa} = 496.9 \text{ GPa},$$

$$V_s = \frac{2L}{t_4 - t_3} = \frac{2 \times 4 \times 10^{-3}}{(2.15 - 1.10) \times 10^{-6}} = 7.62 \times 10^3 \text{m/s}$$

$$V_s = \sqrt{\frac{G}{\rho}},$$

$$G = \rho V_s^2 = 3.18 \times 10^3 \times (7.62 \times 10^3)^2 = 184.6 \times 10^9 \text{ Pa} = 184.6 \text{ GPa}.$$

Since, according to Table 2.2,

$$G = \frac{E}{2(1 + \nu)},$$

where ν is Poisson's ratio, explained in Section 2.4, it follows that

$$\nu = \frac{E}{2G} - 1 = \frac{496.9}{2 \times 184.6} - 1 = 0.346.$$

(*Note:* The preceding calculations were conducted assuming uniaxial stress and without the dispersion correction; hence, the results are only approximate. A correct equation from the elastic modulus would be

$$V_\ell = \sqrt{\frac{\bar{\bar{E}}}{\rho}}, \qquad \bar{E} = \frac{(1 - \nu)}{(1 + \nu)(1 - 2\nu)} E.$$

[1]For more details, see M. A. Meyers, *Dynamic Behavior of Materials* (New York: Wiley, 1994).

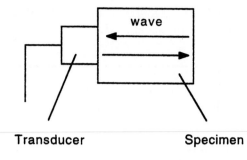

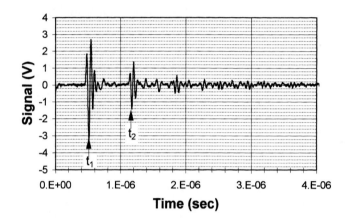

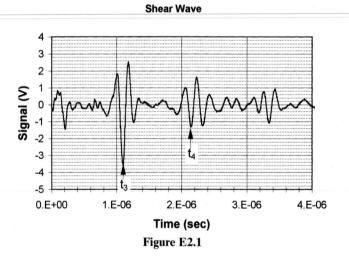

Figure E2.1

This is due to the fact that the length of the pulse is much shorter than the lateral dimension of the specimen, and therefore, the specimen is stressed in uniaxial strain. [2]

---

[2]The interested student can obtain more information in M. A. Meyers, *Dynamic Behavior of Materials* (New York: Wiley, 1994).

2.3 *SHEAR STRESS AND STRAIN*

Imagine the loading arrangement shown in Figure 2.3(a). The specimen is placed between a punch and a base having a cylindrical orifice; the punch compresses the specimen. The internal resistance to the external forces now has the nature of a shear. The small cube in Figure 2.3(b) was removed from the region being sheared (between punch and base). It is distorted in such a way that the perpendicularity of the faces is lost. The shear stresses and strains are defined as

$$\tau = \frac{F}{A}, \qquad \gamma = \frac{dl}{l} = \tan\theta \cong \theta. \tag{2.4}$$

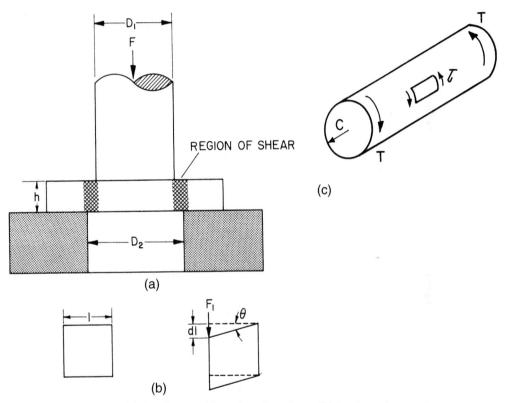

REGION OF SHEAR

(c)

(a)

(b)

Figure 2.3 (a) Specimen subjected to shear force. (b) Strain undergone by small cube in shear region. (c) Specimen (cylinder) subjected to torsion.

The sign convention for shear stresses is given in Section 2.6. The area of the surface that undergoes shear is

$$A \cong \pi\left(\frac{D_1 + D_2}{2}\right)h.$$

The average of the two diameters is taken because D_2 is slightly larger than D_1.

A mechanical test commonly used to find shear stresses and strains is the torsion test. The equations that give the shear stresses and strains in terms of the torque are given in texts on the mechanics of materials. Figure 2.3(c) shows a cylindrical specimen subjected to a torque T. The relationship between the torque and the shear stresses that are generated is given by[3]

$$\tau_{max} = \frac{Tc}{J},$$

where c is the radius of the cylinder and $J = \pi c^4/2$ is the polar moment of inertia. Tubular specimens are preferred over solid cylinders because the shear stress can be approximated as constant over the cross section of the cylinder. For a hollow cylinder with b and c as inner and outer radii, respectively, we subtract out (the hollow part to obtain)

$$J = \frac{\pi c^4}{2} - \frac{\pi b^4}{2}.$$

For metals, ceramics, and certain polymers (the Hookean solids), the proportionality between τ and γ is observed in the elastic regime. In analogy with Young's modulus, a transverse elasticity, called, the *rigidity*, or *shear modulus*, is defined as

$$G = \frac{\tau}{\gamma}. \tag{2.5}$$

G, which is numerically less than E, is related to E by Poisson's ratio, discussed in the next section. Values of G for different materials are given in Table 2.5; it can be seen that G varies between one-third and one-half of E.

EXAMPLE 2.2

A cylindrical steel specimen (length = 200 mm, diameter = 5 mm), is subjectd to a torque equal to 40 N · m.
a) What is the deflection of the specimen end, if one end is fixed?
b) Will the specimen undergo plastic deformation?

Given: $E = 210$ GPa

$\nu = 0.3$

$\sigma_y = 300$ MPa

Solution:

a) $\tau_{max} = \frac{T \cdot c}{J}$ \hfill (1)

Given

$$T = 40 \text{ N} \cdot \text{m}, \qquad c = \frac{d}{2} = 2.5 \text{ mm}$$

[3]See E. P. Popov, *Engineering Mechanics of Solids* (Englewood Cliffs, NJ: Prentice Hall, 1990).

To calculate τ_{max}, we need to know J.

$$J = \pi \frac{c^4}{2} \qquad (2)$$

Substitute (2) into (1).

$$\tau_{max} = \frac{T \cdot c}{\pi \left(\dfrac{c^4}{2} \right)} = \frac{2T}{\pi c^3} = \frac{2 \cdot 40}{\pi \cdot (2.5)^3} \frac{\text{N} \cdot \text{m}}{\text{mm}^3}$$

$$= 1630 \text{ MPa}$$

$$= 1.63 \text{ GPa}$$

Shear stress and shear strain are related as

$$\tau = G\gamma$$

G can be calculated from E and ν.

$$G = \frac{E}{2(1 + \nu)} = \frac{210}{2(1 + 0.3)} = 81 \text{ GPa}$$

$$\gamma = \frac{\tau_{max}}{G} = \frac{1.63}{81} = 0.02$$

But,

$$\gamma = \frac{c\theta}{L}$$

where θ is the angle of rotation.

$$\text{Torsional deflection} = \frac{\text{angle of rotation}}{\text{length}}$$

$$= \frac{\theta}{L}$$

$$= \frac{\gamma}{c} = \frac{0.02}{2.5} = 0.008 \, \frac{\text{radians}}{\text{mm}}.$$

b) $\qquad\qquad\qquad\qquad \tau_{max} = 1.63 \text{ GPa}.$

The shear stress required to cause permanent deformation is related to the yield stress as follows:

$$\tau_y = \frac{\sigma_y}{2}.$$

When the stress is σ_y,

$$\tau = \frac{300}{2} = 150 \text{ MPa}.$$

But from (a), $\tau_{max} = 1.63$ GPa > 150 MPa. Therefore, the specimen will undergo plastic deformation.

EXAMPLE 2.3

What is the strain energy density in a low-carbon steel sample loaded to its elastic limit of 500 MPa?

Solution: Take E for a low-carbon steel to be 210 GPa. For such a material under a stress σ, we have a strain energy density given by

$$U = \frac{1}{2}\frac{\sigma^2}{E} = \frac{1}{2}\frac{(500 \times 10^6)^2}{210 \times 10^9}$$

$$= 595 \text{ kJ/m}^3.$$

2.4 POISSON'S RATIO

A body, upon being pulled in tension, tends to contract laterally. The cube shown in Figure 2.4 exhibits this behavior. The stresses are now defined in a tridimensional body, and they have two indices. The first indicates the plane (or the normal to the plane) on which they are acting; the second indicates the direction in which they are pointing. These stresses are schematically shown acting on three faces of a unit cube in Figure 2.4a. The normal stresses have two identical subscripts: $\sigma_{11}, \sigma_{22}, \sigma_{33}$. The shear stresses have two different subscripts: $\sigma_{12}, \sigma_{13}, \sigma_{23}$. These subscripts refer to the reference system $Ox_1x_2x_3$. If this notation is used, both normal and shear stresses are designated by the same letter, lower case sigma. On the other hand, in more simplified cases where we are dealing with only one normal and one shear stress component, σ and τ will be used, respectively; this notation will be maintained throughout the text. In Figure 2.4, the stress σ_{33} generates strains $\varepsilon_{11}, \varepsilon_{22}, \varepsilon_{33}$. (The same convention is used for stresses and strains.) Since the initial dimensions of the cube are equal to 1, the changes in length are equal to the strains. Poisson's ratio is defined as the ratio between the lateral and the longitudinal strains. Both ε_{11} and ε_{22} are negative (signifying a decrease in length), and ε_{33} is positive. In order for Poisson's ratio to be positive, the negative sign is used. Hence,

$$\nu = -\frac{\varepsilon_{11}}{\varepsilon_{33}} = -\frac{\varepsilon_{22}}{\varepsilon_{33}}. \tag{2.6}$$

In an isotropic material, ε_{11} is equal to ε_{22}. We can calculate the value of ν for two extreme cases: (1) when the volume remains constant and (2) when there is no lateral contraction. When the volume is constant, the initial and final volumes, V_0 and V, respectively, are equal to

$$V_0 = 1,$$
$$V = (1 + \varepsilon_{11})(1 + \varepsilon_{22})(1 + \varepsilon_{33}).$$

Neglecting the cross products of the strains, because they are orders of magnitude smaller than the strains themselves, we have

$$V = 1 + \varepsilon_{11} + \varepsilon_{22} + \varepsilon_{33}.$$

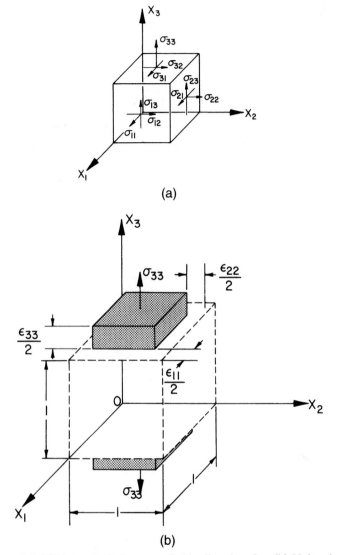

Figure 2.4 (a)Unit cube being extended in direction Ox_3. (b) Unit cube in body subjected to tridimensional stress; only stresses on the three exposed faces of the cube are shown.

Since $V = V_0$,

$$\varepsilon_{11} + \varepsilon_{22} + \varepsilon_{33} = 0.$$

For the isotropic case, the two lateral contractions are the same ($\varepsilon_{11} = \varepsilon_{22}$). Hence,

$$2\varepsilon_{11} = -\varepsilon_{33}. \tag{2.7}$$

Substituting Eq. 2.7 into Eq. 2.6, we arrive at

$$\nu = 0.5.$$

For the case in which there is no lateral contraction, ν is equal to zero. Poisson's ratio for metals is usually around 0.3. (See Table 2.5.) The values given in the table apply to the elastic regime; in the plastic regime, ν increases to 0.5, since the volume remains constant. Figure 2.2 shows the variation of Poisson's ratio with strain.

Poisson's ratio for cork (and other cellular materials) is approximately zero, which means that we can push cork into a glass bottle without expanding the bottle. The student should try to do this with a *rubber* cork ($\nu \sim 0.5$)!

2.5   MORE COMPLEX STATES OF STRESS

The relationships between stress and strain described in Sections 2.2 and 2.3 are unidimensional or uniaxial stress states, and do not apply to bidimensional and tridimensional states of stress. The most general state of stress can be represented by the unit cube of Figure 2.4(b). The generalized Hooke's law (as the set of equations relating tridimensional stresses and strains is called) is derived next, for an isotropic solid. It is assumed that shear stresses can generate *only* shear strains. Thus, the longitudinal strains are produced exclusively by the normal stresses. σ_{11} generates the following strain:

$$\varepsilon_{11} = \frac{\sigma_{11}}{E}. \tag{2.8}$$

Since $\nu = -\varepsilon_{22}/\varepsilon_{11} = -\varepsilon_{33}/\varepsilon_{11}$ for stress σ_{11}, we also have

$$\varepsilon_{22} = \varepsilon_{33} = -\frac{\nu\sigma_{11}}{E}.$$

The stress σ_{22}, in its turn, generates the following strains:

$$\varepsilon_{22} = \frac{\sigma_{22}}{E} \quad \text{and} \quad \varepsilon_{11} = \varepsilon_{33} = -\frac{\nu\sigma_{22}}{E}. \tag{2.9}$$

For σ_{33},

$$\varepsilon_{33} = \frac{\sigma_{33}}{E} \quad \text{and} \quad \varepsilon_{11} = \varepsilon_{22} = -\frac{\nu\sigma_{33}}{E}. \tag{2.10}$$

In this treatment, the shear stresses generate only shear strains:

$$\gamma_{12} = \frac{\sigma_{12}}{G}, \qquad \gamma_{13} = \frac{\sigma_{13}}{G}, \qquad \gamma_{23} = \frac{\sigma_{23}}{G}.$$

The second simplifying assumption is called the "principle of superposition." The total strain in one direction is considered to be equal to the sum of the strains generated by the various stresses along that direction. Hence, the total ε_{11} is the sum of ε_{11} produced by σ_{11}, σ_{22}, and σ_{33}. Adding strains from Equations 2.8 through 2.10, we obtain

$$\varepsilon_{11} = \frac{1}{E}[\sigma_{11} - \nu(\sigma_{22} + \sigma_{33})].$$

Similarly,

$$\varepsilon_{22} = \frac{1}{E}[\sigma_{22} - \nu(\sigma_{11} + \sigma_{33})],$$

$$\varepsilon_{33} = \frac{1}{E}[\sigma_{33} - \nu(\sigma_{11} + \sigma_{22})],$$

$$\gamma_{12} = \frac{\sigma_{12}}{G}, \qquad \gamma_{13} = \frac{\sigma_{13}}{G}, \qquad \gamma_{23} = \frac{\sigma_{23}}{G}. \qquad (2.11)$$

Applying these equations to a hydrostatic stress situation ($\sigma_{11} = \sigma_{22} = \sigma_{33} = -p$), we can see perfectly that there are no distortions in the cube ($\gamma_{12} = \gamma_{13} = \gamma_{23} = 0$) and that $\varepsilon_{11} = \varepsilon_{22} = \varepsilon_{33}$.

The triaxial state of stress is difficult to treat in elasticity (and even more difficult in plasticity). In the great majority of cases, we try to assume a more simplified state of stress that resembles the tridimensional stress. This is often justified by the geometry of the body and by the loading configuration. The example discussed in Section 2.2 is the simplest state (uniaxial stress). It occurs when beams are axially loaded (in tension or compression). In sheets and plates (where one dimension can be neglected with respect to the other two), the state of stress can be assumed to be bidimensional. This state of stress is also known as *plane stress,* because normal stresses (normal to the surface) are zero at the surface, as are shear stresses (parallel to the surface) at the surface. In Figure 2.4(a), one would be left with $\sigma_{11}, \sigma_{12}, \sigma_{22}$ if Ox_1x_2 were the plane of the sheet. Since the sheet is thin, there is no space for buildup of the stresses that are zero at the surface. The solution to this problem is approached graphically in Section 2.6. The opposite case, in which one of the dimensions is infinite with respect to the other two, is treated under the assumption of plane strain. If one dimension is infinite, strain in it is constrained; hence, one has two dimensions left. This state is called *bidimensional* or, more commonly, *plane strain.* It also occurs when strain is constrained in one direction by some other means. A long dam is an example in which deformation in the direction of the dam is constrained. Yet another state of stress is pure shear, when there are no normal stresses.

EXAMPLE 2.4

Consider a plate under uniaxial tension that is prevented from contracting in the transverse direction. Find the effective modulus along the loading direction under this condition of plane strain.

Solution: Take

$$E = \text{Young's modulus}, \qquad \nu = \text{Poisson's ratio}$$

Let the loading and transverse directions be 1 and 2, respectively. There is no stress normal to the free surface, i.e., $\sigma_3 = 0$. Although the applied stress is uniaxial, the constraint on contraction in direction 2 results in a stress in that direction also.

The strain in direction 2 can be written in terms of Hooke's law as

$$\varepsilon_2 = 0 = (1/E)[\sigma_2 - \nu\sigma_1].$$

Thus, $\sigma_2 = \nu\sigma_1$.

In direction 1, we can write, for the strain,

$$\varepsilon_1 = (1/E)[\sigma_1 - \nu\sigma_2] = (1/E)[\sigma_1 - \nu^2\sigma_1]$$
$$= (\sigma_1/E)(1 - \nu^2).$$

Hence, the plane strain modulus in direction 1 is

$$E' = (\sigma_1/\varepsilon_1) = E/(1 - \nu^2).$$

If we take $\nu = 0.33$, then the plane strain modulus $E' = 1.12E$.

EXAMPLE 2.5

An isotropic, linear, elastic material is compressed by a force P by means of a punch in a rigid die. The material has a Young's modulus E and a Poisson's ratio ν. The displacement of the material is Δ, and the cavity has a height h and a square base of side a. (See Figure E2.5.) Determine the stress and strain components. Also, determine the relationship between P and the displacement Δ of the material.

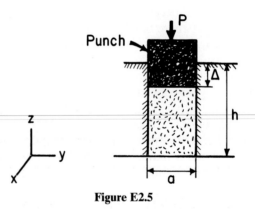

Figure E2.5

Solution: This is a three-dimensional problem. There are no shear shear strains, and the only nonzero normal strain component is ε_z as shown in the figure. The normal strains in the x- and y-directions are zero, because the rigid die does not allow deformation in these directions. However, the stress components in these directions are not zero. We use the generalized Hooke's law to obtain the three stress components. We can write, for the strain components,

$$\varepsilon_z = -\Delta/h, \qquad \varepsilon_x = \varepsilon_y = \varepsilon_{xy} = \varepsilon_{yz} = \varepsilon_{zx} = 0.$$

Now we can write the following constitutive relationships by inverting Equation (2.11) and using x, y, and z instead of 1, 2, and 3:

$$\sigma_x = E/[(1 + \nu)(1 - 2\nu)][(1 - \nu)\varepsilon_x + \nu(\varepsilon_y + \varepsilon_z)]$$
$$= E/[(1 + \nu)(1 - 2\nu)] \cdot [0 + \nu(-\Delta/h)],$$

or

$$\sigma_x = -[E\nu/(1 + \nu)(1 - 2\nu)][\Delta/h].$$

Similarly,

$$\sigma_y = E/[(1 + \nu)(1 - 2\nu)][(1 - \nu)\varepsilon_y + \nu(\varepsilon_x + \varepsilon_z)]$$
$$= E/[(1 + \nu)(1 - 2\nu)] \cdot [0 + \nu(-\Delta/h)],$$

or

$$\sigma_y = -E\nu/[(1 + \nu)(1 - 2\nu)][\Delta/h].$$

Finally,

$$\sigma_z = E/[(1 + \nu)(1 - 2\nu)][(1 - \nu)\varepsilon_z + \nu(\varepsilon_y + \varepsilon_x)]$$
$$= E/[(1 + \nu)(1 - 2\nu)] \cdot [(1 - \nu)(-\Delta/h) + 0],$$

or

$$\sigma_z = -E(1 - \nu)/[(1 + \nu)(1 - 2\nu)][\Delta/h].$$

The load–displacement relationship is obtained by writing

$$P = \sigma_z a^2,$$
$$\sigma_z = P/a^2 = -E(1 - \nu)/[(1 + \nu)(1 - 2\nu)][\Delta/h],$$

or

$$P = -Ea^2(1 - \nu)(\Delta/h)/[(1 + \nu)(1 - 2\nu)]$$

Note the linear relationship between P and Δ, as it should be because the material in the cavity is linear elastic.

2.6 GRAPHICAL SOLUTION OF A BIAXIAL STATE OF STRESS: THE MOHR CIRCLE

Figure 2.5(a) shows a biaxial (or bidimensional) state of stress. The graphical scheme developed by O. Mohr allows the determination of the normal and shear stresses in any orientation in the plane. The reader should be warned, right at the onset, that a *change in sign convention* for the shear stresses has to be introduced here. The former sign convention—positive shear stresses pointing toward the positive direction of axes in faces shown in Figure 2.4a—has to be *temporarily* abandoned and the following convention adopted: Positive shear stresses produce counterclockwise rotation of a cube (or square), and negative shear stresses produce clockwise rotation. The sign convention for normal stresses remains the same. Figure 2.5b shows Mohr's construction. The normal stresses are plotted on the abcissa, while the shear stresses are plotted on the ordinate axis. Point A in the diagram corresponds to a state of stress on the face of the cube perpendicular to Ox_1; point B represents the state of stress on the face perpendicular to Ox_2. From A and B, we construct a circle with center in the axis of the abcissa and passing through A and B. The center is the point where the segment AB intersects the abcissa. Note that the center occurs at $(\sigma_{11} + \sigma_{22})/2$. The stress states for all orientations of the cube (in the same plane) correspond to points diametrically opposed in Mohr's circle. Hence, we can determine the state of stress for any orientation.

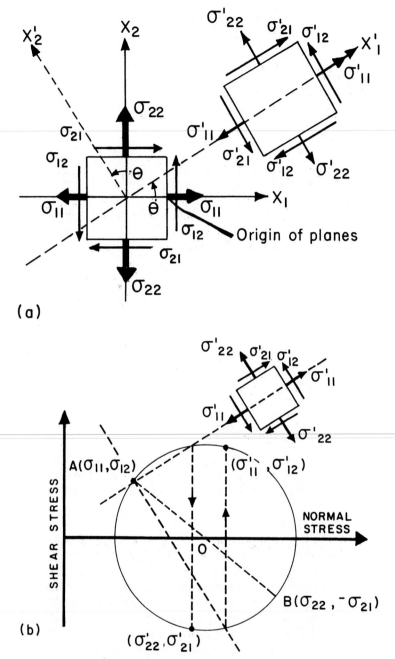

Figure 2.5 (a)Biaxial (or bidimensional) state of stress. (b) Mohr circle and construction of general orientation $0x_1'x_2'$. (c) Mohr circle and construction of principal stresses and maximum shear stresses.

There are different methods of operating on the Mohr circle. One of these is as follows: Point A (the stress system on the right-hand of the cube face) is called the "origin of planes." We will always start from it on the Mohr circle. We solve two problems.

(a) First, we determine the stresses on a general coordinate direction ox_1', ox_2'. This is shown in Figure 2.5b. Lines are drawn through A (the origin of planes) parallel to ox_1' and ox_2'. We seek the intersection of the axes with the circle. We draw lines perpendicular to

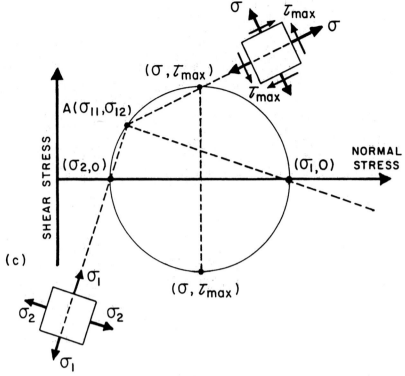

Figure 2.5 (Continued)

the normal stress axis and find the new intersection. Thus, $(\sigma'_{11}, \tau'_{12})$ represents stresses on the face perpendicular to ox'_1, and $(\sigma'_{22}, \tau'_{12})$ represents stresses on face perpendicular to ox'_2. These stresses are drawn in Figure 2.5b. Remember that the clockwise–counterclockwise convention has to be used and that shear stresses are such that the summation of moments is zero.

  (b) Now we determine the maximum normal stresses (principal stresses) and maximum shear stresses. From point A (the origin of planes), we draw lines to the points corresponding to the maximum and minimum principal stresses (Figure 2.5c). Notice that these planes make an angle of 90°. Since we are on a normal stress axis, the intersection of the perpendicular to this axis corresponds to the initial point. We draw a square and place the stresses (σ_1, σ_2) on the square. This represents the orientation and values of the principal stresses. For the maximum shear stresses, we repeat the procedure $(\tau_{max} = (\sigma_1 - \sigma_2)/2)$. At points of intersection, (Figure 2.5c), we go to the opposite intersection with respect to the normal stress axis) and obtain the values. We draw these on the square, with the convention that clockwise is positive. This represents the maximum shear stress value and orientation. Note that the normal stresses for this orientation (and the one 90° from it) are nonzero. Note also that τ_{max} occurs in orientations that make 45° with the principal stress orientations.

EXAMPLE 2.6

  Elisabeth S., a bright, but somewhat nerdy, graduate student, went skiing in her brand-new boots. She had an unfortunate mishap on the slopes, and her right ski twisted beyond the strength of her femur, resulting in a fracture. The doctor took some X-rays and informed Elisabeth that she had a "spiral fracture." This triggered a spirited dialogue between Elisabeth and the doctor. Elisabeth claims that her fracture ("peeking" through the ruptured skin) is helical. With whom do you agree? Why? Show, using your knowledge of engineering, what is the maximum torque? The tensile strength of bone is 80 MPa and the diameter of the femur is 25 mm.

The Mohr circle construction (see Figure E2.6) shows that the torsion T applied to the bone leads to a state of simple shear in the cross section. If the material were ductile, the failure plane would be the plane of maximum shear. Since bone is brittle, however, failure will occur along the surface where the tensile stresses are maximum. This surface is at angle of 45° with the cross-sectional plane. Thus, the fracture is helical, and not spiral. (Students should

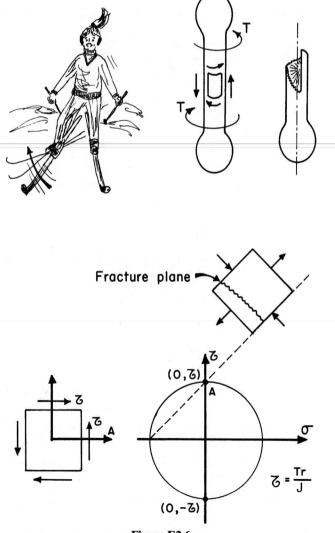

Figure E2.6

repeat this analysis by using a piece of chalk and subjecting it to torsion.) The maximum torque that the bone will withstand is

$$T = \frac{\tau J}{r},$$

where J is the polar moment on inertia. (The student should consult a text on the mechanics of materials). Now, since $\tau_{max} = \sigma_{1max}$, it follows that $\tau_{max} = 80$ MPa. Also,

$$J = \frac{\pi d^4}{32}.$$

Thus,

$$T = \frac{\tau_{max} \pi d^3}{16} = 245 \text{ N} \cdot \text{m}.$$

The weight of a normal person is 750N. Here is a ski tip, then: A distance of 1 meter from the axis of the leg can easily generate a torque of sufficient magnitude for a helical bone fracture to occur. Skiers, beware!

EXAMPLE 2.7

A state of stress is given by

$$\sigma_{11} = 350 \text{ MPa},$$
$$\sigma_{12} = 70 \text{ MPa},$$
$$\sigma_{22} = 210 \text{ MPa}.$$

Determine the principal stresses, the maximum shear stress, and their angle with the given direction by the Mohr circle.

Solution: Figure E2.7 shows the desired quantities.

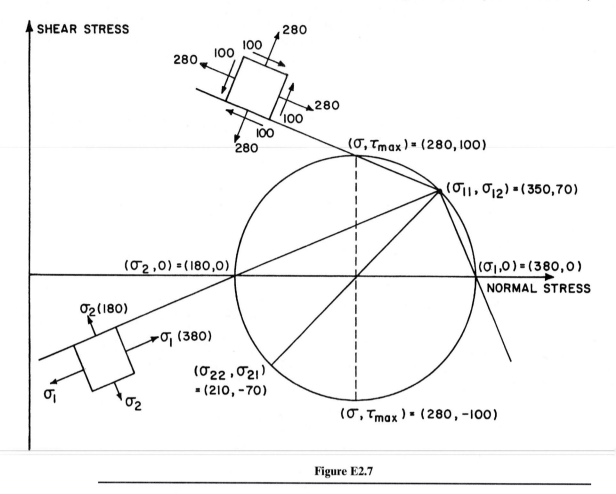

Figure E2.7

2.7 PURE SHEAR: RELATIONSHIP BETWEEN G AND E

There is a special case of bidimensional stress in which $\sigma_{22} = -\sigma_{11}$. This state of stress is represented in Figure 2.6a. It can be seen that $\sigma_{12} = 0$, implying that σ_{11} and σ_{22} are principal stresses. Hence, we can use the special subscripts for principal stresses and write $\sigma_2 = -\sigma_1$. In Mohr's circle of Figure 2.6b, the center coincides with the origin of the axes. We can see that a rotation of 90° (on the circle) leads to a state of stress in which the normal stresses are zero. This rotation is equivalent to a 45° rotation in the body (real space). The magnitude of the shear stress at this orientation is equal to the radius of the circle. Hence, the square shown in Figure 2.6c is deformed to a lozenge under the combined effect of the shear stresses. Such a state of stress is called *pure shear*.

It is possible, from this particular case, to obtain a relationship between G and E; furthermore, the relationship has a general nature. The strain ε_{11} is, for this case,

$$\varepsilon_{11} = \frac{1}{E}(\sigma_1 - \nu\sigma_2) = \frac{\sigma_1}{E}(1 + \nu). \tag{2.12}$$

We have, for the shear stresses (using the normal, and not the Mohr, sign convention),

$$\tau = -\sigma_1. \tag{2.13}$$

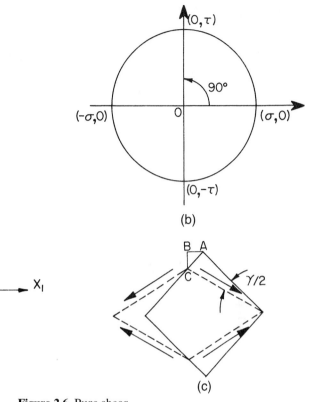

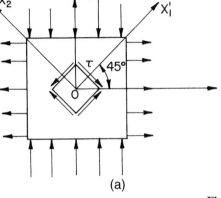

Figure 2.6 Pure shear.

But we also have,

$$\tau = G\gamma \tag{2.14}$$

Substituting Equations 2.13 and 2.14 into Equation 2.12 yields

$$\varepsilon_{11} = -\frac{G\gamma}{E}(1 + \nu)$$

It is possible, by means of geometrical considerations on the triangle ABC in Fig. 2.6(c), to show that

$$2\epsilon_{11} = -\gamma$$

The reader should do this, as an exercise. Hence,

$$G = \frac{E}{2(1 + \nu)}$$

Consequently, G is related to E by means of Poisson's ratio. This theoretical relationship between E and G is in good agreement with experimental results. For a typical metal having $\eta = 0.3$, we have $G = E/2.6$. The maximum value of G is $E/2$.

The state of *simple shear* should not be confused with *pure shear;* simple shear involves an additional rotation, so that two faces remain parallel after deformation.

2.8 ANISOTROPIC EFFECTS

Figure 2.4 shows that a general stress system acting on a unit cube has nine components and is a symmetrical tensor. (The off-diagonal components are equal, i.e., $\sigma_{13} = \sigma_{31}$, $\sigma_{12} = \sigma_{21}$, and $\sigma_{23} = \sigma_{32}$.) We can therefore write

$$\begin{pmatrix} \sigma_{11} & \sigma_{12} & \sigma_{13} \\ \sigma_{21} & \sigma_{22} & \sigma_{23} \\ \sigma_{31} & \sigma_{32} & \sigma_{33} \end{pmatrix} \equiv \begin{pmatrix} \sigma_{11} & \sigma_{12} & \sigma_{13} \\ \sigma_{12} & \sigma_{22} & \sigma_{23} \\ \sigma_{13} & \sigma_{23} & \sigma_{33} \end{pmatrix}.$$

When the unit cube in Figure 2.4 is rotated, the stress state at that point does not change; however, the components of the stress change. The same applies to strains. A general state of strain is described by

$$\begin{pmatrix} \varepsilon_{11} & \varepsilon_{12} & \varepsilon_{13} \\ \varepsilon_{21} & \varepsilon_{22} & \varepsilon_{23} \\ \varepsilon_{31} & \varepsilon_{32} & \varepsilon_{33} \end{pmatrix} \equiv \begin{pmatrix} \varepsilon_{11} & \varepsilon_{12} & \varepsilon_{13} \\ \varepsilon_{12} & \varepsilon_{22} & \varepsilon_{23} \\ \varepsilon_{13} & \varepsilon_{23} & \varepsilon_{33} \end{pmatrix}.$$

We can also use a matrix notation for stresses and strains, replacing the indices by the following:

$$\begin{aligned} 11 &\rightarrow 1 & 23 &\rightarrow 4 \\ 22 &\rightarrow 2 & 13 &\rightarrow 5 \\ 33 &\rightarrow 3 & 12 &\rightarrow 6 \end{aligned}$$

$$\begin{pmatrix} 11 & 12 \leftarrow 13 \\ & 22 & 23 \\ & & 33 \end{pmatrix} \equiv \begin{pmatrix} 1 & 6 \leftarrow 5 \\ & 2 & 4 \\ & & 3 \end{pmatrix}$$

We now have the stress and strain, in general form, as

$$\begin{pmatrix} \sigma_1 & \sigma_6 & \sigma_5 \\ \sigma_6 & \sigma_2 & \sigma_4 \\ \sigma_5 & \sigma_4 & \sigma_3 \end{pmatrix} \quad \text{and} \quad \begin{pmatrix} \varepsilon_1 & \varepsilon_6/2 & \varepsilon_5/2 \\ \varepsilon_6/2 & \varepsilon_2 & \varepsilon_4/2 \\ \varepsilon_5/2 & \varepsilon_4/2 & \varepsilon_3 \end{pmatrix}.$$

It should be noted that $\varepsilon_1 = \varepsilon_{11}$, $\varepsilon_2 = \varepsilon_{22}$, and $\varepsilon_3 = \varepsilon_{33}$, but

$$\varepsilon_4 = 2\varepsilon_{23} = \gamma_{23},$$
$$\varepsilon_5 = 2\varepsilon_{13} = \gamma_{13},$$
$$\varepsilon_6 = 2\varepsilon_{12} = \gamma_{12}.$$

These differences in notation are important to preserve the equations (see shortly) that relate stresses to strains.

The foregoing transformation is easy to remember: One proceeds first along the diagonal ($1 \rightarrow 2 \rightarrow 3$) and then back ($4 \rightarrow 5 \rightarrow 6$). It is now possible to correlate the stresses and strains for a general case, in which the elastic properties of a material are dependent on its orientation. We use two elastic constants: C (stiffness) and S (compliance), or

$$C \rightarrow \text{Stiffness.}$$

$$S \rightarrow \text{Compliance}$$

We have

$$\begin{pmatrix} \sigma_1 \\ \sigma_2 \\ \sigma_3 \\ \sigma_4 \\ \sigma_5 \\ \sigma_6 \end{pmatrix} = \begin{pmatrix} C_{11} & C_{12} & C_{13} & C_{14} & C_{15} & C_{16} \\ C_{21} & C_{22} & C_{23} & C_{24} & C_{25} & C_{26} \\ C_{31} & C_{32} & C_{33} & C_{34} & C_{35} & C_{36} \\ C_{41} & C_{42} & C_{43} & C_{44} & C_{45} & C_{46} \\ C_{51} & C_{52} & C_{53} & C_{54} & C_{55} & C_{56} \\ C_{61} & C_{62} & C_{63} & C_{64} & C_{65} & C_{66} \end{pmatrix} \begin{pmatrix} \varepsilon_1 \\ \varepsilon_2 \\ \varepsilon_3 \\ \varepsilon_4 \\ \varepsilon_5 \\ \varepsilon_6 \end{pmatrix}.$$

In short notation, noting that repeated indices in one term imply summation, we have:

$$\sigma_i = C_{ij}\varepsilon_j,$$

$$\varepsilon_i = S_{ij}\sigma_j.$$

The elastic stiffness and compliance matrices are symmetric, and the 36 components (6×6) are reduced to 21. We now apply this general expression to crystals having different structures and, therefore, different symmetries to obtain successive simplifications. In the isotropic case, the elastic constants are reduced from 21 to 2.

The different crystal systems can be characterized exclusively by their symmetries. The proof of this is beyond the scope of the book; however, it is sufficient to say that the cubic system can be perfectly described by four threefold rotations. The seven crystalline systems can be perfectly described by their axes of rotation.

Table 2.1 presents the different symmetry operations defining the seven crystal systems. For example, a threefold rotation is a rotation of 120° ($3 \times 120° = 360°$); after 120°, the crystal system comes to a position identical to the initial one. The hexagonal system exhibits a sixfold rotation around the c axis; after each 60°, the structure superimposes upon itself. In terms of a matrix, we have the following:

Orthorhombic

$$\begin{bmatrix} 11 & 12 & 13 & 0 & 0 & 0 \\ \cdot & 22 & 23 & 0 & 0 & 0 \\ \cdot & \cdot & 33 & 0 & 0 & 0 \\ \cdot & \cdot & \cdot & 44 & 0 & 0 \\ \cdot & \cdot & \cdot & \cdot & 55 & 0 \\ \cdot & \cdot & \cdot & \cdot & \cdot & 66 \end{bmatrix},$$

Tetragonal

$$\begin{bmatrix} 11 & 12 & 13 & 0 & 0 & 16 \\ \cdot & 11 & 13 & 0 & 0 & -16 \\ \cdot & \cdot & 33 & 0 & 0 & 0 \\ \cdot & \cdot & \cdot & 44 & 0 & 0 \\ \cdot & \cdot & \cdot & \cdot & 44 & 0 \\ \cdot & \cdot & \cdot & \cdot & \cdot & 66 \end{bmatrix},$$

Hexagonal

$$\begin{bmatrix} 11 & 12 & 13 & 0 & 0 & 0 \\ & 11 & 13 & 0 & 0 & 0 \\ & & 33 & 0 & 0 & 0 \\ & & & 44 & 0 & 0 \\ & & & & 44 & x \end{bmatrix}$$

where

$$x \rightarrow 2(S_{11} - S_{12}), \qquad \text{or}$$

$$x \rightarrow \frac{1}{2}(C_{11} - C_{12}),$$

TABLE 2.1 Minimum Number of Symmetry Operations in Various Systems

System	Rotation
Triclinic	None (or center of symmetry)
Monoclinic	1 twofold rotation
Orthorhombic	2 perpendicular twofold rotations
Tetragonal	1 fourfold rotation around [001]
Rhombohedral	1 threefold rotation around [111]
Hexagonal	1 sixfold rotation around [0001]
Cubic	4 threefold rotations around <111>

Laminated composites made by the consolidation of prepregged sheets, with individual plies having different fiber orientations, have orthotropic symmetry with nine independent elastic constants. Orthotropic symmetry is analogous to orthorhombic symmetry: There are three mutually perpendicular axes of symmetry, and the elastic constants along these three axes are different. For the cambric system, the elastic matrix is the following configuration:

$$
\begin{bmatrix}
11 & 12 & 12 & 0 & 0 & 0 \\
\cdot & 11 & 12 & 0 & 0 & 0 \\
\cdot & \cdot & 11 & 0 & 0 & 0 \\
\cdot & \cdot & \cdot & 44 & 0 & 0 \\
\cdot & \cdot & \cdot & \cdot & 44 & 0 \\
\cdot & \cdot & \cdot & \cdot & \cdot & 44
\end{bmatrix}.
$$

The number of independent elastic constants in a cubic system is three.
For isotropic materials (most polycrystalline aggregates can be treated as such):

$$
C_{44} = \frac{C_{11} - C_{12}}{2}. \tag{2.15}
$$

The stiffness matrix is

$$
\begin{bmatrix}
C_{11} & C_{12} & C_{12} & 0 & 0 & 0 \\
\cdot & C_{11} & C_{12} & 0 & 0 & 0 \\
\cdot & \cdot & C_{11} & 0 & 0 & 0 \\
\cdot & \cdot & \cdot & \dfrac{C_{11} - C_{12}}{2} & 0 & 0 \\
\cdot & \cdot & \cdot & \cdot & \dfrac{C_{11} - C_{12}}{2} & 0 \\
\cdot & \cdot & \cdot & \cdot & \cdot & \dfrac{C_{11} - C_{12}}{2}
\end{bmatrix}. \tag{2.16}
$$

For cubic systems, Equation (2.15) does not apply, and we define an anisotropy ratio (also called the Zener anisotropy ratio, in honor of the scientist who introduced it):

$$A = \frac{2C_{44}}{C_{11} - C_{12}} \neq 0. \tag{2.17}$$

Some metals have high anisotropy ratios, whereas others, such as aluminum and tungsten, have values of A very close to 1. For the latter, even single crystals are almost isotropic.

For the elastic compliances, we have, for the isotropic case:

$$\begin{bmatrix} S_{11} & S_{12} & S_{12} & 0 & 0 & 0 \\ \cdot & S_{11} & S_{12} & 0 & 0 & 0 \\ \cdot & \cdot & S_{11} & 0 & 0 & 0 \\ \cdot & \cdot & \cdot & 2(S_{11} - S_{12}) & 0 & 0 \\ \cdot & \cdot & \cdot & \cdot & 2(S_{11} - S_{12}) & 0 \\ \cdot & \cdot & \cdot & \cdot & \cdot & 2(S_{11} - S_{12}) \end{bmatrix}. \tag{2.18}$$

Hence, for the isotropic system, the 81 components of the elastic constants have been reduced to three independent ones while for the isotropic case, only two independent elastic constants are needed. However, it is not under this form that the elastic constants are usually known. Several parameters are used to describe the elastic properties of isotropic materials. They are related to the elastic stiffness and compliance constants by the following equations:

Young's modulus:

$$E = \frac{1}{S_{11}}. \tag{2.19}$$

Rigidity or shear modulus:

$$G = \frac{1}{2(S_{11} - S_{12})}.$$

Compressibility (B) and bulk modulus (K):

$$B = \frac{1}{K} = \frac{\varepsilon_{11} + \varepsilon_{22} + \varepsilon_{33}}{-\dfrac{1}{3}(\sigma_{11} + \sigma_{22} + \sigma_{33})}.$$

Poisson's ratio:

$$\nu = -\frac{S_{12}}{S_{11}}.$$

Lamé's constants:

$$\mu = C_{44} = \frac{1}{2}(C_{11} - C_{12}) = \frac{1}{S_{44}} = G,$$

$$\lambda = C_{12}.$$

Table 2.2 gives the various equations interrelating the foregoing parameters.

TABLE 2.2 Relations among the Elastic Constants for Isotropic Materials

Elastic Constants	In Terms of:				
	E, ν	E, G	K, ν	K, G	λ, μ
E	$= E$	$= E$	$= 3(1 - 2\nu)K$	$= \dfrac{9K}{1 + 3K/G}$	$= \dfrac{\mu(3 + 2\mu/\lambda)}{1 + \mu/\lambda}$
ν	$= \nu$	$= -1 + \dfrac{E}{2G}$	$= \nu$	$= \dfrac{1 - 2G/3K}{2 + 2G/3K}$	$= \dfrac{1}{2(1 + \mu/\lambda)}$
G	$= \dfrac{E}{2(1 + \nu)}$	$= G$	$= \dfrac{3(1 - 2\nu)K}{2(1 + \nu)}$	$= G$	$= \mu$
K	$= \dfrac{E}{3(1 - 2\nu)}$	$= \dfrac{E}{9 - 3E/G}$	$= K$	$= K$	$= \lambda + \dfrac{2\mu}{3}$
λ	$= \dfrac{E\nu}{(1 + \nu)(1 - 2\nu)}$	$= \dfrac{E(1 - 2G/E)}{3 - E/G}$	$= \dfrac{3K\nu}{1 + \nu}$	$= K - \dfrac{2G}{3}$	$= \lambda$
μ	$= \dfrac{E}{2(1 + \nu)}$	$= G$	$= \dfrac{3(1 - 2\nu)K}{2(1 + \nu)}$	$= G$	$= \mu$

The relationships between stresses and strains for isotropic materials become

$$\varepsilon_1 = S_{11}\sigma_1 + S_{12}\sigma_2 + S_{12}\sigma_3 = \frac{1}{E}[\sigma_1 - \nu(\sigma_2 + \sigma_3)],$$

$$\varepsilon_2 = S_{12}\sigma_1 + S_{11}\sigma_2 + S_{12}\sigma_3 = \frac{1}{E}[\sigma_2 - \nu(\sigma_1 + \sigma_3)],$$

$$\varepsilon_3 = S_{12}\sigma_1 + S_{12}\sigma_2 + S_{11}\sigma_3 = \frac{1}{E}[\sigma_3 - \nu(\sigma_1 + \sigma_2)],$$

$$\varepsilon_4 = 2(S_{11} - S_{12})\sigma_4 = \frac{1}{G}\sigma_4,$$

$$\varepsilon_5 = 2(S_{11} - S_{12})\sigma_5 = \frac{1}{G}\sigma_5,$$

$$\varepsilon_6 = 2(S_{11} - S_{12})\sigma_6 = \frac{1}{G}\sigma_6.$$

Expressing the strains as function of the stresses, we have

$$\sigma_1 = C_{11}\varepsilon_1 + C_{12}\varepsilon_2 + C_{12}\varepsilon_3 = (2\mu + \lambda)\varepsilon_1 + \lambda\varepsilon_2 + \lambda\varepsilon_3,$$
$$\sigma_2 = C_{12}\varepsilon_1 + C_{11}\varepsilon_2 + C_{12}\varepsilon_3 = \lambda\varepsilon_1 + (2\mu + \lambda)\varepsilon_2 + \lambda\varepsilon_3,$$
$$\sigma_3 = C_{12}\varepsilon_1 + C_{12}\varepsilon_2 + C_{11}\varepsilon_3 = \lambda\varepsilon_1 + \lambda\varepsilon_2 + (2\mu + \lambda)\varepsilon_3,$$
$$\sigma_4 = \frac{1}{2}(C_{11} - C_{12})\varepsilon_4 = \mu\varepsilon_4,$$
$$\sigma_5 = \frac{1}{2}(C_{11} - C_{12})\varepsilon_5 = \mu\varepsilon_5,$$
$$\sigma_6 = \frac{1}{2}(C_{11} - C_{12})\varepsilon_6 = \mu\varepsilon_6.$$

Note that $\mu = G$.

A great number of materials can be treated as isotropic, although they are not microscopically so. The individual grains exhibit the crystalline anisotropy and symmetry, but when they form a polycrystalline aggregate and are randomly oriented, the material is macroscopically isotropic (i.e., the elastic constants are the same in all directions). Often, a material is not completely isotropic; if the elastic modulus E is different along three perpendicular directions, the material is orthotropic; composites are a typical case.

In a cubic material, the elastic moduli can be determined along any orientation, from the elastic constants, by application of the following equations:

$$\frac{1}{E_{ijk}} = S_{11} - 2\left(S_{11} - S_{12} - \frac{1}{2}S_{44}\right)(l_{i1}^2 l_{j2}^2 + l_{j2}^2 l_{k3}^2 + l_{i1}^2 l_{k3}^2), \qquad (2.20\text{a})$$

$$\frac{1}{G_{ijk}} = S_{44} + 4\left(S_{11} - S_{12} - \frac{1}{2}S_{44}\right)(l_{i1}^2 l_{i2}^2 + l_{j2}^2 l_{k3}^2 + l_{i1}^2 l_{k3}^2). \qquad (2.20\text{b})$$

E_{ijk} and G_{ijk} are the Young's and shear modulus, respectively, in the [ijk] direction; l_{i1}, l_{j2}, and l_{k3} are the direction cosines of the direction [ijk].

Figure 2.7a illustrates the dependence on orientation of elastic Young's modulus for copper. The [100], [010], and [001] directions are "softer," whereas the [111], [1$\bar{1}$1], and

[001]

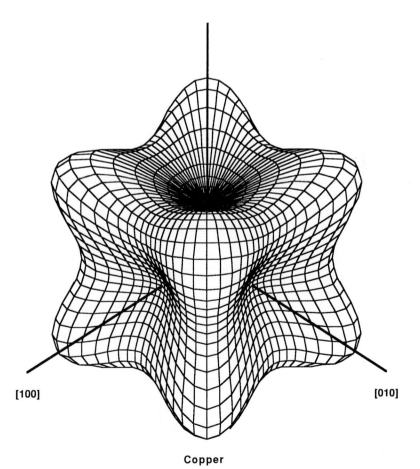

Copper

Figure 2.7 Dependence on orientation of Young's modulus for monocrystalline (a) copper and (b) cubic zirconia. (Courtesy of R. Ingel).

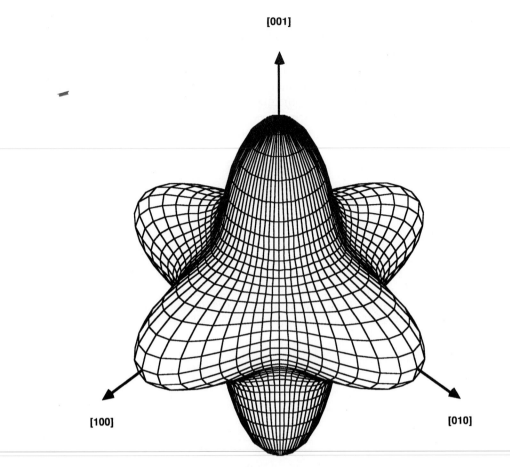

[001]

[100]

[010]

Cubic Zirconia - 20 w/o Y2O3
Figure 2.7 (Continued)

[11$\bar{1}$] directions are stiffer. For cubic zirconia (Figure 2.7b), the opposite occurs: The coordinate axes correspond to the stiff directions. These diagrams illustrate very well the importance of anisotropy of elastic properties. For a cubic material that has the same Young's modulus along all directions (an isotropic material), we have the relationship

$$2(S_{11} - S_{12}) = S_{44}. \tag{2.21}$$

EXAMPLE 2.8

A hydrostatic compressive stress applied to a material with cubic symmetry results in a dilation of -10^{-5}. The three independent elastic constants of the material are $C_{11} = 50$ GPa, $C_{12} = 40$ GPa, and $C_{44} = 32$ GPa. Write an expression for the generalized Hooke's law for this material, and compute the applied hydrostatic stress.

Solution: Dilation is the sum of the principal strain components:

$$\varepsilon = \varepsilon_1 + \varepsilon_2 + \varepsilon_3 = -10^{-5}.$$

Cubic symmetry implies that

$$\varepsilon_1 = \varepsilon_2 = \varepsilon_3 = -3.33 \times 10^{-6}.$$

and

$$\varepsilon_4 = \varepsilon_5 = \varepsilon_6 = 0.$$

From Hooke's law,

$$\sigma_i = C_{ij}\varepsilon_j$$

and

$$\sigma_1 = C_{11}\varepsilon_1 + C_{12}\varepsilon_2 + C_{13}\varepsilon_3.$$

The applied hydrostatic stress is

$$\sigma_p = \sigma_1 = (50 + 40 + 40)(-3.33)\ 10^3\ \text{Pa}$$
$$= -130 \times 3.33 \times 10^3\ \text{Pa}$$
$$= -433\ \text{kPa}.$$

EXAMPLE 2.9

From the elastic stiffnesses for a cubic material, Nb (C_{11} = 242 GPa; C_{12} = 129 GPa; C_{44} = 286 GPa), find the elastic compliances.

The relationship between stiffnesses and compliances is given by the product of their two matrices, which is an identity matrix:

$$
\begin{vmatrix}
S_{11} & S_{12} & \cdots\cdots & S_{16} \\
S_{21} & S_{22} & \cdots\cdots & S_{26} \\
\vdots & & & \\
\vdots & & & \\
S_{61} & S_{62} & \cdots\cdots & S_{66}
\end{vmatrix}
\begin{vmatrix}
C_{11} & C_{12} & \cdots\cdots & C_{16} \\
C_{21} & C_{22} & \cdots\cdots & C_{26} \\
\vdots & & & \\
\vdots & & & \\
C_{61} & C_{62} & \cdots\cdots & C_{66}
\end{vmatrix} = (I).
$$

For materials with cubic symmetry,

$$
\begin{vmatrix}
S_{11} & S_{12} & S_{12} & 0 & 0 & 0 \\
S_{12} & S_{11} & S_{12} & 0 & 0 & 0 \\
S_{12} & S_{12} & S_{11} & 0 & 0 & 0 \\
0 & 0 & 0 & S_{44} & 0 & 0 \\
0 & 0 & 0 & 0 & S_{44} & 0 \\
0 & 0 & 0 & 0 & 0 & S_{44}
\end{vmatrix}
\begin{vmatrix}
C_{11} & C_{12} & C_{12} & 0 & 0 & 0 \\
C_{12} & C_{11} & C_{12} & 0 & 0 & 0 \\
C_{12} & C_{12} & C_{11} & 0 & 0 & 0 \\
0 & 0 & 0 & C_{44} & 0 & 0 \\
0 & 0 & 0 & 0 & C_{44} & 0 \\
0 & 0 & 0 & 0 & 0 & C_{44}
\end{vmatrix} = (I).
$$

All the off-diagonal terms of the identity matrix are zero. The diagonal terms are equal to 1.

Row 1 and column 1 give

$$S_{11}C_{11} + S_{12}C_{12} + S_{12}C_{12} = 1. \tag{1}$$

From row 6 and column 6, we have

$$S_{44}C_{44} = 1.$$

Therefore,

$$S_{44} = \frac{1}{C_{44}}.$$

Row 1 and column 2 yield

$$S_{11}C_{12} + S_{12}C_{11} + S_{12}C_{12} = 0. \tag{2}$$

From equations 1 and 2, we get, for row 1 and column 1,

$$S_{12} = \frac{-C_{12}}{C_{11}^2 + C_{11}C_{12} - 2C_{12}^2} = \frac{-C_{12}}{(C_{11} + 2C_{12})(C_{11} - C_{12})}. \qquad (3)$$

Substituting Equation 3 into Equation 1 yields

$$S_{11} = \frac{1}{C_{11}} + \frac{2C_{12}^2}{C_{11}(C_{11} + 2C_{12})(C_{11} - C_{12})} = \frac{C_{11} + C_{12}}{(C_{11} + 2C_{12})(C_{11} - C_{12})}$$

Thus,

$$S_{44} = 3.5 \times 10^{-2} \text{ GPa}^{-1},$$
$$S_{12} = -0.22 \times 10^{-2} \text{ GPa}^{-1},$$
$$S_{11} = 0.66 \times 10^{-2} \text{ GPa}^{-1}.$$

These values are fairly close to the values given in Table 2.4.

2.9 ELASTIC PROPERTIES OF POLYCRYSTALS

The elastic constants of materials are determined by the bonding between the individual atoms. While monocrystals have the elastic properties dictated by the crystalline symmetry, most metals and ceramics are polycrystalline. In polycrystals, the properties are determined from the individual grains by an averaging process.

In a polycrystalline aggregate, the deformation of one grain is not independent of the deformation of its neighbor. The compatibility requirements are such that we have to apply either one of two simplifying assumptions:

1. The local strain is equal to the mean strain (all grains undergo the same strain); this is called the *Voigt average*. The Young's modulus is then

$$E = \frac{(F - G + 3H)(F + 2G)}{2F + 3G + H},$$

where

$$F = \frac{1}{2}(C_{11} + C_{22} + C_{33}),$$

$$G = \frac{1}{3}(C_{12} + C_{23} + C_{13}),$$

$$H = \frac{1}{3}(C_{44} + C_{55} + C_{66}).$$

2. The local stress is equal to the mean stress (all grains are under the same stress); this is called the *Reuss average*. The inverse of the Young's modulus is then

$$\frac{1}{E} = \frac{1}{5}(3F' + 2G' + H')$$

$$F' = \frac{1}{3}(S_{11} + S_{22} + S_{32}),$$

$$G' = \frac{1}{3}(S_{12} + S_{23} + S_{13}),$$

$$H' = \frac{1}{3}(S_{44} + S_{55} + S_{66}),$$

$$H' = \frac{1}{3}(S_{44} + S_{55} + S_{66}).$$

The actual stress and strain configuration is probably between the two assumptions. There are more advanced methods, such as the Hashin–Shtrikman upper and lower bound method; however, this will not be treated here.

EXAMPLE 2.10

Determine the Young's moduli along [100], [110], and [111] of copper, tungsten, and ZrO_2. We use Equation 2.20a:

$$\frac{1}{E_{ijk}} = S_{11} - 2\left(S_{11} - S_{12} - \frac{1}{2}S_{44}\right)(\ell_{i1}^2\ell_{j2}^2 + \ell_{j2}^2\ell_{k3}^2 + \ell_{i\ell}\ell_{k3}^2).$$

The direction consines are as follows:

	ℓ_{i1}	ℓ_{j2}	ℓ_{k3}	$(\ell_{i1}^2\ell_{j2}^2 + \ell_{j2}^2\ell_{k3}^2 + \ell_{i1}^2\ell_{k3}^2)$
[100]	1	0	0	0
[110]	$\sqrt{2}/2$	$\sqrt{2}/2$	0	1/4
[111]	$1/\sqrt{3}$	$1/\sqrt{3}$	$1/\sqrt{3}$	1/3

The compliances for Cu and W are given in Table 2.4; Table 2.6 provides the stiffnesses for cubic ZrO_2. We have:

W	Cu
$S_{11} = 0.257 \times 10^{-2}\ GPa^{-1}$	$S_{11} = 1.498 \times 10^{-2}\ GPa^{-1}$
$S_{44} = 0.66 \times 10^{-2}\ GPa^{-1}$	$S_{44} = 1.326 \times 10^{-2}\ GPa^{-1}$
$S_{12} = -0.073 \times 10^{-2}\ GPa^{-1}$	$S_{12} = -0.629 \times 10^{-2}\ GPa^{-1}$

This yields

Cu: $E_{100} = 66\ GPa,\qquad E_{110} = 130\ GPa,\qquad E_{111} = 191\ GPa,$

W: $E_{100} = E_{110} = E_{111} = 389\ GPa.$

For ZrO_2, we have to use the equations derived in Example 2.9 to obtain the elastic compliances:

$$C_{11} = 410\ GPa,\qquad C_{12} = 110\ GPa,\qquad C_{44} = 60\ GPa,$$

$$S_{44} = \frac{1}{C_{44}} = 1.6 \times 10^{-2}\ GPa^{-1},$$

$$S_{12} = \frac{-C_{12}}{(C_{11} + 2C_{12})(C_{11} - C_{12})} = -0.058 \times 10^{-2} \text{ GPa}^{-1},$$

$$S_{11} = \frac{1}{C_{11}} + \frac{2C_{12}^2}{C_{11}(C_{11} + 2C_{12})(C_{11} - C_{12})} = 0.275 \times 10^{-2} \text{ GPa}^{-1}.$$

These yield

$$E_{100} = 363.5 \text{ GPa},$$

$$E_{110} = 196.7 \text{ GPa},$$

$$E_{111} = 171 \text{ GPa}.$$

EXAMPLE 2.11

Determine the elastic anisotropy ratios of Ag, Al, Cu, Ni, Fe, Ta, and W. Which one of these metals has the greatest dependence on orientation for Young's modulus? Which one has the smallest?

Solution: First, we have

$$A = \frac{2C_{44}}{C_{11} - C_{12}}.$$

From Table 2.3, we obtain the following results:

$$\text{Ag:} \quad A = \frac{46.1 \times 2}{124 - 93.4} = 3.01.$$

$$\text{Al:} \quad A = \frac{28.5 \times 2}{108.2 - 61.3} = 1.22.$$

$$\text{Cu:} \quad A = \frac{75.4 \times 2}{168.4 - 121.4} = 3.21.$$

$$\text{Ni:} \quad A = \frac{124.7 \times 2}{246.5 - 147.3} = 2.51.$$

$$\text{Fe:} \quad A = \frac{116.5 \times 2}{228 - 132} = 2.43.$$

$$\text{Ta:} \quad A = \frac{82.5 \times 2}{267 - 161} = 1.56.$$

$$\text{W:} \quad A = \frac{151.4 \times 2}{501.0 - 198} = 1.00.$$

Copper has the highest and W the lowest anisotropy ratio. Elastic properties should therefore be most orientation dependent for Cu and orientation independent for W.

EXAMPLE 2.12

Determine the Young's modulus for polycrystalline iron, using Reuss' and Voigt's averages. From Tables 2.3 and 2.4, we get the elastic stiffnesses and compliances:

$$C_{11} = 228 \text{ GPa}, \qquad S_{11} = 0.762 \times 10^{-2} \text{ GPa}^{-1},$$

$$C_{44} = 116.5 \text{ GPa}, \qquad S_{44} = 0.858 \times 10^{-2} \text{ GPa}^{-1},$$

$$C_{12} = 132 \text{ GPa}, \qquad S_{12} = -0.279 \times 10^{-2} \text{ GPa}^{-1}.$$

Voigt method:
We first calculate the parameters, taking into account cubic symmetry:

$$F = \frac{1}{2}(C_{11} + C_{22} + C_{33}) = \frac{3}{2}C_{11},$$

$$G = \frac{1}{3}(C_{12} + C_{23} + C_{13}) = C_{12},$$

$$H = \frac{1}{3}(C_{44} + C_{55} + C_{66}) = C_{44}.$$

Then

$$E = \frac{(F - G + 3H)(F + 2G)}{2F + 3G + H} = \frac{\left(\frac{3}{2}C_{11} - C_{12} + 3C_{44}\right)\left(\frac{3}{2}C_{11} + 2C_{12}\right)}{3C_{11} + 3C_{12} + C_{44}}$$

$$= \frac{559.5 \times 606}{1196.5} = 283.3 \text{ GPa}.$$

Reuss method:

$$F' = \frac{1}{3}(S_{11} + S_{22} + S_{32}) = \frac{1}{3}(2S_{11} + S_{12}),$$

$$G' = \frac{1}{3}(S_{12} + S_{23} + S_{13}) = S_{12},$$

$$H' = \frac{1}{3}(S_{44} + S_{55} + S_{66}) = S_{44},$$

$$\frac{1}{E} = \frac{1}{5}(3F' + 2G' + H') = \frac{1}{5}(2S_{11} + S_{12} + 2S_{12} + S_{44}),$$

$$= \frac{1}{5}(1.545 \times 10^{-2}).$$

so

$$E = 323.6 \text{ GPa}.$$

2.10 ELASTIC PROPERTIES OF MATERIALS

Figure 2.8 presents a comparison of the elastic constants of different classes of materials. At the top, we have diamond (with covalent bonding). For metals, there is a correlation between the melting point (indicative of the bonding energy between atoms) and the Young's modulus. Thus, the metals with the highest bonding energies have the highest melting

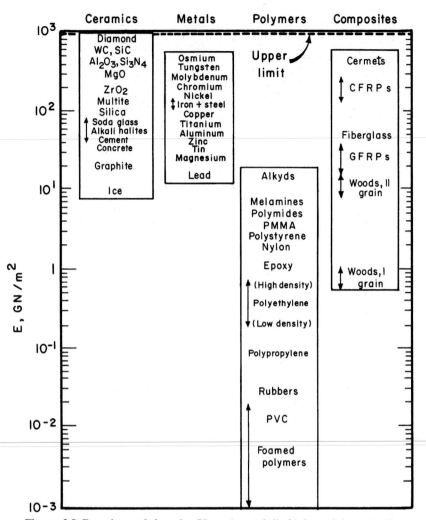

Figure 2.8 Bar chart of data for Young's moduli. (Adapted from M. F. Ashby and D. R. H. Jones, *Engineering Materials* (Oxford: Pergamon Press, 1980), p. 32)

points, interatomic forces, and Young's modulus. The ranking of the metals in the second column of the figure shows this relationship; at the top are osmium and tungsten, and at the bottom is lead. The third column of the figure shows the polymers, which have elastic constants that are much lower than those of the metals. The composites in the last column show a wide variation in elastic constants. The carbon-fiber reinforced polymers (CFRPs) can have a very high modulus.

2.10.1 Elastic Properties of Metals

Tables 2.3 and 2.4 give the elastic stiffnesses and compliances, respectively, of metallic monocrystals. One of the most complete compilations of elastic constants for crystals is that by Simmons and Wang. (See suggested reading.) The elastic constants for a number of polycrystalline metals are given in Table 2.5. We can also determine the polycrystalline (isotropic) elastic constants from the monocrystalline ones, using equations given earlier.

TABLE 2.3 Elastic Stiffnesses of Monocrystals at Ambient Temperature (GPa)

Element	Structure	C_{11}	C_{44}	C_{12}	C_{33}	C_{66}	C_{13}	C_{14}
Ag	FCC	124.0	46.1	93.4				
Al	FCC	108.2	28.5	61.3				
Au	FCC	186.0	42.0	157.0				
Cu	FCC	168.4	75.4	121.4				
Ni	FCC	246.5	124.7	147.3				
Pb	FCC	49.5	14.9	42.3				
Fe	BCC	228.0	116.5	132.0				
Mo	BCC	460.0	110.0	176.0				
Ta	BCC	267.0	82.5	161.0				
W	BCC	501.0	151.4	198.0				
Co	HCP	307.0	75.3	165.0	358.1		103.0	
Zn	HCP	161.0	38.3	34.2	61.0		50.1	
Ti	HCP	162.4	46.7	92.0	180.7	69.0		
Be	HCP	292.3	162.5	26.7	336.4	14.0		
Zr	HCP	143.4	32.0	72.8	164.8	65.3		
Mg	HCP	59.7	16.7	26.2	61.7	21.7		
Sn	Tetragonal	73.5	22.0	23.4	87.0	22.6	28.0	
In	Tetragonal	44.5	06.6	39.5	44.4	12.2	40.5	
Hg	Rhombohedral	36.0	12.9	28.9	50.5		30.3	05.0

TABLE 2.4 Elastic Compliances for Monocrystalline Metals at Ambient Temperature $(10^{-2} \text{ GPa}^{-1})$

Element	Structure	S_{11}	S_{44}	S_{12}	S_{23}	S_{13}
Ag	FCC	2.29	2.17	−0.983		
Al	FCC	1.57	3.51	−0.568		
Au	FCC	2.33	2.38	−1.065		
Cu	FCC	1.498	1.326	−0.629		
Ni	FCC	0.734	0.802	−0.274		
Pb	FCC	9.51	6.72	−4.38		
Fe	BCC	0.762	0.858	−0.279		
Mo	BCC	0.28	0.91	−0.078		
Nb	BCC	0.69	3.42	−0.249		
Ta	BCC	0.685	1.21	−0.258		
W	BCC	0.257	0.66	−0.073		
BI	HCP	0.348	0.616	−0.030	0.298	−0.031
Mg	HCP	2.20	6.1	−0.785	1.97	−0.50
Ti	HCP	0.958	2.14	−0.462	0.698	−0.189
Zr	HCP	1.013	3.13	−0.404	0.799	−0.241

2.10.2 Elastic Properties of Ceramics

The elastic properties of ceramic monocrystals possess the symmetry of the crystal (see Table 2.6). As an example, consider the stiffnesses and compliances for MgO at room temperature. Magnesia is a cubic crystal, and alumina has the rhombohedral structure. The corresponding Young and shear moduli, computed along the three crystallographic axes of the monocrystal from Equations 2.20 are given in Table 2.7. Table 2.8 presents the elastic moduli for a number of ceramics and glasses. The largest elastic constant is that for diamond and is equal to 1,000 GPa. This is due to the C–C bonds, as is explained in Chapter 4.

The elastic moduli of ceramics are strongly dependent on porosity. Ceramics are porous due to their fabrication, and one should be aware of the effect of porosity.

TABLE 2.5 Elastic and Shear Moduli and
Poisson Ratios for Polycrystalline Metals[a]

Metal (20°C)	E (GPa)	G (GPa)	ν
Aluminum	70.3	26.1	0.345
Cadmium	49.9	19.2	0.300
Chromium	279.1	115.4	0.210
Copper	129.8	48.3	0.343
Gold	78.0	27.0	0.440
Iron	211.4	81.6	0.293
Magnesium	44.7	17.3	0.291
Nickel	199.5	76.0	0.312
Niobium	104.9	37.5	0.397
Silver	82.7	30.3	0.367
Tantalum	185.7	69.2	0.342
Titanium	115.7	43.8	0.321
Tungsten	411.0	160.6	0.280
Vanadium	127.6	46.7	0.365

[a] Adapted with permission from R. W. Hertzberg, *Deformation and Fracture Mechanics of Engineering Material,* New York: John Wiley, 1976, p. 8.

TABLE 2.6 Elastic Constants for Ceramics (S_{ij} in 10^{-10} Pa^{-4}; C_{ij} in GPa)

Material	C_{11}	C_{12}	C_{44}	C_{14}	C_{13}	C_{33}	S_{11}	S_{12}	S_{44}
MgO	289.2	88.0	154.6				4.03	−0.94	6.47
Al$_2$O$_3$	497.1	162.3	147.7	−23	117	502			
ZrO$_2$	410	110	60						
MgAl$_2$O$_4$	279	153	153				5.83	−2.08	6.54
TiC	513	106	178		0.2	2.1	−0.36	5.61	
Diamond	1076	125	576						
LiF	112	46	63						
NaCl	49	13	13						
ThO$_2$	367	106	797				3.13	−0.70	12.5
LIO$_2$	395	121	64.1				2.96	−0.70	15.6
SiC (hexagonal)	500	186	168		176	521			
SiC (cubic)	352	140	233						

TABLE 2.7 Orientation Dependence of Young's Modulus and Shear Modulus for MgO and Al$_2$O$_3$ at 25°C.

Crystal Orientation	Young's Modulus, Al$_2$O$_3$ (GPa)	Young's Modulus, MgO (GPa)	Shear Modulus, MgO (GPa)
<100>	299	248.2	154.6
<110>	330	316.4	121.9
<111>	344	348.9	113.8

TABLE 2.8 Modulus of Elasticity of Some Ceramic Materials

Material	E (GPa)
Aluminum oxide crystals	378
Sintered alumina*	365
Alumina porcelain (90–95% Al_2O_3)	365
Sintered beryllia	310
Hot-pressed boron nitride*	82.7
Hot-pressed boron carbide*	289
Graphite*	9
Sintered magnesia*	210
Sintered molybdenum silicide*	406
Sintered spinel*	238
Dense silicon carbide (cubic or hexagonal)	280–510
Sintered titanium carbide*	310
Sintered stabilized zirconia*	152
Silica glass	72.3
Vycor glass	72.3
Pyrex glass	68.9
Superduty fire-clay brick	96.4
Magnesite brick	172.2
Bonded silicon carbide**	345
Silicon nitride	320–365
Aluminum nitride	
Mullite (aluminosilicate) porcelain	69
Steatite (magnesia aluminosilicate)	69
Diamond	450–650
Tungsten carbide	400–530
Cobalt/tungsten carbide cermets	379
Titanium dioxide	290
Titanium diboride	440

*(ca. 5% porosity)
**(ca. 20% porosity)

(Adapted from W. D. Kingery, H. K. Bowen, and D. R. Uhlmann, *Introduction to Ceramics,* 2d ed., 1976 (New York: John Wiley, 1976), p. 777.

Figure 2.9 shows the variations in the Young modulus of alumina with volume fraction of pores. For 10% porosity (a common value for commercial alumina), the Young's modulus is decreased by 20%.

The change in Young's modulus with porosity has been empirically expressed by Wachtman and MacKenzie.[4]

$$E = E_0(1 - f_1 p + f_2 p^2), \qquad (2.22)$$

where p is the porosity and f_1 and f_2 are constants. For spherical voids, MacKenzie found that f_1 and f_2 are equal to 1.9 and 0.9, respectively, for a Poisson's ratio of 0.3. The data of Coble and Kingery[5] may be compared with the prediction of Eq. 2.22. If one assumes the law of mixtures for the porosity, then, as a first approximation, one has

$$E = E_A(1 - f_B) + E_B f_B. \qquad (2.23)$$

where f is the volume fraction of a phase and the subscripts A and B denote the two phases.

[4] See J. B. Wachtman, in *Mechanical and Thermal Properties of Ceramics,* ed. J. B. Wachtman, NBS Special Publication 303, NBS Washington, 1963, p. 139; and J. K. MacKenzie, *Proc. Phys. Soc.,* B63 (1950) 2.

[5] R. L. Coble and W. D. Kingery, *J. Am. Cer. Soc. 39* (1956) 377.

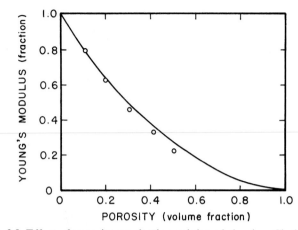

Figure 2.9 Effect of porosity on elastic modulus of alumina. Circles represent experimental measurements. (After R. L. Coble and W. D. Kingery, *J. Am. Ceramic Soc,* 39 (1956) 377).

However, if phase B is the pore, one has

$$E = E_0(1 - p). \qquad (2.24)$$

For relatively low porosity, the quadratic term in Equation 2.22 can be neglected, leaving

$$E = E_0(1 - 1.9p). \qquad (2.25)$$

If E varied linearly with p, the form would be $E = E_0 (1 - p)$. Thus, the physical significance of MacKenzie's equation is that porosity has an effect of E equal to approximately double the volume of pores.

Another effect of considerable importance on Young's modulus for ceramics is the presence of microcracks, which decrease the stored elastic energy and reduce the effective Young's modulus. Figure 2.10 shows schematically how the presence of microcracks would affect the slope of the stress–strain curve. The initial slope, E_0, is decreased by microcracking. Microcracks can also form during the cooling of the ceramic, due to thermal expansion (or contraction) anisotropy. Different grains contract by different amounts along different orientations, resulting in a buildup of elastic stress in the boundary area. Elastic stress can generate microcracks. Similarly, the anisotropy of elastic constants can generate elastic stress concentrations at the grain boundaries, where the neighboring grains undergo different strains (due to differences in crystallographic orientation). The change in the Young's modulus with microcracking has been computed by a number of investigators. The formulations give predictions that vary with the orientation of the cracks with respect to the tensile axis, among other parameters. An expression developed by Salganik[6] is

$$\frac{E}{E_0} = \left[1 + \frac{16(10 - 3\nu_0)(1 - \nu_0^2)}{45(2 - \nu_0)} Na^3 \right]^{-1} = (1 + ANa^3)^{-1}, \qquad (2.26)$$

where E is the Young's modulus of the cracked ceramic, ν_0 and E_0 are, respectively, Poisson's ratio and Young's modulus of the uncracked material, a is the radius of a mean crack, and N is the number of cracks per unit volume. The factor

[6] R. L. Salganik, Izv. Akad. Nauk SSR Mekh. Tverd. Tela, 8 (1973) 149.

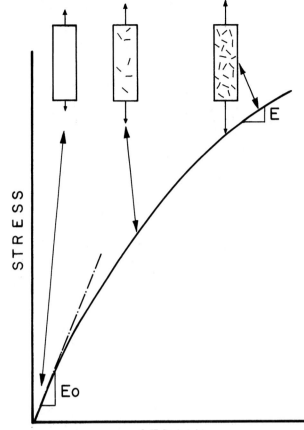

Figure 2.10 Effect of microcracks on Young's modulus for ceramics.

$$A = \frac{16(10 - 3\nu_0)(1 - \nu_0^2)}{45(2 - \nu_0)} \tag{2.27}$$

varies between 1.77 and 1.5 when ν_0 varies between 0 and 0.5. To a first approximation, one can say that

$$\frac{E}{E_0} = [1 + 1.63\, Na^3]^{-1}. \tag{2.28}$$

O'Connell and Budiansky arrived at a slightly different expression:[7]

$$\frac{E}{E_0} = 1 - \frac{16(10 - 3\nu)(1 - \nu^2)}{45(2 - \nu)}\, f_s. \tag{2.29}$$

Here, f_s is defined as the volume fraction of cracks. (i.e., the number of cracks per unit volume, N, multiplied by the cube of the mean crack radius, a^3) and ν is Poisson's ratio of the porous material, which is related to Poisson's ratio of the fully dense material by

$$\nu = \nu_0\left(1 - \frac{16\, f_s}{9}\right). \tag{2.30}$$

[7] R. J. O'Connell and B. Budiansky, J. Geol. Res. 79 (1974) 5412.

By applying the same approximation as in Salganik's equation, we arrive at

$$\frac{E}{E_0} = 1 - 1.63\, Na^3. \tag{2.31}$$

Note that Na^3 is a measure of the fraction of the material that is under the effect of the cracks. Figure 2.11 shows the effect of microcracks on the Young's modulus of alumina. This effect is substantial. For $f_s = 0.1$, the Young's modulus is reduced by 20%. Both Salganik's and O'Connell and Budiansky's predictions are plotted, and it can be seen that they are in fairly close agreement for values of f_s smaller than 0.1. For higher values, O'Connell and Budiansky's equation predicts a more rapid decrease in E.

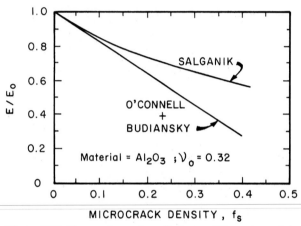

Figure 2.11 Comparison of predictions of Young's modulus.

2.10.3 Elastic Properties of Polymers

Polymers have elastic constants that range from the lower end of the metallic elastic constants to values even lower by several orders of magnitude. As an example, melamines have elastic constants of 6–7 GPa (E (lead) = 14 GPa), while the elastic constant of polymeric foams is between 3 and 10 MPa. Table 2.9 lists the elastic constants of a number of polymers. The bar chart of Figure 2.8 provides a comparison of the elastic constants of the different classes of materials. The elastic behavior of polymeric materials is more difficult to describe than that of metals or ceramics, because it is strongly dependent on both temperature and time. This behavior, called *viscoelastic* or *anelastic,* is described separately in Section 2.11. Here we merely introduce the subject briefly. In most polymers, there are dramatic changes in E between $-20°C$ and $200°C$; for most metals and ceramics, the changes in E in this range can be neglected. The glass transition temperature T_g plays an important role in polymers. Above T_g, E is considerably low, and the behavior of the polymer can be described as rubbery and viscous. Below T_g, the modulus of elasticity is considerably higher, and the behavior is closer to linear elastic. Figure 2.12 shows schematically the elastic behavior of a linear polymer as a function of temperature. The modulus of elasticity ranges from 10^3 to 10^{-1} MPa.

TABLE 2.9 Elastic Constants of Some Polymers

Material	E (GPa)
Phenolformaldehyde	8
Melamines	6–7
Polyimides	3–5
Polyesters	1.3–4.5
Acrylics	1.6–3.4
Nylon	2–4.5
PMMA	3.4
Polystyrene	3–3.4
Polycarbonate	2.1
Epoxies	2.1–5.5
Polypropylene	1.2–1.7
Polyethylene, high density	0.15–0.24
Foamed polyurethane	0.01–0.06
Polyethylene, low density	0.15–0.24
Rubbers	0.01–0.1
PVC (unplasticized)	2.4–3.0
Foamed polymers	0.001–0.01

Adapted from M. F. Ashby and D. R. H. Jones, *Engineering Materials* (Oxford: Pergamon Press, 1986), p. 31, Table 3.1.

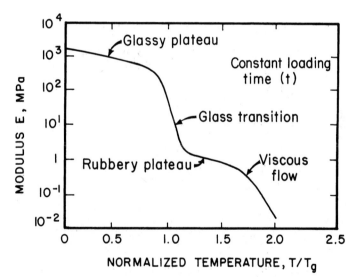

Figure 2.12 Schematic variation in the modulus of elasticity of a linear polymer with temperature.

2.11 VISCOELASTICITY

Glasses or amorphous materials show the phenomenon of time-dependent strain, called *viscoelasticity* or *anelasticity*. The deformation of an amorphous material does not involve atomic displacements on specific crystallographic planes, as is the case in crystalline metals. Rather, a continuous displacement of atoms or molecules takes place with time at a constant load. This flow mechanism of noncrystalline materials is associated with the diffusion of atoms or molecules within the material; that is, it is a thermally activated process and is thus described by an Arrhenius-type equation. Of course, at sufficiently high temperatures, where diffusion becomes important, crystalline as well as amorphous materials show a large amount of thermally activated plastic flow. Liquids and even fluids in general show a characteristic resistance to flow called *viscosity*. The viscosity of a fluid results in a frictional energy loss, which appears as heat. The more viscous a fluid, the higher is the frictional energy loss.

Over a range of temperatures, the viscosity η can be described by the Arrhenius-type relationship

$$1/\eta = A \exp(-Q/RT), \tag{2.32}$$

or

$$\eta = A \exp(Q/RT),$$

where Q represents the activation energy for the atomic or molecular process responsible for the viscosity, R is the universal gas constant, and T is the temperature in kelvin. The S. I. units of the viscosity η are Nm^{-2} s or Pa s. Another common unit of viscosity is poise, P; 1 P = 0.1 Pa s.

A purely viscous material shows stress proportional to strain rate. Thus, if we apply a shear stress τ to a glassy solid above its glass transition temperature, then we can write, for the rate of shear deformation,

$$\dot{\gamma} = \frac{d\gamma}{dt} = \frac{\tau}{\eta} = \phi\tau, \tag{2.33}$$

where ϕ is the *fluidity* (the reciprocal of viscosity) of the material.

Equation (2.33) can be written as

$$\tau = \eta\dot{\gamma}. \tag{2.34}$$

If the viscosity of a material does not change with the strain rate (i.e., if the stress is linearly proportional to the strain rate), then we call the viscosity a *Newtonian viscosity* and such a material a *Newtonian material*. Figure 2.13 shows a Newtonian (or linear) response curve. If the stress is not directly proportional to the strain rate, we have a non-Newtonian response, which can be written as

$$\tau = \eta\dot{\gamma}^n. \tag{2.35}$$

This is shown by the curve marked "nonlinear" in the figure. If the stress is independent of the strain rate, we have a plastic material. A special case is that of a material whose viscosity decreases when subjected to high strain rates. Such a material is called a *thixotropic* material, a good example of which is a latex paint. When we apply the paint to a vertical wall, it does not sag, because its viscosity is very high on the wall. However, we can stir and brush the paint easily because its viscosity decreases when subjected to shear stress in the stirring action.

Polymers, polymer solutions and dispersions, metals at very high temperatures, and amorphous materials (organic and inorganic) show viscoelastic behavior—that is, characteristics intermediate between perfectly elastic and perfectly viscous behavior. Commercial silica-based glasses have a high proportion of additives: about 30% in soda–lime glass

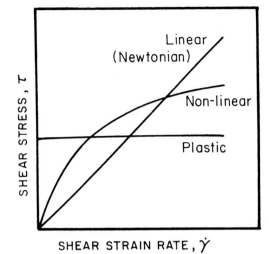

Figure 2.13 Linear or Newtonian response (stress proportional to deformation rate), nonlinear response, and plastic response (stress independent of deformation rate).

and 20% in high-temperature glasses such as Pyrex. The main purpose of the additives is to lower the viscosity by breaking up the silica network, thus making the processing of glass easy.

Conventionally, glasses are formed by melting an appropriate composition and then casting or drawing the melt into a desired form. It is interesting to compare the viscosity values of liquid metals with glasses. Molten metals have about the same viscosity as that of water ($\sim 10^{-3}$ Pa s) and transform to a crystalline solid state in a discontinuous manner when cooled. The viscosity of glasses, however, falls slowly and continuously with temperature. The shaping of glass is carried out in the viscosity range of 10^3–10^6 Pa s. Polymers are formed in the range 10^3–10^5 Pa s. Perhaps the most important characteristic of a viscoelastic material is that its rheological properties are dependent on time. This characteristic is manifested very markedly by amorphous or noncrystalline materials such as polymers.

A viscoelastic substance has a viscous and an elastic component. Figure 2.14a shows the stress–strain curve of an ideal elastic material. The load and unload curves are the same, and the energy lost as heat per cycle is zero in this case. In practice, there is always present an anelastic (i.e., a time-dependent) component, with the result that the unload curve does not in fact follow the load curve. Energy equal to the shaded area in Figure 2.14b is dissipated in each cycle. This phenomenon is exploited in damping out vibrations. Some polymers and soft metals (e.g., lead) have a high damping capacity. In springs and bells, a high damping capacity is undesirable. For such applications, one uses materials such as bronze, spring steel, etc., which have a low damping capacity.

2.11.1 Storage and Loss Moduli

In order to characterize the viscoelastic behavior of a material, the material is sinusoidally deformed, and the resulting stress is recorded. For an ideal elastic material, the stress and strain are *in phase,* and the phase shift $\delta = 0$. For an ideal viscous material, the stress and strain are 90° *out of phase* (i.e., $\delta = 90°$). As pointed out before, a viscoelastic behavior—a combination of an ideal elastic response and an ideal viscous response—is more common. Figure 2.15 shows a viscoelastic response with a phase lag between the stress and the strain. Dynamic (commonly sinusoidal) perturbations are used to study the viscoelastic

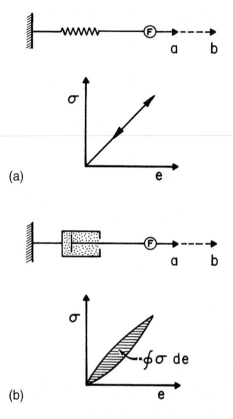

(a)

(b)

Figure 2.14 Stress–strain plots for (a) elastic behavior (no energy is lost during a load–unload cycle) and (b) viscoelastic behavior (energy equal to the shaded area is lost in a load–unload cycle).

behavior of a material. The material is subjected to an oscillatory strain with frequency ω. From the figure, we can write the following expressions for strain and stress:

$$e = e_0 \sin \omega t,$$

$$\sigma = \sigma_0 \sin(\omega t + \delta).$$

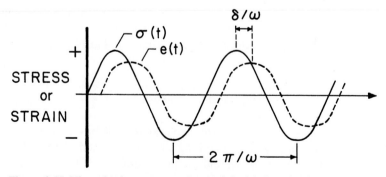

Figure 2.15 Viscoelastic response of material with time lag between stress and strain.

In the latter equation, δ is the phase angle or phase lag between the stress and strain. From these expressions, we can define two moduli,

$$E' = \left(\frac{\sigma_0}{e_0}\right)\cos \delta$$

and

$$E'' = \left(\frac{\sigma_0}{e_0}\right)\sin \delta,$$

where E' is the tensile storage modulus and E'' is the tensile loss modulus.

Alternatively, we can use complex variables and write

$$e = e_0 \exp i(\omega t),$$

$$\sigma = \sigma_0 \exp i(\omega t + \delta),$$

$$E = \frac{\sigma}{e} = \frac{\sigma_0}{e_0} \exp i\delta = \frac{\sigma_0}{e_0}(\cos \delta + i \sin \delta)$$

$$= E' + iE'',$$

where i is the imaginary number $\sqrt{-1}$.

Figure 2.16 shows graphically the relationship among these quantities. Proceeding in a manner similar to that for deriving the tensile modulus, we can obtain the shear modulus. (Experimentally, this is generally obtained by means of a torsion pendulum.) The complex modulus

$$G = G' + iG'',$$

where G' is the shear storage modulus and G'' is the shear loss modulus. The storage modulus is a measure of the stored energy, i.e., the elastic part. The loss modulus is a measure of the energy lost as heat, i.e., the viscous part. These two modulus components can be written in terms of the phase shift as

$$G'' = G \sin \delta, \qquad E'' = E \sin \delta,$$
$$G' = G \cos \delta, \qquad E' = E \cos \delta.$$

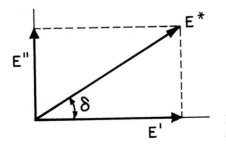

Figure 2.16 Relationship between tensile storage and tensile loss modulus.

We can now define a term called the *loss tangent* as follows:

$$\text{Loss tangent} = \tan \delta = \frac{\text{energy loss}}{\text{energy stored}} = \frac{G''}{G'} = \frac{E''}{E'}.$$

Sometimes, a related term called the *logarithmic decrement* Δ is used, which is defined as

$$\text{Logarithmic decrement } \Delta = \pi \tan \delta = \frac{\pi E''}{E'}.$$

The logarithmic decrement is the natural logarithm of the amplitude ratio between successive vibrations; that is,

$$\Delta = \ln \frac{\theta_n}{\theta_{n+1}},$$

where θ_n and θ_{n+1} are the amplitudes of two successive vibrations.

Both the loss tangent and the logarithmic decrement are proportional to the ratio of the maximum energy dissipated per cycle to the maximum energy stored in the cycle.

EXAMPLE 2.13

In a free-vibration test, a polymer showed a drop of 50% in two successive amplitudes. Compute the logarithmic decrement for this polymer.

Solution: If θ_n and θ_{n+1} are the successive amplitudes, then the logarithmic decrement, $\Delta = \ln(\theta_n/\theta_{n+1}) = \ln 2 = 0.69$.

EXAMPLE 2.14

Recall that the stress–strain relationship involving real and imaginary moduli is given by

$$\sigma = (E' + iE'')\varepsilon = E^*\varepsilon.$$

Derive an expression for the complex modulus E^* in terms of E' and $\tan \delta$. Show that for small values of $\tan \delta$, $E^* \approx E'$.

Solution: The magnitude of the complex modulus is given by

$$E^* = \frac{\sigma}{\varepsilon} = (E'^2 + E''^2)^{1/2} = E'[1 + \tan^2 \delta]^{1/2}.$$

For $\tan \delta < 0.2$, E^* will be within 2% of E'.

2.12 RUBBER ELASTICITY

A polymeric molecule is generally not rigid like a straight rod, although there are some special liquid crystal polymers that do have a rigid, rodlike molecule (e.g., the aramid fibers). Barring these special cases, the polymeric molecule is a very long and flexible chain that can change form easily because many independent vibrations and rotations of the individual atoms that compose the molecular chain are possible. Long, flexible polymeric chains can change their configuration and lengths rather easily when a stress is applied. When the number of configurations available is very large and the chains are cross-linked to form a network, we get a special polymer called an *elastomer*. Elastomers characteristically show very high reversible, nonlinear extensions (5–700%) in response to an applied stress. The requirement of cross-linking (i.e., the existence of a network) is established to avoid chains slipping past one another in a *permanent* manner. High chain mobility is also required. Glassy and crystalline polymers will not have enough chain mobility, and therefore, the reversible strains are not very large. In crystalline materials such as metals and ceramics, the deformation involves a change in equilibrium interatomic distance, which requires the application of rather large forces. This is why the elastic modulus values of metals and ceramics are very high.

The first law of thermodynamics says that the internal energy of a system is given by

$$dU = dQ + dW, \tag{2.36}$$

where dQ is the heat absorbed and dW is the work done on the system by the surroundings. Also, for a reversible process, we can write, from the second law of thermodynamics,

$$dQ = TdS \tag{2.37}$$

and

$$dW = Fd\ell - PdV, \tag{2.38}$$

where T is the temperature, V is the volume, P is the external pressure, S is the entropy, and F is the tensile force causing a change in the length ℓ.

From Eqs. (2.36) and (2.38), we get the following for the internal energy:

$$dU = TdS + Fd\ell - PdV.$$

For conditions of constant temperature and volume, we can write

$$F = \left(\frac{\delta U}{\delta \ell}\right)_{T,V} - T\left(\frac{\delta S}{\delta \ell}\right)_{T,V} \tag{2.39}$$

$$= F_e + F_s,$$

where F_e is the energy contribution and F_s is the entropy contribution to the tensile force.

In the case of crystalline metals, the first term in Eq. (2.39) is predominant, while the second terms is negligible. This is because the crystalline structure of a metal remains essentially unchanged with deformation. Such is not the case with amorphous polymers, especially the polymers that are rubberlike and that show rather large elastic deformations. On deforming these kinds of polymers, the form of the molecular chains can change considerably, and the entropy contribution $F_s = -T(\delta S/\delta \ell)_{T,V}$ becomes considerably large. (See Figure 2.17.) In fact, the first term (i.e., the energy term) in Equation 2.39 is equal to zero for an ideal rubbery material. The rubber elasticity thus has its origins in the entropy effects. For such polymers, one can write an expression for the entropy of the form

$$S = k \ln p$$

where k is Boltzmann's constant and p is the probability of finding a particular chain configuration for which the entropy effects will be very important.

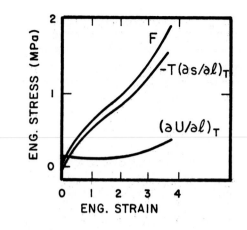

Figure 2.17 Changes in internal energy, U, and entropy S, accompanying the extension of rubber.

When an elastomer is stretched, the distance between cross-linked points increases, and the number of possible chain configurations decreases. Consider a piece of rubber of dimensions ℓ_1, ℓ_2, and ℓ_3 in the deformed state. (See Figure 2.18) Let us subject the rubber to a uniaxial force F. Then we can write the following expressions for this stretched piece of rubber:

Extension or draw ratio $= \lambda_1 = \ell_1/\ell_0$, or $\ell_1 = \lambda_1\ell_0$.

Volume in the undeformed state $= \ell_0^3$.

Volume in the deformed state $= \ell_1\ell_2\ell_3$.

In rubberlike materials, the deformation does not cause a change in volume (i.e., $\nu \simeq \frac{1}{2}$). This condition of constancy of volume means that

$$\ell_0^3 = \ell_1\ell_2\ell_3. \tag{2.40}$$

The strains in the lateral directions are equal, i.e.,

$$\ell_2 = \ell_3.$$

Putting the values of ℓ_1 and ℓ_3 into Eq. 2.40, we get

$$(\lambda_1\ell_0)\ell_2^2 = \ell_0^3,$$

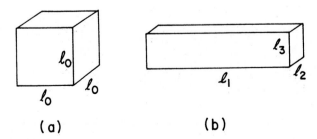

(a)   (b)

Figure 2.18 Deformation at constant volume. (a) Unstrained state. (b) Strained state.

or

$$\ell_2 = \ell_3 = \ell_0/\sqrt{\lambda_1}. \tag{2.41}$$

Equation 2.41 tells us that the rubber sample in directions 2 and 3 is reduced to $\ell_0/\sqrt{\lambda_1}$, or

$$\lambda_2 = \lambda_3 = 1/\sqrt{\lambda_1}. \tag{2.42}$$

If the applied force produces distortions λ_1, λ_2, and $\lambda_3(\lambda_1 = 1 + d\ell_1/\ell_1)$, then it can be shown[8] that the change in entropy of the network is

$$\Delta S = -(1/2)Nk(\lambda_1^2 + \lambda_2^2 + \lambda_3^2 - 3), \tag{2.43}$$

where N is the number of chains and k is Boltzmann's constant. From Equation 2.42, the entropy change in Equation 2.43 then becomes

$$\Delta S = -(1/2)[Nk(\lambda_1^2 + 2\lambda_1^{-1} - 3)]. \tag{2.44}$$

If the rubber is subjected to an isothermal extension, the $\delta U/\delta\ell = 0$, and we have, from Equation (2.39),

$$F = -T(\partial S/\partial\ell)_{T,V}. \tag{2.45}$$

Differentiating Equation 2.44 with respect to λ_1 ($= \ell_1/\ell_0$) and substituting into Equation 2.45, we get

$$F = NkT/\ell_0[\lambda_1 - \lambda_1^{-2}]. \tag{2.46}$$

Rearranging and writing force divided by area as the tensile stress, we obtain

$$\sigma = F/A = \frac{F\ell}{A_0\ell_0},$$

or

$$\sigma = F/A = (NkT/A_0\ell_0)[\lambda_1^1 - \lambda_1^{-2}]. \tag{2.47}$$

Denoting the number of chain segments per unit volume $(N/A_0\ell_0)$ by n and we can rewrite Eq. (2.47) as

$$\sigma = nkT[\lambda_1^2 + \lambda_1^{-1}]. \tag{2.48}$$

Equation (2.48) shows a linear dependence of stress, at a given strain, on temperature. This follows from the dominance of the entropic elasticity. Any deviation from this linear relationship between stress and temperature of a rubbery or elastomeric materials can be taken as a measure of its deviation from thermodynamic ideal behavior. For an ideal rubbery behavior, the energetic component of force is zero. Also, the stress is not linearly dependent on strain, i.e., the Hooke's law is not obeyed in tension for an elastomer. Up to ~400% strain, the theoretical stress-strain curve is in quite good accord with experimental values as shown in Fig. 2.19. At very large strains, i.e., at strains >400% ($\lambda = 5$)

---

[8] See L. R. G. Treloar, *The Physics of Rubber Elasticity,* 3d ed. (Oxford: Clarendon Press, 1975).

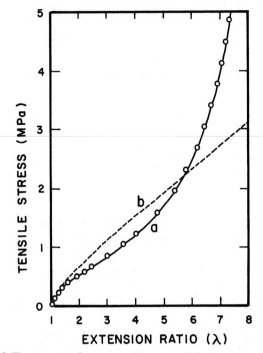

Figure 2.19 Force–extension curve for cross-linked rubber. (a) Experimental. (b) Theoretical. (After L. R. G. Treloar, *The Physics of Rubber Elasticity,* 3d ed. (Oxford: Clarendon Press, 1975), p. 87.)

secondary bonds form between the partially aligned chains, i.e., strain induced crystallization occurs. At such large strain values, the chains begin to align themselves and stretching of the primary bonds in the chain becomes important.

Because the tensile stress–strain curve of rubber is nonlinear, Young's modulus cannot be defined for rubber, as it can be for crystalline metals and ceramics. One can, however, define a secant modulus at a given strain. Another important thing that a perceptive reader may have noticed is that the number of network chains per unit volume and, correspondingly, the modulus of an elastomer increases as the degree of cross-linking increases. This is as expected if we just compare a lightly cross-linked rubber band with a highly cross-linked bowling ball.

EXAMPLE 2.15

Make a schematic plot of the internal energy and entropy as a function of strain for a crystalline solid (e.g., a metal) and for a rubbery solid (e.g., an elastomer). Make a drawing showing the structure before and after deformation in the two cases.

Solution: Figure E2.14 shows the requested plots.

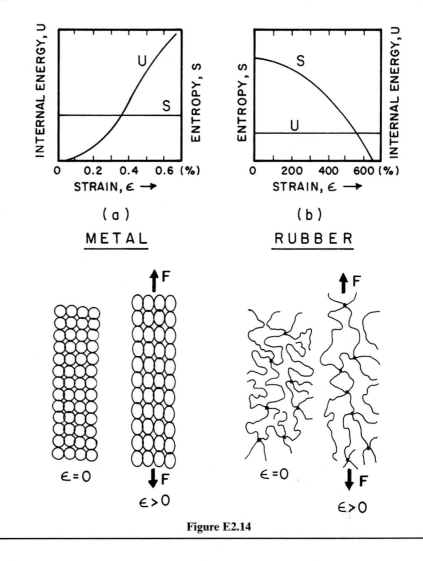

Figure E2.14

EXAMPLE 2.16

It is frequently said that elastic deformation on loading and the recovery of strain on unloading involves the stretching of atomic bonds. Would this statement be true of the large elastic deformation that is observed in rubbery, or elastomeric, materials?

Solution: No. The large elastic deformation observed in elastomeric materials involves the *uncoiling* of randomly coiled polymeric chains. When we deform an elastomeric material, the end-to-end distance of the chains increases. When the material is unloaded, the chains return to the original random configuration. This uncoiling of chains results in the entropy effects discussed in the text. Such entropy effects are insignificant in metals and other nonelastomeric materials.

SUGGESTED READINGS

Y. C. FUNG. *A First Course in Solid Mechanics,* 2d. ed. Upper Saddle River, NJ: Prentice Hall, 1997.

H. B. HUNTINGTON. *The Elastic Constants of Crystals.* New York: Academic Press, 1958.

A. KELLY and G. W. GROVES. *Crystallography and Crystal Defects.* Reading, MA: Addison-Wesley, 1970.

J. LEMAITRE, and J.-L. CHABOCHE. *Mechanics of Solid Materials.* Cambridge, U.K.: Cambridge U. Press, 1990.

A. E. H. LOVE. *The Mathematical Theory of Elasticity.* New York: Dover, 1952.

F. A. McCLINTOCK and A. S. ARGON (eds.). *Mechanical Behavior of Materials.* Reading, MA: Addison-Wesley, 1966.

J. F. NYE. *Physical Properties of Crystals.* London: Oxford University Press, 1957.

G. SIMMONS and H. WANG. *Single Crystal Elastic Constants.* Cambridge, MA: MIT Press, 1971.

I. S. SOKOLNIKOFF. *Mathematical Theory of Elasticity,* 2d ed. New York: McGraw-Hill, 1956.

S. TIMOSHENKO and J. N. GOODIER. *Theory of Elasticity.* New York: McGraw-Hill, 1951.

L. R. G. TRELOAR. *The Physics of Rubber Elasticity,* 3d ed. Oxford, U.K.: Clarendon Press, 1975.

EXERCISES

2.1 The rubber specimens, having an initial length of 5 cm, are tested, one in compression and one in tension. If the engineering strains are -1.5 and $+1.5$, respectively, what will be the final lengths of the specimens? What are the true strains, and why are they numerically different?

2.2 An aluminum polycrystalline specimen is elastically compressed in plane strain. If the true strain along the compression direction is -2×10^{-4}, what are the other two longitudinal strains?

2.3 Determine K, λ, and G for polycrystalline niobium, titanium, and iron, from E and γ.

2.4 A state of stress is given by

$$\sigma_{11} = 350 \text{ MPa},$$

$$\sigma_{12} = 70 \text{ MPa},$$

$$\sigma_{22} = 210 \text{ MPa}.$$

Determine the principal stresses and the maximum shear stress, as well as their angle with the given direction.

2.5 Calculate the anisotropy ratio for the cubic metals in Table 2.3.

2.6 Show that a uniaxial hydrostatic compressive stress can be decomposed into a hydrostatic pressure and two states of pure shear. Use sketches if necessary.

2.7 Determine the principal stresses and the maximum shear stress, as well as their angles with the system of reference given by the following stresses:

$$[\sigma_{ij}] = \begin{pmatrix} 3 & 2 \\ 2 & 0 \end{pmatrix} \text{MPa}$$

2.8 Extensometers attached to the external surface of a steel pressure vessel indicate that $\varepsilon_t = 0.002$ and $\varepsilon_t = 0.005$ along the longitudinal and transverse directions, respectively. Determine the corresponding stresses. What would be the error if Poisson's ratio were not considered?

2.9 Calculate Young's and shear moduli for monocrystalline iron along [100], [110], and [111].

2.10 From the values obtained in Exercise 2.9, obtain a rough estimate of the Young's modulus of a polycrystalline aggregate, assuming that there are only three orientations for the grains ([100], [110], and [111]) and that they occur proportionally to their multiplicity factors. Compare your result with the predictions of Voigt averages (isostrain) and Reuss averages (isostress).

2.11 A silver monocrystal is extended along [100]. Obtain the values for the Young's and shear moduli, as well as Poisson's ratio.

2.12 (a) For Figure 2.18, plot the curve of true stress vs. true strain. (b) Taking the slopes of the curve at various strains, plot the elastic modulus of rubber as a function of strain. (c) Schematically draw polymer chains at different positions in the curve.

2.13 A steel specimen is subjected to elastic stresses represented by the matrix

$$\sigma_{ij} = \begin{pmatrix} 2 & -3 & 1 \\ -3 & 4 & 5 \\ 1 & 5 & -1 \end{pmatrix} \text{MPa.}$$

Calculate the corresponding strains.

Section 2.2

2.14 Ultrasonic equipment was used to determine the longitudinal and shear sound velocities of a metallic specimen having a density of 7.8 g/cm³. The values obtained are

$$V_\ell = 5{,}300 \text{ m/s,}$$

$$V_s = 3{,}300 \text{ m/s.}$$

Determine the Young's and shear moduli and Poisson's ratio for this material. What is the material?

2.15 A tubular specimen is being subjected to a torsional moment $T = 600\text{N} \cdot \text{m}$. If the shear modulus of the material (Al) is equal to 26.1 GPa, what is the total angular deflection if the length is 1 m? The tube has a diameter of 5 cm and a wall thickness of 0.5 cm. Assume the process to be elastic.

2.16 Using the Mohr circle construction, calculate the principal stresses and the maximum shear stresses, as well as their orientation, for the sheet subjected to the stresses shown in Figure Ex.2.16.

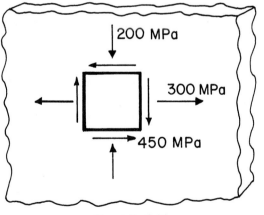

Figure Ex.2.16

2.17 A state of stress is given by

$$\sigma_{11} = -500 \text{ MPa,}$$

$$\sigma_{22} = 300 \text{ MPa,}$$

$$\sigma_{12} = 150 \text{ MPa.}$$

Determine the principal stresses and the maximum shear stress, as well as their orientation, using the Mohr circle construction.

2.18 From the elastic stiffnesses for copper (see Table 2.3), determine the elastic compliances.

2.19 From the elastic compliances S_{11}, S_{12}, and S_{14} for chromium and tungsten, determine the Young's moduli along [111], [110], and [100].

2.20 Determine the elastic Young's moduli for tungsten and ZrO_2 along [112], [122], and [123].

2.21 Determine the polycrystalline Young's modulus for molybdenum using Reuss' and Voigt's averages. Use elastic stiffnesses and compliances from Tables 2.3 and 2.4.

Section 2.11

2.22 Plot Young's modulus as a function of porosity for alumina, and show what the value should be for a specimen having 5% porosity ($E_{Al_2O_3} = 378$ GPa).

2.23 A specimen of Al_2O_3 contains microcracks that are approximately equal to its grain size (20 μm). One grain in each 10 grains contains cracks. If the uncracked materials has $E_0 = 378$ GPa, determine Young's modulus for the cracked material using Budiansky and O'Connell's and Salganik's equations.

Section 2.8

2.24 Young's modulus (E) of a cubic single crystal as a function of orientation is given by

$$\frac{1}{E_{hkl}} = \frac{1}{E_{100}} - 3\left(\frac{1}{E_{100}} - \frac{1}{E_{111}}\right)(\ell_1^2\ell_2^2 + \ell_2^2\ell_3^2 + \ell_3^2\ell_1^2),$$

where l_1, l_2, and l_3 are the direction cosines between the direction hkl and [100], [010], and [001], respectively. This is another version of the expression given in Example 2.9. For copper, $E_{111} = 19$ GPa and $E_{100} = 66$ GPa. Calculate Young's modulus for a copper single crystal in the [110] direction, and check your answer against the one in Example 2.9.

Section 2.12

2.25 A polymer has a viscosity of 10^{12} Pas at 150°C. If this polymer is subjected to a tensile stress of 100 MPa at that temperature, compute the deformation after 10 h. Assume the polymer to behave as a Maxwell solid. Take $E = 5$ GPa, and use the equation

$$\varepsilon_t = \frac{\sigma}{E} + \frac{1}{3\eta}\sigma t.$$

2.26 For a given polymer, the activation energy for stress relaxation was measured to be 10 kJ/mole. If the stress relaxation time for this polymer at room temperature is 3,600 s, what would be the relaxation time at 100°C?

Section 2.13

2.27 For an elastomeric material, we have the constitutive equation

$$\sigma = G\left(\lambda - \frac{1}{\lambda^2}\right) = \frac{E}{3}\left[\lambda - \frac{1}{\lambda^2}\right],$$

where E is the elastic modulus at zero elongation. Show that, for very small strains, this equation reduces to $\sigma = E\varepsilon$.

2.28 A cylindrical steel specimen (length = 200 mm, diameter = 5 mm) is subjected to a torque equal to 40 Nm. If one end of the specimen is fixed what is the deflection of the other end? Take E = 210 GPa and ν = 0.3.

2.29 A material is subjected to the following state of stress:

$$\sigma_{11} = 350 \text{ MPa},$$

$$\sigma_{12} = 70 \text{ MPa},$$

$$\sigma_{22} = 210 \text{ MPa}.$$

Determine the maximum and minimum principal stresses, the maximum shear stress, and the angle between the maximum principal stress and the maximum shear stress.

3

Plasticity

3.1 INTRODUCTION

Upon being mechanically stressed, a material will, in general, exhibit the following sequence of responses: elastic deformation, plastic deformation, and fracture. This chapter addresses the second response: plastic deformation. A sound knowledge of plasticity is of great importance because:

1. Many projects are executed in which small plastic deformations of the structure are accepted. The "theory of limit design" is used in applications where the weight factor is critical, such as space vehicles and rockets. The rationale for accepting a limited plastic deformation is that the material will work-harden at that region, and plastic deformation will cease once the flow stress (due to work-hardening) reaches the applied stress.
2. It is very important to know the stresses and strains involved in deformation processing, such as rolling, forging, extrusion, drawing, and so on. All these processes involve substantial plastic deformation, and the response of the material will depend on its plastic behavior during the processes. The application of plasticity theory to such processes is presented later in this chapter.
3. The mechanism of fracture involves plastic deformation at the tip of a crack. The way in which the high stresses that develop at the crack can be accommodated by the surrounding material is of utmost importance in the propagation of the crack. A material in which plastic deformation can take place at the crack is "tough," while one in which there is no such deformation is "brittle."
4. The stress at which plastic deformation starts is dependent upon the stress state. A material can have a much greater strength when it is confined—that is, when

it is not allowed to flow laterally—than when it is not confined. This will be discussed in detail later. A number of criteria for plastic deformation and fracture will be examined in this chapter.

The mechanical strength of a material under a steadily increasing load can be determined in uniaxial tensile tests, compression (upsetting) tests, bend tests, shear tests, plane-strain tensile tests, plane-strain compression (Ford) tests, torsion tests, and biaxial tests. The uniaxial tensile test consists of extending a specimen whose longitudinal dimension is substantially larger than the two lateral dimensions (Figure 3.1a). The upsetting test consists of compressing a cylinder between parallel platens; the height/diameter ratio has to be lower than a critical value in order to eliminate the possibility of instability (buckling) (Figure 3.1b). After a certain amount of strain, "barreling" takes place, destroying the state of uniaxial compression. The three-point bend test is one of the most common bending tests. A specimen is simply placed between two supports; a wedge advances and bends it through its middle point (Figure 3.1c). Plane-strain tests simulate the conditions encountered by a metal in, for instance, rolling. Loading is imparted in such a way as to result in zero strain along one direction. The two most common geometries are shown in Figures 3.1d and e. In the tensile mode, two grooves are made parallel to each other, on opposite sides of a plate. The width of the plate is much greater than its thickness in the region of reduced thickness; hence, flow is restricted in the direction of the width. In the compressive mode (Ford test), a parallelepiped of metal is machined and inserted between the groove-and-punch setup of Figure 3.1e. As the top punch is lowered, the specimen is plastically deformed. Strain is restricted in one direction. In the torsion test (Figure 3.1f), the cylindrical (or tubular) specimen is subjected to a torque and undergoes an attendant angular displacement. One of the problems in the analysis of the torsion test is that the stress varies as the distance from the

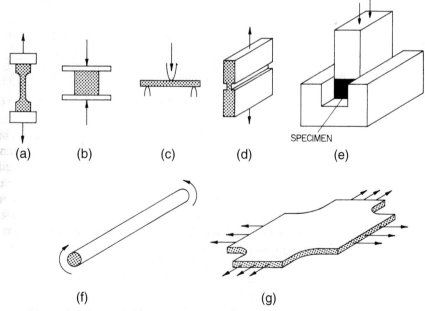

SPECIMEN

(a)   (b)   (c)   (d)   (e)

(f)   (g)

Figure 3.1 Common tests used to determine the monotonic strength of metals. (a) Uniaxial tensile test. (b) Upsetting test. (c) Three-point bending test. (d) Plane-strain tensile test. (e) Plane-strain compression (Ford) test. (f) Torsion test. (g) Biaxial test.

central axis of the specimen. Accordingly, the biaxial test is usually applied to thin sheets, and one of the configurations is shown in Figure 3.1g. Other configurations involve testing a tubular specimen in tension with an internal pressure and testing a tubular specimen in tension with torsion. The results of the tests just described can be expressed graphically as stress-versus-strain curves. They can be compared directly by using effective stresses and effective strains. A machine commonly used to carry out the tests is the so-called universal testing machine. Both screw-driven (Figure 2.1) and servohydraulic machines are very useful for mechanical testing. Figure 3.2 shows a typical servohydraulic testing machine.

Figure 3.2 A servohydraulic universal testing machine linked to a computer. (Courtesy of MTS Systems Corp.)

3.2 PLASTIC DEFORMATION IN TENSION

Figure 3.3 shows a number of stress–strain curves for the same material: AISI 1040 steel. This might look surprising at first, but it merely reflects the complexity of the microstructural-mechanical behavior interactions. Both engineering and true stress–strain curves are shown. (The definitions of these are given in Chapter 2.) Engineering (or nominal) stress is defined as P/A_0, while true stress is P/A, where A_0 and A are the initial and current cross-sectional areas, respectively. Engineering (or nominal) strain is defined as $\Delta L/L_0$, while true strain is $\ln L/L_0$, where L and L_0 are the current and initial lengths, respectively. The yield stress varies from 250 to 1,100 MPa, depending on the heat treatment. Conversely, the total elongation varies from 0.38 to 0.1. The properties of steel are highly dependent upon heat treatment, and quenching produces a hard, martensitic structure, which is gradually softened by tempering treatments at higher temperatures (200, 400, and 600°C). The annealed structure is ductile, but has a low yield stress. The ultimate tensile stresses (the maximum engineering stresses) are marked by arrows. After these points,

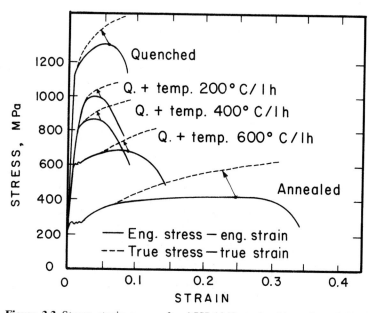

Figure 3.3 Stress–strain curves for AISI 1040 steel subjected to different heat treatments; curves obtained from tensile tests.

plastic deformation becomes localized (called *necking*), and the engineering stresses drop because of the localized reduction in cross-sectional area. However, the true stress continues to rise because the cross-sectional area decreases and the material work-hardens in the neck region. The true-stress–true-strain curves are obtained by converting the tensile stress and its corresponding strain into true values and extending the curve.

We know that the volume V is constant in plastic deformation:

$$V = A_0 L_0 = AL.$$

Consequently,

$$A = \frac{A_0 L_0}{L}. \tag{3.1}$$

In what follows, we use the subscripts e and t for engineering (nominal) and true stresses and strains, respectively. We have

$$\varepsilon_e = \frac{L - L_0}{L_0} = \frac{A_0}{A} - 1, \tag{3.2}$$

$$\frac{\sigma_t}{\sigma_e} = \frac{P}{A} \times \frac{A_0}{P} = \frac{A_0}{A} = 1 + \varepsilon_e, \tag{3.3}$$

$$\sigma_t = (1 + \varepsilon_e)\sigma_e. \tag{3.4}$$

On the other hand, the incremental longitudinal true strain is defined as

$$d\varepsilon_t = \frac{dL}{L}. \tag{3.5}$$

For extended deformations, integration is required:

$$\varepsilon_t = \int_{L_0}^{L} \frac{dL}{L} = \ln \frac{L}{L_0},$$ (3.6)

$$\exp(\varepsilon_t) = \frac{L}{L_0}.$$ (3.7)

Substituting Equations 3.2 and 3.3 into Equation 3.7, we get

$$\sigma_t = \frac{P}{A_0} \exp(\varepsilon_t).$$ (3.8)

Engineering (or nominal) stresses and strains are commonly used in tensile tests, with the double objective of avoiding complications in the computation of σ and ε and obtaining values that are more significant from an engineering point of view. Indeed, the load-bearing ability of a beam is better described by the engineering stress, referred to the initial area A_0. It is possible to correlate engineering and true values.

From Equations 3.4 and 3.8, the following relationship is obtained:

$$\varepsilon_t = \ln (1 + \varepsilon_e)$$ (3.9)

All of the preceding curves, as well as other ones, are represented schematically by simple equations in various ways. Figure 3.4 shows four different idealized shapes for stress–strain curves. Note that these are true-stress–true-strain curves. When we have a large amount of plastic deformation, the plastic strain is large with respect to the elastic strain, and the latter can be neglected. If the material does not work-harden, the plastic curve is horizontal, and the idealized behavior is called perfectly plastic. This is shown in Figure 3.4a. If the plastic deformation is not so large, the elastic portion of the curve cannot be neglected, and one has an ideal elastoplastic material (Figure 3.4b). A further approximation to the behavior of real materials is the ideal elastoplastic behavior depicted in Figure 3.4c; this is a linear curve with two slopes E_1 and E_2 that represent the material's elastic and plastic behavior, respectively. One could represent the behavior of the steels in Figure 3.3 fairly well by this elastoplastic, linear work-hardening behavior. It can be seen that $E_2 \ll E_1$. For example, for annealed steel, $E_2 \cong 70$ MPa, while $E_1 = 210$ GPa. However, a better representation of the work-hardening behavior is obtained by assuming a gradual decrease in the slope of the curve as plastic deformation proceeds (shown in Figure 3.4d). The convex shape of the curve is well represented by an equation of the type

$$\sigma = K\varepsilon^n,$$ (3.10)

where $n < 1$. This response is usually called "parabolic" hardening, and one can translate it upward by assuming a yield stress σ_0, so that Equation 3.10 becomes

$$\sigma = \sigma_0 + K\varepsilon^n.$$ (3.11)

The exponent n is called the work-hardening coefficient.

These equations that describe the stress–strain curve of a polycrystalline metal are known as the Ludwik–Hollomon equations.[1] In them, K is a constant, and the exponent n

[1] See P. Ludwik, *Elemente der Technologischen Mechanik* (Berlin: Springer, 1909), p. 32; and J. H. Hollomon, *Trans. AIME,* 162 (1945) 268.

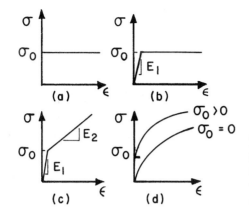

Figure 3.4 Idealized shapes of uniaxial stress–strain curve. (a) Perfectly plastic. (b) Ideal elastoplastic. (c) Ideal elastoplastic with linear work-hardening. (d) Parabolic work-hardening $(\sigma = \sigma_0 + K\varepsilon^n)$.

depends on the nature of the material, the temperature at which it is work-hardened, and the strain. The exponent n generally varies between 0.2 and 0.5, while the value of K varies between $G/100$ and $G/1{,}000$, G being the shear modulus. In Equation 3.11 ε is the true plastic strain, while in Equation 3.10 ε is true total strain. Equations 3.11 and 3.10 describe parabolic behavior. However, such a description is valid only in a narrow stretch of the stress–strain curve. There are two reasons for this. First, the equations predict a slope of infinity for $\varepsilon = 0$, which does not conform with the experimental facts. Second, the equations imply that $\sigma \to \infty$ when $\varepsilon \to \infty$. But we know that this is not correct and that, experimentally, a saturation of stress occurs at higher strains.

Voce[2] introduced a much different equation,

$$\frac{\sigma_s - \sigma}{\sigma_s - \sigma_0} = \exp\left(-\frac{\varepsilon}{\varepsilon_c}\right), \tag{3.12}$$

where σ_s, σ_0, and ε_c are empirical parameters that depend on the material, the temperature, and the strain rate. This equation says that the stress exponentially reaches an asymptotic value of σ_s at higher strain values. Furthermore, it gives a finite slope to the stress–strain curve at $\varepsilon = 0$ or $\sigma = \sigma_0$.

It should be noted that the parameters in the preceding equations (3.10–3.12) depend on the choice of the initial stress and/or strain. For instance, if one prestrained a material, one would affect K in the Ludwik–Hollomon equation.

The fact that some equations reasonably approximate the stress–strain curves does not imply that they are capable of describing the curves in a physically satisfactory way. There are two reasons for this: (1) In the different positions of stress–strain curves, different microscopic processes predominate. (2) Plastic deformation is a complex physical process that depends on the path taken; it is not a thermodynamic state function. That is to say, the accumulated plastic deformation is not uniquely related to the dislocation structure of the material. This being so, it is not very likely that simple expressions could be derived for the stress–strain curves in which the parameters would have definite physical significance.

Some alloys, such as stainless steels, undergo martensitic phase transformations induced by plastic strain. This type of transformation alters the stress–strain curve. (See Chapter 11). Other alloys undergo mechanical twinning beyond a threshold stress (or strain), which affects the shape of the curve. In these cases, it is necessary to divide the plastic regime into stages. It is often useful to plot the slope of the stress–strain curve vs. stress (or strain) to reveal changes in mechanism more clearly.

[2]E. Voce, *J. Inst. Met.,* 74 (1948) 537.

In spite of its limitations, the Ludwik–Hollomon Equation 3.11 is the most common representation of plastic response. When $n = 0$, it represents ideal plastic behavior (no work-hardening). More general forms of this equation, incorporating both strain rate and thermal effects, are often used to represent the response of metals; in that case they are called *constitutive equations*. As will be shown in Chapter 4, the flow stress of metals increases with increasing strain rate and decreasing temperature, because thermally activated dislocation motion is inhibited.

The Johnson-Cook equation

$$\sigma = (\sigma_0 + K\varepsilon^n)\left(1 + C \ln \frac{\dot{\varepsilon}}{\dot{\varepsilon}_0}\right)\left[1 - \left(\frac{T - T_r}{T_m - T_r}\right)^m\right] \qquad (3.13)$$

is widely used in large-scale deformation codes. The three groups of terms in parentheses represent work-hardening, strain rate, and thermal effects, respectively. The constants K, n, C, and m are material parameters, and T_r is the reference temperature, T_m the melting point, and $\dot{\varepsilon}_0$ the reference strain rate. There are additional equations that incorporate the microstructural elements such as grain size and dislocation interactions and dynamics: they are therefore called "physically based." The most common ones are the *Zerilli–Armstrong*[3] and the *MTS* (materials threshold stress, developed at Los Alamos National Laboratory) equations. The basic idea is to develop one equation that represents the mechanical response of a material from 0 K to $0.5T_m$ and from very low strain rates ($\sim 10^{-5}$ s^{-1}) to very high strain rates ($\sim 10^5$ s^{-1}). Nevertheless, three factors throw monkey wrenches into these equations: creep (see Chapter 13), fatigue (Chapter 14), and environmental effects. The effects of these factors are very complex and cannot be simply "plugged into" the equations.

EXAMPLE 3.1

For the stress–strain curve shown in Figure E3.1.1 (tantalum tested at strain rate of 10^{-4} s^{-1}), obtain the parameters of the Ludwik–Hollomon equation. Estimate the duration of the test in seconds.

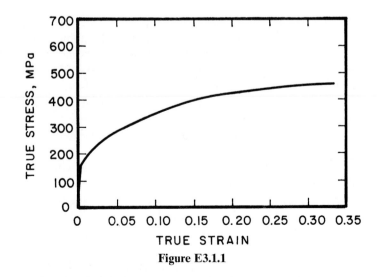

Figure E3.1.1

[3]See F. Zerilli and R. W. Armstrong, *J. Appl. Phys.* 68 (1990) 1580.

Solution: From the σ–ε curve, we have

$$\sigma_0 = 160 \text{ MPa}.$$

We use the Ludwik–Hollomon equation

$$\sigma - \sigma_0 = K\varepsilon^n,$$

so that

$$\log(\sigma - \sigma_0) = \log K + n \log \varepsilon,$$

which is a linear equation. We then make a plot of $\log(\sigma - \sigma_0)$ vs $\log \varepsilon$ (shown in Figure E3.1.2) from the following table of values:

σ	ε	$\log(\sigma - \sigma_0)$	$\log \varepsilon$
280	0.05	2.08	−1.3
345	0.1	2.27	−1
385	0.15	2.35	−0.82
415	0.2	2.41	−0.70
435	0.25	2.44	−0.60
455	0.3	2.47	−0.52

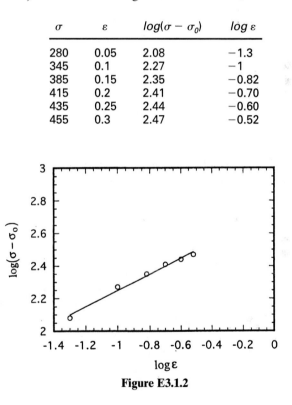

Figure E3.1.2

From the new plot, we have

$$\log K = 2.75,$$
$$K = 562,$$
$$n = \text{slope} \approx 0.5.$$

Substituting K and n into the Ludwik–Hollomon equation yields

$$\sigma = 160 + 562\varepsilon^{0.5} \text{ (in MPa)}.$$

The duration of the test, given that

$$\dot{\varepsilon} = 10^{-4}\,\mathrm{s}^{-1}$$

$$= \frac{d\varepsilon}{dt} \approx \frac{\Delta\varepsilon}{\Delta t},$$

is

$$\Delta t = \frac{\Delta\varepsilon}{\dot{\varepsilon}} \approx \frac{.33}{10^{-4}}$$

$$= 3.3 \times 10^3\,\mathrm{s}.$$

The volume of a material is assumed to be constant in plastic deformation. It is known that such is not the case in elastic deformation. As was shown in Section 2.4, the constancy in volume implies that

$$\varepsilon_{11} + \varepsilon_{22} + \varepsilon_{33} = 0$$

or

$$\varepsilon_1 + \varepsilon_2 + \varepsilon_3 = 0 \tag{3.14}$$

and that Poisson's ratio is 0.5. Figure 3.5 shows that this assumption is reasonable and that ν rises from 0.3 to 0.5 as deformation goes from elastic to plastic.

However, prior to delving into the plasticity theories, we have to know, for a complex state of stress, the stress level at which the body starts to flow plastically. The methods developed to determine this are called *flow criteria* (See Section 3.7). Figure 3.6 shows

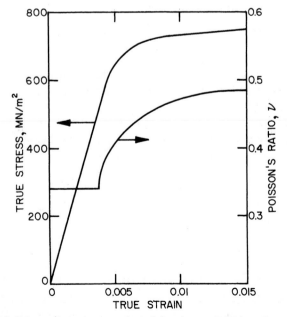

Figure 3.5 Schematic representation of the change in Poisson's ratio as the deformation regime changes from elastic to plastic.

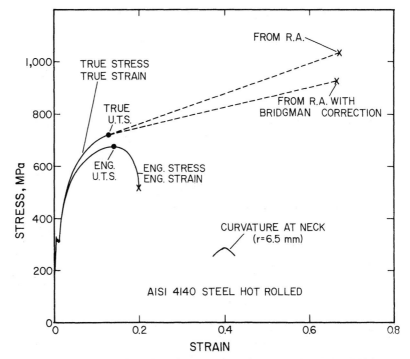

Figure 3.6 True- and engineering-stress–strain curves for AISI 4140 hot-rolled steel.

engineering- and true-stress–strain curves for the same hot-rolled AISI 4140 steel. In the elastic regime the coincidence is exact, because strains are very small (~0.5%). From Equation 3.4, we can see that we would have $\varepsilon_e \simeq \varepsilon_t$. As plastic deformation increases, ε_t and ε_e become progressively different. For $\varepsilon_t = 0.20$ (a common value for metals), we have $\varepsilon_e = 0.221$. For this deformation, the true stress is 22.1% higher than the nominal one. It can be seen that these differences become greater with increasing plastic deformation. Another *basic* difference between the two curves is the decrease in the engineering stress beyond a certain value of strain (~0.14 in Figure 3.6). This phenomenon is described in detail in Section 3.2.2.

3.2.1 Tensile Curve Parameters

Figure 3.7 shows two types of engineering stress–strain curves. The first does not exhibit a yield point, while the second does. Many parameters are used to describe the various features of these curves. First, there is the elastic limit. Since it is difficult to determine the maximum stress for which there is no permanent deformation, the 0.2% offset yield stress (point *A* in the figure) is commonly used instead; it corresponds to a permanent strain of 0.2% after unloading. Actually, there is evidence of dislocation activity in a specimen at stress levels as low as 25% of the yield stress. The region between 25 and 100% of the yield stress is called the *microyield* region and has been the object of careful investigations. In case there is a drop in yield, an *upper* (*B*) and a *lower* (*C*) *yield point* are defined in Fig. 3.7b. The lower yield point depends on the machine stiffness. A *proportional limit* is also sometimes defined *(D)*; it corresponds to the stress at which the curve deviates from linearity. The maximum engineering stress is called the *ultimate tensile stress* (UTS); it corresponds to point *D'* in Figure 3.7. Beyond the UTS, the engineering stress drops until the *rupture stress* (*E*) is reached. The *uniform strain* (*F*) corresponds to the plastic strain that takes place uniformly in the specimen. Beyond that point, necking occurs. Necking is treated in detail in Section 3.2.2. *G* is the *strain-to-failure*. Additional parameters can be

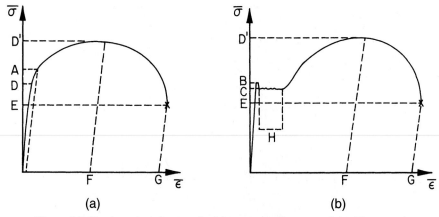

Figure 3.7 Engineering- (or nominal-) stress–strain curves (a) without and (b) with a yield point.

obtained from the stress–strain curve: (1) The elastic energy absorbed by the specimen (the area under the elastic portion of the curve) is called *resilience;* (2) the total energy absorbed by the specimen during deformation, up to fracture (the area under the whole curve), is called *work of fracture.* The strain rate undergone by the specimen, $\dot{\varepsilon}_e = d\varepsilon_e/dt$, is equal to the crosshead velocity, divided by the initial length L_0 of the specimen.

The *reduction in area* is defined as

$$q = \frac{A_0 - A_f}{A_0}, \tag{3.15}$$

where A_0 and A_f are the initial area and cross-sectional area in the fracture region, respectively. The true strain at the fracture is defined as

$$\varepsilon_f = \ell n \frac{A_0}{A_f}. \tag{3.16}$$

The true uniform strain is

$$\varepsilon_u = \ell n \frac{A_0}{A_u}, \tag{3.17}$$

where A_u is the cross-sectional area corresponding to the onset of necking (when the stress is equal to the UTS).

3.2.2 Necking

Necking corresponds to the part of the tensile test in which instability exists. The neck is a localized region in the reduced section of the specimen in which the greatest portion of strain concentrates. The specimen "necks" down in this region. Figure 3.8 shows the onset of necking in a tensile specimen; arrows show the region where the cross section starts to decrease.

Several criteria for necking have been developed. The oldest one is due to Considère.[4] According to Considère, necking starts at the maximum stress (UTS), when the increase in strength of the material due to work-hardening is less than the decrease in the load-bearing ability due to the decrease in cross-sectional area. In other words, necking starts when the increase in stress due to the reduction in cross-sectional area starts to exceed the increase in load-bearing ability due to work-hardening. We have, at the onset of necking,

[4]A. Considère, *Ann. Ponts. Chaussées,* Ser. 6. (1885) 574.

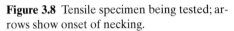

Figure 3.8 Tensile specimen being tested; arrows show onset of necking.

$$\frac{d\sigma_e}{d\varepsilon_e} = 0 \tag{3.18}$$

Substituting Equations 3.4 and 3.9 into 3.18 yields

$$\frac{d\left(\dfrac{\sigma_t}{1 + \varepsilon_e}\right)}{d(e^{\varepsilon_t} - 1)} = \frac{d\left(\dfrac{\sigma_t}{e^{\varepsilon_t}}\right)}{d(e^{\varepsilon_t} - 1)} = 0.$$

Making the transformation of variables

$$e^{\varepsilon_t} - 1 = Z, \quad e^{\varepsilon_t} = Z + 1$$

yields

$$\frac{d\left(\dfrac{\sigma_t}{Z + 1}\right)}{dZ} = \sigma_t \frac{d(Z + 1)^{-1}}{dZ} + (Z + 1)^{-1} \frac{d\sigma_t}{dZ} = 0,$$

$$-\sigma_t(Z + 1)^{-2} + (Z + 1)^{-1} \frac{d\sigma_t}{dZ} = 0,$$

or

$$-\sigma_t e^{-2\varepsilon_t} + e^{-\varepsilon_t} \frac{d\sigma_t}{d(e^{\varepsilon_t} - 1)} = 0,$$

$$\frac{d\sigma_t}{d(e^{\varepsilon_t} - 1)} = \sigma_t e^{-\varepsilon_t},$$

$$\frac{d\sigma_t}{\sigma_t} = e^{-\varepsilon_t} d(e^{\varepsilon_t} - 1) = d\varepsilon_t. \tag{3.19}$$

Using the Hollomon equation, we obtain

$$d\sigma_t = nK\varepsilon^{n-1}d\varepsilon_t$$

and it follows from Equation 3.19 that $\sigma_t = nK\varepsilon^{n-1}$. Finally, applying the Hollomon equation again results in $K\varepsilon^n = nK\varepsilon^{n-1}$, so that

$$\varepsilon_u = n.$$

The work-hardening coefficient is numerically equal to the true uniform strain and can be easily obtained in this way.

It is sometimes useful to present results of tensile tests in plots of $d\sigma/d\varepsilon$ versus σ or $d\sigma/d\varepsilon$ versus ε. An example of a plot of log $(d\sigma/d\varepsilon)$ versus log ε for AISI 302 stainless steel is given in Figure 3.9. It can be seen that $d\sigma/d\varepsilon$ decreases with ε, indicating that the necking tendency steadily increases. For metals that do not exhibit any work-hardening capability, necking should start immediately at the onset of plastic flow. Under certain conditions (predeformation at very low temperature or very high strain rate) some metals can exhibit this response, called *work-softening*.

The formation of the neck results in an accelerated and localized decrease in the cross-sectional area. Figure 3.6 shows how the true-stress–true-strain curve continues to rise after the onset of necking. It can also be seen that the true strain at fracture is much higher than the "total strain." The correct plotting of the true-stress–true-strain curve beyond the UTS requires determination of the cross-sectional area in the neck region continuously after necking. This is difficult to do, and the simplest way is to obtain one single point on the plot, joining it to the point corresponding to the maximum load. For this reason, a dashed line is used in Figure 3.6. The deformation in the neck region is much higher than the one uniformly distributed in the specimen. It can be said that the neck acts as a second tensile specimen. Since its length is smaller than that of the specimen, and the crosshead velocity is constant, the strain rate is necessarily higher.

The onset of necking is accompanied by the establishment of a triaxial state of stress in the neck; the uniaxial stress state is destroyed by the geometrical irregularity. After

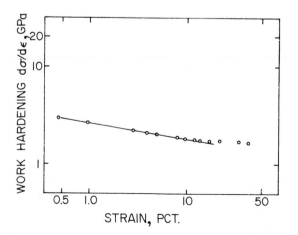

Figure 3.9 Log $d\sigma/d\varepsilon$ versus log ε for stainless steel AISI 302. (Adapted with permission from A. S. de S. e Silva and S. N. Monteiro, *Metalurgia-ABM, 33* (1977) 417, p. 420).

studying flow criteria (see Section 3.7), we will readily see that the flow stress of a material is strongly dependent on the state of stress. Hence, a correction has to be introduced to convert the triaxial flow stress into a uniaxial one. If we imagine an elemental cube aligned with the tensile axis and situated in the neck region, it can be seen that it is subjected to tensile stresses along three directions. (The external boundaries of the neck generate the tensile components perpendicular to the axis of the specimen.) The magnitude of the transverse tensile stresses depends on the geometry of the neck, the material, the shape of the specimen, the strain-rate sensitivity of the material, the temperature, the pressure, and so on. Bridgman[5] introduced a correction from a stress analysis in the neck. His analysis applies to cylindrical specimens. The equation that expresses the corrected stress is

$$\sigma = \frac{\sigma_{av}}{(1 + 2R/r_n)\, \ell n\, (1 + r_n/2R)},\qquad(3.21)$$

where R is the radius of curvature of the neck and r_n is the radius of the cross section in the thinnest part of the neck. Thus, one has to continuously monitor the changes in R and r_n during the test to perform the correction.

Figure 3.10 presents a plot in which the corrections have already been computed as a function of strain beyond necking. There are three curves, for copper, steel, and aluminum. The correction factor can be read directly from the plot shown. ε_u is the true uniform strain (the strain at onset of necking). In Figure 3.6, the true-stress–true-strain curve that was corrected for necking by the Bridgman technique lies slightly below the one determined strictly from the reduction in area at fracture and the load at the breaking point. This is consistent with Figure 3.10; σ is always lower than σ_{av}.

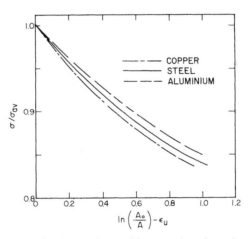

Figure 3.10 Correction factor for necking as a function of strain in neck, $\ln(A_0/A)$, minus strain at necking, ε_u. (Adapted with permission from W. J. McGregor Tegart, *Elements of Mechanical Metallurgy*, New York: MacMillan, 1964), p. 22.

Necking is a characteristic of tensile stresses; compressive stresses are not characterized by necking. Barreling is the corresponding deviation from the uniaxial state in compressive tests. Hence, metals will exhibit necking during deformation processing only if the state of stress is conducive to it (tensile). Figure 3.11 shows plainly how the work-hardening capacity of a metal greatly exceeds that in an individual tensile test. Wire was drawn to different strains: Drawing the wire consists of pulling it through a conical die; at

[5]P. W. Bridgman, *Trans. ASM*, 32 (1974) 553.

each pass, there is a reduction in cross section. Tensile tests were conducted after different degrees of straining (0 to 7.4) by wire drawing; it can be seen that the wire work hardens at each step. However, the individual tensile tests are interrupted by necking and fracture. In wire drawing, necking and fracture are inhibited by the state of stress in the deformation zone (compressive). The individual true-stress–true-strain curves were corrected for necking by Bridgman's technique; in each case, the individual curve fits fairly well into the overall work-hardening curve. It can be concluded that the individual tensile test gives only a very limited picture of the overall work-hardening response of a metal; for the wire in Figure 3.11, the total strain can exceed 7.4.

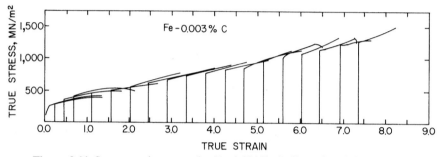

Figure 3.11 Stress–strain curves for Fe–0.003% C alloy wire, deformed to increasing strains by drawing; each curve is started at the strain corresponding to the prior wire-drawing reduction. (Courtesy of H. J. Rack)

3.2.3 Strain Rate Effects

For many materials, the stress–strain curves are sensitive to the strain rate $\dot{\varepsilon}$. The lowest range of strain rates corresponds to creep and stress-relaxation tests. The tensile tests are usually conducted in the range $10^{-4}\,s^{-1} < \dot{\varepsilon} < 10^{-2}\,s^{-1}$. At strain rates on the order of $10^2\,s^{-1}$, inertial and wave-propagation effects start to become important. The highest range of strain rates corresponds to the passage of a shock wave through the material.

More often than not, the flow stress increases with strain rate; the work-hardening rate is also affected by it. A parameter defined to describe these effects

$$m = \frac{\partial \ln \sigma}{\partial \ln \dot{\varepsilon}}\bigg|_{\varepsilon, T}, \tag{3.22}$$

is known as the *strain rate sensitivity*. Equation 3.22 can also be expressed as

$$\sigma = K\dot{\varepsilon}^m. \tag{3.23}$$

where K is a constant. Note that this K is different from the Ludwik-Hollomon parameter.

Materials can be tested over a wide range of strain rates; however, standardized tensile tests require well-characterized strain rates that do not exceed a critical value. High-strain-rate tests are often used to obtain information on the performance of materials under dynamic impact conditions. The cam plastometer is one of the instruments used. In certain industrial applications, metals are also deformed at high strain rates. Rolling mills generate bar velocities of 180 km/h; the attendant strain rates are extremely high. In wire drawing, the situation is similar.

Figure 3.12a shows the effect of different strain rates on the tensile response of AISI 1040 steel. The yield stress and flow stresses at different values of strain increase with strain rate. The work-hardening rate, on the other hand, is not as sensitive to strain rate. This il-

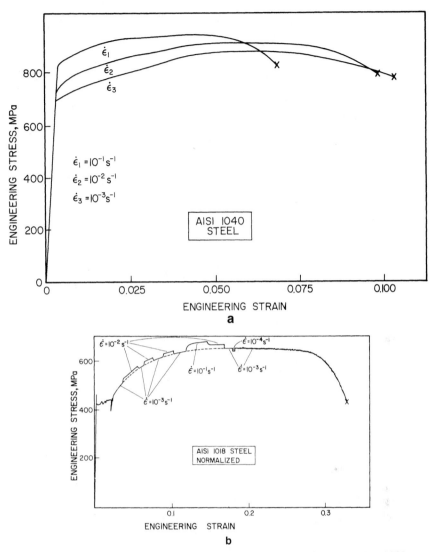

Figure 3.12 (a) Effect of strain rate on the stress–strain curves for AISI 1040 steel. (b) Strain-rate changes during tensile test. Four strain rates are shown: 10^{-1}, 10^{-2}, 10^{-3}, and $10^{-4}\,\mathrm{s}^{-1}$.

lustrates the importance of correctly specifying the strain rate when giving the yield stress of a metal. Not all metals exhibit a high strain rate sensitivity: Aluminum and some of its alloys have either zero or negative m. In general, m varies between 0.02 and 0.2 for homologous temperatures between 0 and 0.9 (90% of the melting point in K). Hence, one would have, at the most, an increase of 15% in the yield stress by doubling the strain rate. It is possible to determine m from tensile tests by changing the strain rate suddenly and by measuring the instantaneous change in stress. This technique is illustrated in Figure 3.12b. Applying Equation 3.23 to two strain rates and eliminating K, we have

$$m = \frac{\ln\,(\sigma_2/\sigma_1)}{\ln\,(\dot{\varepsilon}_2/\dot{\varepsilon}_1)} \tag{3.24}$$

The reader can easily obtain m from the strain-rate changes in the figure.

Some alloys show a peculiar plastic behavior and are called *superplastic*. When neck-ing starts, the deformation concentrates itself at the neck. Since the velocity of deforma-tion is constant, and the effective length of the specimen is reduced during necking, the strain rate increases ($\dot{\varepsilon} = v/L$). If a material exhibits a positive strain-rate sensitivity, the flow stress in the neck region will increase due to the increased strain rate; hence, necking is inhibited. This topic is treated in greater detail in Section 15.8—Superplasticity; it is what takes place in superplastic alloys, which can undergo plastic strains of up to 5,000%.

EXAMPLE 3.2

Can the necking phenomenon be observed in *any* kind of mechanical test? Point out some of the problems that this phenomenon can cause during tensile testing.

Solution: No, necking is an artifact of the tensile test only. A reduction in cross-sectional area at any irregularities along the length of the specimen occurs in the tension mode only, and therefore, the phenomenon of necking occurs in tension only. In compression, the specimen bulges out.

After necking starts, the plastic deformation is concentrated in a very nar-row region of the sample. Thus, one must not compare the total deformation corresponding to failure for two specimens that have different gage lengths. In order to avoid such complications, one should only compare the uniform elon-gation or use the reduction in area, i.e., the true-strain definition of the final strain. Strain gages and clip-on extensometers will not function properly or give accurate results after necking has begun.

EXAMPLE 3.3

Tensile testing of brittle materials such as ceramics is not very common, but is being resorted to in many laboratories. Why? Comment on the problems of do-ing tensile testing on ceramics.

Solution: Direct tensile testing of a sample results in a simple stress state over the whole volume of the sample gage length. All the volume and surface flaws in the gage length of the specimen are called into play and lead to a true measure of the ma-terial strength. Hence, there is increasing interest in tensile testing of ceramics. One major problem, however, is that of alignment of the sample. Any off-center application of the load or loading at an angle can result in a combined state of bending and tension in the specimen. Stresses induced in such a state are called *parasitic bending stresses* and can lead to errors in the computed ten-sile strength values or even fracture the sample while it is being aligned in the machine. Some self-aligning grips have been designed to take care of these problems. This leads to rather long specimens and rather complex machining of the specimen. All of this makes tensile testing of ceramics very expensive!

EXAMPLE 3.4

Determine, for the curve shown in Figure E3.4.1,
  a) Young's modulus
  b) the UTS
  c) the yield stress (with a 0.2% offset)
  d) the uniform elongation
  e) the total elongation
  f) the engineering stress–strain curve

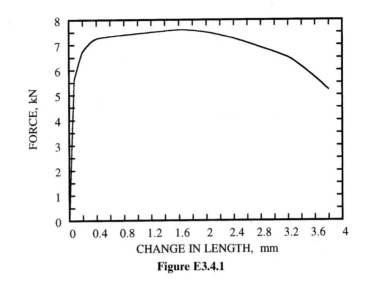

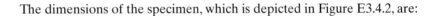

Figure E3.4.1

The dimensions of the specimen, which is depicted in Figure E3.4.2, are:

$$L_0 = 20 \text{ mm},$$

$$D_0 = 4 \text{ mm}.$$

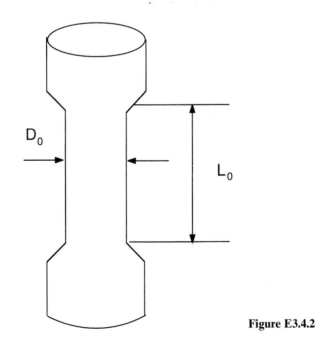

Figure E3.4.2

Solution: (a) The elastic region is the straight line of the stress–strain curve. Taking both ends of this line, we obtain

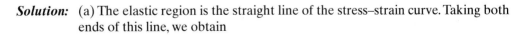

Point 1: $F_1 = 0$ kN, $\Delta l_1 = 0$,

Point 2: $F_2 = 5.5$ kN, $\Delta l_2 \approx 0.175$ mm.

To calculate Young's modulus ($E = \Delta\sigma/\Delta\varepsilon$), we have to change F, Δl in terms of σ,ε:

Point 1: $\sigma_1 = \dfrac{F_1}{A_0} = 0,$

$\varepsilon_1 = \dfrac{\Delta l_1}{L_0} = 0.$

Point 2: $\sigma_2 = \dfrac{F_2}{A_0} = \dfrac{5.5}{\pi(2)^2} \dfrac{\text{kN}}{\text{mm}^2} \approx 0.44 \text{ kN/mm}^2 = 440 \text{ MPa},$

$\varepsilon_2 = \dfrac{\Delta l_2}{L_0} \approx \dfrac{0.175}{20} \approx 0.009.$

So

$$E = \frac{\Delta\sigma}{\Delta\varepsilon} = \frac{\sigma_2 - \sigma_1}{\varepsilon_2 - \varepsilon_1} \approx \frac{440}{0.009} \approx 49000 \text{ MPa} \approx 49 \text{ GPa}$$

(b) The UTS is the maximum value of the stress reached just before necking. Therefore, from the stress–strain curve, the UTS is equal to the stress corresponding to $F \approx 7.5$ kN. So

$$\text{UTS} = \frac{7.5}{\pi(2)^2} \approx 0.6 \text{ kN/mm}^2 \approx 600 \text{ MPa}$$

(c) The 0.2%-offset yield stress is

$$\varepsilon = \frac{\Delta l}{\Delta l_0}, \qquad \begin{aligned} \varepsilon &= 0.2\% = 0.002, \\ l_0 &= 20 \text{ mm}. \end{aligned}$$

Therefore,

$$\Delta l = \varepsilon \cdot l_0 = 0.002 \times 20 = 0.04 \text{ mm}.$$

If you draw a line parallel to the elastic region calculated in part (a), from $\Delta l = 0.04$ mm, you will find that the point of intersection with the stress–strain curve is at $F \approx 6$ kN. At that point,

$$\sigma_y = \frac{6}{\pi(2)^2} \approx 0.48 \text{ kN/mm}^2 = 480 \text{ MPa}.$$

(d) For uniform elongation, make a parallel line from the UTS point to the stress axis. You will then find that

$$\Delta l_u \approx 1.5 \text{ mm}.$$

The percent uniform strain is

$$\frac{\Delta l_u}{L_0} \times 100\% = \frac{1.5}{20} \times 100\% = 7.5\%.$$

(e) To find the total strain, we repeat (d) from the failure point. We have

$$\Delta l_t \approx 3.7 \text{ mm.}$$

The percent total strain is

$$\frac{\Delta l_t}{L_0} \times 100\% = \frac{3.7}{20} \times 100\% = 18.5\%.$$

(f) The engineering stress–strain curve is as shown in Figure E3.4.3.

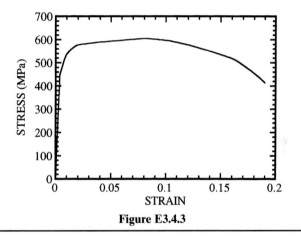

Figure E3.4.3

EXAMPLE 3.5

The load-extension curve of an aluminum alloy, shown in Figure E3.5.1 was taken directly from a testing machine. A strain-gage extensometer was used, so machine stiffness effects can be ignored. From this curve, obtain the true and engineering stress–strain curves. Also, calculate the following parameters:
a) Young's modulus
b) the UTS
c) the 0.2%-offset yield stress
d) the uniform elongation
e) the total elongation
f) the reduction in area at the fracture

Solution: We first change the coordinates to stress and strain. For engineering stresses, this is easily done:

$$\sigma_e = \frac{P}{A_0} \qquad (A_0 = 28.26 \text{ mm}^2),$$

$$\varepsilon_e = \frac{\Delta L}{L_0} \qquad (L_0 = 54 \text{ mm}).$$

The shape of the curve remains the same. For true stresses and true strains, we have to convert the engineering values into true values using the equations

$$\sigma_t = \sigma_e(1 + \varepsilon_e),$$

$$\varepsilon_t = \ln(1 + \varepsilon_e).$$

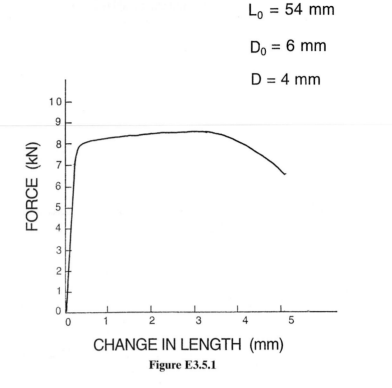

$$L_0 = 54 \text{ mm}$$

$$D_0 = 6 \text{ mm}$$

$$D = 4 \text{ mm}$$

Figure E3.5.1

This is valid up to the onset of necking. Beyond necking (which starts at the UTS), we have only one point: that corresponding to failure. We can establish the true strain in the neck from the equation

$$\varepsilon_f = \ln \frac{A_0}{A_f} = \ln \frac{\pi \times 9}{\pi \times 4} = 0.81.$$

The corresponding true stress is

$$\sigma_t = \frac{P}{A} = \frac{6.5}{\pi \times 4} \frac{\text{kN}}{\text{mm}^2},$$

$$\sigma_t = 515 \text{ MPa}.$$

The other parameters are determined as follows:
a) Young's modulus:

$$E = \text{slope of elastic part}$$

$$= \frac{\Delta \sigma}{\Delta \varepsilon}$$

$$= \frac{250}{0.004} \text{ MPa}$$

$$\approx 63 \text{ GPa}.$$

b) UTS $\approx$ 300 MPa (σ_{max}).
The corresponding true stress is

$$\sigma_t = 300(1 + 0.056) = 317 \text{ MPa}.$$

c) 0.2%-offset yield stress:

$$\sigma_{ys} \approx 280 \text{ MPa}.$$

d) The uniform elongation is approximately equal to 0.056.
The corresponding true strain is

$$\varepsilon_t = \ln(1 + 0.056) = 0.054.$$

e) The total elongation is approximately equal to 9%.
f) Reduction in area at the fracture:

$$q = \frac{A_0 - A_f}{A_0} = \frac{\pi \times 3^2 - \pi \times 2^2}{\pi \times 3^2} = 0.55, \quad \text{or } 55\%.$$

The true and engineering stress–strain curves are shown in Figure E3.5.2a. The engineering curve is shown blown up in Figure E3.5.2b..

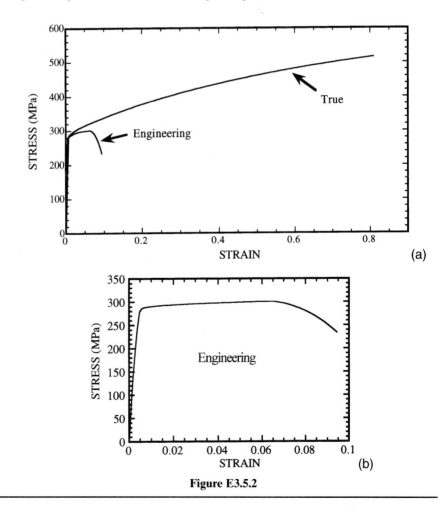

Figure E3.5.2

3.3 PLASTIC DEFORMATION IN COMPRESSION TESTING

In compression testing, a cylinder or a parallelepiped cube (with one side—the one parallel to the loading direction—longer than the other two) is subjected to compression between two parallel plates. The plates should have a self-alignment system, and they often ride on one or two hemispherical caps, as shown in Figure 3.13a. If ceramics are being tested, it is also common to use special ceramic (WC, for instance) inserts between the specimen and the hemispherical caps. This eliminates indentation and plastic deformation of platens. Lubrication between the specimen and the plate is also very desirable, to decrease barreling (nonuniform deformation) of the specimen. (Barreling will be discussed shortly.) The use of a thin Teflon™ coating, molybdenum disulfide, or graphite is recommended. It is also very important to ensure homogeneous loading of the specimen. This is particularly critical for ceramics, which often fail in the elastic range. It is easy to calculate stresses that arise when one of the parallel sides of a specimen is longer than the other. Figure 3.13b shows a specimen with a height difference Δh. The right side will experience a stress $\sigma = E(\Delta h/h_2)$ before the left side is loaded. For a typical ceramic, it is a simple matter to calculate the relationship between $\Delta\sigma$, the difference in stress from one side to the other, from Δh. For example, consider alumina, for which $E = 400$ GPa and $h = 10$ mm. The compressive strength of alumina can be as high as

$$\sigma_c = 4 \text{ GPa.}$$

Therefore, the failure strain is

$$\varepsilon_f = \frac{\sigma}{E} = 10^{-2}.$$

The corresponding displacement is

$$\Delta h = \varepsilon h = 0.1 \text{ mm.}$$

If the difference in height in the specimen is greater than 0.1 mm, the right side will fail as the left side starts to experience loading. This inhomogeneous loading is eliminated by the hemispherical caps, which can rotate to accommodate differences in height. However, if

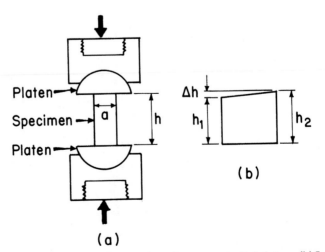

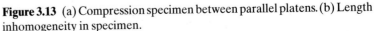

Figure 3.13 (a) Compression specimen between parallel platens. (b) Length inhomogeneity in specimen.

the surfaces of the specimen are not flat, stress inhomogeneities will arise, which can cause significant differences in the stress–strain response.

In reality, the platens also undergo elastic deformation, and a more uniform stress state is reached. Nevertheless, it is not a good practice to have the stresses on the two sides vary significantly, as this will result in erroneous strength determinations. The use of Teflon or thin metallic shims (stainless steel foil) also helps to alleviate the problem. This example illustrates the care that has to be exercised in choosing the dimensions of the specimen. In the case of ductile materials, it is not so critical, because plastic deformation will "homogenize" stresses.

Figure 3.14a shows a typical compressive stress–strain curve for a metal (70–30 brass). The engineering-stress–engineering-strain curve $(\sigma_e, \varepsilon_e)$ is concave, whereas it is convex in a tensile test. (See, for instance, Figure 3.3.) The true-stress–true-strain curve is obtained by means of Equations 3.4 and 3.9. (See also Section 2.2.) The translation of five points by using these equations is shown in Figure 3.14a. After conversion to

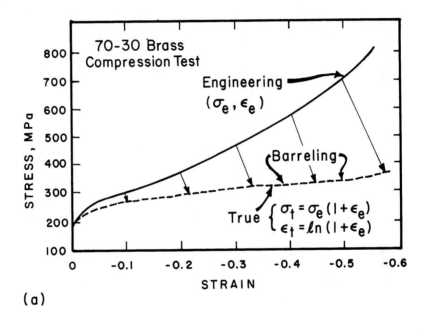

(a)

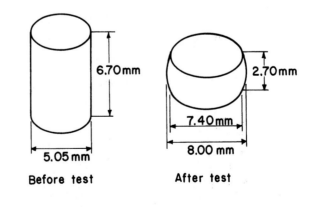

(b)

Figure 3.14 (a) Stress–strain (engineering and true) curves for 70–30 brass in compression. (b) Change of shape of specimen and barreling.

true-stress–true-strain values, the concavity of the curve is, for the most part, lost. In contrast, the true stress–strain curves in tension are displaced to the left (on the strain axis) and up (on the stress axis) from the engineering stress–strain curves. (See Figure 3.6.) The phenomenon of necking is absent in compression testing, and much higher strains are reached. However, necking is replaced by barreling, a nonuniform plastic deformation resulting from friction between the specimen and the platen. Figure 3.14b shows the barreling of the brass specimen after the test. This barreling is responsible for some concavity in the true stress–strain curve (at a strain greater than -0.4) and limits the range of strain in compression testing of ductile materials to approximately -0.3–-0.4. It will be shown, through a stress analysis, that frictional effects play an increasing role as the length/diameter ratio is decreased. This can significantly affect the results of a test. The compression of a cylindrical specimen under an engineering strain of -0.5, as simulated by finite elements under sticking conditions (i.e., there is no sliding at the specimen–platen interface), is shown in Figure 3.15. The distortion of the initially perpendicular grid is visible. This is an extreme case; strain inhomogeneities in the specimen are evident by differences in distortion of the grid. Barreling also can be seen.

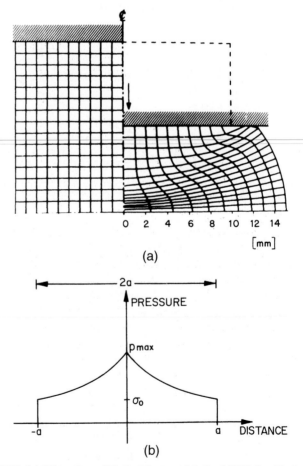

Figure 3.15 (a) Distortion of Finite Element Method (FEM) grid after 50% reduction in height h of specimen under sticking-friction conditions. (Reprinted with permission from H. Kudo and S. Matsubara, *Metal Forming Plasticity* (Berlin: Springer, 1979, p. 395.) (b) Variation in pressure on surface of cylindrical specimen being compressed.

The pressure or compressive stress is not uniform over the top and bottom surfaces of the specimen. Pressure differences can be calculated from an equation derived by Meyers and Chawla:[6]

$$p = \sigma_0 e^{2\mu(a-r)/h}.$$

This is the equation for the "friction hill." The compressive stress at the outside ($r = a$) is equal to σ_0, the material flow stress. In the center, it rises to p_{max}. The greater the ratio a/h, the more severe the problem is. The "friction hill" is schematically plotted in Figure 3.15b. The pressure rises exponentially toward the center of the cylinder. The greater the coefficient of friction, the greater is p_{max}. A friction coefficient $\mu = 0.15$ is a reasonable assumption. It is instructive to calculate the maximum pressure for three a/h ratios:

$$a/h = 2, \qquad p_{max} = 1.82\sigma_0;$$
$$a/h = 1, \qquad p_{max} = 1.34\sigma_0;$$
$$a/h = 0.5, \qquad p_{max} = 1.16\sigma_0.$$

A specimen with an initial length/diameter ratio of 2 would have a maximum pressure of $1.07\sigma_0$. However, after a 50% reduction in length, the ratio a/h is changed to $1.23\sigma_0$. The calculation is left as a challenge to student; remember that the volume is constant. This can cause significant differences between the actual strength values of materials and stress readings. It is therefore recommended that these effects be considered. On the other hand, if a/h is too small, the specimen will tend to buckle under the load.

3.4 THE BAUSCHINGER EFFECT

In most materials, plastic deformation in one direction will affect subsequent plastic response in another direction. The translation of the von Mises ellipse (kinematic hardening; see Section 3.7.4) is a manifestation of this relationship. The ellipse will move toward the direction in which the material is stressed. In one-dimensional deformation, the phenomenon is known as the *Bauschinger effect*. A material that is pulled in tension, for example, shows a reduction in compressive strength. Figure 3.16 illustrates the effect. A stress–strain curve is drawn, and the sequence 0–1–2–3 or 0–1–2–4 represents the loading direction. The material is first loaded in tension and yields at 1. At 2, the loading direction is reversed. Unloading occurs along the elastic line until the stresses become compressive. If there were no directionality effect, the material would start flowing plastically at a stress equal to σ_2. The idealized reverse curve is also shown in the figure. If the material did not exhibit a dependence on the stress direction, the compressive curve would be symmetrically opposite to the tensile curve. This idealized curve is drawn in dashed lines. The sequence is 0–1R–2R. Thus, compressive plastic flow, after the 0–1–2 tensile sequence, should occur at $\sigma_3 = \sigma_2R = -\sigma_2$. If the material exhibits a Bauschinger effect, this stress is decreased from σ_3 to σ_4. Hence, the material "softens" upon inversion of the loading direction.

An actual example is shown in Figure 3.17. The 0.2% proof stress (the stress at which 0.2% plastic strain occurs) in compression is divided by the tensile flow stress that preceded it. These values are marked in the figure, which shows three plain carbon steels and one alloy steel. The change in flow stress is indeed highly significant and increases with plastic strain in tension. Thus, this factor cannot be ignored in design considerations when a component is to be subjected to compression stresses in service after being plastically deformed in tension.

[6]M. A. Meyers and K. K. Chawla, *Mechanical Metallurgy* (Englewood Cliffs, NJ: Prentice-Hall, 1984), p. 122.

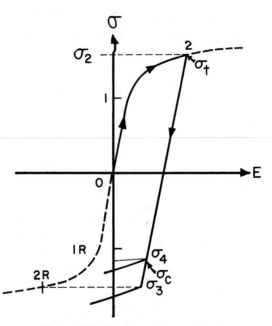

Figure 3.16 The Bauschinger effect.

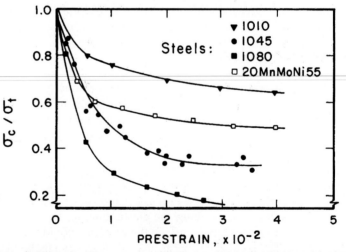

Figure 3.17 Ratio of compressive flow stress (0.2% plastic strain) and tensile flow stress at different levels of plastic strain for different steels. (After B. Scholtes, O. Vöhringer, and E. Macherauch, *Proc. ICMA6,* Vol. 1 (New York: Pergamon, 1982), p. 255.)

3.5 PLASTIC DEFORMATION OF POLYMERS

3.5.1 Stress–Strain Curves

At a microscopic level, deformation in polymers involves stretching and rotating of molecular bonds. More commonly, one distinguishes the deformation mechanisms in polymers as brittle, ductile (with or without necking), and elastomeric. Figure 3.18 schematically shows the curves that correspond to these mechanisms. Clearly, factors such as the strain rate and temperature affect the shape of stress–strain curves, much more so in polymers

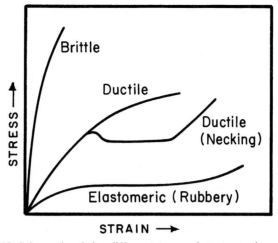

Figure 3.18 Schematic of the different types of stress–strain curves in a polymer.

than in ceramics or metals. This is because the polymers are viscoelastic; that is, their stress–strain behavior is dependent on time. Temperature and strain rate have opposite effects. Increasing the strain rate (or decreasing the temperature) will lead to higher stress levels, but lower values of strain. Figure 3.19 shows this schematically.

Polymers (especially, linear, semicrystalline polymers), in a manner superficially similar to metals, can show the phenomena of yielding and necking. The necking condition for polymers can be represented, again in a manner similar to that for metals (see Section 3.2.2), by the equation

$$\frac{d\sigma_t}{d\varepsilon_t} = \sigma_t. \tag{3.25}$$

This equation says that necking occurs when the work-hardening rate $d\sigma_t/d\varepsilon_t$ attains a value equal to σ. At that point, the increase in strength due to work-hardening cannot compensate for the loss in strength due to a decrease in cross-sectional area, and necking ensues.

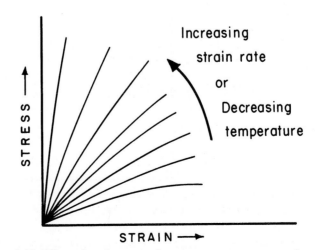

Figure 3.19 Effect of strain rate and temperature on stress–strain curves.

3.5.2 Glassy Polymers

In a manner similar to its occurrence in metals, plastic deformation occurs inhomogeneously in polymers. Two forms of inhomogeneous deformation are observed in glassy polymers: shear bands and crazes. Shear bands form at about 45° to the largest principal stress. The polymeric molecular chains get oriented within the shear bands without any accompanying change in volume. The process of shear band formation can contribute to a polymer's toughness because it is an energy-dissipating process. Shear yielding can take two forms: diffuse shear yielding and localized shear band formation. In localized shear, the shear is concentrated in thin planar regions, and the process involves a "cooperative" movement of molecular chains. The bands form at about 45° to the stress axis. Crazes are narrow zones of highly deformed polymer containing voids; the zones are oriented perpendicular to the stress axis. In the crazed zone, the molecular chains get aligned along the stress axis, but they are interspersed with voids. The void content in a craze can be as much as 55%. Unlike shear band formation, craze formation does not require the condition of constancy of volume. Generally, crazing occurs in brittle polymers. It can also occur to some extent in ductile polymers, but the dominant mode of deformation in these polymers is shear yielding. The phenomena of shear yielding and crazing are discussed further in Chapter 8.

Like ceramics, glassy or amorphous polymers show different stress–strain behaviors in tension and compression. The reason for this is that the surface flaws are much more dangerous in tension than in compression.

3.5.3 Semicrystalline Polymers

Semicrystalline polymers containing spherulites show a highly complex mode of deformation. Characteristically, these materials exhibit a ductile stress–strain curve with necking. Figure 3.20 shows such a stress–strain curve. Also illustrated is the process of transformation of a spherulitic structure to a fibrillar structure under the action of a tensile stress. Such orientation of polymeric chains parallel to the direction of stress increases the strength in that direction. Figure 3.21a shows a picture of the neck propagating in a linear polyethylene tensile sample while Figure 3.21b shows a schematic of the neck formation and propagation.

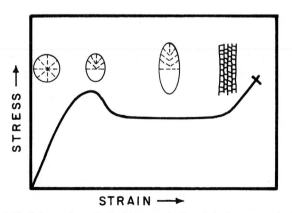

Figure 3.20 Schematic of necking and drawing in a semicrystalline polymer.

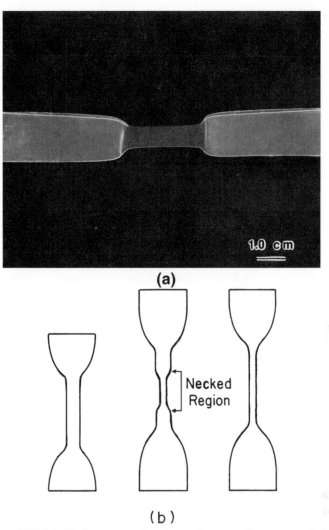

(a)

(b)

Figure 3.21 (a) Neck propagation in a sheet of linear polyethylene. (b) Neck formation and propagation in a specimen, shown in schematic fashion.

3.5.4 Viscous Flow

At high temperatures ($T \geq T_g$, the glass transition temperature), polymers undergo a viscous flow. Under these conditions, the stress is related to the strain rate, rather than the strain. Thus,

$$\tau = \eta \frac{d\gamma}{dt},\tag{3.26}$$

where τ is the shear stress, η is the viscosity, and t is the time. (The derivation of Equation 3.26 is given in Section 3.6.2.)

Viscous flow is a thermally activated process. It occurs by molecular motion, which increases as the temperature increases. The reader can appreciate the fact that such a viscous flow would involve the local breaking and re-forming of the polymeric network structure. The thermal energy for this is available above the glass transition temperature T_g. Below T_g, the thermal energy is too low for breaking and re-forming bonds, and the material does not flow so easily. At very high temperatures, the viscosity η is given by the Arrhenius-type relationship

$$\eta = A \exp\left(\frac{Q}{RT}\right), \tag{3.27}$$

where A is a constant, Q is the activation energy, R is the universal gas constant, and T is the temperature in kelvin.

3.5.5 Adiabatic Heating

There is a unique feature associated with the plastic deformation of polymers. Most of the work done during the plastic deformation of any material is converted into heat. In metals, this is not very important, because metals are good conductors, and except at extremely high rates of deformation, the heat generated is dissipated to the surroundings rather quickly, so that the temperature rise of the metal is insignificant. Polymers are generally poor conductors of heat. Thus, any heat generated in localized regions of a specimen due to plastic deformation can cause local softening. In the case of fatigue, heat may be dissipated rather easily at low strains and at low frequencies, even in polymers. A significant amount of softening, however, can result under conditions of high strain rates and high-frequency cyclic loading. This phenomenon is called *adiabatic heating*.

EXAMPLE 3.6

Polyethylene is a linear-chain thermoplastic; that is, relatively speaking, it is easy to crystallize by stretching or plastic deformation. An extreme case of this is the high degree of crystallization obtained in a gel-spun polyethylene fiber. Describe a simple technique that can be used to verify the crystallization in polyethylene.

Solution: An easy way would be to use an X-ray diffraction technique. Unstretched polyethylene will consist mostly of amorphous regions. Such a structure will give diffuse halos. A diffuse halo indicates an irregular atomic arrangement—that is, an amorphous structure. A polyethylene sample that has been subjected to stretching or a gel-spun polyethylene fiber will have highly crystalline regions aligned along the draw axis. There may also be some alignment of chains in the amorphous regions. An X-ray diffraction pattern of such a sample would show regular spots and/or regular rings. The discrete spots indicate regular spacing characteristic of an orderly arrangement in a single crystal. Well-spaced regular rings indicate a polycrystalline region. Regular rings result from overlapping spots due to random crystalline orientations.

3.6 PLASTIC DEFORMATION OF GLASSES

The unique mechanical properties exhibited by metallic glasses are connected to their structure. Table 3.1 lists the hardnesses, yield stresses, and Young's moduli for several metallic glasses. The unique compositions correspond to regions in the phase diagram that have a very low melting point. The low melting points aid in the retention of the "liquid" structure. Metallic glasses are primarily formed by rapid cooling from the molten state, so

TABLE 3.1 Mechanical Properties of Some Metallic Glasses[a]

Alloy	HV (GPa)	σ_ν (GPa)	H/σ_ν	E_g (GPa)	E_g/σ_ν
$Ni_{36}Fe_{32}CR_{14}P_{12}B_6$ (Metglas 286AA)	6.1	1.9 (tension)	3.16	99.36	52
$Ni_{49}Fe_{29}P_{14}B_6S_2$ (Metglas 286B)	5.5	1.7 (tension)	3.26	91.1	54
$Fe_{80}P_{16}C_2B_1$ (Metglas 2615)	5.8	1.7 (tension)	3.35		
$Pd_{77.5}Cu_6Si_{36.5}$	3.4	1.08 (compression)	3.17	61.9	57
$Pd_{64}Ni_{16}P_{20}$	3.1	1 (compression)	3.17	61.9	57
$Fe_{80}B_{20}$ (Metglas 2605)	7.6	2.55 (tension)	2.97	116.6	45

[a]Adapted with permission from (L. A. Davis in *Rapidly Quenched Metals,* N. J. Grant and B. C. Giessen (eds.) Cambridge, MA: MIT Press, 1976, p. 401), p. 369, Table 1.

that the atoms do not have time to form crystals. The Metglas group is commercially produced in wire and ribbon form. Young's modulus for glasses varies between 60 and 70% of the Young's modulus of the equilibrium crystalline structure. Li[7] has proposed a relationship between the shear modulus of the glassy and crystalline states, namely,

$$G_g = \frac{0.947}{1.947 - \nu} G_c \qquad (3.28)$$

where G_g and G_c are the shear moduli of the glassy and crystalline states, respectively, and ν is Poisson's ratio. The crystalline Young's modulus of glasses is recovered when the material is annealed and crystallinity sets in. The yield stresses of metallic glasses are high, as can be seen in Table 3.1. For Fe–B metallic glasses, strength levels over 3.5 GPa were achieved. This is close to the highest yield strengths achieved in polycrystalline metals. (See Section 1.4.) The yield stresses of the metallic glasses are usually 10 to 30 times higher than the yield stress of the same alloy in the crystalline state.

The micromechanical deformation mechanisms responsible for the unique mechanical properties of metallic glasses are still not very well understood. The absence of crystallinity has a profound effect on the mechanical properties. Grain boundaries, dislocations, mechanical twinning, and other very important components of the deformation of crystalline metals are not directly applicable to metallic glasses. Although the dislocations are not fully described until Chapter 4 (a brief description is given in Section 1.4), the concept is used in this section in an attempt to rationalize the mechanical response of metallic glasses. The lower Young's modulus is probably due to the less efficient packing of atoms, with a consequent larger average interatomic distance. The plastic part of the stress–strain curve also differs from the crystalline one. Here we have to distinguish between the behavior of the metallic glass above and below T_g, the glass transition temperature. As in silicate glasses, a temperature is defined above which the glass becomes viscous and deformation occurs by a viscous flow that is homogeneous. Only the deformation at temperatures below T_g will be discussed here. Curves for small cylindrical specimens under compression are shown in Figure 3.22. There is little evidence of work-hardening, and the

[7]J. C. M. Li, in *Frontiers in Materials Science—Distinguished Lectures,* L. E. Murr and C. Stein (eds.) (New York: Marcel Dekker, 1976), p. 527.

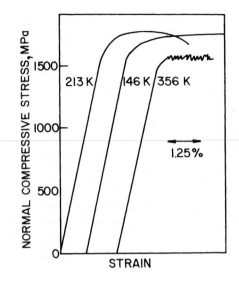

Figure 3.22 Compression stress–strain curves for $Pd_{77.5}CU_6Si_{16.5}$. (Adapted with permission from C. A. Pampillo and H. S. Chen, *Mater. Sci. Eng., 13* (1974) 181).

plastic range is close to horizontal. The surface of the specimens usually exhibits steps produced by shear bands. These shear bands have been found to be 20 nm thick, and the shear offset (step) has been found to be around 200 nm. This shows that deformation is highly inhomogeneous in metallic glasses and that, once shear starts on a certain plane, it tends to continue there. The plane of shear actually becomes softer than the surrounding regions. We can compute the amount of shear strain in a band by dividing the band offset by the thickness. In the preceding case, it is equal to 10. This behavior is termed work-softening. The curves of Figure 3.22 provide macroscopic support for the absence of work-hardening. The equivalent of a dislocation can exist in a glass. The slip vector of the dislocation would fluctuate in direction and magnitude along the dislocation line, but its mean value would be dictated by some structural parameter.

Figure 3.23 shows slip lines and steps produced after bending and after unbending. We can see the slip lines terminating inside the metallic glass. The slips decrease in height on unbending. These observations tend to confirm the relevance of some kind of shear localization in the plastic deformation of metallic glasses.

3.6.1 Microscopic Deformation Mechanism

Of the theories explaining the microscopic aspects of plastic deformation of metallic glasses, the best known are the dislocation theory of Gilman and the strain ellipsoid theory of Argon.

Figure 3.24(a) shows dislocation lines in crystalline and vitreous silica. Dislocations in crystalline solids will be studied in Chapter 4. The two-dimensional picture in the figure is analogous to the Zachariasen model for silica in Figure 1.18. The dislocation line is shown in the two cases, and we are looking at the dislocation "from the top down"; that is, the extra atomic plane is perpendicular to the surface of the paper. For the regular crystalline structure, all Burgers vectors are parallel and have the same magnitude. For the glassy structure, b fluctuates both in magnitude and direction. The dislocation line is not forced to remain in a crystallographic plane (there are no such planes in glasses), but can fluctuate. This is the *Gilman* mechanism for plastic deformation of glasses.

Experiments using "bubble rafts" and computational simulations indicate that there are localized regions of approximately ellipsoidal shape that undergo larger distortions than the bulk of the material and that are the main entities responsible for the plastic deformation of glasses. The ellipsoidal regions do not move, but undergo gradual distortion. Figure 3.24(b) shows the result of a computer simulation, including the positions and displacements of individual atoms. The lengths of the lines represent the displacements of the

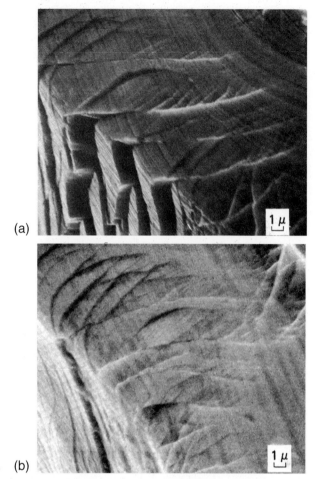

(a)

(b)

Figure 3.23 Shear steps terminating inside material after annealing at 250°C/h, produced by (a) bending and decreased by (b) unbending. Metglas $Ni_{82.4}Cr_7Fe_3Si_{4.5}B_{3.1}$ strip. (Courtesy of X. Cao and J. C. M. Li).

atoms. One can see regions of the material where the displacements of the atoms are larger. The ellipses become distorted, and the entire body deforms. This is the so-called *Argon* model for deformation of glasses, named after a renowned MIT professor (and not after a gas!).

3.6.2 Temperature Dependence and Viscosity

The mechanical response of glasses is often represented by their viscosity, which is a property of liquids. The viscosity, η is defined as the velocity gradient that will be generated in a liquid when it is subjected to a specific shear stress, or

$$\tau = \eta \frac{dv}{dy}, \tag{3.29}$$

where τ is temperature, v is the velocity and dv/dy is the velocity gradient. For temperature $T > T_m$, the viscosity is very low and the glass is a fluid. A characteristic value is $\eta \cong 10^{-3}$ Pas. For $T \sim T_g$, (the glass transition temperature), the viscosity is between 10^{10} and 10^{15} Pas. A common unit of viscosity is the Poise (P). Note that 1 P = 0.1 Pas. For $T < T_g$, the viscosity is $\eta > 10^{15}$ Pas. Mechanically speaking, the material is solid.

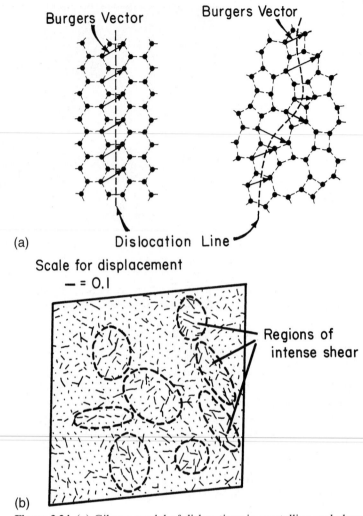

(a)

(b)

Figure 3.24 (a) Gilman model of dislocations in crystalline and glassy silica, represented by two-dimensional arrays of polyhedra. (Adapted from J. J. Gilman, *J. Appl. Phys.* 44 (1973) 675) (b) Argon model of displacement fields of atoms (indicated by magnitude and direction of lines) when assemblage of atoms is subjected to shear strain of 5×10^2, in molecular dynamics computation. (Adapted from D. Deng, A. S. Argon, and S. Yip, *Phil. Trans. Roy. Soc. Lond.* A329 (1989) 613.)

Figure 3.25 shows these different regimes of mechanical response as a function of temperature, for soda–lime–silica glass and for some metallic glasses ($Au_{77}Si_{14}Ge_{19}$, $Pd_{77.5}Cu_6Si_{16.5}$, $Pd_{80}Si_{20}$, and $Co_{75}P_{25}$). The temperature is normalized by dividing it by T_g. The viscosity decreases at $T > T_g$, as

$$\eta = \eta_0 e^{Q/RT}, \tag{3.30}$$

where Q is the activation energy for viscous flow. This is a classic Arrhenius response. The shear strength of the material can be related to the viscosity by

$$\tau = \eta \frac{dv}{dy} = \eta \frac{ds/dt}{dy} = \eta \frac{ds/dy}{dt} = \eta \frac{d\gamma}{dt} = \eta \dot{\gamma},$$

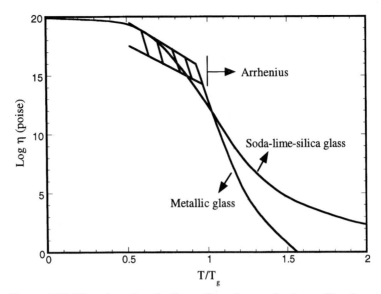

Figure 3.25 Viscosity of soda–lime–silica glass and of metallic glasses (Au–Si–Ge, Pd–Cu–Si, Pd–Si, C_0P) as a function of normalized temperature. (Adapted from J. F. Shakelford, *Introduction to Materials Science for Engineers,* 4th ed. (Englewood Cliffs, NJ: Prentice Hall, 1991), p. 331, and F. Spaepen and D. Turnbull in *Metallic Glasses,* ASM.) 1P = 0.1 Pas.

where v is the velocity of one part of the material with respect to the other. The velocity is the displacement with time, ds/dt. By changing the order of differentiation, we obtain $d\gamma = ds/dy$. The change of strain with time is $\dot{\gamma} = d\gamma/dt$. A general relationship between shear stress, shear strain, and shear strain rate is

$$\tau = \tau_0 \gamma^n \dot{\gamma}^m, \tag{3.31}$$

where n is the work-hardening coefficient and m is the strain rate sensitivity. Since glasses do not work harden, $n = 1$. When τ is proportional to $\dot{\gamma}$, the strain rate sensitivity is equal to unity, and the material will be resistant to necking in tension. This is why glass can be pulled in tension to extremely high strains. Such behavior is discussed in greater detail in Chapter 13. Another class of materials, called *superplastic* materials, also exhibits this response when the grain size of the material is very small.

EXAMPLE 3.7

Viscosity is a very important characteristic of glassy materials. On the viscosity-versus-temperature curve of a given glassy material, one can identify certain important points. The *strain point* of glass is the temperature at which internal stresses are reduced significantly in a few hours. This corresponds to $\eta = 10^{13.5}$ Pas. The *annealing point* of a glass is the temperature at which the internal stresses are reduced in a few minutes such that $\eta = 10^{12}$ Pas. The *softening point* of a glass corresponds to $\eta = 10^{6.65}$ Pas. At this viscosity, the glass deforms rapidly under its own weight. The *working point* of glass corresponds to $\eta = 10^3$ Pas. At this viscosity, the glass is soft enough to be worked.

Consider a glass with a strain point of 500°C and a softening point of 800°C. Using the preceding viscosity values for the strain point and softening point, estimate the activation energy for the deformation of this glass.

Solution: We can write the viscosity as a function of temperature as

$$\eta = A \exp [Q/RT].$$

At the softening point,

$$10^{6.65} = A \exp \left[\frac{Q}{8.314 \times 1073} \right],$$

while at the strain point,

$$10^{13.5} = A \exp \left[\frac{Q}{8.314 \times 773} \right].$$

From these two expressions, we obtain, by division

$$10^{6.85} = \exp \left[\left(\frac{Q}{8.314} \right) \left(\frac{1}{773} - \frac{1}{1073} \right) \right],$$

or

$$Q = 362 \text{ kJ/mol}.$$

3.7   FLOW, YIELD, AND FAILURE CRITERIA

The terms *flow criterion, yield criterion,* and *failure criterion* have different meanings. *Failure criterion* has its historical origin in applications where the onset of plastic deformation indicated failure. However, in deformation-processing operations this is obviously not the case, and plastic flow is desired. *Yield criterion* applies only to materials that are in the annealed condition. It is known that, when a material is previously deformed by, for instance, rolling, its yield stress increases due to work-hardening. (See Chapter 6.) The term *flow stress* is usually reserved for the onset of plastic flow in a previously deformed material. *Failure criterion* is applied to brittle materials, in which the limit of elastic deformation coincides with failure. To be completely general, a flow criterion has to be valid for any stress state. In a uniaxial stress state, plastic flow starts when the stress–strain curve deviates from its initial linear range. Uniaxial stress–strain curves are very easily obtained experimentally, and the deformation response of a material is usually known for this situation. The main function of flow criteria is to predict the onset of plastic deformation in a complex state of stress when one knows the flow stress (under uniaxial tension) of the material. Note that the value of the flow stress is strongly dependent on the state of stress, and if this effect is not considered, it can lead to potentially dangerous errors in design. We next present some of these criteria.

3.7.1   Maximum-Stress Criterion (Rankine)

According to the maximum-stress criterion, plastic flow takes place when the greatest principal stress in a complex state of stress reaches the flow stress in uniaxial tension. Since $\sigma_1 > \sigma_2 > \sigma_3$, we have

$$\sigma_0(\text{tension}) < \sigma_1 < \sigma_0(\text{compression}),$$

where σ_0 is the flow stress of the material. Later (Section 3.7.5) we will see the situation where the compressive strength is greater than the tensile strength. The great weakness of this criterion is that it predicts plastic flow of a material under a hydrostatic state of stress; however, this is impossible, as shown by the following example. It is well known that tiny shrimp can live at very great depths. The hydrostatic pressure due to water is equivalent to 1 atm (10^5 N/m²) for every 10 m; at 1,000 m below the surface, the shrimp would be subjected to a hydrostatic stress of 10^7 N/m². Hence

$$-p = \sigma_1 = \sigma_2 = \sigma_3 = -10^7 \, \text{N/m}^2.$$

A quick experiment to determine the yield stress of the shrimp could be conducted by carefully holding it between two fingers and pressing it. By doing the test with a live shrimp, one can define the flow stress as the stress at which the amplitude of the tail wiggling will become less than a critical value. This will certainly occur at a stress of about 0.1 MPa. Hence,

$$\sigma_0 = 0.1 \, \text{MPa}.$$

Rankine's criterion would produce shrimp failure at

$$p \equiv -\sigma_0 = -0.1 \, \text{MPa}.$$

This corresponds to a depth of only 10 m. Fortunately for all lovers of crustaceans, this is not the case, and hydrostatic stresses do not contribute to plastic flow.

3.7.2 Maximum-Shear-Stress Criterion[8] (Tresca)

Plastic flow starts when the maximum shear stress in a complex state of deformation reaches a value equal to the maximum shear stress at the onset of flow in uniaxial tension (or compression). The maximum shear stress is given by (see Section 2.6)

$$\tau_{\text{max}} = \frac{\sigma_1 - \sigma_3}{2}. \tag{3.32}$$

For the uniaxial stress state, we have, at the onset of plastic flow,

$$\sigma_1 = \sigma_0, \quad \sigma_2 = \sigma_3 = 0;$$

so

$$\tau_{\text{max}} = \frac{\sigma_0}{2}.$$

Therefore,

$$\sigma_0 = \sigma_1 - \sigma_3. \tag{3.33}$$

This criterion corresponds to taking the differences between σ_1 and σ_3 and making it equal to the flow stress in uniaxial tension (or compression). It can be seen that it does

[8]See H. Tresca, *Compt. Rend. Acad. Sci. Paris,* 59 (1864) 754; 64 (1867) 809.

not predict failure under hydrostatic stress, because we would have $\sigma_1 = \sigma_3 = p$ and no resulting shear stress.

3.7.3 Maximum-Distortion-Energy Criterion (von Mises)[9]

This criterion was originally proposed by Huber as "When the expression

$$\frac{\sqrt{2}}{2}[(\sigma_1 - \sigma_2)^2 + (\sigma_2 - \sigma_3)^2 + (\sigma_1 - \sigma_3)^2]^{1/2} > \sigma_0 \qquad (3.34)$$

then the material will plastically flow." The above expression is known as *effective stress.* The criterion was stated by von Mises without a physical interpretation. It is now accepted that it expresses the critical value of the distortion (or shear) component of the deformation energy of a body. Based on this interpretation, a body flows plastically in a complex state of stress when the distortional (or shear) deformation energy is equal to the distortional (or shear) deformation energy in uniaxial stress (tension or compression). This will be shown shortly. This criterion is also called J_2, which is the second invariant of the stress deviator. Students will learn about this in advanced "Mechanic of Materials" courses.

3.7.4 Graphical Representation and Experimental Verification of Rankine's, Tresca's, and von Mises' Criteria

There is a convenient way to represent Rankine's, Tresca's, and von Mises' criteria for a plane state of stress. For this, one makes $\sigma_3 = 0$ and has σ_1 and σ_2. It will be necessary to momentarily forget the convention that $\sigma_1 > \sigma_2 > \sigma_3$, because it would not be obeyed for $\sigma_2 < 0$; we have $\sigma_2 < \sigma_3 = 0$. Figure 3.26 shows a plot of σ_1 versus σ_2. According to Tresca's criterion, plastic flow starts when

$$\tau_{max} = \frac{\sigma_0}{2}.$$

The four quadrants have to be analyzed separately. In the first quadrant, there are two possible situations. For σ_1 greater than σ_2, $\tau_{max} = (\sigma_1 - \sigma_3)/2$ and $\sigma_1 = \sigma_0$. This is a line passing through $\sigma_1 = \sigma_0$ and parallel to $O\sigma_2$. For σ_2 greater than σ_1, we have the converse situation and a line passing through $\sigma_2 = \sigma_0$ and parallel to σ_1.

In the second quadrant, $\sigma_2 > 0$ and $\sigma_1 < 0$. We have

$$\tau_{max} = \frac{\sigma_1 - \sigma_2}{2} \quad \text{and} \quad \sigma_1 - \sigma_2 = \sigma_0.$$

This equation represents a straight line intersecting the $O\sigma_1$ axis at σ_0 and the $O\sigma_2$ axis at $-\sigma_0$.

[9]See R. von Mises, *Göttinger Naehr. Math. Phys. Klasse,* 1913, p. 582.

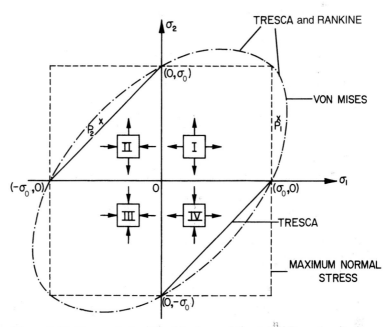

Figure 3.26 Comparison of Rankine's, von Mises', and Tresca's criteria.

The flow criteria for quadrants III and IV are found in a similar way. For von Mises' criterion, we have, from Equation 3.34 and $\sigma_3 = 0$,

$$\sigma_0 = \frac{\sqrt{2}}{2} [(\sigma_1 - \sigma_2)^2 + \sigma_2^2 + \sigma_1^2]^{1/2},$$

$$\sigma_1^2 - \sigma_1\sigma_2 + \sigma_2^2 = \sigma_0^2.$$

This is the equation of an ellipse whose major and minor axes are rotated 45° from the orthogonal axes $O\sigma_1$ and $O\sigma_2$, respectively. It can be easily shown by applying a rotation of axes to the equation of an ellipse referred to its axes:

$$\left(\frac{\sigma_1}{a}\right)^2 + \left(\frac{\sigma_2}{b}\right)^2 = k^2. \tag{3.35}$$

From Equation 3.35, it can be seen that Tresca's criterion is more conservative than von Mises'. Tresca's criterion would predict plastic flow for the stress state defined by point P_1, whereas von Mises' would not. However, both criteria are fairly close. It can be seen from Figure 3.26 that plastic flow may require a stress σ_1 greater than σ_0 for a combined state of stress. (See point P_2). However, there are regions (when one stress is tensile and another is compressive) where plastic flow starts when both stresses are within the interval

$$\sigma_0 < \sigma_1, \qquad \sigma_2 < \sigma_0.$$

This occurs in the second and fourth quadrants. Point P_2 shows the situation very clearly. The conclusion is that the correct application of a yield criterion is very important for design purposes. For comparison purposes, the maximum-normal stress (Rankine) criterion is also drawn in Figure 3.26. It is just a square with sides parallel to the $O\sigma_1$ and $O\sigma_2$ axes and intersecting them at $(\sigma_0, 0)$, $(-\sigma_0, 0)$ $(0, \sigma_0)$, and $(0, -\sigma_0)$. We see that there is a considerable difference between Rankine's criterion, on the one hand, and Tresca's and von Mises' criteria, on the other hand, for quadrants II and IV. This difference is readily explained by the fact that Rankine's criterion applies to brittle solids (including cast irons and steel below the ductile–brittle transition temperature), in which failure (or fracture) is produced by tensile stresses.

Figure 3.27 shows the three criteria, together with experimental results for copper, aluminum, steel, and cast iron. While copper and aluminum tend to follow von Mises' criterion (and, in a more conservative way, Tresca's), cast iron clearly obeys Rankine's criterion. This is plainly in line with the low ductility exhibited by cast iron. The reader is warned that the ratio $\sigma/\sigma_{\text{ult}}$, and not σ/σ_0, is used in the figure. Nevertheless, it serves to illustrate the difference in response.

The determination of the flow locus is usually conducted in biaxial testing machines, which operate in a combined tension–torsion or tension–hydrostatic-pressure mode. These two modes use tubular specimens, and one has to use the appropriate calculations to find the principal stresses. As the material is plastically deformed, we have an expansion of the flow locus. For von Mises' criterion, we can envision concentric ellipses having increasing major and minor axes. This is illustrated in Figure 3.28a. When the ellipse expands in a symmetric fashion, the hardening is the same in all directions and is called *isotropic*. Often, however, hardening in one direction (the loading direction) causes a change in flow stress in

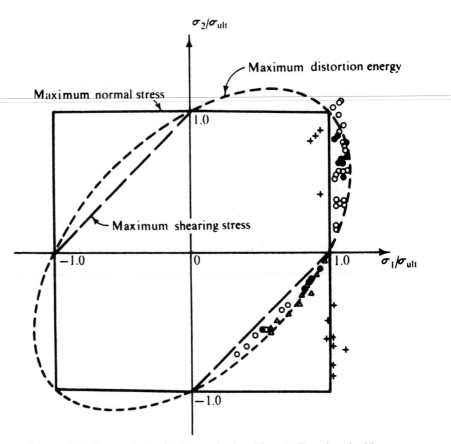

Figure 3.27 Comparison of failure criteria with test. (Reprinted with permission from E. P. Popov, *Mech. of Mat'ls,* 2d ed., (Englewood Cliffs, NJ: Prentice-Hall, 1976), and G. Murphy, *Adv. Mech. of Mat'ls,* (New York: McGraw-Hill, 1964), p. 83.

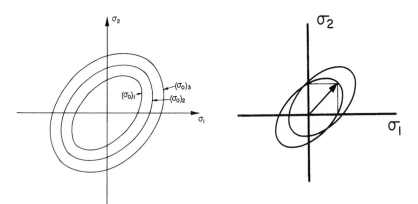

Figure 3.28 Displacement of the yield locus as the flow stress of the material due to plastic deformation. (a) Isotropic hardening. (b) Kinematic hardening.

other directions that is different. This is very important in plastic-forming operations (stamping, deep drawing). The extreme case where the ellipse is just translated is shown in Figure 3.28(b). This case is called *kinematic hardening*. (See Section 3.4.)

EXAMPLE 3.8

A region on the surface of a 6061-T4 aluminum alloy component has strain gages attached, which indicate the following stresses:

$$\sigma_{11} = 70 \text{ MPa},$$

$$\sigma_{22} = 120 \text{ MPa},$$

$$\sigma_{12} = 60 \text{ MPa}.$$

Determine the yielding for both Tresca's and von Mises' criteria, given that $\sigma_0 = 150$ MPa (the yield stress).

Solution: We first have to establish the principal stresses. This is easily accomplished by a Mohr circle construction or by its analytical expression (the equation of a circle):

$$\sigma_{1,2} = \frac{\sigma_{11} + \sigma_{22}}{2} \pm \left[\left(\frac{\sigma_{11} - \sigma_{22}}{2}\right)^2 + \sigma_{12}^2\right]^{1/2},$$

$$\sigma_1 = 160 \text{ MPa}; \qquad \sigma_2 = 30 \text{ MPa}; \qquad \sigma_3 = 0.$$

According to Tresca, $\tau_{\text{max}} = 160 - 0/2 = 80$ MPa.

The value $\tau_{\text{max}} = 80$ MPa exceeds Tresca's criterion ($\sigma_0/2 = 75$ MPa) and would be unsafe, according to it. The von Mises criterion gives

$$J_2 = \frac{1}{6}[(\sigma_1 - \sigma_2)^2 + (\sigma_1 - \sigma_3)^2 + (\sigma_2 - \sigma_3)^2]$$

$$= \frac{1}{6}[130^2 + 160^2 + 30^2]$$

$$= 7233 \text{ MPa}^2.$$

The maximum value of $J_2^M = (1/3)\sigma_0^2 = (1/3)150^2 = 7{,}500$ MPa^2.

So $J_2 < J_2^M$, and the material does not yield. Plainly, Tresca's criterion is more conservative than von Mises'.

3.7.5 Failure Criteria for Brittle Materials

As shown in Figure 3.29, the tensile strength of Al_2O_3 is approximately one-tenth of its compressive strength. Such is also the case for many brittle materials, such as concrete, rock, etc. Therefore, the Rankine, Tresca, and von Mises criteria have to be modified to incorporate this behavior. This will be done in the rest of the section, with the presentation of the Mohr–Coulomb, Griffith, and McClintock–Walsh criteria. There are also other criteria (e.g., Babel–Sines), which will *not* be presented here.

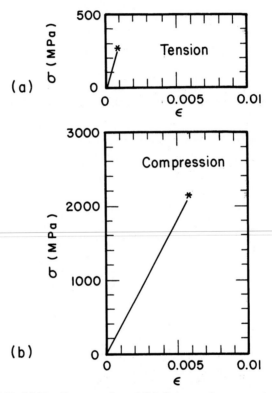

Figure 3.29 (a) Tensile strength and (b) Compressive strength of Al_2O_3.

Mohr–Coulomb Failure Criterion. This is simply the equivalent of the Tresca criterion with different tensile and compressive strengths. Figure 3.30 shows the Mohr–Coulomb criterion in a schematic fashion. The criterion for failure is a maximum shear stress; the compressive strength σ_c is much higher than the tensile strength σ_t.

Griffith Failure Criterion.[10] This criterion simply states that failure will occur when the tensile stress tangential to an ellipsoidal cavity and at the cavity surface reaches a critical level σ_0. The criterion is a classic spin-off of Griffith's work of 1919. Griffith recognized that brittle materials contained flaws and that failure would occur at a specific level of stress at the flaw surfaces. He considered an elliptical crack oriented in a general

[10]See A. A. Griffith, *Proc. 1st Int'l. Congress in Appl. Mech.*, 1925, p. 55.

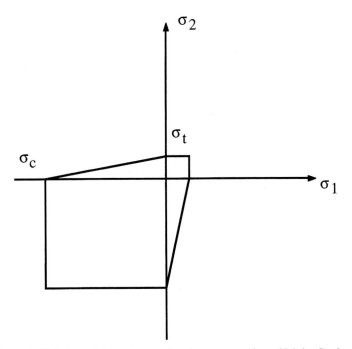

Figure 3.30 Schematic two-dimensional representation of Mohr-Coulomb failure criterion.

direction with respect to the compression axis and calculated the stresses generated at the surface of the crack. Tensile stresses are generated by compressive loading; this might appear surprising at first sight, but will become clear in Chapter 7. If σ_0 is the tensile strength of the material, the following relationship is obtained:

$$(\sigma_1 - \sigma_2)^2 + 8\sigma_0(\sigma_1 + \sigma_2) = 0 \qquad \text{if } \sigma_1 + 2\sigma_2 > 0,$$
$$\sigma_2 = \sigma_0 \qquad \text{if } \sigma_1 + 2\sigma_2 < 0. \qquad (3.36)$$

The criterion proposed by Griffith is shown in Figure 3.31. The compressive failure stress is eight times the tensile failure stress, as is evident from Equation 3.36. This very important result is consistent with the experimental results observed for brittle materials.

McClintock–Walsh Criterion. McClintock and Walsh[11] extended Griffith's criterion by considering a frictional component acting on the flaw faces that had to be overcome in order for them to grow. This term is a function of the applied stress. The frictional stress f was considered equal to the product of the frictional coefficient μ and the normal stress σ_0 acting on the flaw surface. McClintock and Walsh assumed that there was a stress σ_c at infinity necessary to close the flaw so that the opposite surfaces would touch each other. This approach led to the following expression:

$$\sigma_1[(\mu^2 + 1)^{1/2} - \mu] - \sigma_2[(\mu^2 + 1)^{1/2} + \mu] = 4\sigma_0\left(1 + \frac{\sigma_c}{\sigma_0}\right)^{1/2} - 2\mu\sigma_c. \qquad (3.37)$$

[11]See F. A. McClintock and J. B. Walsh, *Proc. 4th U.S. Nat'l. Cong. of Appl. Mech.* (1962), p. 1015.

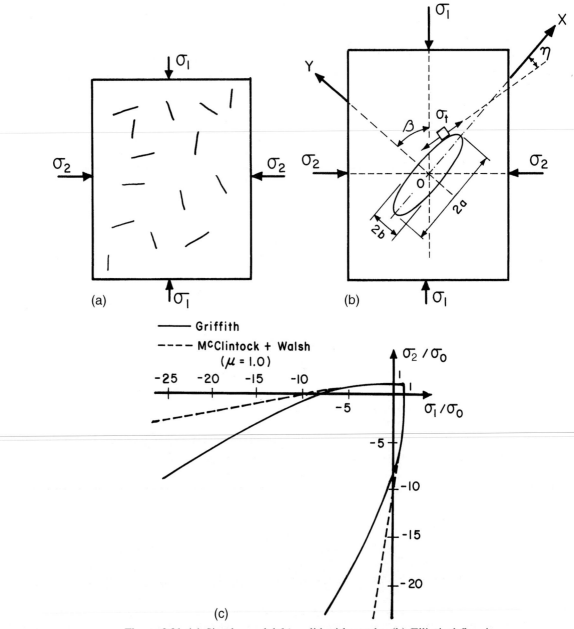

Figure 3.31 (a) Simple model for solid with cracks. (b) Elliptical flaw in elastic solid subjected to compression loading. (c) Biaxial fracture criterion for brittle materials initiated from flaws without (Griffith) and with (Mc-Clintock and Walsh) crack friction.

Assuming that $\sigma_c = 0$, we get the following simple version of this criterion:

$$\sigma_1[(\mu^2 + 1)^{1/2} - \mu] - \sigma_2[(\mu^2 + 1)^{1/2} + \mu] = 4\sigma_0. \tag{3.38}$$

McClintock and Walsh's criterion is shown in Figure 3.31 for $\mu = 1$. Griffith's criterion is more conservative, and the compressive strength is 10 times the tensile strength for Mc-Clintock and Walsh. The frictional forces retard failure in compression.

EXAMPLE 3.9

Determine the fracture stress for SiC in compression in a complex loading situation in which $\sigma_1/\sigma_2 = 2$ if σ_0 in tension is 400 MN/m². Perform all calculations assuming (a) no friction between crack surfaces and (b) a friction coefficient of 0.5.

Solution: Applying Equation 3.36 (with no friction), we have

$$\left(\sigma_1 - \frac{\sigma_1}{2}\right)^2 + 8 \times 400\left(\sigma_1 + \frac{\sigma_1}{2}\right) = 0,$$

$$\frac{\sigma_1}{4} + 4{,}800 = 0,$$

$$\sigma_1 = -19{,}200 \text{ MPa},$$

$$\sigma_1 = -19.2 \text{ GPa}.$$

Applying Equation 3.38 (with friction), we obtain

$$\sigma_1[(0.5^2 + 1)^{1/2} - 0.5] - 0.5\sigma_1[(0.5^2 + 1)^{1/2} + 0.5] = 4 \times 400,$$

$$\sigma_1(0.618) - \sigma_1(0.809) = 1{,}600,$$

$$\sigma_1 = \frac{-1{,}600}{0.272},$$

$$\sigma_1 = -5.88 \text{ GPa}.$$

The very high compressive strengths are due to the confinement. If the ceramic were not confined (i.e., if $\sigma_2 = 0$), the compressive strengths would be −3.2 GPa (Griffith) and −2.5 GPa (McClintock–Walsh).

3.7.6 Yield Criteria for Ductile Polymers

Brittle polymers such as epoxies fail at the end of their linear elastic stage without any significant plastic deformation. Ductile polymers such as thermoplastics undergo plastic deformation. Does this mean that we can use Tresca's or von Mises' criterion to describe their yielding? The answer is no, because, unlike the yield strength of metals, that of polymers depends on the hydrostatic component of stress. Tresca's and von Mises' criteria, on the other hand, do not show any such dependence. This dependence on hydrostatic stress in polymers stems from the more liquidlike structure of polymers. Specifically, the polymers have some free volume, which makes them highly compressible.

Let us consider von Mises' criterion for isotropic metals. According to this criterion, yielding occurs when the condition

$$(\sigma_1 - \sigma_2)^2 + (\sigma_2 - \sigma_3)^2 + (\sigma_3 - \sigma_1)^2 \geq 6\,k^2 = \text{constant}$$

is satisfied, where σ_1, σ_2, and σ_3 are the principal stresses and k is constant equal to the yield stress in torsion τ_0. For metals, we take k or τ_0 to be a constant at room temperature, equal to $\sigma_0/\sqrt{3}$ for uniaxial stress, with σ_0 the uniaxial yield stress. This equation also implicitly assumes that the tensile and compressive yield strengths are numerically the same, equal to $\sqrt{3}\,k$ or $\sqrt{3}\,\tau_0$. It turns out that for polymers, yield stress in compression is greater than that in tension by 10 to 20%.[12] This stems from the fact that, again unlike yielding in metals, yielding in polymers shows a strong dependence on any superimposed hydrostatic pressure. That is,

$$k = k(\dot{\varepsilon}, T, \sigma_p),$$

where $\dot{\varepsilon}$ is the strain rate, T is the temperature, and σ_p is the hydrostatic pressure. As we mentioned, in molecular terms, this dependence of yield stress on hydrostatic pressure can be traced to the fact that polymers have some *free volume* associated with them, which is diminished by hydrostatic compression. We can modify the yield criterion to take into account this dependence on the hydrostatic compressive stress σ_p by using the expression

$$k = k_0 + A\sigma_p,$$

where k_0 is a constant and A is another constant that represents the dependence of yield stress on hydrostatic pressure. As σ_p increases, the free volume decreases, and molecular motion becomes more difficult. The presence of a hydrostatic component translates the von Mises ellipse from quadrant I to quadrant III, as shown in Figure 3.32.

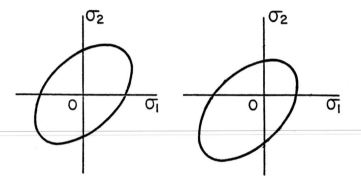

Figure 3.32 Translation of von Mises ellipse for polymer due to hydrostatic stresses.

Several glassy polymers, such as polystyrene, polycarbonate, and PMMA, show the phenomenon of crazing. (See Section 8.4.2.) Crazing involves the formation of microvoids and stretched chains or fibrils under tension. The fibril formation depends on shear flow and free volume. A yield criterion that takes crazing into account is

$$\sigma_1 - \sigma_2 = A + \frac{B}{(\sigma_1 + \sigma_2)},$$

where $(\sigma_1 - \sigma_2)$ represents the shear, $(\sigma_1 + \sigma_2)$ represents the hydrostatic component, and A and B are adjustable constants that depend on temperature. Note that as the hydrostatic component $(\sigma_1 + \sigma_2)$ increases, the shear stress $(\sigma_1 - \sigma_2)$ required for yielding decreases. The yield envelopes for a polymer or metal that does not show yield stress dependence on the hydrostatic component is shown in Figure 3.32a, while that for a polymer showing crazing takes the shape shown in Figure 3.32b. Note that crazing occurs only in tension, not in compression. The yield envelope in Figure 3.32b has been translated with respect to that in Figure 3.32a.

[12]K. Matsushige, S. V. Radcliffe, and E. Baer, *J. Polymer Sci., Polymer Phys.*, 14 (1965) 703.

A better and more complete scenario for yielding in polymers is as follows. Under multiaxial stress, glassy polymers can undergo yielding by shear or crazing. Figure 3.33 shows schematically the yield envelope under a biaxial stress condition. The constants A and B can be chosen to fit the curve to experimental data. The pure-shear line, $\sigma_1 = -\sigma_2$, is the boundary between hydrostatic compression and hydrostatic tension. Below the pure-shear line, crazing (a void-forming process) does not occur because hydrostatic pressure reduces the volume. Above this line, crazing is the main mechanism of failure. The curves for crazing are asymptotic to the pure shear line. The yield envelope shown in the figure also shows the pressure-dependent shear yielding; that is, the envelope has been translated with respect to the conventional von Mises criterion. Note that in the first quadrant the crazing envelope is completely inside the shear yield envelope. This means that for all combinations of biaxial tensile stresses, crazing will precede shear yielding. In the second and fourth quadrants, the two envelopes intersect. The heavy line indicates the overall yielding or failure envelope.

A word of caution is in order here. Crazing in *air* does not occur in pure shear or under conditions of compressive hydrostatic stress. The modified criterion just described requires a dilative component of the applied stress for crazing in air. In the presence of an appropriate environmental agent, crazing can be observed under conditions of simple tension and hydrostatic pressure.

3.7.7 Yield and Failure Criteria for Anisotropic Materials

The criteria discussed above become completely invalid for anisotropic materials. The source of anisotropy in a material can be any of the following:

(i)   A single crystal can have different properties in different directions due to its inherent crystal symmetry.

(ii)   A cold-rolled sheet, tube, or wire of a metal or alloy can show a very high degree of preferred orientation of grains. Polymers are also frequently processed by drawing, extrusion, or injection molding techniques. Such techniques impart a high degree of anisotropy to the polymer. Figure 3.34 shows the change in shape of the yield surface as a function of anisotropy, where $R = \sigma_2/\sigma_1$. For $R = 1$, we have isotropy, and a classical von Mises curve is obtained.

(iii)   In an aligned fiber-reinforced composite, the properties vary markedly as the function of the angle θ of the fiber direction.

     While the most anisotropic crystal would render the plasticity treatment prohibitively complex, there is one type of anisotropy that can be studied

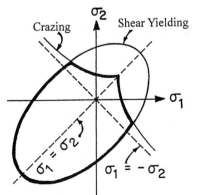

Figure 3.33 Envelopes defining shear yielding and crazing for an amorphous polymer under biaxial stress. (After S. S. Sternstein and L. Ongchin, *Am. Chem. Soc., Div. of Polymer Chem., Polymer Preprints*, 10 (1969), 1117.)

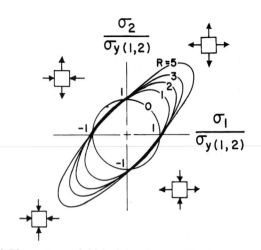

Figure 3.34 Plane-stress yield loci for sheets with planar isotropy or textures that are rotationally symmetric about the thickness direction, x_3. (Values of R indicate the degree of anisotropy = σ_2/σ_1.)

without excessive complications. The type of response displayed by wood is a good illustration of this anisotropy. Wood has different yield stresses along the three directions defined by the wood fibers and by the normals to the fibers. Similarly, a rolled sheet or slab of metal will exhibit orthotropic plastic properties; the rolling direction, transverse direction, and thickness direction define the three axes.

3.8 HARDNESS

The simplest way of determining the resistance of a metal to plastic deformation is through the hardness test. Indentation tests constitute the vast majority of metallurgical tests. They are essentially divided into two classes, commonly called *microindentation* and *macroindentation* tests, but improperly referred to as microhardness and macrohardness tests. The division between the two occurs for a load of approximately 200 gf (~2 N). The indentation tests in metals measure the resistance to plastic deformation; both the yield stress and the work-hardening characteristics of the metal are important in determining the hardness. In spite of the theoretical studies done on hardness, hardness cannot be considered a fundamental property of a metal. Rather, it represents a quantity measured on an arbitrary scale.[13] Hardness measurements should not be taken to mean more than what they are: an empirical, comparative test of the resistance of the metal to plastic deformation. Any correlation with a more fundamental parameter, such as the yield stress, is valid only in the range experimentally determined. Similarly, comparisons between different hardness scales are meaningful only through experimental verification. For steels, Table 3.2 gives a fair conversion of hardnesses and the UTS equivalents.

The most important macro- and microindentation tests are described in Sections 3.8.1 and 3.8.2.

---

[13]M. C. Shaw, in *The Science of Hardness Testing and its Research Applications*, J. H. Westbrook and H. Conrad, eds., (Metals Park, OH: ASM, 1973), p. 1.

TABLE 3.2 Approximate Hardness Conversions for Steels[†]

Vickers HV	HB (Brinell) (10-mm ball, 3,000 kgf)		Rockwell						Shore Scleroscope	Approximate Tensile Strength (MPa)
	Stand.	WC	A (Brale 60 kgf)	B (1/16-in. ball, Brale 100 kgf)	C (Brale 150 kgf)	D (Brale 100 kgf)	Rockwell Superficial 15N	45N		
940			85.6		68.0	76.9	93.2	75.4	97	
900			85.0		67.0	76.1	92.9	74.2	95	
860		757	84.4		65.9	75.3	92.5	73.1	92	
820		733	83.8		64.7	74.3	92.1	71.8	90	
780		710	83.0		63.3	73.3	91.5	70.2	87	
740		684	82.2		61.8	72.1	91.0	68.6	84	
700		656	81.3		60.1	70.8	90.3	66.7	81	
660		620	80.3		58.3	69.4	89.5	64.7	79	2199
620		582	79.2		56.3	67.9	88.5	62.4	75	2061
580		545	78.0		54.1	66.2	87.5	59.9	72	1923
540	496	507	76.7		51.7	64.4	86.3	57.0	69	1792
500	465	471	75.3		49.1	62.2	85.0	53.9	66	1655
460	433	433	73.6		46.1	60.1	83.6	50.4	62	1517
420	397	397	71.8		42.7	57.5	81.8	46.4	57	1379
380	360	360	69.8		38.8	54.4	79.8	41.7	52	1241
340	322	322	67.6		34.4	51.1	77.4	36.5	47	1110
300	284	284	65.2		29.8	47.5	74.9	31.1	42	972
260	247	247	62.4		24.0	43.1	71.6	24.3	37	834
220	209	209		95.0					32	696
200	190	190		91.5					29	634
180	171	171		87.1					26	579
160	152	152		81.7					24	517
140	133	133		75.0					21	455
120	114	114		66.7						
100	95	95		62.3						
85	81	81		41						

[†]Tables for other metals and alloys can be found in ASTM 140 (*Standard Hardness Conversion Tables for Metals*).

[a]Adapted with permission from E. R. Petty, in *Techniques of Metals Research*, Vol. 5, Pt. 2, R. F. Bunshah (ed.), (New York: Wiley-Interscience, 1971), p. 180.

3.8.1 Macroindentation Tests

The impressions caused by macroindentation tests are shown in Figure 3.35. The Brinell test produces by far the largest indentation. The Vickers test may produce very small indentations, depending on the load used.

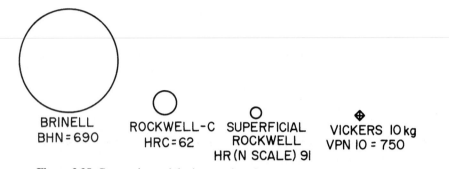

Figure 3.35 Comparison of the impression sizes produced by various hardness tests on material of 750 HV. (Adapted with permission from E. R. Petty, in *Techniques of Metals Research,* Vol. 5, Pt. 2, R. F. Bunshah (ed.) (New York: Wiley-Interscience, 1971), p. 174.

Brinell Hardness Test.　　In this test, a steel sphere is pressed against a metal surface for a specified period of time (10 to 15 s, according to the ASTM), and the surface of the indentation is measured. The load (in kgf) divided by the area (in mm²) of the curved surface gives the hardness HB, or

$$HB = \frac{P}{\pi D \times \text{depth}} \tag{3.39}$$

$$= \frac{2P}{\pi D(D - \sqrt{D^2 - d^2})}, \tag{3.40}$$

where D and d are the diameters of the sphere and impression, respectively. The parameters are indicated in Figure 3.36. Since $d = D \sin \phi$, we have

$$HB = \frac{2P}{\pi D^2(1 - \cos \phi)}. \tag{3.41}$$

Different spheres produce different impressions, and if we want to maintain the same HB, independent of the size of the sphere, the load has to be varied according to the relationship

$$\frac{P}{D^2} = \text{constant}. \tag{3.42}$$

This assures the same geometrical configuration (the same ϕ). The diameter of the impressions between $0.25D$ and $0.5D$ gives good, reproducible results. The target sought is $d = 0.375D$. If the same d/D ratio is maintained (constant ϕ), the Brinell test is reliable. Spheres with diameters of 1, 2, 5, and 10 mm have been used, and some of the ratios P/D^2 that provide good d/D ratios for different metals are: steels and cast irons (30), Cu and Al

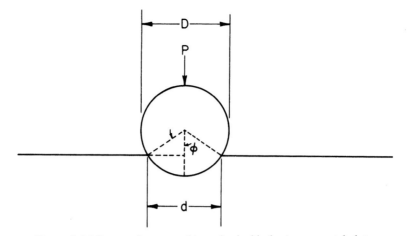

Figure 3.36 Impression caused by spherical indenter on metal plate.

(5), Cu and Al alloys (10), and Pb and Sn alloys (1). The softer the material, the lower is the P/D^2 ratio required to produce $d/D = 0.375$.

One of the problems of the Brinell test is that HB is dependent on the load P for the same sphere. In general, HB decreases as the load is increased. ASTM standard E10-78 provides details and specifications for Brinell hardness tests. It states that the standard Brinell test is conducted under the following conditions:

> Ball diameter: 10 mm
>
> Load: 3000 kgf
>
> Duration of loading: 10 to 15 s

In this case, 360 HB indicates a Brinell hardness of 360 under the foregoing testing conditions. For different conditions, the parameters have to be specified. For example, 63 HB 10/500/30 indicates a Brinell hardness of 63, measured with a ball of 10 mm diameter and a load of 500 kgf applied for 30 s. Brinell tables and additional instructions are provided in ASTM E10–78. Meyer[14] was aware of this problem and proposed a modification of the Brinell formula. He found out that the load divided by the *projected* area of the indentation ($\pi d^2/4$) was constant. Hence, he proposed, in place of Equation 3.39, the equation

$$\text{Meyer} = \frac{4P}{\pi d^2},\tag{3.43}$$

where P is expressed in kilograms force and d in millimeters. The Meyer hardness never gained wide acceptance, in spite of being more reliable than the Brinell hardness. For work-hardened metals, it seems to be independent of P.

Rockwell Hardness Test. The most popular hardness test is also the most convenient, since there is no need to measure the depth or width of the indentation optically. This testing procedure is illustrated in Figure 3.37. A preload is applied prior to the application of the main load. The dial of the machine provides a number that is related to the depth of the indentation produced by the main load. Several Rockwell scales are used, and the numbers refer to arbitrary scales and are not directly related to any fundamental parameter of the material. Two different types of indenters are used. The A, C, D, and N

[14]E. Meyer, *Z. Ver. Dtsch. Ing.,* 52 (1980) 645, 740, 835.

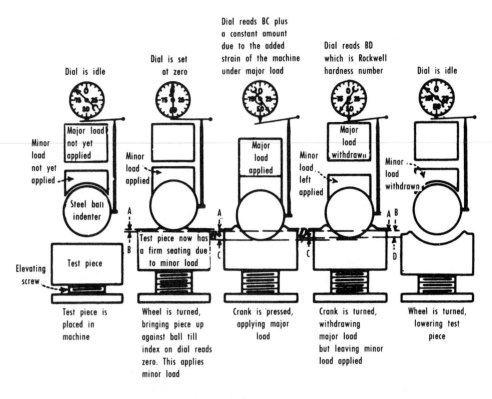

AB = Depth of hole made by minor load.

AC = Depth of hole made by major load and minor load combined.

DC = Recovery of metal upon withdrawal of major load. This is caused by
elastic recovery of the metal under test, and does not enter the hardness reading.

BD = Difference in depth of holes   Rockwell hardness number.

Figure 3.37 Procedure in using Rockwell hardness tester. (Reprinted with permission from H. E. Davis, G. E. Troxel, and C. T. Wiscocil, *The Testing and Inspection of Engineering Materials,* (New York: McGraw-Hill, 1941), p. 149.)

scales use the Brale indenter, which is a diamond cone with a cone angle of 120°. The other scales use either 1/8-in. (3.175-mm) or 1/16-in. (1.587-mm)-diameter steel spheres. The loads also vary, depending on the scale. Table 3.3 shows the various loads and typical applications. Usually, the C scale is used for harder steels and the B scale for softer steels; the A scale covers a wider range of hardness. Because of the nature of the measurement, any sagging of the test piece will produce changes in hardness. Therefore, it is of utmost importance to have the sample well supported; specimens embedded in Bakelite cannot be tested. The Brinell and Vickers tests, on the other hand, which are based on optical measurements, are not affected by the support.

For very thin samples, there is a special superficial Rockwell test. The testing procedure is described in detail in the ASTM Standard E18–74, and conversion tables for a number of alloys are given in ASTM Standard E140–78. The symbol used to designate this hardness is, according to the ASTM, HR; 64HRC corresponds to Rockwell hardness number 64 on the C scale.

---

[15]See G. E. Dieter, *Mechanical Metallurgy,* 2d ed. (New York: McGraw-Hill, 1976), p. 398.

TABLE 3.3 Details of the More Important Scales Available for the Rockwell Hardness Tester

Scale Designation	Type of Indenter	Major Load (kgf)	Typical Field of Application
A	Brale	60	The only continuous scale from annealed brass to cemented carbide, but is usually used for harder materials
B	1/16-in.-diameter steel ball	100	Medium-hardness range (e.g., annealed steels)
C	Brale	150	Hardened steel > HRB100
D	Brale	100	Case-hardened steels
E	1/8-in.-diameter steel ball	100	Al and Mg alloys
F	1/16-in.-diameter steel ball	60	Annealed Cu and brass
L	1/4-in.-diameter steel ball	60	Pb or plastics
N	N Brale	15, 30, or 45	Superficial Rockwell for thin samples or small impressions

The following precautions are recommended for reproducible results in Rockwell testing.[15]

1. The indenter and anvil should be clean and well seated.
2. The surface to be tested should be clean, dry, smooth, and free from oxide. A rough-ground surface is usually adequate for the Rockwell test.
3. The surface should be flat and perpendicular to the indenter.
4. Tests on cylindrical surfaces will give low readings, the error depending on the curvature, load, indenter, and hardness of the material. Corrections are given in ASTM E140-78.
5. The thickness of the specimen should be such that a mark or bulge is not produced on the reverse side of the piece. It is recommended that the thickness be at least 10 times the depth of the indentation. Tests should be made on only a single thickness of material.
6. The spacing between indentations should be three to five times the diameter of the indentation.
7. The speed of application of the load should be standardized. This is done by adjusting the dashpot on the Rockwell tester. Variations in hardness can be appreciable in very soft materials, unless the rate of application of the load is carefully controlled. For such materials, the operating handle of the Rockwell tester should be brought back as soon as the major load has been fully applied.

Vickers (or Diamond Pyramid) Hardness Test.   This test uses a pyramidal indenter with a square base, made of diamond. The angle between the faces is 136°. The test was introduced because of the problems encountered with the Brinell test. One of the known advantages of the Vickers test is that one indenter covers all the materials, from the softest to the hardest. The load is increased with hardness, and there is a continuity in scale. The angle of 136° was chosen on the basis of results with spherical indenters. For these, the best results were obtained when $d/D = 0.375$. If we take the points at which the sphere touches the surface of the specimen and the point of highest deformation (the center of the depression in metal) and calculate this angle, we obtain 136°. This exercise is left to the student. The description of the procedures used in testing is given in ASTM Standard E92–72. The Vickers hardness (HV) is computed from the equation

$$\text{HV} = \frac{2P\sin(\alpha/2)}{d^2} = \frac{1.8544P}{d^2}, \tag{3.44}$$

where P is the applied load (in kgf), d is the average length of the diagonals (in mm), and α is the angle between the opposite faces of the indenter (136°). Conversion to MPa is accomplished by multiplying this value by 9.81. The Vickers test described by ASTM E92–72 uses loads varying from 1 to 120 kgf. For example, 440HV30 represents a Vickers hardness number of 440, measured with a load of 30 kgf. Vickers testing requires a much better preparation of the material's surface than does Rockwell testing; hence, it is more time consuming. The surface has to be ground and polished, care being taken not to work-harden it. After the indentation, both diagonals of impression are measured, and their average is taken. If the surface is cylindrical or spherical, a correction factor has to be introduced. ASTM Standard E92 (Tables 4 through 6) provide correction factors. As with other hardness tests, the distance between the indentations has to be greater than two-and one-half times the length of the indentation diagonal, to avoid interaction between the work-hardening regions.

The manner in which the material flows and work-hardens (or work-softens) beneath the indenter affects the shape of the impression. The sides of the square impression can be deformed into concave or convex curves, depending on the nature of the deformation process, and this results in reading errors.

3.8.2 Microindentation Tests

Microindentation hardness tests—or microhardness tests—utilize a load lighter than 200 gf, and very minute impressions are thus formed; a load of 200 gf produces an indentation of about 50 μm for a medium-hardness metal. These tests are ideally suited to investigate changes in hardness at the microscopic scale. One can measure the hardness of a second-phase particle and identify regions within a grain where differences in hardness occur. Microhardness tests are also used to perform routine tests on very small precision components, such as parts of watches.

The results shown in Figure 3.38 illustrate well an application of microindentation testing. When a metal is alloyed, the distribution of the solute is not even throughout the grain, due to the stress fields produced by the solute atom. (See Chapter 7.) The solute atoms often tend to segregate at the grain boundaries. Figure 3.38a shows how the addition of aluminum to zinc is reflected by an increase in the hardness in the grain-boundary region, and the addition of gold results in a lowering of the grain-boundary hardness. This effect can be noted at extremely low concentrations of solute (a few parts per million). Figure 3.38b shows how this "excess" hardening increases with the concentration of aluminum.

In spite of the attempts made, several problems have arisen in the standardization of microindentation testing and its extrapolation to macroindentation results. There are several reasons for this. First, almost invariably, the microhardness of any material is higher than its standard macrohardness. Additionally, the microhardness varies with load. Second, there is a tendency for the microhardness to increase (up to a few grams); then the hardness value drops with load. At very low loads, one is essentially measuring the hardness of a single grain; the indenter "sees" a single crystal, and the plastic deformation produced by the indentation is contained in this grain. As the load is increased, plastic deformation of adjoining grains is involved, and a truly polycrystalline deformation regime is achieved. As we know well (see Chapter 5), the grain size has a marked effect on the yield strength

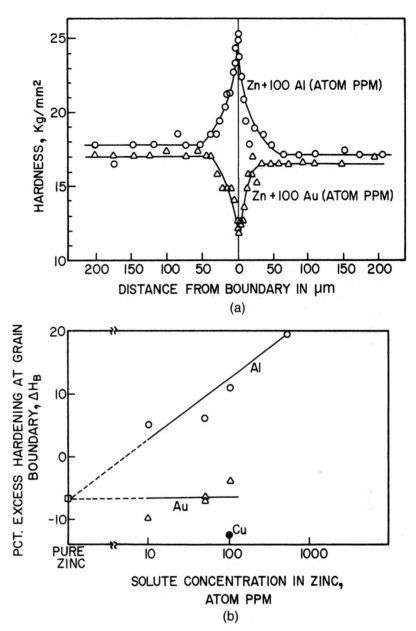

Figure 3.38 (a) Hardness–distance profiles near a grain boundary in zinc with 100-atom ppm of Al and zinc with 100-atom ppm of Au (1-gf load). (b) Solute concentration dependence of percent excess boundary hardening in zinc containing Al, Au, or Cu (3-gf load). (Adapted with permission from K. T. Aust, R. E. Hanemann, P. Niessen, and J. H. Westbrook, *Acta Met.,* 16 (1968) 291.

and work-hardening characteristics of metals. Yet another source of error is the work-hardening introduced in the surface by polishing. The effect of crystallographic orientation, when the impression is restricted to a single grain, is of utmost importance. It is well known that both the yield stress and the work-hardening are dependent on the crystallographic orientation of the material. The Schmid law relates the applied stress to the shear stress "seen" by the various slip systems. The Schmid relation is discussed in Section 6.2.2.

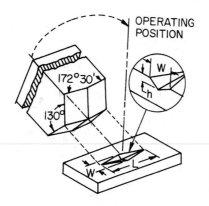

Figure 3.39 Some of the details of the Knoop indenter, together with its impression.

The two most common microindentation tests are the Knoop and Vickers tests. The Knoop indenter is an elongated pyramid, shown in Figure 3.39. The hardness is obtained from the surface area of the impression and is given by

$$KHN = \frac{14.228P}{d^2},$$

where P is the load of kgf and d is the length of the major diagonal, in mm. The ratio between the dimensions of the impression is

$$h/W/L = 1:4.29:30.53.$$

This results in an especially shallow impression, making the technique very helpful for testing brittle materials. Indeed, that was the purpose of introducing the test. The ratio between the major and minor diagonal of the impression is approximately 7:1, resulting in a state of strain in the material that can be considered to be plane strain; the strain in the L direction can be neglected. This subject is treated in Section 3.3. The very shallow Knoop impression is also helpful in testing thin components, such as electrodeposits or hardened layers. The Vickers microhardness test uses the same 136° pyramid with loads of a few grams. Both Knoop and Vickers indenters require prepolishing of the surface to a microscopic grade.

EXAMPLE 3.10 (Inspired by M. F. Ashby and D. R. H. Jones)

Obtain, for a simple two-dimensional case, a relationship between the hardness H and flow stress σ_0 of a material.

Solution: We assume a flat indenter and deformation on one plane only, as shown in Figure E3.10,1. Deformation is assumed to occur by the movement of blocks. The

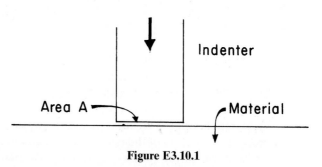

Figure E3.10.1

shear yield strength is τ_0. We set the work done by the punch, Fu, equal to the energy dissipated at the various interfaces. Student should compute the areas of triangles and assume that the resistance to motion is τ_0. The frictional forces between blocks is τ_0 times the areas (A or $A/\sqrt{2}$). We have

$$Fu = 2 \times \frac{A\tau_0}{\sqrt{2}} \times u\sqrt{2} + 2 \times A\tau_0 \times u + 4 \times \frac{A\tau_0}{\sqrt{2}} \times \frac{u}{\sqrt{2}},$$

$$\underbrace{}_{\text{(block 1)}} \quad \underbrace{}_{\text{(block 2,3)}} \quad \underbrace{}_{\text{(block 4,5)}}$$

central triangle   two lateral   two end
                triangles   triangles

where F is the applied force, u is the displacement of the punch, and A is the area of the indentation (Fig. E.3.10.1).

$$Fu = u6A\tau_0,$$

$$\frac{F}{A} = 6\tau_0.$$

But $\tau_0 = \sigma_0/2$; hence,

$$\boxed{\frac{F}{A} = H = 3\sigma_0.}$$

We assume a total displacement u of the punch, shown in Figure E 3.10.2. Block ① moves down by u. Blocks ② and ③ move sideways by u. Blocks ④ and ⑤ are pushed upward by u/2 and we compute the forces on two of their surfaces.

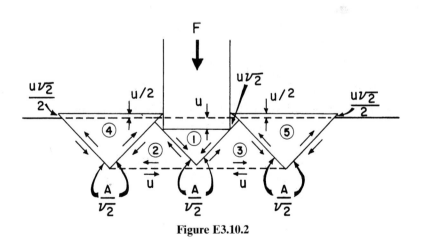

Figure E3.10.2

EXAMPLE 3.11

Estimate the flow stress of the material shown in Figure E3.11 if the indentation was done with a load of 1,000 g and the magnification of the photograph is 100×.

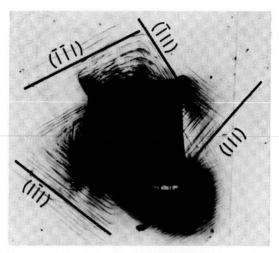

Figure E3.11 Indentation into iron-nickel single crystal; notice {111} traces of slip planes with specimen surface.

Solution: This is a Vickers microindentation. We measure the sides of the square, which are more visible (in this case) than the diagonal:

$$a = \frac{26 + 28}{2} = 27 \text{ mm.}$$

The diagonal is $d = a\sqrt{2} = 38.2$ mm. Dividing this value by the magnification, we obtain $d' = 0.382$ mm. So

$$\text{HV} = \frac{1.8544P}{d^2} = \frac{1.8544}{0.145},$$

$$\text{HV} = 12.80 \text{ kg/mm}^2.$$

We will convert this value to a yield stress, assuming that the material does not work-harden. We have (see Example 3.10)

$$\text{HV} = 3\sigma_0, \quad \text{so} \quad \sigma_y = 4.25 \text{kg/mm}^2.$$

But $1 \text{ kg/mm}^2 = 9.8 \times 10^6$ Pa; thus,

$$\sigma_y = 41.8 \text{ MPa.}$$

3.9 FORMABILITY: IMPORTANT PARAMETERS

An excellent introductory overview on sheet-metal forming is provided by Hecker and Ghosh.[16] *Deep drawing* and *stretching* are the two main processes involved in most sheet-metal-forming operations. In a stamping operation, one part of the blank might be subjected to a deformation process similar to deep drawing (thickness increasing with time). In deep

[16]See S. S. Hecker and A. K. Ghosh, *Sci. Am.,* Nov. (1976), p. 100.

drawing the material is required to contract circumferentially, while in stretching the stresses applied on the sheet are tensile in all directions. Sheet-metal forming has evolved from an art into a science, and important material parameters have been identified. These material properties are obtained in special tests and allow a reasonable prediction of the blank in the actual sheet-forming operation.

The work-hardening rate n is important, because it determines the onset of necking (tensile instability), an undesired feature. According to Considère's criterion (see Section 3.2.2), n is equal to ε_u, the uniform strain. Hence, the higher n, the higher ε_u. The strain-rate sensitivity m is an important parameter, too, because it also helps to avoid necking. If m is positive, the material becomes stronger at incipient necks because the strain rate in the neck region is higher. (see Section 3.2.3). The parameter R (the through-thickness plastic anisotropy) is also important; it is equal to the ratio between the strain in the "stretching" direction and the strain in the thickness direction. The greater the resistance to "thinning" in stretching, the better is the formability of the metal. This resistance to thinning corresponds to a value of R larger than 1: The strength in the thickness direction is greater than the strength in the plane of the sheet. The three parameters n, m, and R are readily obtained in a tensile test. (See Sections 2.2 and 3.2.)

Additional important information on the workability of sheets is provided by the yield and flow loci. Section 3.7.8 gives a description of yield criteria and how they are graphically presented in a plane-stress situation. The experimental determination of the yield locus and its expansion as plastic deformation takes place is conducted in biaxial tests. (See Section 3.7.4, Fig. 3.28).

Figure 3.40 show the most simple formability tests applied to metals. In the simple bending test, the specimen is attached to a die, and one end is clamped at a vise. The other end is bent to a specific radius. Specimens are bent to 180° using bending dies with smaller and smaller bending radii. Observations are made to see whether cracks are formed. In the free-bending test, the specimen is first bent between two rollers until an angle between 30° and 45° is achieved. It is then further bent between two grips, such as a vise.

The Olsen and Erichsen tests are typical stretch tests. A hardened steel sphere (diameters of 22.2 mm for the Olsen test, 20 mm for the Erichsen test) is pushed into the clamped metal, forming a bulge. The depth of the bulge at the fracture point is measured. The clamp-down pressure is very high (>70 kN), to minimize the drawing of the material.

The Swift and Fukui tests (Figures 3.41d and e) are drawing tests. The clamp-down pressure in the Swift test allows the sheet to slip inward. The overall diameter of the part is decreased in the process. This test simulates the deep drawing of parts. The drawability is expressed as the limiting draw ratio

$$\text{LDR} = \frac{\text{maximum blank diameter}}{\text{punch diameter}} = \frac{D}{d}.$$

There are two geometries for the Swift test, shown in Figure 3.41d: the round-bottomed cup test and the cup test. The latter test causes stretching of the center of the cup in addition to drawing. The Fukui test (Figure 3.41e) is the Japanese (JIS Z 2249) equivalent of the US stretch-drawing Swift test. A sphere 12.5 to 27 mm in diameter is pushed into a disk and advanced until either failure results or necking occurs in the cup. A hold-down ring maintains the specimen in place. The ratio between the diameters of the base of the deformed cup and the original disk provides the Fukui conical cup value. The modern counterpart of these older, but reliable, tests is the forming-limit curve, described in Section 3.9.2. The circle-grid analysis, which consists of applying a circle grid to the blank and measuring the strains in the critical regions of the stamped part, is also described in that section.

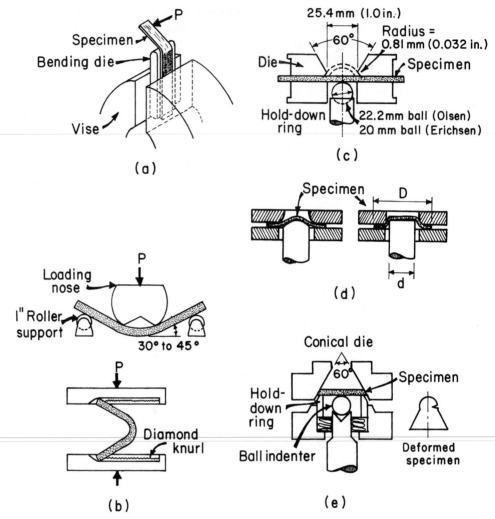

Figure 3.40 Simple formability tests for sheets. (a) Simple bending test. (b) Free-bending test. (c) Olsen cup test. (d) Swift cup test. (e) Fukui conical cup test.

3.9.1 Plastic Anisotropy

Elastic deformation under anisotropic conditions is described by elastic constants, whose number can vary from 21 for the most anisotropic solid to 3 for one exhibiting cubic symmetry. (For isotropic solids, the number of independent elastic constants is 2.) In a similar way, plasticity increases in complexity as the anisotropy of the solid increases. Sections 3.2–3.3 cover only the isotropic case, and even that in a very superficial way. In polycrystals, anisotropy in plasticity is more the rule than the exception. Essentially, there are two sources of anisotropy. First is texture, in which the grains are not randomly oriented, but have one or more preferred orientations. Texturing is often introduced by deformation processing. Well-known and well-characterized textures accompany cold rolling, wire drawing, and extrusion. This type of anisotropy is also called *crystallographic anisotropy*. Second, anisotropy is produced by the alignment of inclusions or second-phase particles along specific directions. When steel is produced, the inclusions existing in the ingot take the shape and orientation of the deformation process (rolling). These inclusions, such as MnS, produce mechanical effects called fibering. This type of anisotropy is also known as *mechanical anisotropy*. Whereas crystallographic anisotropy can strongly affect the yield stress, mechanical anisotropy usually manifests itself only in the later stages of deformation, influencing fracture.

Figure 3.41 shows the effect of texture on a deep-drawn cup. This effect is known as "earing." Prior to drawing, the sheet exhibited different yield stresses along different directions. The orientation in which the sheet is softer is drawn in faster than the harder direction, resulting in "ears." The number of ears (four) actually shows the type of texture. Figure 3.42 on the other hand, illustrates the effect of inclusions on the formability of an alloy. Fracture is much more probable if the sheet is bent along the second-phase strings than if it is bent perpendicular to them.

Section 3.7.7 shows the yield locus for anisotropic materials; this equation is an ellipse essentially identical to that described by the von Mises yield criterion in plane stress. (See Section 3.7.4.) The ellipse is distorted, however.

3.9.2 Punch–Stretch Tests and Forming-Limit Curves (or Keeler–Goodwin Diagrams)

An ideal test is the one that predicts exactly the performance of a material. The m, n, and R values are insufficient to predict the formability, and tests more closely resembling the actual plastic-forming operations have been used for a long time. The main parameter that they can provide is the strain to fracture. These tests are called punch–stretch tests, or simply, "cupping" tests.

The punch–stretch test consists of clamping a blank firmly on its edges between two rings or dies; the next step is to force a plunger or punch through the center area of the specimen enclosed by the area of the ring, until the blank fractures. Several punch–stretch tests have been developed throughout the years, including the Olsen, Erichsen, Guillery, and Wazau tests. These "cupping" tests are routinely used for inspection purposes, since they provide a quick indication of ductility; they also show the change in surface appearance of the sheet upon forming. Two important defects appear in stamping:

1. The orange-peel effect (surface rugosity) is due to the large grain size of the blank. The anisotropy of plastic deformation of the individual grains results in an irregular surface, perfectly visible to the naked eye, when the grain size is large.

Figure 3.41 "Ears" formed in deep-drawn cups due to in-plane anisotropy. (Reprinted with permission from R. L. Whiteley, W. A. Backofen, J. J. Burke, L. F. Coffin, Jr., N. L. Reed, V. Weiss (eds), *Fundamentals of Deformation Processing*, Sagamore Army Materials Research Conf. Proc., Volume 9, New York: Syracuse University Press, 1964, 150, 644.

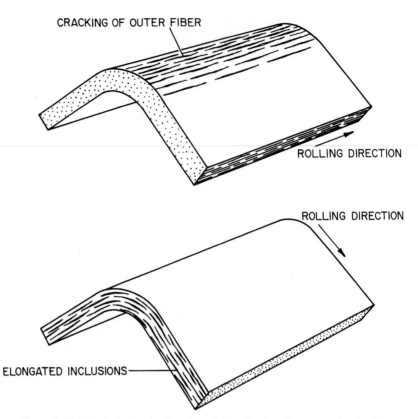

Figure 3.42 Effect of "fibering" on formability. The bending operation is often an integral part of sheet-metal forming, particularly in making flanges so that the part can be attached to another part. During bending, the fibers of the sheet on the outside of the bend are under tension, and the inside ones are under compression. Impurities introduced in the metal as it was made become elongated into "stringers" when the metal is rolled into sheet form. During bending, the stringers can cause the sheet to fail by cracking if they are oriented perpendicular to the direction of bending (top). If they are oriented in the direction of the bend (bottom), the ductility of the metal remains normal. (Adapted with permission from S. S. Hecker and A. K. Ghosh, *Sci. Am.*, Nov. (1976), p. 100).

2. Stretcher strains are produced when Lüders bands appear in the forming process. The interface between the Lüders band and undeformed materials exhibits a step perfectly visible to the naked eye. This is an undesirable feature that can be eliminated either by prestraining the sheet prior to forming (beyond the Lüders band region) or by alloying the material in such a way as to eliminate the yield drop and plateau from the stress–strain curve. In low-carbon steels, Lüders bands are formed by the interactions of carbon and nitrogen atoms with dislocations. After a process called temper rolling, the susceptibility is eliminated; however, it can return following aging. This problem is easily solved by flexing the sheet by effective roller leveling just prior to forming.[17]

The poor correlation between the common "cupping" test and the actual performance of the metal led investigators to look at some more fundamental parameters. The first break-

[17]H. E. McGannon (ed.), *The Making, Shaping, and Treating of Steel*, U.S. Steel, 9th ed., Pittsburgh, Pa., 1971, pp. 1126, 1260.

through came in 1963, when Keeler and Backofen[18] found that the localized necking required a critical combination of major and minor strains (along two perpendicular directions in the sheet plane). This concept was extended by Goodwin to the negative strain region, and the resulting diagram is known as the Keeler–Goodwin,[19] or forming-limit, curve (FLC). The FLC is an important addition to the arsenal of techniques for testing formability and is described after the description of Hecker's testing technique, presented next.[20]

Hecker developed a punch–stretch apparatus and technique well suited for the determination of FLC. The device consists of a punch with a hemispherical head with a 101.6-mm (4-in.) diameter. The die plates are mounted in a servohydraulic testing machine with the punch mounted on the actuator. The hold-down pressure on the die plates (rings) is provided by three hydraulic jacks. (The hold-down load is 133 kN.) The bead-and-groove arrangement in the rings eliminates any possible drawing in. The specimens are all gridded with 2.54-mm circles by a photoprinting technique. The load versus displacement is measured and recorded during the test, and the maximum load is essentially coincident with localized instability and the onset of fracture. A gridded specimen after failure is shown in Figure 3.43. The circles become distorted into ellipses. The clear circumferential mark is due to necking. The strains ε_1 and ε_2 are called meridian and circumferential strains, respectively, and are measured at various points when the test is interrupted. Figure 3.44a shows how these strains vary with distance from the axis of symmetry of the punch, at the point where the punch has advanced a total distance of $h = 27$ mm. ε_1, the meridional strain, is highest at about 25 mm from the center ($\varepsilon_1 \simeq 0.25$); ε_2, the circumferential strain, shows a definite plateau. By using sheets with different widths and varying lubricants between the sheet and the punch, different strain patterns are obtained. (Figure 3.44b shows the geometry of the deformed sheet.) The tests are conducted to obtain different combinations of minor–major strains leading to failure. Figure 3.45 shows how the FLC curve is obtained. The minor strain (circumferential) is plotted on the abscissa, and the major strain (meridional) is plotted on the ordinate axis. Four different specimen geometries are shown. The V-shaped curve (FLC) marks the boundary of the safe–fail zone. The region above the line corresponds to failure; the region below is safe. In order to have both major and minor strains positive, we use a full-sized specimen. By increasing lubrication, the major strain is increased; a polyurethane spacer is used to decrease friction. The drawings on the lower

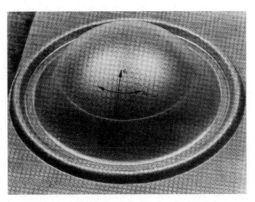

Figure 3.43 Sheet specimen subjected to punch–stretch test until necking; necking can be seen by the clear line. (Courtesy of S. S. Hecker).

[18]S. P. Keeler and W. A. Backofen, *Trans. ASM, 56* (1963) 25.

[19]G. M. Goodwin, "Application of Strain Analysis to Sheet Metal Forming Problems in the Press Shop," *SAE Automotive Eng. Congr.,* Detroit, Jan. 1968, SAE Paper No. 680093.

[20]See S. S. Hecker, *Metals Eng. Quart., 14* (1974) 30.

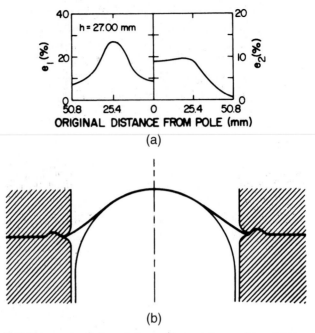

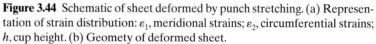

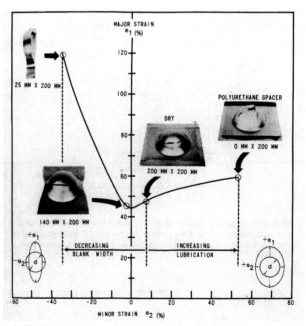

Figure 3.44 Schematic of sheet deformed by punch stretching. (a) Representation of strain distribution: ε_1, meridional strains; ε_2, circumferential strains; h, cup height. (b) Geomety of deformed sheet.

Figure 3.45 Construction of a forming-limit curve (or Keeler–Goodwin diagram). (Courtesy of S. S. Hecker)

left- and right-hand corners of the figure show the deformation undergone by a circle of the grid. When both strains are positive, there is a net increase in area. Consequently, the thickness of the sheet has to decrease proportionately. On the left-hand side of the plot, negative strains are made possible by reducing the lateral dimension of the blank. This allows free contraction in this dimension. The strains in an FLC diagram are obtained by carefully measuring the dimensions of the ellipses adjacent to the neck-failure region. It is interesting to notice that diffuse necking (thinning) starts immediately after deformation, whereas localized necking occurs only after substantial forming. Semiempirical criteria for localized necking that agree well with experimental results have been developed.

FLCs provide helpful guidelines for press-shop formability. Coupled with circle-grid analysis, they can serve as a guide in modifying the shape of stampings. Circle-grid analysis consist of photoprinting a circle pattern on a blank and stamping it, determining the major and minor strains in its critical areas. The strain pattern in the stamping is then compared with the FLC to verify the available safety margin. The strain pattern can be monitored with changes in lubrication, hold-down pressure, and size and shape of draw-beads and the blank; such monitoring can lead to changes in the experimental procedure. Circle-grid analysis also serves, in conjunction with the FLC, to indicate whether a certain alloy might be replaced by another one, possibly cheaper or lighter. During production, the use of occasional circle-grid stampings provides a valuable help with respect to wear, faulty lubrication, and changes in hold-down pressure. Hecker and Ghosh[21] claim that the circle-grid analysis has replaced the craftsman's "feel" for the proper flow of the metal.

The strain pattern undergone by a stamped part is shown schematically in Figure 3.46. Different portions exhibit different strains, and this is evident by observing the distortion of circles at different regions.

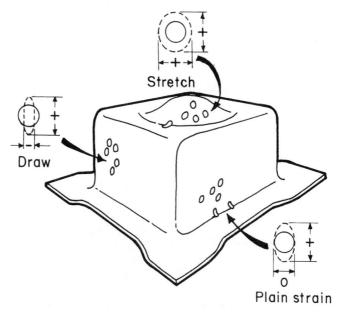

Figure 3.46 Different strain patterns in stamped part. (Adapted from W. Brazier, *Closed Loop*, 15, No. 1 (1986) 3)

[21]S. S. Hecker and A. K. Ghosh, *Sci. Am.*, Nov. (1976), p. 100.

SUGGESTED READINGS: PLASTICITY

J. E. Gordon, *The New Science of Strong Materials, or Why You Don't Fall Through the Floor*. Princeton, NJ: Princeton University Press, 1976.

E. P. Popov, *Engineering Mechanics of Solids*. Englewood Cliffs, NJ: Prentice Hall, 1990.

D. Roylance, *Mechanics of Materials* New York: J. Wiley, 1996.

J. B. Wachtman, *Mechanical Properties of Ceramics*. New York: J. Wiley, 1996.

R. H. Wagoner and J. L. Chenot, *Fundamentals of Metal Forming*. New York: J. Wiley, 1996.

SUGGESTED READINGS: HARDNESS

H. E. Boyer, ed. Hardness Testing. Metals Park, OH: ASM Intl., 1987. *Metals Handbook, Vol. 8: Mechanical Testing*. Metals Park, OH: ASM Int., 1985.

M. C. Shaw, *Mechanical Behavior of Materials*. F. A. McClintock and A. S. Argon (eds.). Reading, MA: Addison-Wesley, 1966, p. 443.

D. Tabor, *The Hardness of Metals*. London: Oxford University Press, 1951.

J. H. Westbrook and H. Conrad (eds.), *The Science of Hardness Testing and Its Research Applications*. Metals Park, OH: ASM Intl., 1973.

SUGGESTED READINGS: FORMABILITY

W. Brazier, *Closed Loop (MTS Journal)*, 15, No. 1 (1986) 3.

K. S. Chan, *J. of Metals*. Feb. (1990) 6.

S. S. Hecker, in *Constitutive Equations in Viscoplasticity: Computational and Engineering Aspects*. New York: ASME, 1976, p. 1.

S. S. Hecker and A. K. Ghosh, *Sci. Am.*, Nov. 1976, p. 100.

S. S. Hecker, A. K. Ghosh, and H. L. Gegel (eds.), *Formability: Analysis, Modeling, and Experimentation*. New York: TMS-AIME, 1978.

W. F. Hosford and R. M. Caddell, Metal Forming—Mechanics and Metallurgy, Englewood Cliffs, NJ: Prentice-Hall, 1983.

EXERCISES

3.1 A polycrystalline metal has a plastic stress–strain curve that obeys Hollomon's equation,

$$\sigma = K\varepsilon^n.$$

Determine n, knowing that the flow stresses of this material at 2% and 10% plastic deformation (offset) are equal to 175 and 185 MPa, respectively.

3.2 You are traveling in an airplane. The engineer who designed it is, casually, on your side. He tells you that the wings were designed using von Mises' criterion. Would you feel safer if he had told you that Tresca's criterion had been used? Why?

3.3 A material is under a state of stress such that $\sigma_1 = 3\sigma_2 = 2\sigma_3$. It starts to flow when $\sigma_2 = 140$ MPa.

   (a)   What is the flow stress in uniaxial tension?

   (b)   If the material is used under conditions in which $\sigma_1 = -\sigma_3$ and $\sigma_2 = 0$, at which value of σ_3 will it flow, according to Tresca's and von Mises' criteria?

3.4 A steel with a yield stress of 300 MPa is tested under a state of stress where $\sigma_2 = \sigma_1/2$ and $\sigma_3 = 0$. What is the stress at which yielding occurs if it is assumed that:

   (a)   The maximum-normal-stress criterion holds?

   (b)   The maximum-shear-stress criterion holds?

   (c)   The distortion-energy criterion holds?

3.5 Determine the maximum pressure that a cylindrical gas reservoir can withstand, using the three flow criteria. Use the following information:

Material: AISI 304 stainless steel—hot finished and annealed, $\sigma_0 = 205$ MPa

Thickness: 25 mm

Diameter: 500 mm

Length: 1 mm

Hint: Determine the longitudinal and circumferential (hoop) stresses by the method of sections.

3.6 Determine the value of Poisson's ratio for an isotropic cube being plastically compressed between two parallel plates.

3.7 A low-carbon-steel cylinder, having a height of 50 mm and a diameter of 100 mm, is forged (upset) at 1,200°C and a velocity of 1 m/s, until its height is equal to 15 mm. Assuming an efficiency of 60%, and assuming that the flow stress at the specified strain rate is 80 MPa, determine the power required to forge the specimen.

3.8 Obtain the work-hardening exponent n using Considère's criterion for the curve of Example 3.4.

3.9 The stress–strain curve of a 70–30 brass is described by the equation

$$\sigma = 600\varepsilon_p^{0.35} \text{ MPa}$$

until the onset of plastic instability.

(a) Find the 0.2% offset yield stress.

(b) Applying Considère's criterion, find the real and engineering stress at the onset of necking.

3.10 The onset of plastic flow in an annealed AISI 1018 steel specimen is marked by a load drop and the formation of a Lüders band. The initial strain rate is 10^{-4} s^{-1}, the length of the specimen is 5 cm, and the Lüders plateau extends itself for a strain equal to 0.1. Knowing that each Lüders band is capable of producing a strain of 0.02 after its full motion, determine:

(a) The number of Lüders bands that traverse the specimen.

(b) The velocity of each Lüders band, assuming that only one band exists at each time.

3.11 A tensile test on a steel specimen having a cross-sectional area of 2 cm^2 and length of 10 cm is conducted in an Instron universal testing machine with stiffness of 20 MN/m. If the initial strain rate is 10^{-3} s^{-1}, determine the slope of the load-extension curve in the elastic range ($E = 210$ GN/m^2).

3.12 Determine all the parameters that can be obtained from a stress–strain curve from the load-extension curve (for a cylindrical specimen) shown in Figure Ex 3.12, knowing that the initial cross-sectional area is 4 cm^2, the crosshead velocity is 3 mm/s, the gage length is 10 cm, the final cross-sectional area is 2 cm^2, and the radius of curvature of the neck is 1 cm.

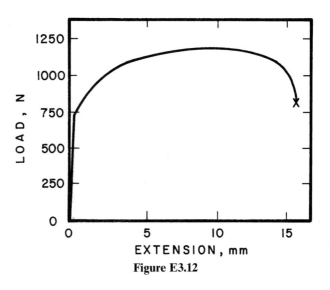

Figure E3.12

3.13 Draw the engineering-stress–engineering-strain and true-stress–true-strain (with and without Bridgman correction) curves for the curve in Exercise 3.12.

3.14 What is the strain-rate sensitivity of AISI 1040 steel at a strain of 0.02 and a strain of 0.05 (Obtain your data from Figure 3.12(a).)

Section 3.2

3.15 From the load-extension curve shown in Example 3.12, draw the true-stress–true-strain and the engineering-stress–engineering-strain curves.

Section 3.4

3.16 For an AISI 1045 steel obeying the relationship

$$\sigma(\text{MPa}) = 300 + 450e^{0.5}$$

in tension, obtain the compressive stress–strain curve, considering the Bauschinger effect. Use data from Figure 3.17 for making the correction.

3.17 The PMMA specimens shown in Figure Ex. 3.17 were deformed in uniaxial tension. (a) Plot the total elongation, ultimate tensile stress, and Young's modulus as a function of temperature. (b) Discuss changes in these properties in terms of the internal structure of the specimen.

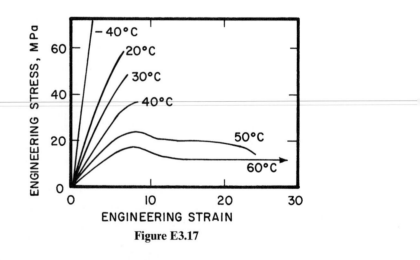

Figure E3.17

3.18 For the force–displacement curve of Figure Ex 3.18, obtain the engineering and true-stress–strain curves if the specimen were tested in compression.

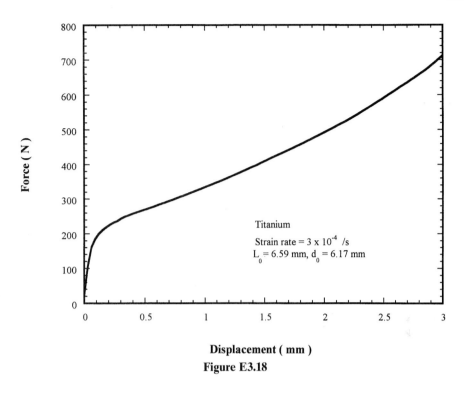

Figure E3.18

Section 3.7

3.19 Calculate the softening temperature for a soda–lime silica glass at which the viscosity is equal to $10^{13.4}$ P if the activation energy for viscous flow is 2.5 kJ/mole and the viscosity at 1,000°C is 10^{14} P. Note: $1P = 0.1$ Pas.

3.20 The viscosity of a SiO_2 glass is 10^{14} P at 1,000°C and 10^{11} P at 1,300°C. What is the activation energy for viscous flow in this glass? Note: $1\ P = 0.1$ Pas.

Section 3.8

3.21 When tested at room temperature, a thermoplastic material showed a yield strength of 51 MPa in uniaxial tension and 55 MPa in uniaxial compression. Compute the yield strength of this polymer when tested in a pressure chamber with a superimposed hydrostatic pressure of 300 MPa.

3.22 From Equation 3.35, obtain Equation 3.34. Then prove that Equation 3.34 represents an ellipse rotated 45° from its principal axis.

3.23 An annealed sheet of AISI 1040 steel (0.85 mm thick and with in-plane isotropy) was tested in uniaxial tension until the onset of necking, to determine its formability. The initial specimen's length and width were 20 and 2 cm, respectively. At the onset of necking, the length and width were 25 and 1.7 cm, respectively.

   (a) Determine the ratio between the through-thickness and the in-plane yield stress, assuming that R does not vary with strain.

   (b) Draw the flow locus of this sheet, assuming that $\sigma_{y(1,2)} = 180$ MN/m².

3.24 Repeat Exercise 3.23 if the final width of the specimen is 1.9 cm, and explain the differences. Which case has a better formability?

3.25 Imagine that you want to perform a circle-grid analysis, but you do not have the facilities for photoprinting. Hence, you decide to make a grid of perpendicular and equidistant lines. After plastic deformation of the material, can you still determine the major and minor strains from the distorted grid? (*Hint*: Use the method for determining principal strains.)

4

Imperfections: Point
and Line Defects

The mechanical properties of materials are often limited by their imperfections. We saw in Chapter 1 that the theoretical cleavage and shear strengths of materials are given by Equations 1.9 and 1.16, namely,

$$\sigma_{th} = \sqrt{\frac{E\gamma}{a}} \approx \frac{E}{\pi} \quad \text{and} \quad \tau_{th} = \frac{Gb}{2\pi a} \approx \frac{G}{2\pi},$$

where E and G are the Young's and shear moduli, respectively; a is the interatomic spacing, and γ is the surface energy of the material. These equations predict exceedingly high strengths (on the order of GPa's), and few materials reach such strengths. (See Chapter 1.) Indeed, this is somehow the Holy Grail of materials science: If materials were perfect, those values could be reached. However, all materials contain imperfections, either by design or inadvertently during processing. These defects are reviewed in this and subsequent chapters. They are classified, according to their dimensions, into four kinds, each discussed in a separate section as follows:

Point defects (Section 4.1)
Line (or one-dimensional) defects (Section 4.2)
Interfacial (or two-dimensional) defects (Chapter 5, Section 5.1)
Volume (or three-dimensional) defects (Chapter 5, Section 5.2).

Cracks are discussed in Chapters 7 and 8, on fracture.

Imperfections determine the mechanical response of materials, and the manner in which the response is used to enhance performance in a material will be analyzed in considerable detail in Chapters 5 through 9. Note that the dimensional scale of defects covers a wide spectrum, 10^{14} m, as shown schematically in Figure 4.1. Electronic point defects do not affect mechanical properties significantly and will therefore not be discussed in this text.

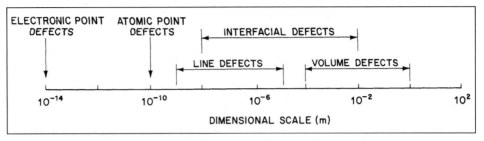

Figure 4.1 Dimensional ranges of different classes of defects.

4.1 *POINT DEFECTS*

Point defects exist on an atomic scale. These defects can have a diameter of approximately 10^{-10} m. Although relatively small compared to other imperfections, atomic defects do generate a stress field in the crystal lattice and affect the properties of the material. Figure 4.2 shows the following three types of atomic point defects:

1. *Vacancy.* When an atomic position in the Bravais lattice is vacant.
2. *Interstitial point defect.* When an atom occupies an interstitial position. This interstitial position can be occupied by an atom of the material itself or by a foreign atom; the defect is called a self-interstitial and an interstitial impurity, respectively, for the two cases.
3. *Substitutional point defect.* When a regular atomic position is occupied by a foreign atom.

The vacancy concentration in pure elements is very low at low temperatures. The probability that an atomic site is a vacancy is approximately 10^{-6} at low temperatures, rising to 10^{-3} at the melting point. In spite of their low concentration, vacancies have a very important effect on the properties of a material, because they control the self-diffusion and substitutional diffusion rates. The movement of atoms in the structure is coupled to the movement of vacancies. In Section 4.1.1, the equilibrium concentration of vacancies is calculated.

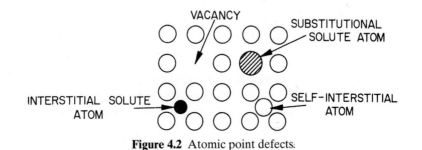

Figure 4.2 Atomic point defects.

In compounds (ceramics and intermetallics), defects cannot occur as freely as in metals, because we have additional requirements, such as electrical neutrality. Two types of defects are prominent in compounds and are shown in Figure 4.3: the Schottky defect, which is a pair of vacancies that have opposite sign (one cation and one anion); and the Frenkel defect, which consists of a vacancy–self-interstitial pair.

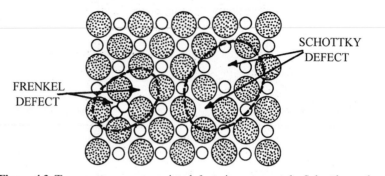

Figure 4.3 Two most common point defects in compounds: Schottky and Frenkel defects.

The self-interstitial and interstitial impurities lodge themselves in the "holes" that the structure has. There is more than one type of hole in the FCC, BCC, and HCP structures, and their diameters and positions will be determined in what follows.

The FCC structure, shown in Figure 4.4 has two types of voids: the larger, called octahedral, and the smaller, called tetrahedral. The names are derived from the nearest neighbor atoms; they form the vertices of the polyhedra shown. If we consider the atoms as rigid spheres, we can calculate the maximum radius of a sphere that would fit into the void without straining the lattice. The reader is encouraged to engage in this exercise; with some luck, he or she will find radii of 52 and 28 pm for octahedral and tetrahedral voids, respectively, in γ-iron. Hence, carbon ($r = 80$ pm) and nitrogen ($r = 70$ pm) produce distortions in the lattice when they occupy the voids.

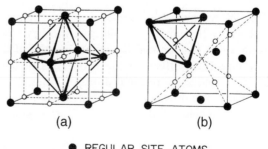

(a)   (b)

● REGULAR SITE ATOMS

Figure 4.4 Interstices in FCC structure. (a) Octahedral void. (b) Tetrahedral void.

In BCC metals there are also octahedral and tetrahedral voids, as shown in Figure 4.5. In this case, however, the larger void is tetrahedral. For rigid spheres in α-iron, the void radii are 36 and 19 pm for tetrahedral and octahedral interstices, respectively. Hence, a solute atom is accommodated in an easier way in FCC than in BCC iron, in spite of the fact that the FCC structure is more closely packed.

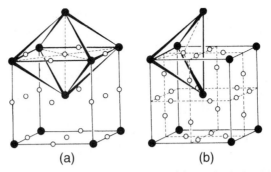

Figure 4.5 Interstices in the BCC structure. (a) Octahedral void. (b) Tetrahedral void.

Analogously, the HCP structure presents tetrahedral and octahedral voids, shown in Figure 4.6; the reader is reminded of the similarity between the FCC and HCP structures (see Section 4.3), which explains the presence of the same voids.

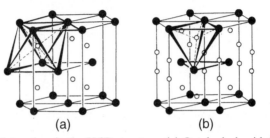

Figure 4.6 Interstices in the HCP structure. (a) Octahedral void. (b) Tetrahedral void.

4.1.1 Equilibrium of Point Defects

A very important characteristic of vacancies and self-interstitial atoms, in contrast to line and surface defects, is that they can exist in thermodynamic equilibrium at temperatures above 0 K. The thermodynamic equilibrium in a system of constant mass, at a constant pressure and temperature, and that does not execute any work in addition to the work against pressure, is reached when the Gibbs free energy is minimum. The formation of point defects in a metal requires a certain quantity of heat δq (as there is no work being executed, except against pressure). Hence, if $dH = \delta q$, the enthalpy of the system increases. The configurational entropy S also increases, because there are a certain number of different ways of putting the defects into the system.

The Gibbs free energy is, by definition,

$$G = H - TS. \tag{4.1}$$

One can thus see that the free energy will reach a minimum for a certain value of n (the number of point defects) different from zero; at 0 K, the entropic term is zero and the equilibrium concentration is zero.

The equilibrium concentration of point defects can be calculated from statistical considerations and is equal to

$$\boxed{\frac{n}{N} = \exp(-\,G_f/kT)} \tag{4.2}$$

where n and N are the number of point defects and sites, respectively, H_f is the free energy of formation of the defects, and k is Boltzmann's constant. For copper, the formation of vacancies and interstitials are

$$G_v = 83 \text{ kJ/mol}, \qquad G_i = 580 \text{ kJ/mol}.$$

We have, approximately, the following ratio:

$$\frac{G_i}{G_v} \approx 7.$$

Therefore, for copper, the enthalpy of formation of a vacancy is approximately seven times lower than that of a self-interstitial defect. Using Equation 4.2, we can obtain the ratio between the vacancy (X_v) and interstitial (X_i) concentrations:

$$\frac{X_v}{X_i} \approx \exp\left(\frac{G_i - G_v}{kT}\right). \tag{4.3}$$

For copper at 1,000 K (we have to convert molar quantities or use $R = 8.314 \text{ J (mole} - \text{k)}$:

$$\frac{X_v}{X_i} \approx 10^{26}.$$

It can be concluded that, at least in close-packed structures, the concentration of interstitials is negligible with respect to that of the vacancies. Using Equation 4.2 for copper at 1,000 K, we obtain

$$X_v \approx 4.5 \times 10^{-5}.$$

Hence, there is only one vacancy for each 2×10^4 copper atoms at 1,000 K. This number is very small; in spite of this, it corresponds to approximately 10^{14} vacancies/cm^3. The low concentration of self-interstitials in close-packed structures is a consequence of the small diameter of the interstitial voids. (See Section 4.1.) In more open structures these concentrations can be higher. Even so, high interstitial concentrations are not observed in equilibrium structures.

EXAMPLE 4.1

If, at 400°C, the concentration of vacancies in aluminum is 2.3×10^{-5}, what is the excess concentration of vacancies if the aluminum is quenched from 600°C to room temperature? What is the number of vacancies in one cubic μm of quenched aluminum?

We are given:

$$G_v = 0.62 \ eV,$$

$$k = 86.2 \times 10^{-6} \ eV/K,$$

$$r_{Al} = 0.143 \text{ nm}.$$

Solution: We have

$$\frac{n_v}{N} = e^{-G_v/kT}.$$

At 400°C (= 673 K),

$$2.3 \times 10^{-5} = e^{-0.62/86.2 \times 10^{-6} \times 673},$$

Thus,

$$\frac{n_v}{n} = e^{-0.62/86.2 \times 10^{-6} \times 873} = 2.6 \times 10^{-4}.$$

Aluminum has the FCC structure, with four atoms per unit cell. The lattice parameter a is related to the unit cell by

$$a = 2\sqrt{2}r = 0.404 \text{ nm}.$$

The corresponding volume is

$$V = a^3 = 0.0662 \text{ nm}^3.$$

In one μm^3, the number of atoms is

$$n = \frac{4 \times 10^9}{0.0662} = 6.04 \times 10^{10},$$

$$n_v = (2.6 \times 10^{-4})n = 1.6 \times 10^7.$$

Hence, there are about 1.6×10^7 vacancies per cubic μm of the quenched aluminum.

Point defects can group themselves in more complex arrangements (for instance, two vacancies form a divacancy, two interstitials form a diinterstitial, etc). The enthalpy of formation of divacancies has been determined for several metals. For example, for copper (with $G_f = 5.63 \times 10^{-19}$ J), it is: 0.96×10^{-19} J. The enthalpy of formation of divacancies in noble metals in on the order of 0.48×10^{-19} J. It is thought that divacancies are stable, in spite of the fact that their enthalpies of bonding are not very well known.

Diinterstitials also exist, and their energies can be calculated by the same processes as mono-interstitials. Similarly, the vacancies can bind themselves to atoms of impurities when the binding energy is positive.

4.1.2 Production of Point Defects

Intrinsic point defects in a metal—either vacancies or self-interstitials—exist in well-established equilibrium concentrations. (See Section 4.1.1.) By appropriate processing, the concentration of these defects can be increased. Quenching, or ultra-high-speed cooling, is one of these methods. The concentration of vacancies in BCC, FCC, and HCP metals is greatly superior to that of interstitials and on the order of 10^{-3} when the metal is at a temperature close to the melting point; it is only 10^{-6} when the metal is at a temperature of about half the melting point. Hence, if a specimen is cooled at a high enough rate, the high-temperature concentration can be retained at low temperatures. For this to occur, the rate of cooling has to be such that the vacancies cannot diffuse to sinks—grain boundaries, dislocations, the surface, and so on. Theoretically, gold would have to be cooled from 1,330 K to ambient temperature at a rate of 10^{11} K/s to retain its high-temperature vacancy concentration. The fastest quenching technique to cool thin wires produces cooling rates lower than 10^5 K/s; nevertheless, a significant portion of the high-temperature point defects is retained.

Another method of increasing the concentration of point defects is by plastic deformation. The movement of dislocations, as will be seen in Section 4.2.8, generates point defects by two mechanisms: the nonconservative motion of jogs, and the annihilation of parallel dislocations of opposite sign, producing a line of vacancies or interstitials. *Jogs* are created by dislocation intersections; since they cannot glide with dislocations, they have to climb as the dislocation moves. In a screw dislocation, they are small segments having the character of an edge. The slip plane of this segment is not compatible with that of the dislocation. The climb is possible only by continuous emission of vacancies or interstitials. The second mechanism is depicted schematically in Figure 4.7. When the two dislocations cancel each other, they create rows of interstitials or vacancies if their slip planes do not coincide.

Quenching produces mostly vacancies and vacancy groups. The concentrations obtained are lower than 10^{-4}. Deformation, on the other hand, can introduce higher concentrations of vacancies and equivalent ones of interstitials; the problem is that it also introduces a number of other substructural changes that complicate the situation. Dislocations are introduced, and they interact strongly with point defects. One method of producing point defects does not present these problems: *Radiation* of the metal by high-energy particles allows the introduction of a high concentration of point defects. The radiation displaces the electrons, or ionizes, displaces atoms by elastic collisions, and produces fission and thermal spikes. This subject is treated in greater detail in Section 4.1.4. The displacement of atoms is produced by the elastic collision of the bombarding particles with the lattice atoms, transferring the kinetic energy of the particles to the atoms. This may cause the atoms to travel through the lattice. In the majority of cases, the atom travels a few atomic distances and enters an interstitial site. Consequently, a vacancy is produced, together with a self-interstitial. The energy transferred in the collision has to be well above the energy required to form an interstitial–vacancy pair in a reversible thermodynamic process (3 to 6 eV, or 4.8×10^{-19} to 9.6×10^{-19} J). It is believed that the energy transferred

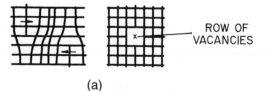

(a)

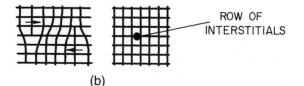

(b)

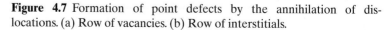

Figure 4.7 Formation of point defects by the annihilation of dislocations. (a) Row of vacancies. (b) Row of interstitials.

to the atom has to be approximately 25 eV (40×10^{-19} J). Different particles can be used in the bombardment process: neutrons, electrons, γ rays, and α particles.

4.1.3 Effect of Point Defects on Mechanical Properties

Point defects have a marked effect on the mechanical properties of a material. For this reason, the effect of radiation is of great importance. Maddin and Cottrell[1] used aluminumonocrystals with various purity levels, observing that the yield stress increased with quenching. Quenching was accomplished by taking the specimens from 600°C and throwing them into a water–ice mixture, while annealed material was slowly cooled in the furnace. The yield stress increased from 550 to 5,900 kPa, on average. The effect of impurity atoms could be neglected because the increase in yield stress was consistent throughout the specimens. The effect of possible residual stresses due to quenching was also neglected. With the purpose of obtaining evidence that was still more convincing, a crystal was tested immediately after quenching, while another was tested after staying a few days at ambient temperature. The yield stress increased from 5.9 GPa to 8.4 GPa in the aged condition. The strengthening in quenching is due to the interaction of dislocations and vacancies or groups thereof. The effect of jogs, formed by the condensation of vacancies on the dislocations, can also be considerable. During aging, the excess concentration of vacancies forms groups and/or annihilates preexisting dislocations.

There are also alterations in the plastic portion of the stress-versus-strain curve seen in Figure 4.8. The initial work-hardening rate of the quenched aluminum is lower than that of slowly (furnace) cooled aluminum. At greater strains, however, the two work-hardening rates become fairly similar. Hence, the effect of quenching disappears at higher strains. This is thought to be due to the fact that the excess concentrations are eliminated during plastic deformation; at the same time, excess vacancies are generated by dislocation motion, so that the concentrations in the quenched and furnace cooled materials become the same.

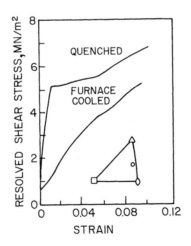

Figure 4.8 Stress-versus-strain curves for aluminum monocrystals. The crystallographic orientation is shown in the stereographic triangle. (Adapted with permission from R. Maddin and A. H. Cottrell, *Phil. Mag.*, 46 (1955), p. 737.)

The increase in hardness in many quenched metals is negligible, in spite of the obvious changes in the stress-versus-strain curve. This is explained by the fact that the effect of quenching disappears after a certain amount of plastic deformation. Since the indenter deforms the metal plastically (in an extensive way), the effect of quenching is minimal.

[1]R. Maddin and A. H. Cottrell, *Phil. Mag.*, 46 (1955) 735.

4.1.4 Radiation Damage

Irradiation of solids by high-energy particles may produce one or more of the following effects:

1. Displaced electrons (i.e., ionization).
2. Displaced atoms by elastic collision.
3. Fission and thermal spikes.

Ionization has a much more important role in nonmetals than it has in metals. The high electrical conductivity of metals leads to a very quick neutralization of ionization, and there is no observable change in properties due to this phenomenon. Electronic excitations in metals are also eliminated almost instantaneously. Such would not be the case in semiconductors and dielectrics, where electronic excitation configurations are almost permanent. Thus, in the case of metals, only collisions among incident particles and atomic nuclei are of importance. The basic mechanism in all processes of radiation damage is the transfer of energy and motion from the incident particle beams to the atoms of the material. The incident particle beam may consist of positive particles (protons, for example), negative particles (which are invariably electrons), or neutral particles (X rays, γ rays, neutrons, etc.). Irradiation by neutrons results in a large spectrum of constant energy until the maximum energy that a particle can transmit to an atom which suffered the impact. A neutron of 1 MeV (0.16 pJ) can transfer about 10^5 eV (0.016 pJ) to an atom. High-energy transfers can also be obtained by means of positive particles, but such energy transfers are less common. In the case of electrons, only low-energy transfers are possible. We shall consider here mainly the effects of neutron radiation on metals. The primary collision has the function of transferring energy to the atomic system. The subsequent events that occur are as follows:

1. Displacement of an atom from its normal position in the lattice to a position between the normal lattice sites.
2. Creation of defects by displacements and their migrations and interactions.

When an atom is displaced from its normal lattice site, two defects are created: an interstitial atom referred to as autointerstitial or self-interstitial, and a vacant lattice site called a vacancy. More complex configurations can be regarded as having started from this fundamental step. When an atom receives an energy impulse greater than a certain value E_e, called the *effective displacement energy,* some atom is displaced from its normal position to an interstitial position. In the most simple case, if an atom receives the primary impact of energy E_e, the atom itself is displaced. This, however, is not inevitable; sometimes another atom, a neighboring one, is displaced. With an increase in the energy imparted to the affected atom, various events can occur. At low energies, but higher than E_e, only an interstitial and its connected vacancy are possible. At high energies, the affected atom becomes an important particle for creating more damage. This leads to cascade elements.

Near the end of its trajectory, an energetic atom displaces all the atoms that it encounters; this is called a "displacement spike." Through a cascade effect, damage propagates through the lattice. Many atoms that spread about by displacement spikes will become situated along the atomic packing lines, and thus these lines will be a most efficient manner of transporting energy far away from the spike. The impact transferred along a crystallographic direction is called a *Focuson* (analogous to *photon* and *phonon*). If the energy is not well above the energy required for atomic displacement, it will be transferred into a chain of exchange collisions that makes the atom travel far away from the spike before it comes to a stop as an interstitial. The efficiency of this process is much higher in the close-packed directions (the ⟨110⟩ directions in FCC crystals). The atomic configuration in the ⟨110⟩ direction in which an interstitial is propagated along a line is called a *dynamic crowdion*. The efficiency of the focusing processes is directly proportional to the interatomic potential, being higher for heavy metals and lower for light metals (such as Al). According to the Seeger model, at zero kelvin, for each initially displaced atom, one would have one or more regions in which a good fraction of atoms (about 30%) disappear. These regions are surrounded by interstitial clouds that extend a few hundreds of atomic distances in noble metals and perhaps a few atomic distances in a metal such as Al. Seeger called the region of lost atoms in the center of a cascade a "depleted zone" and estimated that its typical size would be less than 1 nm. Figure 4.9 shows the Seeger model of damage produced by irradiation.

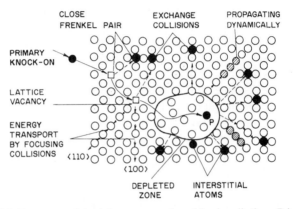

Figure 4.9 Seeger model of damage produced by irradiation. *P* indicates the position where the first "knock-on" terminates. (Reprinted with permission from A. Seeger, in *Proc. Symp. Radiat. Damage Solids React.*, Vol. 1, IAEA, Vienna, 1962, pp. 101, 105)

Vacancies generated during exposure to radiation often condense and form voids inside the material. An illustration of this is provided in Figure 4.10, which shows Ni irradiated by a high dosage of N_2^+ ions. A high concentration of voids is produced. The voids have polyhedral shapes because the surface energy is anisotropic and this shape, rather than a sphere, minimizes the overall surface energy.

Figure 4.10 Voids formed in nickel irradiated using 400 keV $^{14}N_2^+$ ions to a dose of 40 dpa at 500°C; notice the voids with polyhedral shape. (Courtesy L. J. Chen and A. J. Ardell)

In any event, a major portion of radiation damage in common metals caused by neutrons in reactors consists of a large number of interstitials and vacancies produced in a cascade process that follows after a primary knock-on impact. These point defects act as small obstacles to dislocation movement and result in a hardening of the metals. Besides this direct effect on mechanical properties, some indirect effects are possible. These indirect effects, which arise from the fact that irradiation by neutrons changes the rates and mechanisms of atomic interchange, are as follows:

1. Destruction of order of lattice
2. Fractionating of precipitates
3. Acceleration of nucleation
4. Acceleration of diffusion

These processes have their origin, directly or indirectly, in the kinetic energy exchanges between energetic neutrons and atoms. According to Seeger's model, atoms can be transported long distances by "cooperative" focalization along the more densely packed directions, and the collision processes create simple defects, such as interstitials and vacancies, and complex defects, such as displacement spikes. If an alloy is ordered, focalization and displacement spikes may destroy the order. If the alloy contains precipitates, a displacement spike may break the precipitates if they are smaller than the spike and thus return the precipitates into solution. In an alloy that can have precipitates, the damaged regions caused by spikes can serve as nucleation sites. The excess of vacancies produced by irradiation can accelerate the diffusion rate. All these effects influence significantly the mechanical properties, as described in earlier chapters. At ordinary temperatures (i.e., ambient or slightly above) one or both the defects (interstitials and vacancies) are mobile, and thus, the ones that survive the annihilation, due to recombination or loss of identity at sinks such as dislocations or interfaces, group together. It is well established that in a majority of metals, irradiation at low temperatures ($<0.2T_m$, where T_m is the melting point, in kelvins) results in joining of vacancies and interstitials to form groups that are surrounded by dislocations (i.e., loops and tetrahedral packing defects). These groups impede dislocation motion, as well as increase the strength and reduce the ductility of the material. At high temperatures, the vacancies can group together to form voids. The formation of such groups of defects can cause important and undesirable changes in mechanical properties and result in a dimensional instability of the material. Damage accumulated during irradiation by neutrons (and other particles) can cause significant changes in important properties. For example, the yield stress or the flow stress increases, and frequently there is a loss of ductility.

The problem of mechanical and dimensional stability is a very serious one for structural components in fast reactors. In 1967, it was discovered that fuel cladding consisting of austenitic stainless steel, when exposed to high doses of fast neutrons, showed internal cavities (~10 nm). These cavities, called *voids,* result in an increase in the dimensions of the material. It is estimated that the maximum possible dilation in the structural components is on the order of 10%. However, as neutron flux and the temperature of the sodium coolant are not uniform in the core, the swelling of the component will be nonuniform. This nonuniformity can influence the component's behavior.

Irradiation by neutrons causes marked changes in the properties of the zirconium alloys Zircaloy-2 and Zircaloy-4 (both very much used in light water reactors) and in 304 and 316 stainless steels (used in liquid metal fast-breeder reactors). Figure 4.11 shows the increase in strength (yield strength and ultimate tensile strength) of Zircaloy after neutron radiation. The exact nature of the defects introduced by radiation that are responsible for these changes in Zircaloy are not yet very well characterized. There is a considerable variation in the observed microstructures. One of the few observations about which there exists general agreement is the absence of radiation-induced vacancies in Zircaloy, which is a significant

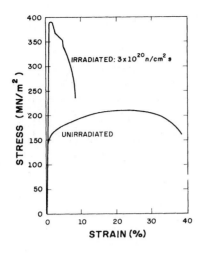

Figure 4.11 Stress–strain curves for irradiated and unirradiated Zircaloy. (Adapted with permission from J. T. A. Roberts, *IEEE Trans. Nucl. Sci.,* NS-22, (1975) 2219, p. 2223)

difference compared with, say, the behavior of stainless steels. Stainless steels show swelling due to neutron irradiation. The dilatation induced by neutron irradiation in stainless steel depends on the neutron flux and the temperature, as shown in Figure 4.12. It is believed that the vacancies introduced by irradiation combine to form voids, while the interstitials are preferentially attracted to dislocations. According to Shewmon[2] this dilatation of stainless steel does not affect the viability or security of breeder-type reactors, but will have a significant effect on core design and economy of reproduction. Preliminary results indicate that, in spite of not being able to eliminate the effect completely, cold work, heat treatments, or changes in composition can reduce the swelling by a factor of two or more. Figure 4.13 shows the change in dilatation of stainless steel as a function of Cr and Ni content.

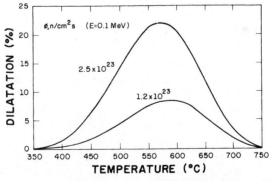

Figure 4.12 Stress-free dilatation in AISI 316 steel (20% cold worked). (Adapted with permission from J. T. A. Roberts, *IEEE Trans. Nucl. Sci.,* NS-22, (1975) 2219, p. 2223)

4.1.5 Ion Implantation

An interesting technological application using charged particles is called *ion implantation.* Charged ions are accelerated in an electric field (e.g., in a linear accelerator) to very high energies ($\sim$200 keV) and allowed to strike the target solid in a moderate vacuum ($\sim$1 mPa). It is worth emphasizing that the selected species of ions gets *implanted* into, and not deposited on, the target surface. The technique, originally developed for preparing semiconductor devices in a controlled fashion, has been made into a sophisticated tool for

[2] P. G. Shewmon, *Science,* 173 (1971) 987.

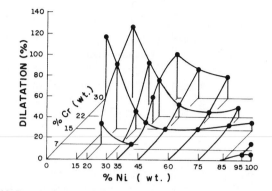

Figure 4.13 Dependence of fast neutron-induced dilatation in stainless steel (Fe–Cr–Ni) as a function of Ni and Cr amounts. (Adapted with permission from W. B. Hillig, *Science,* 191 (1976) 733)

altering the composition and structure of surfaces for any number of purposes—for example, modifying the surface chemistry for better corrosion and oxidation resistance, tribological properties, and superconductivity. The reader can well imagine the power of the technique by the fact that it allows one to introduce elements into a surface, which may not be possible in conventional heat treatment because of low diffusivity. Depending on the dose, B^+, N^+, and Mo^+ ions implanted into steel can reduce the wear of a tool by an order of magnitude. As an illustration, Figure 4.14 shows a gear being ion implanted.

Figure 4.14 Ion implantation of a gear. (Courtesy of G. Dearnaley)

The ion implantation technique of modifying the composition and structure of surfaces has a number of advantages over conventional techniques:

1. The process is essentially a cold one; therefore, there is no loss of surface finish and dimensions (i.e., the process can be applied to finished parts).
2. One can implant a range of metallic and nonmetallic ions, individually or combined.
3. One can implant selected critical areas.

Ion implantation is particularly suited for the selected modification of small, critical parts. Oil burners used for injecting a mixture of fuel oil and air into boilers of oil-fired power plants face rather severe erosion conditions. Ti and B implantation of oil-burner tips improved erosion properties and increased the service life of the boilers.

Another very important aspect of ion implantation has to do with the fact that it is basically a nonequilibrium process. There are thus no thermodynamic constraints, such as solubility limits. In other words, we are able to produce metastable alloys with new and unusual characteristics, amorphous alloys, and so on. Hence, the technique offers a novel way of producing surfaces, in a controlled manner, for scientific studies. This aspect of ion implantation, together with its obvious technological importance, should not be ignored.

4.2 LINE DEFECTS

Bands in the surface of plastically deformed metallic specimens were reported as early as the 19th century. With the discovery of the crystalline nature of metals, these bands were interpreted as being the result of the shear of one part of the specimen with respect to the other. Similar slip bands (or markings) were observed by geologists in rocks. However, calculations of the theoretical strength of crystals based on the simultaneous motion of all atoms along the slip band showed systematic deviations of several orders of magnitude with respect to the experimental values. (See Section 1.4.) This discrepancy led to the concept of line imperfections in crystals called *dislocations* later, the actual existence of such imperfections was verified by a variety of techniques.

Figure 4.15 presents two analogies that help us to visualize dislocations. The displacement of a rug can be accomplished by applying a much lower force if a wave is created

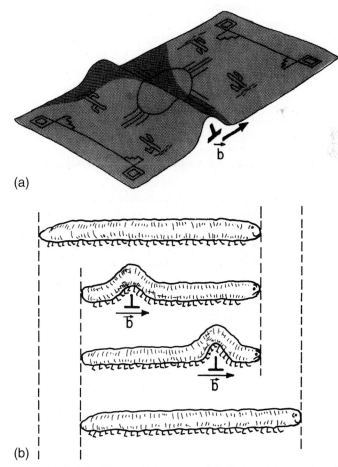

Figure 4.15 Models for a dislocation. (a) Rug with moving fold. (b) Caterpillar.

in the rug and moved from the back to the front. This displacement, **b,** is indicated in Figure 4.15a. In a similar manner, caterpillars move by creating a "dislocation" and displacing it from the back to the front (See Figure 4.15b). Sidewinders use a similar principle: These snakes generate "waves" along their bodies. The movement of the wave propels the snake sideways. Having understood this concept, the diligent student can readily comprehend how the movement of a dislocation in a body can produce plastic deformation.

Figure 4.16 shows two distinct types of dislocations encountered in crystalline solids: edge and screw dislocations. The atomic arrangement surrounding these dislocations is distorted from the regular periodicity of the lattice. The edge dislocation (Figure 4.16a and b) can be visualized as an extra half plane of atoms terminating at the dislocation line **l** (perpendicular to the plane of the paper and passing through the symbol "⊥"). The screw dislocation is best visualized as a "parking garage": A car, driving around the dislocation line **l,** will go up or down the building. Another analogy is the screw. Figure 4.16c shows the atomic arrangement. The distortion of the periodic atomic arrangement is represented by the Burgers vector **b.** A circuit is created around the dislocation line. In Figure 4.16b, this circuit is indicated by $ABCDE$. AB and CD correspond to $4a$, where a is the interatomic spacing. BC and DE correspond to $3a$. The failure of the circuit to close represents the vector **b.** A Burgers circuit is also represented around the screw dislocation in Figure 4.16c. The essential difference between these two types of dislocation is that in the edge dislocation, **b** is perpendicular to **l,** whereas in the screw dislocation, **b** is parallel to **l.**

Edge dislocations were proposed by Orowan, Taylor, and Polanyi in 1934.[3] Screw dislocations were proposed by Burgers in 1939.[4] Figure 4.17 shows how the shearing of the lattice can generate edge and screw dislocations. Imagine a cut made along $ABCD$ in Figure 4.17a. If the shearing direction is as marked in Figure 4.17b, the Burgers vector is perpendicular to line AB. Thus, **b** ⊥ **l,** and the resultant dislocation is of edge character. If the shearing direction, defined by **b,** is parallel to AB, then **b** // **l,** and the resulting dislocation is of screw character. (See Figure 4.17c.) The movement of an edge dislocation under an applied shear stress τ is shown in Figure 4.18. The perfect lattice (Figure 4.18a) is broken (Figure 4.18b), and the dislocation is formed. This edge dislocation (**b** ⊥ **l**) moves from left to right, and the final, deformed configuration is shown in Figure 4.18f. The relationship between the applied shear stress, the direction of movement of dislocation, and the plastic strain generated is quite different for the two types of dislocation. Figure 4.19 shows how a hypothetical crystal subjected to a shear stress τ undergoes a plastic deformation by means of the propagation of (a) an edge dislocation and (b) a screw dislocation. The direction of motion of the dislocations is indicated by **s.** This is parallel to **b** for the edge dislocation and perpendicular to **b** for the screw dislocation. The final shear is the same, but the motion of the two dislocations is completely different. There is also a mixed dislocation that possesses both screw and edge character. Figure 4.20 shows such a dislocation, together with a "cut." It can be seen that the shear direction is neither parallel (screw) nor perpendicular (edge) to the direction of the cut. The relationship between line AB and **b** defines a mixed dislocation.

Another type of dislocation is called a *helical dislocation.* It forms a large spiral and is sometimes observed in crystals that were heat treated to produce climb. "Climb" is the movement of a dislocation perpendicular to its slip plane. "Glide" is the movement along the slip plane. Climb is described in Chapter 15 (Creep). These dislocations are of mixed character; the reader should not confuse them with screw dislocations.

Dislocations will be studied in detail in this chapter, since they are the building blocks for the understanding of the mechanical response of metals. The treatment, however, still

[3] E. Orowan, *Z. Phys.,* 89 (1934) 603, 634.

[4] J. M. Burgers, *Proc. Kon. Ned. Akad. Wetenschap.,* 42 (1939) 293, 378.

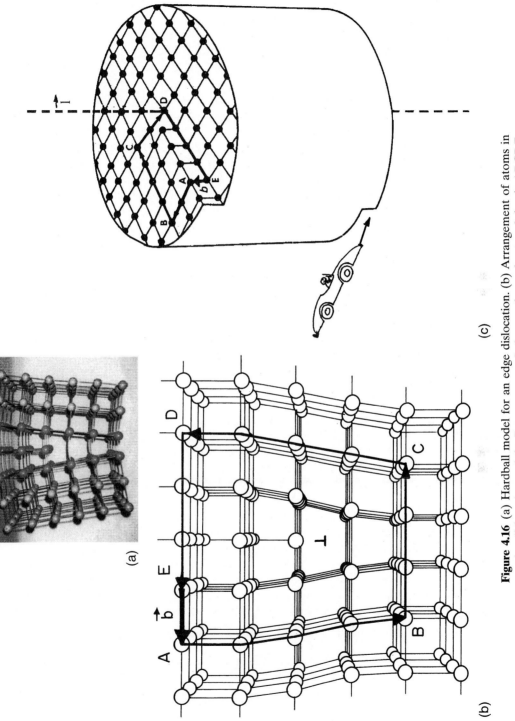

Figure 4.16 (a) Hardball model for an edge dislocation. (b) Arrangement of atoms in an edge dislocation and the Burgers vector **b** that produces closure of circuit *ABCDE*. (c) Arrangement of atoms in screw dislocation with "parking garage" setup. (Notice car entering garage).

(a)

(b)

(c)

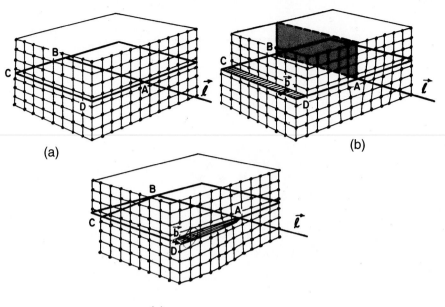

(a)   (b)

(c)

Figure 4.17 Geometrical production of dislocations. (a) Perfect crystal. (b) Edge dislocation. (c) Screw dislocation.

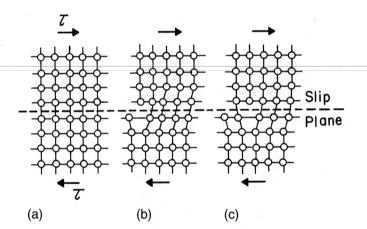

(a)   (b)   (c)

Figure 4.18 The plastic deformation of a crystal by the movement of a dislocation along a slip plane.

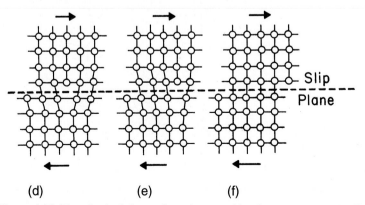

(d)   (e)   (f)

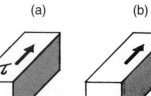

(a)　　(b)

Unsheared

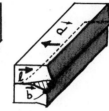

Partially sheared

Completely sheared

Figure 4.19 Plastic deformation (shear) produced by the movement of (a) edge dislocation and (b) screw dislocation. Note d is the direction of dislocation motion; ℓ is the direction of dislocation line.

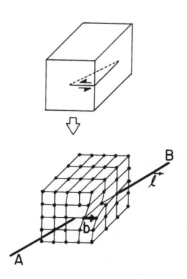

Figure 4.20 Mixed dislocation obtained from cut-and-shear operation; notice the angle between **b** and **l.**

is far from comprehensive. For further details, the reader is referred to the suggested readings at the end of the chapter.

4.2.1 Experimental Observation of Dislocations

It took 20 years to prove, beyond any doubt, the existence of dislocations experimentally, and this period (1935–1955) was surrounded by skepticism and harsh polemics. Nevertheless, the existence of dislocations is nowadays universally recognized, and the "lunatic" theories and models have been proven to be remarkably correct. A number of techniques have allowed the observation of dislocations, including etch pitting, X-ray diffraction (Berg–Barrett topography), and, most importantly, transmission electron microscopy (TEM). The last has established itself as the principal method for observing dislocations.

In TEM, the foil has to be thinned to a thickness between 0.1 and 0.3 μm, becoming transparent to electrons when the accelerating voltage is in the 100–300-kV range. Dislocations produce distortions of the atomic planes. Hence, for certain orientations of the foil with respect to the beam, the region around a dislocation diffracts the beam. The dislocations can then be seen as dark, thin lines under a bright field. TEMs with higher operating voltages (in the megavolt range) are available and allow thicker specimens to be observed. Figure 4.21 shows dislocations in titanium, nickel, and silicon rendered visible by this technique. Figure 4.21a shows dislocation segments in nickel. A variety of orientations is observed. The dislocations in titanium (Figure 4.21b) appear as sets of parallel segments; the segments are parallel because the dislocations minimize their energy by being along certain crystallographic planes. The same phenomenon is observed in silicon. (See Figure 4.21c.) A hardness indentation (lower right-hand corner) generated a profusion of dislocation loops. These loops are not circular, but consist of segments that are crystallographically aligned because of energy minimization considerations. The dislocation configurations in materials are highly varied and depend on a number of parameters, such as total strain, strain rate, strain state, deformation temperatures, crystallographic structure, etc. Note that the dislocations in silicon (Figure 4.21c) appear as white lines, whereas in Figure 4.21a and b they are dark lines. This is due to the fact that Figure 4.21c is a dark-field image, in which the grain diffracts, and the dislocation transmits, the electron beam. The figure is opposite to the normal bright-field transmission images (Figures 4.21a and b).

Dislocations are also present in ceramics, although they are less mobile. They can be produced by plastic deformation at high temperatures, by thermal stresses during cooling, or by applying very high stresses, made possible by, for instance, impacts at several hundred meters per second. Figure 4.22 shows dislocations observed in alumina and titanium carbide. The dislocations in the alumina were generated by impact at 600 m/s. The dislocations in the titanium carbide were produced by plastic deformation above the material's ductile-to-brittle transition temperature ($\sim$2,000°C). At room temperature, this ceramic would simply undergo brittle fracture.

High-resolution TEM can resolve the individual atoms and identify the lattice distortions around a dislocation. Figure 4.23 shows molybdenum imaged in such a fashion. The dark spots represent one atom each. Mo has the BCC structure, and the foil plane imaged is (100). The right-hand side of the picture shows a unit cell. A Burgers circuit is drawn around an edge dislocation, which has a line **l** perpendicular to the plane of the foil. The closure gap represents the Burgers vector of the dislocation. A comparison of the figure with the unit cell establishes the magnitude of the Burgers vector; it is equal to the lattice parameter a. This is clearly indicated in the figure. The presence of the dislocation can also be felt by noticing the break in the [110] planes, making 45° with the cube axes.

The electron micrographs of Figures 4.21–4.23 illustrate the presence and variety of dislocation configurations observed in crystalline materials.

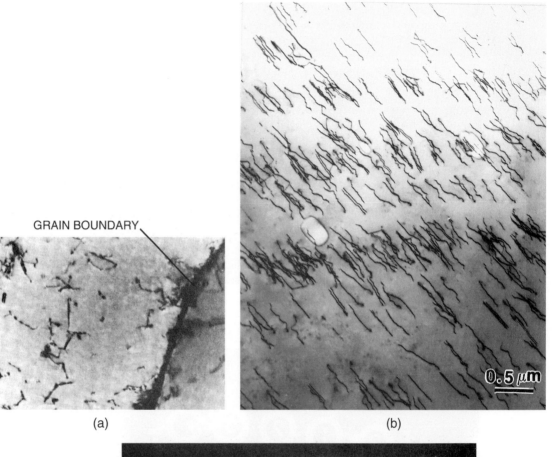

GRAIN BOUNDARY

(a)

(b)

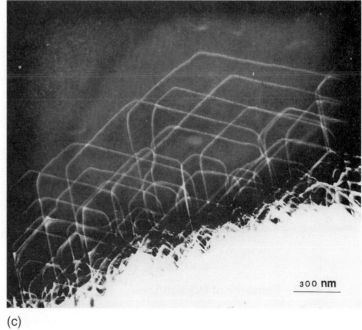

(c)

Figure 4.21 Dislocations in metals. (a) Nickel. (b) Titanium. (Courtesy of B. K. Kad)
(c) Silicon.

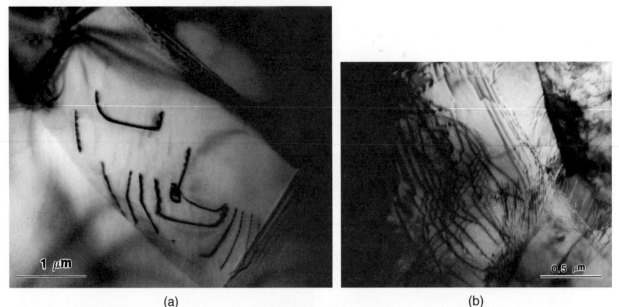

(a)   (b)

Figure 4.22 Dislocations in (a) Al_2O_3 and (b) TiC. (Courtesy J. C. LaSalvia)

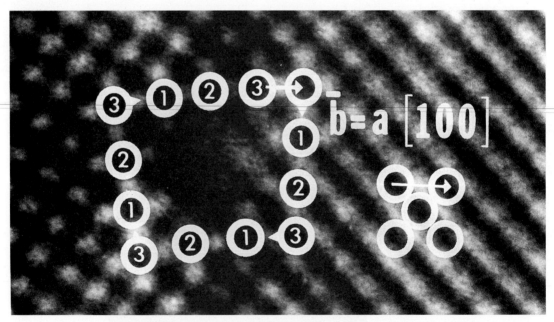

Figure 4.23 Atomic resolution transmission electron micrograph of dislocation in molybdenum with a Burgers circuit around it. (Courtesy of R. Gronsky).

4.2.2 Behavior of Dislocations

Dislocation Loops.   A dislocation line can form a closed loop, instead of extending until it reaches an interface or the surface of the crystal. This is illustrated in Figure 4.24, where a square loop is sketched. Two cuts, along perpendicular sections, were made: *AAAA* and *BBBB*. Figure 4.24(b) and (c) show these sections. It can clearly be seen that the dislocation segments *CF* and *DE* (Figure 4.24b) are of edge character, while segments

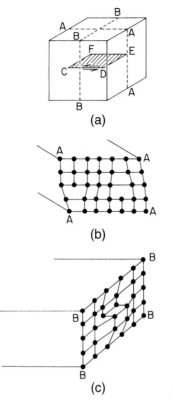

Figure 4.24 Square dislocation loop.

CD and *FE* (Figure 4.24c) are of screw character. This is due to the direction of the shear. The loop can be imagined as a cut made in the interior of the crystal (an impossible feat, of course); the edges of the cut form the dislocation line, after shear is applied to the crystal. Dislocations *CF* and *DE* are of the same type, with opposite signs; the same applies to *CD* and *FE*. The sign convention used for edge dislocations is the following: If the extra semiplane (wedge) is on the top portion, it is positive; if on the bottom, it is negative. Hence, *CF* is positive and *DE* is negative. For screw dislocations, a similar convention is used. If the helix turns in accord with a normal screw, it is positive. If not, it is negative. According to this convention, *CD* is positive and *FE* is negative.

The actual dislocation loops are not necessarily square. An elliptical shape would be more favorable energetically than a square. For an elliptical or circular shape, the character of the dislocation changes continuously along the line. Figure 4.25a shows this situation; the regions that are edge and screw are shown by appropriate symbols. The symbols most commonly used are an inverted T ($\perp$) for a positive edge, and an S for a positive screw dislocation. The negative signs can be described by a correct T and by an inverted S (ς). In Figure 4.25a, all the portions of the loop between the short segments of pure screw and edge character are mixed. These loop segments move as shown in the figure. The loop expands and eventually "pops out" of the parallelepiped, creating the shear shown in Figure 4.25b.

There is another type of loop, called a *prismatic loop,* that should not be confused with a common loop. A prismatic loop is created when a disk of atoms is either inserted or removed from the crystal. Figure 4.26 shows this situation; cuts *AAAA* and *BBBB* are indicated. A disk having the thickness of one atomic layer was introduced and it can be seen that sections *AAAA* (Figure 4.26b) and *BBBB* (Figure 4.26c) are identical. They are edge dislocations with opposite signs. This configuration is very different from that encountered in normal loops. One can also *remove* a disk of atoms, instead of adding it. These loops do not have the same ability to move as do normal loops.

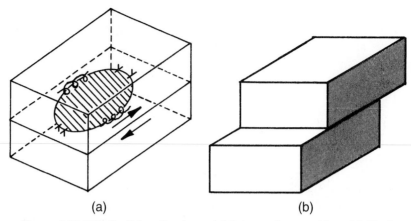

Figure 4.25 Elliptic dislocation loop. (a) Intermediate position. (b) Final (sheared) position.

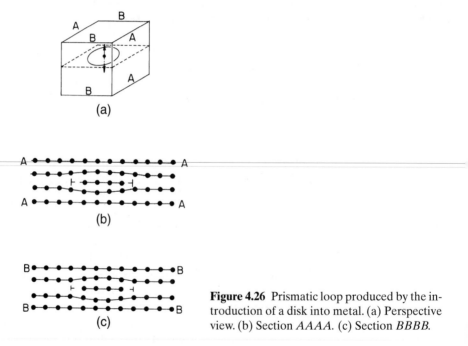

Figure 4.26 Prismatic loop produced by the introduction of a disk into metal. (a) Perspective view. (b) Section *AAAA*. (c) Section *BBBB*.

Movement of Dislocations.   The plastic deformation of metals is normally accomplished by the movement of dislocations. The elements of dislocation motion are reviewed in this section, together with the resulting deformations. In actual deformation and for elevated strains, complex interactions occur between dislocations. These interactions can be broken down into simple basic mechanisms that will be described next. Two edge dislocations are shown in Figure 4.27a. After the passage of one of them, one part of the lattice is displaced in relation to the other part by a distance equal to the Burgers vector. Both a positive and a negative dislocation can generate the same shear; however, they have to move in opposite directions in order to accomplish this. The reader is reminded (see Figure 4.19a) that the shear and motion directions are the same for edge dislocation.

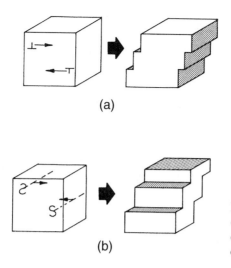

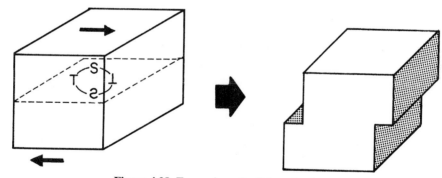

Figure 4.27 Slip produced by the movement of dislocation. (a) Positive and negative edge dislocations. (b) Positive and negative screw dislocations.

Screw dislocations can produce the same lattice shear (Figure 4.27b). However, in this case the shear takes place perpendicular to the direction of motion of the dislocations; positive and negative screw dislocations have to move in opposite directions in order to produce the same shear strain.

The plane in which a dislocation moves is called a *slip plane*. The slip plane and the loop plane coincide in Figure 4.28. A loop will eventually be ejected from a crystal upon expanding if there is no barrier to its motion. The expansion of a loop will produce an amount of shear in the crystal equal to the Burgers vector of the dislocation. It is worth noting that the shears of the different dislocations are all compatible; there is no incompatibility of movement.

Figure 4.28 Expansion of a dislocation loop.

The prismatic loops, consisting totally of edge dislocations, cannot expand like the normal loops. Thus, because the plane of the dislocation does not coincide with the loop plane, the coupled movement of the edge dislocations will force the loop to move perpendicular to its plane, maintaining the same diameter. Upon being ejected from the crystal, a step will be formed at the surface. Figure 4.29 shows a succession of vacancy loops.

4.2.3 Stress Field Around Dislocations

Dislocations are defects; hence, they introduce stresses and strains in the surrounding lattice of a material. The mathematical treatment of these stresses and strains can be substantially simplified if the medium is considered to be isotropic and continuous. Under

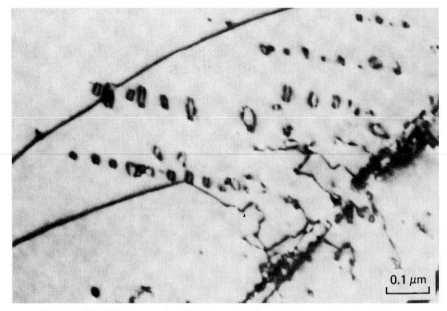

Figure 4.29 Nickel heated at 600°C for 10 min and quenched in liquid nitrogen. Strings of vacancy loops can be clearly seen. (Courtesy of L. E. Murr)

conditions of isotropy, a dislocation is completely described by the line and Burgers vectors. With this in mind, and considering the simplest possible situation, dislocations are assumed to be straight, infinitely long lines. Figure 4.30 shows hollow cylinders sectioned along the longitudinal direction. Different deformations are applied in the two cases. The one in Figure 4.30a portrays the deformation around a screw dislocation, while Figure 4.30b is an idealization of the strains around an edge dislocation. The cylinders, with external radii R, were longitudinally and transversally displaced by the Burgers vector b, which is parallel (perpendicular) to the cylinder axis in the representation of a screw (an edge) dislocation. In either case, an internal hole with radius r_0 is made through the center. This is done to simplify the mathematical treatment. In a continuous medium, the

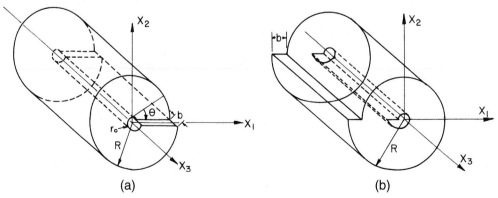

(a)   (b)

Figure 4.30 Simple models for (a) screw and (b) edge dislocations; the deformation fields can be obtained by cutting a slit longitudinally along a thick-walled cylinder and displacing the surface by **b** parallel (screw) and perpendicular (edge) to the dislocation line.

stresses on the center would build up and become infinite in the absence of a hole; in real dislocations the crystalline lattice is periodic, and this does not occur. In mechanics terminology, this is called a *singularity,* A "singularity" is a spike, or a single event. For instance, the Kilimanjaro is a singularity in the African plains. Therefore, we "drill out" the central core, which is a way of reconciling the continuous-medium hypothesis with the periodic nature of the structure. To analyze the stresses around a dislocation, we use the formal theory of elasticity. For that, one has to use the relationships between stresses and strains (constitutive relationships), the equilibrium equations, the compatibility equations, and the boundary conditions. Hence, the problem is somewhat elaborate. We do not present the derivation of these relationships here; both Kuhlmann-Wilsdorf, and Weertman and Weertman (see the suggested readings), provide easy-to-follow derivations.

The strains are obtained from the elastic deformation of the material. The stresses around a screw dislocation are

$$\sigma_{13} = \sigma_{31} = -\frac{Gbx_2}{2\pi(x_1^2 + x_2^2)}, \tag{4.4}$$

$$\sigma_{23} = \sigma_{32} = \frac{Gbx_1}{2\pi(x_1^2 + x_2^2)}. \tag{4.5}$$

The stresses around an edge dislocation are

$$\sigma_{11} = -\frac{Gbx_2(3x_1^2 + x_2^2)}{2\pi(1 - \nu)(x_1^2 + x_2^2)^2}, \tag{4.6a}$$

$$\sigma_{12} = \frac{Gbx_1(x_1^2 - x_2^2)}{2\pi(1 - \nu)(x_1^2 + x_2^2)^2}, \tag{4.6b}$$

$$\sigma_{22} = \frac{Gbx_2(x_1^2 - x_2^2)}{2\pi(1 - \nu)(x_1^2 + x_2^2)^2}. \tag{4.6c}$$

It then follows that

$$\sigma_{33} = \nu(\sigma_{11} + \sigma_{22}) = -\frac{Gb\nu x_2}{\pi(1 - \nu)(x_1^2 + x_2^2)}. \tag{4.6d}$$

These stresses are shown in Figure 4.31 through isostress lines.

4.2.4 Energy of Dislocations

The elastic deformation energy of a dislocation can be found by integrating the elastic deformation energy over the whole volume of the deformed crystal. The deformation energy is given by

$$U = \frac{1}{2}\sigma_{ij}\varepsilon_{ij}. \tag{4.7}$$

For an isotropic material, converting the strain to stresses, we have

$$U = \frac{1}{2G}\left[\frac{1}{2(1 + \nu)}(\sigma_{11}^2 + \sigma_{22}^2 + \sigma_{33}^2) + (\sigma_{12}^2 + \sigma_{13}^2 + \sigma_{23}^2)\right.$$

$$\left. - \frac{\nu}{(1 + \nu)}(\sigma_{11}\sigma_{33} + \sigma_{11}\sigma_{22} + \sigma_{22}\sigma_{33})\right].$$

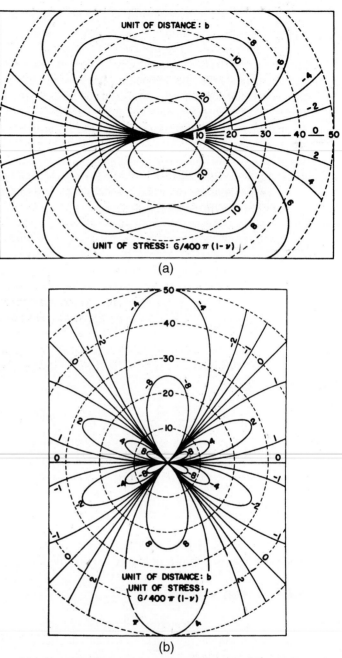

(a)

(b)

Figure 4.31 Stress fields around an edge dislocation. (The dislocation line is Ox_3), (a) σ_{11}; (b) σ_{22}; (c) σ_{33}; (d) σ_{12}. (Adapted with permission from J. C. M. Li, in *Electron Microscopy and Strength of Crystals,* G. Thomas and J. Washburn (eds.) New York: Interscience Publishers, 1963.)

Using Equations 4.4 and 4.5 we have, for a screw dislocation,

$$U_s = \frac{1}{2G}\left[\frac{G^2 b^2 x_2^2}{4\pi^2(x_1^2 + x_2^2)^2} + \frac{G^2 b^2 x_1^2}{4\pi^2(x_1^2 + x_2^2)^2}\right] \qquad (4.8)$$

$$= \frac{Gb^2}{8\pi^2(x_1^2 + x_2^2)}.$$

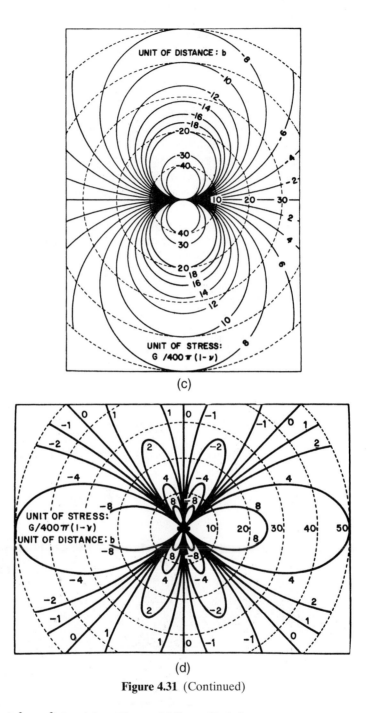

(c)

(d)

Figure 4.31 (Continued)

Substituting $(x_1^2 + x_2^2)$ by r^2 (see Figure 4.30), we find that

$$U_s = \frac{Gb^2}{8\pi^2 r^2}.$$ (4.9)

Integrating Equation 4.9 between r_0 and R, we get

$$U_s = \int_{r_0}^{R} \frac{Gb^2}{8\pi^2 r^2} 2\pi r dr = \frac{Gb^2}{4\pi} \ln \frac{R}{r_0}.$$ (4.10)

In a similar way, the energy of a straight edge dislocation per unit length is equal to

$$U_\perp = \frac{Gb^2}{4\pi(1-\nu)} \ln \frac{R}{r_0}. \tag{4.11}$$

It should be observed that the factor $(1-\nu)$ is approximately equal to $2/3$. Hence, the energy of an edge dislocation is about $3/2$ of that of a screw dislocation.

The schematic drawing of Figure 4.30 removes the core of the dislocation so as to avoid the infinite stresses along the dislocation line. Several methods have been used to estimate r_0. In this book, r_0 will be assumed to be equal to $5b$. Note that the energy given by the foregoing equations become infinite for infinite R; hence, one has to establish an approximate value for R. Dislocations in a metal never occur in a completely isolated manner; they form irregular arrays with mean density ρ. This density is derived as the total length of dislocation line per unit volume. The spaghetti analogy can be used here. Imagine a pot with water and spaghetti. The density of the spaghetti would be obtained by measuring the total length of the spaghetti and dividing it by the volume of the pot. The stress fields of the various dislocations interact, as will be seen in subsequent sections; we generally assume a value of R equal to the average distance between the dislocations. It can be shown, by means of a simplified array, that the average distance or mean free path of dislocations is approximately equal to $\rho^{-1/2}$.

It is possible to calculate the radius of influence of each dislocation line, R, from the dislocation density ρ. This radius of influence is equal to $L/2$, in Figure 4.32. Figure 4.32(a)

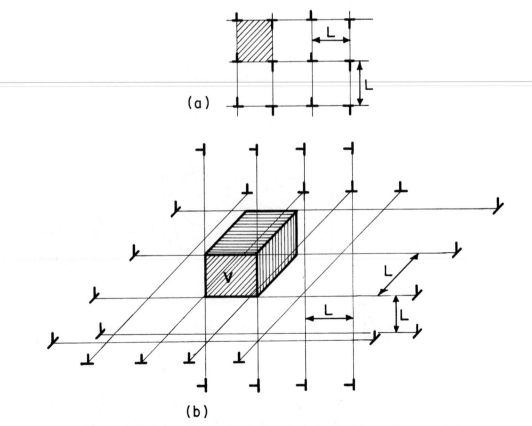

(a)

(b)

Figure 4.32 Schematic representation of an idealized dislocation array (a) in two dimensions and (b) in three dimensions; note that dislocations on three perpendicular atomic planes define a volume V.

shows a two-dimensional array of dislocations; all dislocation lines "poke out" of the plane of the page. The mean spacing is L, and the hatched area is L^2. This area is bounded by four dislocations, and each dislocation is shared by four areas. Thus,

$$L^2 \text{ area} \rightarrow 1 \text{ dislocation,}$$

$$\text{unit area} \rightarrow \rho \text{ dislocations.}$$

As a result,

$$\rho = L^{-2}. \tag{4.12}$$

The tridimensional calculation is slightly more complicated. Figure 4.32(b) shows a tridimensional array of dislocations. The hatched volume is $V = L^3$. This volume is composed of dislocations that lie along the edges. The total dislocation length can be taken to be $12\,L$. However, each dislocation is shared by four adjacent cubes. Hence,

$$\rho = \frac{12L/4}{L^3} = 3L^{-2}. \tag{4.13}$$

But

$$R = \frac{L}{2},$$

so that

$$\rho = 3(2R)^{-2}$$

and

$$R = \frac{1}{2}\left(\frac{\rho}{3}\right)^{-1/2} = 0.86\rho^{-1/2}.$$

The average dislocation radius is often taken to be

$$R \approx \rho^{-1/2}.$$

We now add the energy of the dislocation core. This energy is taken to be $Gb^2/10$ for metals. Hence, the total energy of a dislocation is

$$U_r = U_{\text{nucleus}} + U_{\text{periphery}}.$$

Equation 4.11 can be generalized to:

$$U_r = \frac{Gb^2}{10} + \frac{Gb^2}{4\pi(1-\nu)}(1 - \nu\cos^2\alpha)\ln\frac{\rho^{-1/2}}{5b}, \tag{4.14}$$

where α is a parameter that describes the nature of the dislocation (edge $\alpha = \pi/2$, screw $\alpha = 0$), which can be mixed.

The energy of dislocations is often taken to be approximately

$$U_r = \frac{Gb^2}{2}.$$

(4.15)

For typical metals, U_r is equal to a few electron volts per atomic plane. The energy of the nucleus is 10% of this total. The energy of a dislocation per atomic plane is high in comparison with that of a vacancy: approximately 3 eV (4.8×10^{-19} J) versus about 1 eV (1.6×10^{-19} J). (See Section 4.1.2.)

EXAMPLE 4.2

Annealed materials have a dislocation density of approximately 10^8 cm^{-2} or 10^{12} m^{-2}. Calculate the total strain energy for copper.

Solution: For copper, the Burgers vector is $b = 0.25$ nm. Inserting these values into Equation 4.14 and using $\alpha = 0$ (for a screw dislocation), we obtain

$$U = 0.1Gb^2 + \frac{Gb^2}{4\pi} \ln \frac{10^{-6}}{5 \times 0.25 \times 10^{-9}}$$

$$= 0.63Gb^2 = \frac{Gb^2}{1.87} \cong \frac{Gb^2}{2}.$$

(4.16)

For this example, the energy per unit length is equal to 1.5×10^{-9} J/m ($G = 48.3$ GPa). The total strain energy is $1.5 \times 10^{-9} \times 10^{12}$ J/m^3.

Two additional equations will be derived next: the force required to curve a dislocation to a radius R and the Peach-Koehler equation. The analogy of a string helps to explain the energy of a dislocation. In the absence of an external stress field, a dislocation will tend to be straight, minimizing its length and overall energy. The same occurs for a string under tension. If the string is pushed by a force, it will exert a force back. Thus, a curved dislocation is said to possess a "line tension," which can be calculated. The energy of a curved dislocation with radius R can be calculated (see Weertman and Weertman, p. 50, in the suggested readings) and is equal to

$$U = \frac{Gb^2}{4\pi} \ln \frac{R}{5b}.$$

(4.17a)

It is possible to calculate the force F required to bend a dislocation into a radius R. Figure 4.33 shows a curved dislocation with radius R. The line tension T is defined as the self-energy per unit length of dislocation. In the figure, the segment of the dislocation ds is "sectioned off," and the remaining dislocation is replaced by two tensions T acting tan-

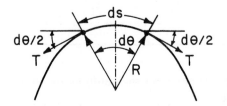

Figure 4.33 Curved dislocation.

gentially to the line at the section points. The line tension is always tangential to the dislocation line. The (downward) vertical force exerted by the line tension on the segment ds is

$$F_1 = 2T \sin(d\theta/2).$$

This is balanced by the force F_2 (per unit length) exerted on the dislocation, multiplied by its length:

$$F_2\, ds = 2T \sin(d\theta/2).$$

Since $d\theta/2$ is a small quantity,

$$F_2 ds = T\, d\theta.$$

But

$$R\, d\theta = ds,$$
$$F_2 R\, d\theta = T\, d\theta,$$
$$F_2 = T/R.$$

Assuming, as a first approximation, that the energy of the curved dislocation is equal to that of a straight dislocation, we have

$$F = Gb^2/2R. \tag{4.17b}$$

The Peach–Koehler equation $F = \tau b$ relates the force applied to a dislocation to a stress. F is the force per unit length of dislocation, and τ is the shear stress acting on the slip plane along the slip direction. This relation can be demonstrated by considering a parallelepiped with dimensions dx, dy, dz. If one dislocation, with length dx, on which a force per unit length is F, moves through the parallelepiped, the work done is

$$W = (F\, dx)\, dy.$$

The change in strain energy of the cube is

$$U = (\tau\gamma)\, dx\, dy\, dz,$$

where $dx\, dy\, dz$ is the volume of the parallelepiped. The shear strain produced by one dislocation is

$$\gamma = b/dz.$$

Since $W \equiv U$,

$$(F\, dx)\, dy = (\tau b/dz)\, dx\, dy\, dz,$$

and

$$F = \tau b. \tag{4.17c}$$

By applying the Peach–Koehler equation to Equation 4.17b, we get

$$\tau = Gb/2R. \tag{4.17d}$$

4.2.5 Dislocations in Various Structures

Dislocations in Face-Centered Cubic Crystals. In Section 1.3.2, we saw that, among the 80 or so metals, 55 are FCC. The FCC structure is the closest packed one, together with the HCP structure. Thus, it is natural that dislocations be more carefully studied for the FCC structure.

When we visualize a dislocation, we generally think of a defect that, upon passing, recomposes the original structure of the crystal. Hence, in a simple cubic structure, the Burgers vector would have the direction [100] and magnitude a (lattice parameter). However, there are cases in which the original structure is not recomposed. This type of dislocation is called *imperfect* or *partial*.

In FCC crystals, the closest packed planes are (111). These planes are usually termed A, B, and C, depending on their order in the stacking sequence. Figure 4.34 shows an atomic plane A. The glide movement of the atoms of the plane A that would recompose the same lattice would be indicated by the Burgers vector b_1. This vector has the direction $[10\bar{1}]$. Its magnitude is

$$d_{hkl} = \frac{a}{\sqrt{h^2 + k^2 + l^2}} = \frac{a}{\sqrt{2}}, \tag{4.18}$$

which is equal to the distance between $(10\bar{1})$; in cubic structures, planes and directions with the same indices are perpendicular. Vector b_1 is expressed with respect to unit vectors $i, j,$ and k of the coordinate system $Ox_1x_2x_3$ as

$$b_1 = \frac{a}{2}\mathbf{i} + 0\mathbf{j} - \frac{a}{2}\mathbf{k} = \frac{a}{2}(\mathbf{i} - \mathbf{k}). \tag{4.18a}$$

It can be seen that the magnitude is

$$|\mathbf{b_1}| = \frac{a}{\sqrt{2}}.$$

This vector is, logically, the same as that of Eq. 4.18. The simplified notation used for Burgers vectors is

$$b_1 = \frac{a}{2}[10\bar{1}] \quad \text{or} \quad b_1 = \frac{1}{2}[10\bar{1}].$$

Hence, the term in brackets gives the direction of the vector, while the term that precedes it is the same fraction as that used in the definition of the unit vectors **i, j,** and **k** (see Eqn. 4.18a). There is also a graphic method to determine this fraction. First, one draws the vec-

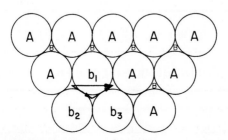

Figure 4.34 Decomposition of dislocation.

tor **b** connecting point $(0, 0, 0)$ to point $(1, 0, -1)$. Then one draws b_1, which will be a fraction of b (in this case, half). The fraction is the term that precedes the bracketed term.

One possibility of decomposition for the dislocation is shown in Figure 4.34, where b_2 and b_3 add up to b_1. Both b_2 and b_3 define partial dislocations, because they change the stacking sequence ABC. But, acting together (or sequentially), they would have the same effect as b_1 and maintain the correct stacking sequence. b_2 and b_3 are:

$$\mathbf{b}_2 = \frac{a}{6}(2\mathbf{i} - \mathbf{j} - \mathbf{k})$$

$$\mathbf{b}_3 = \frac{a}{6}(\mathbf{i} - \mathbf{j} - 2\mathbf{k})$$

and $\mathbf{b}_3 = \mathbf{b}_2 + \mathbf{b}_3$

It is easy to establish whether $\mathbf{b}_1$, $\mathbf{b}_2$, and $\mathbf{b}_3$ belong to (111): the scalar product should be zero, because [111], which is perpendicular to (111), should also be perpendicular to $\mathbf{b}_1$, $\mathbf{b}_2$, and $\mathbf{b}_3$.

$$d_{hkl} = \frac{a}{\sqrt{h^2 + k^2 + l^2}} = \frac{a}{\sqrt{6}}.$$

Hence, we have the following possible reaction:

$$\frac{a}{2}[10\bar{1}] \rightarrow \frac{a}{6}[2\bar{1}1] + \frac{a}{6}[11\bar{2}].$$

From Eq. 4.15, the energy is $Gb^2/2$. Taking the square of the magnitude of the Burgers vectors yields

$$\frac{a^2}{2} > \frac{a^2}{6} + \frac{a^2}{6},$$

and we can see that the total energy decreases with decomposition.

When a perfect dislocation decomposes itself into partials, a region of faulty stacking is created between the partials. This decomposition is shown in Figure 4.35. The dislocations generate a region in which the stacking is $ABC\,AC\,ABC$. Hence, we have four planes in which the stacking is $CACA$. This is exactly the stacking sequence of the HCP structure. This structure has a higher Gibbs free energy than the equilibrium FCC structure, because it is not thermodynamically stable under the imposed conditions. This specific array of planes is called the *stacking fault,* and the energy associated with it determines the separation between the two partial dislocations: The repulsive force between the two partials is balanced by the attraction trying to minimize the region with the stacking fault. The following equations from [Murr[5] and Kelly and Groves, (see the suggested readings) respectively], allow the calculation of the equilibrium separation between the partial dislocations d:

$$\gamma_{SF} = \frac{G|b_p|^2}{8\pi d}\left[\frac{2 - \nu}{1 - \nu}\left(1 - \frac{2\nu\cos 2\theta}{2 - \nu}\right)\right],$$

$$\gamma_{SF} = \frac{Gb_1 b_2}{2\pi d}\left(\cos\theta_1 \cos\theta_2 + \frac{\sin\theta_1 \sin\theta_2}{2 - \nu}\right).$$

Here, γ is the stacking-fault free energy (SFE) per unit area (free energy of HCP minus free energy of FCC), b_p is the Burgers vector of the partial dislocation, and θ is the angle of the

[5]L. E. Murr, *Interfacial Phenomena in Metals and Alloys,* Addison-Wesley, Reading, Massachusetts, 1975, p. 142.

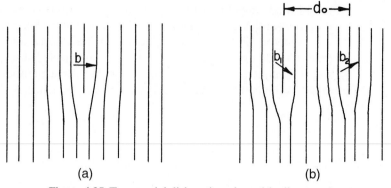

Figure 4.35 Two partial dislocations b_1 and b_2 distance d_0.

TABLE 4.1 Stacking Fault Free Energies and Separation between Shockley Partials for Metals ($\theta = 30°$)[a]

Metal	γ (mJ/m²)	a_0 (nm)	b (nm)	G (GPa)	d (nm)
Aluminum	166	0.41	0.286	26.1	1
Copper	78	0.367	0.255	48.3	3.2
Gold	45	0.408	0.288	27.0	Sec.
Sec. 4.0					
Nickel	128	0.352	0.249	76.0	2.9
Silver	22	0.409	0.289	30.3	9
Stainless steels					

[a] Adapted from L. E. Murr, *Interfacial Phenomena in Metals and Alloys,* Reading, MA: Addison-Wesley, 1975.

Burgers vector with the dislocation line. Table 4.1 presents the SFEs for some materials. From the preceding equations, it can be seen that d is inversely proportional to γ. The effect of alloying elements is generally to decrease the SFE. The addition of aluminum to copper has a drastic effect on the latter's SFE, dropping it from 78 to 6 mJ/m². Aluminum, which has a high SFE (166 mJ/m²), does exhibit a very small separation between partials: 1 nm. On the other hand, in certain alloy systems, the distance can go up to 10 nm or more.

The stacking-fault energy is very sensitive to composition. Usually, alloying has the effect of decreasing the SFE. Hence, brasses have an SFE lower than that of copper, and Al alloys have an SFE lower than that of Al.

Figure 4.36 shows some stacking faults in AISI 304 stainless steels viewed by transmission electron microscopy. The region corresponding to the stacking fault can be clearly seen by the characteristic fringe (////) pattern. The extremities of the fringes are bound by the partial dislocations. In Figure 4.36a, the stacking fault lies parallel to a coherent twin boundary, which is much longer than the stacking fault. The fault can be distinguished from the coherent twin boundary by the differences in fringe contrast. While all the fringes of the stacking fault are dark, the ones in the twin are dark at the top and become successively lighter. Figure 4.36b shows a number of dislocations (probably emitted from the same source) whose segments are trapped on the foil. These segments have decomposed into partials, and one can clearly distinguish the stacking-fault regions by the characteristic fringe contrast.

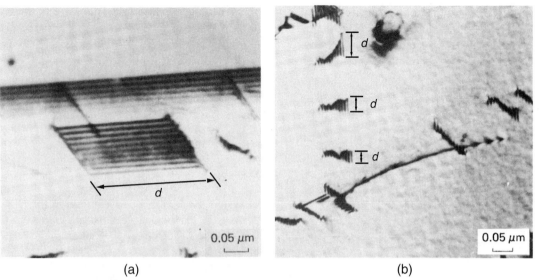

(a)   (b)

Figure 4.36 (a) Short segment of stacking fault in AISI 304 stainless steel overlapping with coherent twin boundary. Differences in the nature of these defects are illustrated by fringe contrast differences. (b) Dislocations in AISI 304 stainless steel splitting into partials bounded by short stacking-fault region. Partials spacing marked as *d*. (Courtesy of L. E. Murr)

The effect of the stacking-fault energy on the deformation substructure can be seen in Figure 4.37. This figure shows (a) an Fe–34.5% Ni and (b) an Fe–15% Cr–15% Ni alloy after deformation by shock loading under identical conditions (7.5 GPa peak pressure, 2 μs pulse duration). The Fe–15% Cr–15% Ni alloy has a significantly lower stacking-fault energy than does the Fe–34.5% Ni alloy, and the resultant deformation substructures seem to be strongly affected by this difference. Low-SFE metals tend to exhibit a deformation substructure characterized by banded, linear arrays of dislocations, whereas high-SFE metals tend to exhibit dislocations arranged in tangles or cells. Cross-slip is more difficult in low-SFE alloys because the dislocations have to constrict in order to change slip planes. (See Chapter 6.) Therefore, the dislocations arrange themselves into parallel bands. The SFE also affects the work-hardening of alloys.

Another type of dislocation in FCC structures is called a *sessile* or *Frank dislocation*, which is immobile. Sessile or Frank dislocations appear under two specific conditions, shown in Figure 4.38. In Figure 4.38a, a disk was removed in plane (111); in Figure 4.38b, a disk was added. It can be seen that in both cases the stacking sequence was changed, to *ABCBCA* and *ABCBABC* for Figures 4.38a and b, respectively. The Burgers vector is determined by applying Equation 4.18a:

$$b = \frac{a}{3}[111].$$

We have a sample of an *intrinsic* stacking fault in Figure 4.39a and an *extrinsic*, or double-stacking fault in Figure 4.38b. Since the Burgers vector is not in the slip plane, the two faults are immobile. Another type of immobile dislocation that can occur in FCC metals is the

Lomer–Cottrell lock. Let us consider two (111) and ($11\bar{1}$) planes. The three perfect dislocations on (111) are

$$b_1 = \frac{a}{2}[1\bar{1}0],$$

$$b_2 = \frac{a}{2}[\bar{1}01],$$

$$b_3 = \frac{a}{2}[01\bar{1}],$$

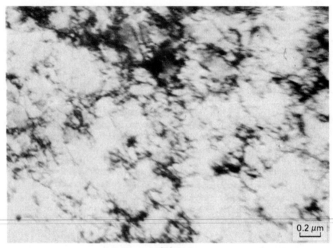

(a)

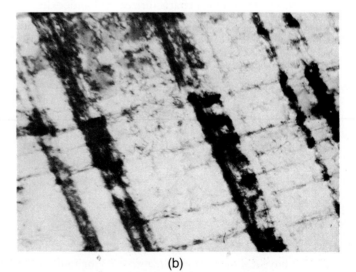

(b)

Figure 4.37 Effect of stacking-fault energy on dislocation substructure. (a) Higher-stacking-fault-energy alloy (Fe–34% Ni). (b) Lower-stacking-fault-energy alloy (Fe–15% Cr–15% Ni). (Both alloys shock-loaded at 7.5 GPa pressure and 2-μs pulse duration).

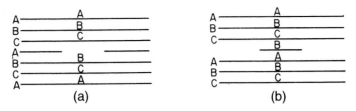

Figure 4.38 Frank or Sessile dislocations. (a) Intrinsic. (b) Extrinsic.

For plane $(11\bar{1})$, we have

$$b_4 = \frac{a}{2}[\bar{1}10],$$

$$b_5 = \frac{a}{2}[101],$$

$$b_6 = \frac{a}{2}[011].$$

One good rule to determine whether a direction belongs to a plane is that the scalar product between the direction **b** and the normal to the plane must be zero (in a cubic structure). This rule comes from vector calculus. Vectors $\mathbf{b}_1$ and $\mathbf{b}_4$ have the same direction and opposite senses; the common direction is also that of the intersection of the two planes. Hence, both dislocations will cancel when they encounter each other. The combination of $\mathbf{b}_2$ and $\mathbf{b}_5$ would result in

$$\mathbf{b}_2 + \mathbf{b}_5 = \frac{a}{2}[\bar{1}01] + \frac{a}{2}[101] = \frac{a}{2}[002] = a[001].$$

The energy of these dislocations is

$$\frac{a^2}{2} + \frac{a^2}{2} = a^2.$$

Therefore, this reaction will not occur, because it would not result in a reduction of the energy. The sole combinations that *would* result in a decrease in the overall energy would be of the type

$$\mathbf{b}_3 + \mathbf{b}_5 = \frac{a}{2}[01\bar{1}] + \frac{a}{2}[101]$$

$$= \frac{a}{2}[110].$$

$$\frac{a^2}{2} + \frac{a^2}{2} > \frac{a^2}{2}.$$

This reaction, which is energetically favorable, is shown in Figure 4.39. The dislocation is not mobile in either the (111) or $(11\bar{1})$ plane; hence, it acts as a barrier for any additional dislocation moving in these planes. Since it impedes slip, it is called a Lomer–Cottrell "lock."

The resultant configuration is shown in Figure 4.40; it resembles a stair and is therefore called a "stair-rod" or "stairway" dislocation. The leading partials react and immobilize

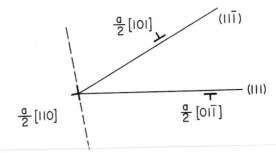

Figure 4.39 Cottrell–Lomer lock.

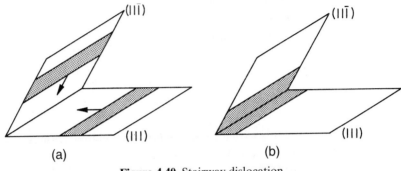

Figure 4.40 Stairway dislocation.

the partials coupled to them (the trailing partials). The bands of stacking faults form a configuration resembling steps on a stairway. These steps are barriers to further slip on the atomic planes involved, as well as in the adjacent planes.

Dislocations in Hexagonal Close-Packed Crystals.   In HCP crystals, the stacking sequence of the most densely packed planes is *ABAB*. (See Section 1.3.1.) These planes are known as basal planes. (See Figure 1.10.) Figure 4.41 shows the main planes in the HCP structure. Perfect dislocations moving in the basal plane can decompose into Shockley partials, just as in the FCC structure. Stacking faults are also formed (only *intrinsic* stacking faults). This analogy can be easily understood if one realizes the similarity between the two structures. The (111) planes in the FCC structure are the equivalent of the basal planes in the HCP structures. A perfect dislocation in the basal plane has the Burgers vector

$$b = \frac{a}{3}[2\ \bar{1}\ \bar{1}\ 0].$$

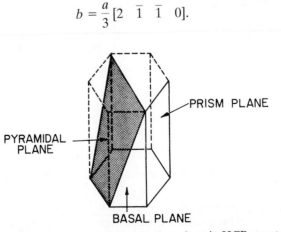

Figure 4.41 Basal, pyramidal, and prism plane in HCP structure.

In an ideal crystal, the c/a ratio is 1.633. However, in real crystals this never happens. It has been experimentally observed that, for crystals with $c/a > 1.633$, slip occurs mainly in the basal plane, while the pyramidal and prism planes are "preferred" in crystals with $c/a < 1.633$. This is due to the dependence of the distance between the atoms upon c/a; it is well known that the dislocations tend to move in the highest packed planes. A detailed treatment of dislocations in HCP metals is given by Teutonico.[6]

Dislocations in Body-Centered Cubic Crystals. In BCC crystals, the atoms are closest to each other along the $\langle 111 \rangle$ direction. (See Figure 1.12b.) Any plane in the BCC crystal that contains this direction is a suitable slip plane. Slip has been experimentally observed in (110), (112), and (123) planes. The slip markings in BCC metals are usually wavy and ill defined. The following reaction has been suggested for a perfect dislocation having its Burgers vector along $\langle 111 \rangle$:

$$\frac{a}{2}[\bar{1}\bar{1}1] \rightarrow \frac{a}{8}[\bar{1}\bar{1}0] + \frac{a}{4}[\bar{1}\bar{1}2] + \frac{a}{8}[\bar{1}\bar{1}0].$$

This corresponds to the equivalent of Shockley partials. Apparently, the stacking-fault energy is very high, because the faults cannot be observed by transmission electron microscopy. The waviness of the slip markings is also indicative of the high stacking-fault energy. If the partials were well separated, slip would be limited to one plane. Cross-slip, which will be treated in Chapter 5, is much easier when the stacking-fault energy is high. If one adds up all the slip systems for BCC, one obtains a number of 48. This is much higher than the number for FCC.

EXAMPLE 4.3

Consider the following body-centered cubic dislocation reaction:

$$\frac{a}{2}[\bar{1}\bar{1}1] \rightarrow \frac{a}{8}[\bar{1}\bar{1}0] + \frac{a}{4}[\bar{1}\bar{1}2] + \frac{a}{8}[\bar{1}\bar{1}0].$$

(a) Prove that the reaction will occur.
(b) What kind of dislocations are the $(a/8)\langle 110 \rangle$ and $(a/4)\langle 112 \rangle$?
(c) What kind of crystal imperfection results from this dislocation reaction?
(d) What determines the distance of separation of the $(a/8)[\bar{1}\bar{1}0]$ and the $(a/4)[\bar{1}\bar{1}2]$ dislocations?

Solution: (a) $U \propto b^2$:

$$\overset{b_1}{\frac{a}{2}[\bar{1}\bar{1}1]} \rightarrow \overset{b_2}{\frac{a}{8}[\bar{1}\bar{1}0]} + \overset{b_3}{\frac{a}{4}[\bar{1}\bar{1}2]} + \frac{a}{8}[\bar{1}\bar{1}0].$$

On the left-hand side:

$$b^2 = \left(\frac{-a}{2}\right)^2 + \left(\frac{-a}{2}\right)^2 + \left(\frac{a}{2}\right)^2 = \frac{3}{4}a^2.$$

[6]L. J. Teutonico, *Mater. Sci & Eng.,* 6 (1970) 27.

On the right-hand side:

$$b_1^2 + b_2^2 + b_3^2$$

$$= \left[\left(\frac{-a}{8}\right)^2 + \left(\frac{-a}{8}\right)^2 + 0^2\right] + \left[\left(\frac{-a}{4}\right)^2 + \left(\frac{-a}{4}\right)^2 + \left(\frac{2a}{4}\right)^2\right]$$

$$+ \left[\left(\frac{-a}{8}\right)^2 + \left(\frac{-a}{8}\right)^2 + 0^2\right]$$

$$= \frac{a^2}{32} + \frac{3a^2}{8} + \frac{a^2}{32}$$

$$= \frac{7a^2}{16}.$$

Since

$$\frac{3}{4}a^2 > \frac{7a^2}{16},$$

the energy is lower after the reaction, and therefore, the reaction will occur.
(b) Partial dislocations.
(c) Stacking fault.
(d) Stacking-fault energy, γ_{SF}.

$$\gamma_{SF} \propto \frac{b_1 b_2}{d}, \qquad \text{with } b_1, b_2 \text{ known from (a)},$$

$$\Rightarrow \gamma_{SF} \alpha \frac{1}{d}$$

$$\Rightarrow d \alpha \frac{1}{\gamma_{SF}}.$$

That is, if γ_{SF} increases, the distance between the dislocations decreases.

EXAMPLE 4.4

Make a table with all 48 slip systems for the BCC structure.

Solution: For each slip system, we have to satisfy the condition $\mathbf{u} \cdot \mathbf{v} = 0$. For $(110)[1\bar{1}1]$, $1 \times 1 + 1 \times (-1) + 0 \times 1 = 0$.

The table of 48 slip systems for the BCC structure is as follows:

Slip Plane {110}		Slip Plane {112}				Slip Plane {123}			
(110)	$[1\bar{1}1]$	(112)	$[\bar{1}\bar{1}1]$	(123)	$[11\bar{1}]$	(123)	$[\bar{1}11]$
(110)	$[11\bar{1}]$	(121)	$[1\bar{1}1]$	(132)	$[1\bar{1}1]$	(132)	$[\bar{1}11]$
(1\bar{1}0)	$[111]$	(211)	$[\bar{1}11]$	(312)	$[\bar{1}11]$	(312)	$[1\bar{1}1]$
(110)	$[111]$	(11\bar{2})	$[111]$	(321)	$[\bar{1}11]$	(321)	$[11\bar{1}]$
(10\bar{1})	$[111]$	(1\bar{2}1)	$[111]$	(213)	$[11\bar{1}]$	(213)	$[1\bar{1}1]$
(101)	$[\bar{1}11]$	(\bar{2}11)	$[111]$	(231)	$[1\bar{1}1]$	(231)	$[111]$
(101)	$[1\bar{1}1]$	(11\bar{2})	$[\bar{1}11]$	(123)	$[111]$	(\bar{1}23)	$[\bar{1}11]$
(101)	$[11\bar{1}]$	(121)	$[\bar{1}\bar{1}1]$	(132)	$[111]$	(1\bar{3}2)	$[111]$
(011)	$[111]$	(211)	$[\bar{1}\bar{1}1]$	(312)	$[111]$	(3\bar{1}2)	$[11\bar{1}]$
(011)	$[1\bar{1}1]$	(112)	$[1\bar{1}1]$	(312)	$[111]$	(32\bar{1})	$[1\bar{1}1]$
(0\bar{1}1)	$[111]$	(\bar{1}21)	$[11\bar{1}]$	(213)	$[111]$	(2\bar{1}3)	$[111]$
(011)	$[111]$	(2\bar{1}1)	$[111]$	(231)	$[111]$	(23\bar{1})	$[111]$

4.2.6 Dislocations in Ceramics

Transmission electron microscopy has revealed dislocations in most nonmetals. Dislocations in semiconductors, minerals, oxide ceramics, and carbides, nitrides, and borides have been described and characterized. Many nonmetals tend to exhibit brittle behavior, in which dislocations play a minor role. However, if the temperature or lateral confinement of the material is sufficiently high, ductile behavior can be observed; in this case, dislocations play an important role. The role of confinement, or externally applied traction on planes parallel to the principal direction of external loading, is described in Chapter 7. The principal effect is to eliminate tensile stresses at the tips of internal flaws, thereby enabling the nonmetal to deform plastically. The temperature provides thermal activation that assists the overcoming of short-range obstacles by dislocations.

Table 4.2 lists the minimum temperatures at which ductile behavior is observed in ceramics. Most ceramics have high ductile-to-brittle transitions, and this has rendered the study of dislocations difficult. These high temperatures also affect the mechanisms of dislocation motion, since diffusion plays an important role at temperatures greater than or equal to $0.5 T_m$, where T_m is the melting point in K. The climb of dislocations is an effective mechanism for overcoming obstacles.

The structures of a number of ceramics are given in Chapter 1. (See Figure 1.17.) In general, ceramics tend to slip along directions that are closest packed. Since ceramics possess ordered structures, and a perfect dislocation has to recompose the original atomic arrangement, the Burgers vectors tend to be large.

Table 4.3 lists slip systems and Burgers vectors for a number of ceramics. For the oxide ceramics, the oxygen atoms (anions) tend to arrange themselves in close-packed structures (FCC or HCP), and this determines the slip systems. For instance, Al_2O_3 (HCP) has basal slip, where the slip lane is (0001) and the slip directions are $<11\bar{2}0>$. Prismatic or pyramidal slip are also possible. (See Table 4.3.) The Burgers vector is given by

$$b = \frac{1}{3} <11\bar{2}0> = \sqrt{3}d_0,$$

where d_0 is the nearest distance between oxygen atoms. Recall that the oxygen atoms form an HCP structure. The arrangement of atoms in the basal plane is shown in

TABLE 4.2 Approximate temperature for macroscopic plasticity in some ceramics

Ceramic	Melting point, T_m (K)	Softening point, $0.4T_m$ (K)
B_4C	2,725	1,090
TiC	3,400	1,360
HfC	4,425	1,770
WC	3,000	1,200
SiC	2,970	1,188
MgO	3,100	1,240
ZrO_2	3,100	1,240
Al_2O_3	2,325	930
TiO_2	2,100	844
SiO_2 (cristobalite)	1,990	796
S_3N_4	2,715	1,086
$MoSi_2$	2,300	920

TABLE 4.3 Crystal structures, slip systems, and Burgers vectors for ceramics (Courtesy of T. E. Mitchell)

Oxide	Slip system	Burgers vector	Other slip systems
MgO	$\{110\} <1\bar{1}0>$	$\frac{1}{2}<1\bar{1}0> = d_0$	$\{001\} <1\bar{1}0>, \{111\} <1\bar{1}0>$
$MgAl_2O_4$	$\{111\} <1\bar{1}0>$	$\frac{1}{2}<1\bar{1}0> = 2d_r$	$\{110\} <1\bar{1}0>$
Al_2O_3	$(0001) <11\bar{2}0>$	$\frac{1}{3}<11\bar{2}0> = \sqrt{3}d_0$	$\{11\bar{2}0\} <10\bar{1}0>, \{\bar{1}102\} <\sim11\bar{2}0>$
TiO_3	$\{001\} <0\bar{1}1>$	$<0\bar{1}1> \approx 2d_0$	$\{110\} [001]$
Mg_2SiO_4	$(100), \{110\} [001]$	$[001] = 2d_0$	$(100) [010], \{0kl\} [100]$
BeO	$(0001) <11\bar{2}0>$	$\frac{1}{3}<11\bar{2}0> = d_0$	$\{1\bar{1}00\} <11\bar{2}0>, [0001] \{10\bar{1}0\}$
UO_2	$\{001\} <1\bar{1}0>$	$\frac{1}{2}<1\bar{1}0> = \sqrt{2}d_0$	$\{110\}, \{111\} <1\bar{1}0>$
SiO_2 (quartz)	$(0001) <11\bar{2}0>$	$\frac{1}{3}<11\bar{2}0>$	$\{11\bar{2}0\}, \{10\bar{1}0\} [\bar{0}001]$

Figure 4.42. The large circles are the oxygen anions, forming a closed-packed hexagonal array. The full circles are the aluminum cations, which stack in the *ABC* sequence (similar to the FCC structure). The empty circles are normally empty octahedral interstices. The vectors $\mathbf{b}_1, \mathbf{b}_2$, and $\mathbf{b}_3$ are the Burgers vectors. They translate interstitial sites in such a manner that they become superposed. The amplitude of the Burgers vectors is equal to $\sqrt{3}$. This can be shown from the triangle *ABC*, where $BC = \sqrt{3}d_0$ and the angle *BCA* is equal to 120°.

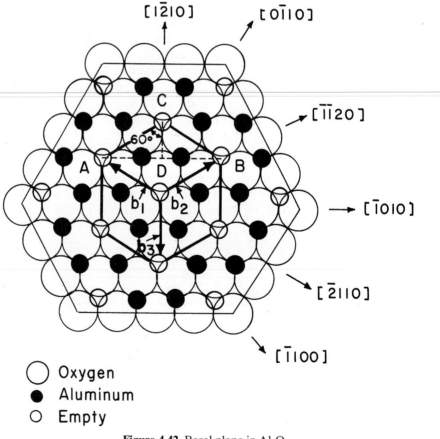

Figure 4.42 Basal plane in Al_2O_3.

For MgO, the anions form an FCC structure, and the Burgers vector has the direction $<1\bar{1}0>$ and a magnitude equal to d_0, the smallest oxygen spacing. Thus,

$$b = \frac{1}{2}<1\bar{1}0> = d_0.$$

The dislocations in ceramics generally have a high energy, due to the large shear modulus and Burgers vector ($U \sim Gb^2/2$). Table 4.4 gives Burgers vectors and self-energies for dislocations in a number of intermetallics and ceramics. For purposes of comparison, the dislocation energy of aluminum is shown. The differences can be dramatic. The Peierls–Nabarro stress (see Section 4.2.10) is very high, in general, because of the directionality of bonding in ionic and covalent structures. For instance, the bond angles of 109° for the carbon atom need very high forces to be distorted. The movement of a dislocation requires the breaking and remaking of bonds, and distortions are produced around the dislocations. Therefore, the movement of dislocations in ceramics is, in general, difficult. There are exceptions, however, such as MgO, which can exhibit significant plasticity at ambient temperature.

Dislocation interactions and reactions occur in a manner similar to that in metals and intermetallics. An example is given in Figure 4.43. Dislocation dipoles are often observed in the deformation of sapphire and are shown in Figure 4.43a (arrows D). These dipoles are parallel edge dislocations of opposite sign that are attracted together into a position of approximately 45° (55° if there is anisotropy) in order to minimize the elastic fields. This is shown in Figure 4.43b. In Figure 4.31d, the elastic (shear stress) fields of edge distortions are shown. The shear stresses σ_{12} are minimized if they place themselves at 45°. These dipoles break down and form loops, as indicated by arrows L in Figure 4.43a. The stress fields of one dislocation are canceled by those of the other dislocation, at 45°, as shown in Figure 4.43(b). Dislocation dissociations and reactions are also observed and can be predicted from energetics. A hexagonal dislocation network is

TABLE 4.4 Elastic energy for dislocations in ceramics and intermetallics (Courtesy of Veyssiere)

	Oxygen Sublattice	b	b(nm)	(GPa)	Gb²/2
Al		½ <110>	0.286	27	1.2
Ni₃Al		<110>	0.356	100	6.4
MgO	fcc	½ <110>	0.298	125	5.1
CoO	fcc	½ <110>	0.301	70	3.2
NiO	fcc	½ <110>	0.296	135	5.9
MgAl₂O₄	fcc	½ <110>	0.57	120	19.5
BeO	hcp	⅓ <11–20>	0.27	160	5.9
Al₂O₃ − α	hcp	⅓ <11–20>	0.476	200	22.6
TiO₂	distorted	<001>	0.296	100	4.4
	hcp	<101>	0.546		14.9
CuO₂	bcc	<001>	0.427	10	0.9
		<011>	0.604		1.8
UO₂	cubic	½ <110>	0.386	94	7.0
Y₂O₃	vacancy—	½ <111>	0.918	65	27.4
	containing cubic	<100>	1.06		31.5
Y₃Fe₅O₁₂	highly	½ <111>	1.072	78	44.8
	distorted	<100>	1.038		42.0

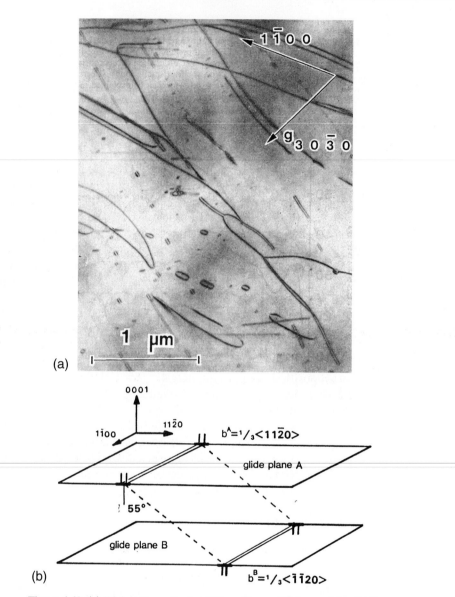

Figure 4.43 (a) Dislocations, dipoles (D), and loops (L) in sapphire. (b) Interaction between dislocations in sapphire. (From K. P. D. Lagerdorf, B. J. Pletka, T. E. Mitchell, and A. H. Heuer, *Radiation Effects*, 74 (1983) 87, p. 90, Figure 3)

shown in Figure 4.44. The total Burgers vector at the nodes has to be equal to zero under equilibrium. This is called *Frank's rule*. For basal dislocations in a hexagonal structure, we have, at the nodes,

$$\frac{1}{3}[11\bar{2}0] + \frac{1}{3}[1\bar{2}10] + \frac{1}{3}[\bar{2}110] = 0.$$

And for the FCC structure,

$$\frac{1}{2}[1\bar{1}0] + \frac{1}{2}[01\bar{1}] + \frac{1}{2}[\bar{1}01] = 0.$$

These structures are often produced during recovery.

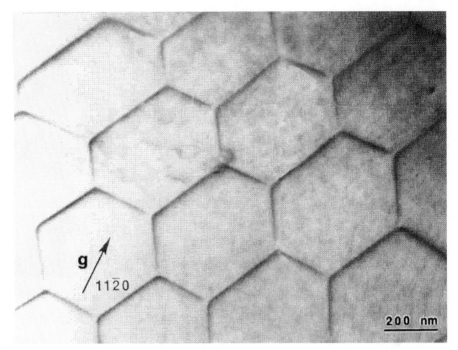

Figure 4.44 Hexagonal array of dislocations in titanium diboride. (Courtesy of D. A. Hoke and G. T. Gray)

The dissociation of a perfect dislocation into partial dislocations is treated in a manner similar to that in metals. The criterion for energy decrease ($U \approx Gb^2/2$) is applied, and dissociation is stable if $b^2 > b_1^2 + b_2^2$.

A few dislocation dissociations have been observed in ceramics. In the spinel structure, the dissociation

$$\frac{1}{2}[1\bar{1}0] \rightarrow \frac{1}{4}[1\bar{1}0] + \frac{1}{4}[1\bar{1}0]$$

was observed, and the following dissociation was suggested to occur in Al_2O_3:

$$\frac{1}{3}[11\bar{2}0] \rightarrow \frac{1}{3}[10\bar{1}0] + \frac{1}{3}[01\bar{1}0].$$

This dissociation has been observed to occur only by climb.

As an illustration of the occurrence of stacking faults in ceramics, Figure 4.45 shows a TEM of gallium phosphide. The large concentration of these faults is evident. They are a common occurrence in thin films deposited on Si substrates by molecular beam epitaxy (MBE), chemical vapor deposition (CVD), or metal–organic CVD (MOCVD). Section 4.2.6 describes the stresses generated in epitaxial growth on a substrate. These mismatch stresses, as well as thermal stresses and growth faults, are responsible for the high concentration of stacking faults, which decreases with distance from the interface. Profuse stacking faults bounded by Shockley partial dislocations and stair-rod dislocations have been observed to occur in SiC grown on Si wafers. The configuration of stacking faults observed in SiC is analogous to that for GaP shown in Figure 4.45.

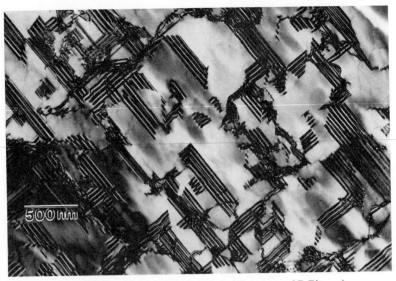

Figure 4.45 Stacking faults in GaP. (Courtesy of P. Pirouz)

4.2.7 Sources of Dislocations

It is experimentally observed that the dislocation density increases with plastic deformation; specifically, the relationship $\tau \propto \rho^{1/2}$ (see Chapter 6 Section 6.3) has been found to be closely obeyed. While the dislocation density of an annealed polycrystalline specimen is typically 10^7 cm^{-2}, a plastic strain of 10% raises this density to 10^{10} cm^{-2} or more, an increase of three orders of magnitude. This is an apparent paradox, because one would think that the existing dislocations would be ejected out of the crystalline structure by the applied stress. If one calculates the strain that the existing dislocations in an annealed metal would be able to produce by their motion until they would leave the crystal, one would arrive at very small numbers. Consequently, the density of dislocations has to increase with plastic deformation, and internal sources have to be activated. Some possible dislocation-generation mechanisms are discussed in the next few paragraphs.

The homogeneous nucleation of a dislocation consists in the rupture of the atomic bonds of a material along a certain line. Figure 4.46 shows schematically the sequence of steps leading to the formation of a pair of edge dislocations (one negative, one positive). In Figure 4.46a the lattice is elastically stressed, until, in Figure 4.46b, an atomic plane is sheared; this generates two dislocations that move in opposite senses. Such a mechanism allows the formation of dislocations from an initially perfect lattice. It can be intuitively seen that the stress required would be extremely high. Calculations were done by Hirth and Lothe (see the suggested readings), and for copper, this stress is on the order of

$$\frac{\tau_{\text{hom}}}{G} = 7.4 \times 10^{-2}.$$

Comparing this with the theoretical strength of crystals, one can see that the difference is not very large. Hence, such values would be obtained only if the applied stresses were very high or there were internal regions of high stress concentration. In conventional deformation, other dislocation-generation mechanisms should become operational at much lower stresses, rendering homogeneous nucleation highly unlikely.

Grain boundaries can serve as sources of dislocation. Irregularities at the boundaries (steps or ledges) could be responsible for the emission of dislocations into the grains.

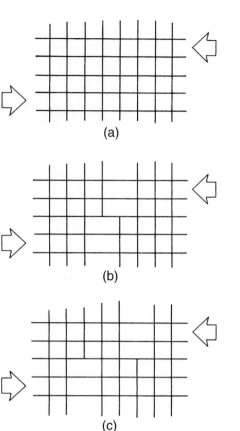

Figure 4.46 Homogeneous nucleation of dislocation in conventional deformation.

Figure 4.47 shows the emission of dislocations from a grain-boundary source; dislocations are seen as they are generated at the ledge. The stress due to heating produced by the electron beam produces the force on the dislocations. It is thought that dislocation emission from grain boundaries can be an important source of dislocations in the first stages of plastic deformation of a polycrystal. (See Section 5.3 for details.)

In monocrystals, the surfaces can act as sources of dislocation. Small steps at the surfaces act as stress concentration sites; hence, the stress can be several times higher than the average stress. At these regions, dislocations can be generated and "pumped" into the monocrystals. The majority of dislocations in monocrystals deformed in tension are generated at the surface. Pangborn et al.[7] investigated the bulk and surface dislocation mechanism in monocrystals. The dislocation density close to the surface was up to six times greater than that in the bulk. The dislocation surface layer (with higher dislocation density) extended for approximately 200 μm into the material at the surface. The surface sources cannot have a significant effect on polycrystal deformation, because the majority of the grains would not be in contact with the free surface. Since dislocation activity is restricted to the grains, the surface sources would not be able to affect the internal grains. Incoherent interfaces between the matrix and precipitates, dispersed phases, or reinforcing fibers (in composites) are also sources of dislocations.

The importance of interfaces in the production of dislocations is seen in the results shown in Figure 4.48. The low-temperature tensile response of BCC metals was

[7]P. N. Pangborn, S. Weissman, and I. R. Kramer, *Met. Trans. 12A* (1981) 109.

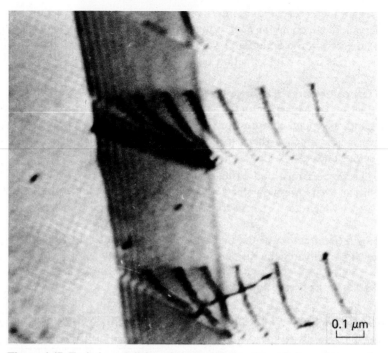

Figure 4.47 Emission of dislocations from ledges in grain boundary, as observed in transmission electron microscopy during heating by electron beam. (Courtesy of L. E. Murr)

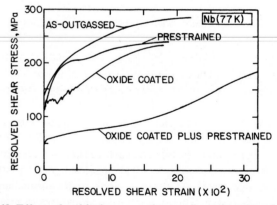

Figure 4.48 Effect of oxide layer on the tensile properties of niobium. (Reprinted with permission from V. K. Sethi and R. Gibala, *Scripta Met.* 9 (1975) 527.)

dramatically affected by the presence of an oxide layer. The figure exemplifies this response for niobium. The flow stress of monocrystalline niobium at 77 K is highly dependent on the state of the surface. The oxide softens the material. Two effects are responsible for the lowering of the flow stress by the introduction of an oxide layer:

1. The oxide puts the surface layers under tensile stresses, because the introduction of oxygen into the lattice expands it. On the other hand, the oxide is under compression. The resultant resolved shear stress at the surface is much higher (in the presence of the oxide layer) than that due exclusively to the externally applied load.

2. The predeformed and oxide-coated specimen (the lowest curve in the figure) has an even lower flow stress because the predeformation introduces surface steps, which act as stress-concentration sites.

Hence, the joint action of the internal stresses generated by the oxide and the surface steps activates the dislocation sources at the surface.

The classic mechanism for dislocation multiplication is called the *Frank–Read Source*. In Figure 4.49a, there is a dislocation *ABCD* with Burgers vector *b*. Only the segment *BC* is mobile in the slip plane α. Segments *AB* and *CD* do not move under the imposed stress. The applied stress will generate a force on segment *BC* equal to

$$F = \frac{T\,ds}{R}.$$

The radius of curvature of the dislocation segment decreases until it reaches its minimum, equal to *BC*/2. At this point, the force is maximum (and so is the stress). Hence, the dislocation reaches a condition of instability beyond that point. The critical position is shown in Figure 4.49c. When *P* approaches *P'*, the dislocation segments have opposite signs; accordingly, they attract each other, forming a complete loop when they touch, and are then pinched off. The stress required to activate a Frank–Read source is equal to that needed to curve the segment *BC* into a semicircle with radius *BC*/2; beyond this point, the stress is decreased. However, as loops are formed, they establish a back stress, so that the stress required to generate successive loops increases steadily. If the loops are expelled from the material, they cease to exert a back stress.

Only a few Frank–Read sources have been observed in metals. However, in a tridimensional array of dislocations, nodes define segments. These segments can bow and effectively act as Frank–Read sources. Another possibility is that the source forms when a screw dislocation cross-slips and returns to a plane parallel to the original slip plane. (See Figure 4.50.) Incidentally, edge dislocations cannot cross-slip because their Burgers vector could not be contained in the cross-slip plane. The Burgers vector of a screw dislocation, on the other hand, is parallel to its line and will be in the cross-slip plane if the

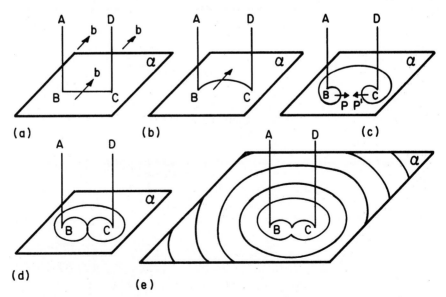

Figure 4.49 Sequence of the formation of dislocation loop by the Frank–Read mechanism.

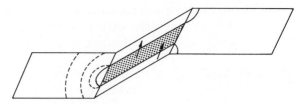

Figure 4.50 Frank–Read source formed by cross-slip.

intersection of the two dislocations is parallel to it. After the segment in the cross-slip plane advances a certain extent, the stress system applied might force it into a plane parallel to the original slip plane. At this point, a Frank–Read source is formed. Although it is thought that the original formulation of the Frank–Read source is not common, its modifications just cited—the node and the cross-slip case—might be the principal mechanism of dislocation generation, after the first few percent of plastic strain.

Crystals formed by growth over a substrate (a technique commonly employed in the production of thin films) show dislocations whose formation can be easily explained. The substrate never has exactly the same lattice parameter as the crystal overgrowth. Figure 4.51 shows the sequence of formation of dislocations as the crystal grows over the substrate. If a_s and a_0 are the lattice parameters of the substrate and overgrowth, respectively, the separation between the dislocations is

$$d = \frac{a_s^2}{|a_s - a_0|}.$$

Often, the impurity content of a crystal varies cyclically due to solidification; this is called *segregation*. The periodic change in composition is associated with changes in the lattice parameter, which can be accommodated by dislocation arrays.

Vacancies can condense and form disks as well as prismatic loops if they are present in a "supersaturated" concentration. In FCC crystals, these disks and loops occur on {111} planes. As seen in Figure 4.38, the dislocations that form the edges of these features are called *Frank dislocations*. Kuhlmann-Wilsdorf (see suggested readings) proposed, that they can act as Frank–Read sources, and this was later experimentally confirmed.

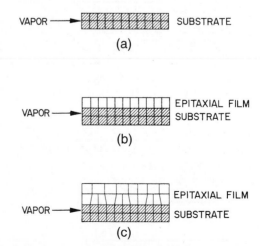

Figure 4.51 Epitaxial growth of thin film. (a) Substrate. (b) Start of epitaxial growth. (c) Formation of dislocations.

4.2.8 Dislocation Pileup

All dislocations generated by a Frank–Read source are in the same slip plane if they do not cross-slip. In metals with low stacking-fault energy, the large separation between the partials renders cross-slip more difficult. In case one of the dislocations encounters an obstacle (a grain boundary, a precipitate, etc.), its motion will be hampered. The subsequent dislocations will "pile up" behind the leading dislocation, after being produced by the Frank–Read source. Figure 4.52 is a schematic diagram of a pileup. The distance between the dislocations increases as their distance from the obstacle increases. On the other hand, if the metal has a very high stacking-fault energy, cross-slip will easily occur, and the planar array will be destroyed; edge dislocations cannot, obviously, cross-slip because of their Burgers vector.

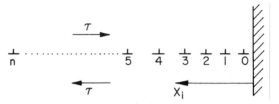

Figure 4.52 Pileup of dislocations against a barrier.

Figure 4.53 shows an example of a pileup, obtained by etch pitting in copper. Observe that the dislocation configurations for a pileup and a grain-boundary source are similar and that many grain-boundary sources have in the past been mistaken for pileups. Figure 4.47 shows a grain-boundary source.

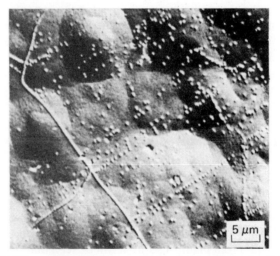

Figure 4.53 Pileup of dislocations against grain boundaries (or dislocations being emitted from grain-boundary sources?) in copper observed by etch pitting.

Each dislocation in a pileup is in equilibrium under the effect of the applied stress and of the stresses due to the other dislocations (in the pileup). Assuming that the dislocations are of edge character and parallel, the resulting force acting on the ith dislocation is obtained by applying the equation that gives the forces between dislocations:

$$\tau b - \sum_{\substack{j=0 \\ i \neq j}}^{n} \frac{Gb^2}{2\pi(1-\nu)(x_i - x_j)} = 0. \tag{4.19}$$

n is the number of dislocations in the pile up.

Solving the n equations with n unknowns $(x_i - x_j)$ for the dislocations behind the lead dislocation, we obtain the positions of the dislocations. This derivation was introduced by Eshelby et al.[8] and is not reproduced here because of its complexity.

The stress acting on the lead dislocation due to the presence of the other dislocations and due to the applied stress is found to be

$$\tau^* = n\tau. \tag{4.20}$$

So the effect of the n dislocations in the pileup is to create a stress at the lead dislocation n times greater than the applied stress. For this reason, the dislocation pileup is sometimes treated as a superdislocation with a Burgers vector nb. The foregoing calculations can also be applied to screw dislocations by removing the term $(1 - \nu)$. The length of the pileup under an applied shear stress τ is given by

$$L = \frac{nGb}{\pi\tau}.$$

4.2.9 Intersection of Dislocations

A dislocation, when moving in its slip plane, encounters other dislocations, moving along other slip planes. If we imagine the first dislocation moving in a horizontal plane, it will "see" the other dislocations as "trees" in a "forest." The latter name designates dislocations in other slip planes. When the dislocation intersects another dislocation, since it shears the material equally (by a quantity b) on the two sides of the slip plane, it will form one or more steps. These steps are of two types: *jogs* if the "tree" dislocation was transferred to another slip plane, and *kink* if the "tree" dislocation remains in the same slip system. Various possible outcomes from dislocation intersections are shown in Figure 4.54. Figure 4.54(a) shows an edge dislocation traversing a "forest" composed of two edge and one screw dislocation. A good rule to determine the direction of jogs and kinks is the following: The direction of the segment is the same as the Burgers vector of the dislocation that is traversing the "forest"; on the other hand, the Burgers vector of the jog or kink is the Burgers vector of the dislocation in which it is located, because the Burgers vector is always the same along the length of a dislocation. Figure 4.54(b) shows a screw dislocation after traversing a "forest." The reader is asked to verify the directions of dislocation segments and Burgers vector; he or she should also verify whether they are jogs or kinks.

The ability of these segments to slip with a dislocation is of great importance in determining the work-hardening of metal. It should be noted that some authors use the name "jog" for both types of segments. Jogs and kinks can have either a screw or an edge character. From Figure 4.55(a), it can be seen that segments on an edge dislocation cannot impede the motion of jogs or kinks, because the segments can slip with the dislocation. On the other hand, in screw dislocations, there are segments that can slip with the dislocations and segments that cannot. When the segment can move with the dislocation, the motion is called *conservative*. When the segment cannot move by slip, the motion is called *nonconservative*. Figure 4.55(b) shows some interactions. At the left there is a conservative motion by slip, and at the right a nonconservative motion. The nonconservative motion of a jog is, in essence, a climb process and requires thermal activation. Vacancies or interstitials are produced as the segment moves. If the temperature is not high enough to provide sufficient thermal activation, the jog does not move, and loops are formed as the dislocation advances; this is shown in Figure 4.56. The dislocation forms a dipole upon

---

[8]J. D. Eshelby, F. C. Frank, and F. R. N. Nabarro, *Phil. Mag.,* 42 (1951) 351.

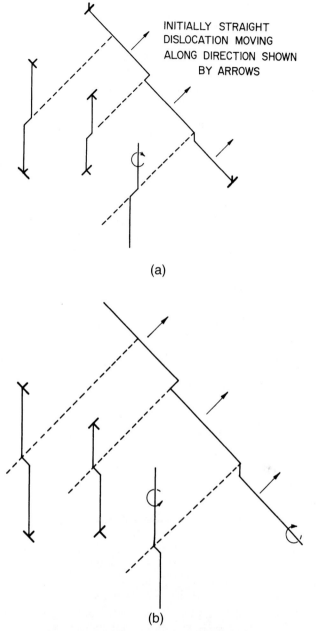

INITIALLY STRAIGHT
DISLOCATION MOVING
ALONG DIRECTION SHOWN
BY ARROWS

(a)

(b)

Figure 4.54 (a) Edge dislocation traversing "forest" dislocation. (b) Screw dislocation traversing "forest" dislocations.

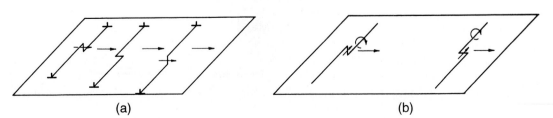

(a)   (b)

Figure 4.55 (a) Kink and jog in edge dislocation. (b) Kink and jog in screw dislocation.

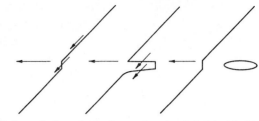

Figure 4.56 Loop being pinched out when jog is left behind by dislocation motion.

advancing, because the jog stays back. At a certain point, the dipole will be pinched out, producing a loop.

4.2.10 Deformation Produced by Motion of Dislocations (Orowan's Equation)

Upon moving, a dislocation produces a certain deformation in a material. This deformation is inhomogeneous. Figure 4.57 shows the steps generated by the passage of dislocations. If we consider a large number of dislocations acting on different systems, we can posit the association of a large number of small steps as creating a homogeneous state of deformation. The deformation is related to both the number of dislocations that move and the distance traveled by them. This equation is known as *Orowan's* or *Taylor–Orowan's equation* and is derived in this section. Figure 4.57 shows a cube dimensions dx_1, dx_2, and dx_3 that was sheared by the passage of N dislocations moving along the plane Ox_1x_2. The plastic shear strain can be expressed as

$$\gamma_{13} = \frac{Nb}{dx_3}.$$ (4.21)

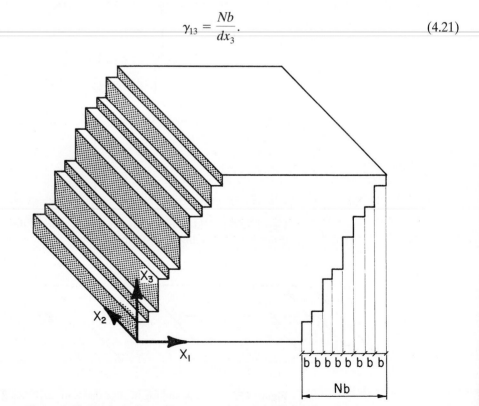

Figure 4.57 Shear produced by the passage of parallel dislocations.

This is so because all dislocations are of edge character and have the same sign, with identical Burgers vector b. The density of dislocations, ρ, is the total length $N\,dx_2$ in the volume $dx_1\,dx_2\,dx_3$. Therefore,

$$\rho = \frac{N\,dx_2}{dx_1\,dx_2\,dx_3} \quad \text{and} \quad N = \rho\,dx_1\,dx_3. \tag{4.22}$$

Substituting Equation 4.22 into Equation 4.21 yields

$$\gamma_{13} = \rho b\,dx_1.$$

A cube isolated in space, in which dislocations are generated on one face and pop out of the opposite face is an idealization. In real situations, dislocations remain within the material, and the deformation generated by each dislocation is related to the distance traveled by it. Assuming that dislocations travel an average distance, $\bar{l}$, we have

$$\gamma_{13} = \rho b\bar{l}.$$

But in a general case of deformation, five independent slip systems are activated. The deformation is not perfectly aligned with the movement of dislocations, and it is necessary to introduce a correction parameter k that takes this into account:

$$\gamma_p = k\rho b\bar{l}. \tag{4.23}$$

This is the Orowan equation. If one assumes that the density of mobile dislocations is not affected by the rate of deformation (strain rate), one would have, taking the time derivative of both sides of Equation 4.23,

$$\frac{d\gamma_p}{dt} = k\rho b\frac{d\bar{l}}{dt}, \quad \text{or} \quad \dot{\gamma}_p = k\rho b\bar{v}. \tag{4.24}$$

where $\bar{v}$ is the mean velocity of the dislocations. We can also use the longitudinal strain ε_{11} if we are applying the situation to a tensile test. It can be shown that $\gamma = 2\varepsilon$ (see Section 6.2.3) for an ideal orientation for slip.

As an illustration, if iron ($b \approx 2.5$ Å) is being deformed at 10^{-3} s^{-1}, and the density of mobile dislocations is around 10^{10} cm^{-2}, their approximate velocity will be 4×10^{-6} cm/s.

Attention should be called to the fact that the density of mobile dislocations is lower than the total density of dislocations in the material. As the dislocation density increases in a deformed material, a greater and greater number of dislocations is locked by various types of barriers, such as grain boundaries, cell walls, or the action of a great number of jogs. The actual density of mobile dislocations is only a fraction of the total dislocation density.

EXAMPLE 4.5

Titanium is deformed by basal slip with edge dislocations. If a cube with one of its sides parallel to the c-axis is being deformed by shear through the passage of dislocations on every fifth (0001) plane, what shear strain γ is the cube undergoing? Take the radius of Ti atom

$$r_{\text{Ti}} = 0.147 \text{ nm}.$$

Solution: We first determine

$$a = 2r = 0.294 \text{ nm}.$$

We assume an ideal c/a ratio equal to 1.633. Thus, $c = 0.48$ nm. The Burgers vector for basal slip is equal to a.

Every fifth atomic plane corresponds to a distance $d = 5c = 2.4$ nm. The shear strain is thus equal to

$$\gamma = \frac{b}{d} = \frac{0.294}{2.4} = 0.1225.$$

EXAMPLE 4.6

An FCC monocrystal of nickel is sheared by $\gamma_{12} = 0.1$. Assuming that the dislocation density is equal to 10^8 cm^{-2} and that it remains constant, what is the average distance each dislocation will have to move? If the shear strain rate is 10^{-4} s^{-1}, what is the mean velocity of the dislocation?

Solution:

$$r_{\text{Ni}} = 0.125 \text{ nm}$$

For FCC, $b = 2r_{\text{Ni}} = 0.250$ nm. Using Orowan's equation taking $k = 1$, $\gamma = \rho b \bar{l}$, we obtain the following:

(i) $\quad \bar{l} = \dfrac{\gamma}{\rho b}$

$$= \frac{0.1}{10^8 \text{ cm}^{-2} \times 0.25 \text{ nm}}$$

$$= \frac{0.1}{10^8 \times (10^4 \ m^{-2}) \times 0.25(10^{-9} \ m)}$$

$$= 4 \times 10^{-4} \text{ m}.$$

(ii) $\quad \dot{\gamma} = \rho b \bar{v}$

so $\bar{v} = \dfrac{\dot{\gamma}}{\rho b}$

$$= \frac{10^{-4} \text{ s}^{-1}}{10^8 \text{ cm}^{-2} \times 0.25 \text{ nm}}$$

$$= 4 \times 10^{-7} \text{ m/s}.$$

4.2.11   The Peierls–Nabarro Stress

The Peierls–Nabarro stress represents the resistance that the crystalline lattice offers to the movement of a dislocation. Figure 4.58 shows the stress that one has to apply to a dislocation to make it move a distance b. When the extra plane is moved away from its equilibrium position (either to the right or to the left), one has to overcome a barrier. The

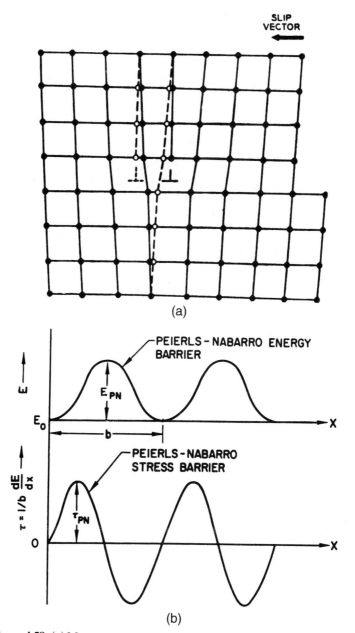

SLIP
VECTOR

(a)

PEIERLS − NABARRO ENERGY
BARRIER

E_{PN}

E_0

b

X

PEIERLS − NABARRO
STRESS BARRIER

τ_{PN}

0

X

(b)

Figure 4.58 (a) Movement of dislocation away from its equilibrium position.
(b) Variation of Peierls–Nabarro stress with distance. (Reprinted with permission from H. Conrad, *J. Metals,* 16 (1964), p. 583.)

difference in energy between the equilibrium (saddle point) and the most unstable position is called the *Peierls–Nabarro energy,* and the stress required to overcome this energy barrier is the *Peierls–Nabarro (P–N) stress.* The dislocation does not advance simultaneously over its entire length. (See Figure 4.59a.) Rather, a small hump, or kink pair is formed, as shown in Figure 4.59b, via what is known as a Seeger mechanism. This kink pair then moves along the dislocation (the parts of the pair move in opposite directions), and when it has covered the entire front, the dislocation has advanced by *b*, the Burgers

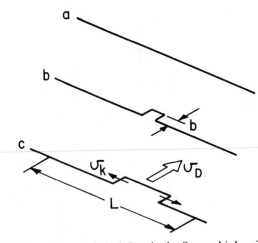

Figure 4.59 Overcoming of Peierls barrier by Seeger kink pair mechanism.
(a) Original straight dislocation. (b) Dislocation with two kinks. (c) Kinks
moving apart at velocity v_k.

vector. In Figure 4.59c, the velocity of movement of a dislocation is v_D, related to the kink
velocity v_k by

$$v_D = v_k \frac{b}{L}.$$

The stress required to overcome the obstacle is known as the Peierls–Nabarro stress. Calculations of this stress are fairly inaccurate because the continuum treatment breaks down
for distances on the order of the atomic spacings. The energy of the dislocation is given by
$U(x)$ as it moves through the barrier. The applied force required to bring this dislocation
to the top of the energy barrier is

$$F = -\frac{dU}{dx}.$$

But from the Peach–Koehler equation ($F = \tau b$), we have

$$\tau = -\frac{1}{b}\frac{dU}{dx}$$

A sinusoidal form for $U(x)$ was assumed by Peierls and Nabarro, leading to the expression

$$\tau_{PN} = \alpha \frac{Gb}{2c} e^{-\pi a/c} \sin \frac{2\pi x}{c},$$

where c is the spacing of atoms in the x direction, a is the lattice parameter, and α is a parameter that depends on the nature of the barrier, for $\alpha = 1$, the barrier is sinusoidal.

4.2.12   The Movement of Dislocations: Temperature and
Strain Rate Effects

The resistance of crystals to plastic deformation is determined by the resolved shear stress
that is required to make the dislocations glide in their slip planes. If no obstacles were present, the dislocations would move under infinitesimally small stresses. However, in real

metals, the nature and distribution of obstacles determines their mechanical response. Becker[9] was the first to point out the importance of thermal energy in helping the applied stress overcome existing obstacles. The stress required for deformation, τ, can be divided into two parts: τ^*, which is dependent on the strain rate and temperature of the material, and τ_G, in which the temperature dependence is equal to that of the shear modulus. Thus,

$$\tau = \tau^* + \tau_G,$$

or, in terms of the normal stresses,

$$\sigma = \sigma^* + \sigma_G.$$

The functional dependence can be expressed as

$$\sigma = \sigma^*(T, \dot{\varepsilon}) + \sigma_G(G).$$

We know that the elastic properties (E, G, ν) are only slightly dependent on temperature. Figure 4.60 shows the temperature dependence of Young's modulus for a number of materials. As the temperature increases, the amplitude of vibration of the atoms increases (but the frequency remains constant at approximately $10^{13}s^{-1}$). This results in thermal dilation, which separates the atoms somewhat and changes their equilibrium positions and interatomic forces. The flow stress of metals, on the other hand, is much more sensitive to temperature and strain rate. Figure 4.61 shows the dependence of the yield stress on temperature for typical BCC and FCC structures. BCC metals (Fe, Cr, Ta, W, etc.) exhibit a greater temperature and strain rate sensitivity. It can be seen that the athermal

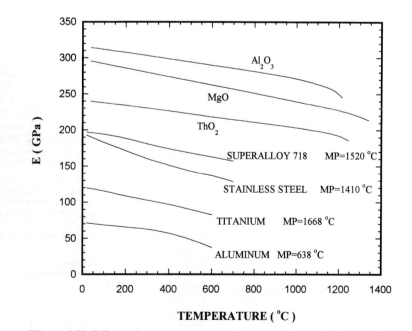

Figure 4.60 Effect of temperature on Young's modulus. (Adapted from J. B. Wachtman Jr., W. E. Tefft, D. G. Lam, Jr., and C. S. Apstein, *J. Res. Natl. Bur. Stand.,* Vol. 64A, 213–228, 1960; and J. Lemartre and J. L. Chaboche, *Mechanics of Solid Materials,* Cambridge: Cambridge U. Press, 1990, 143).

[9]R. Becker, *Z. Phys.* 26 (1925) 919.

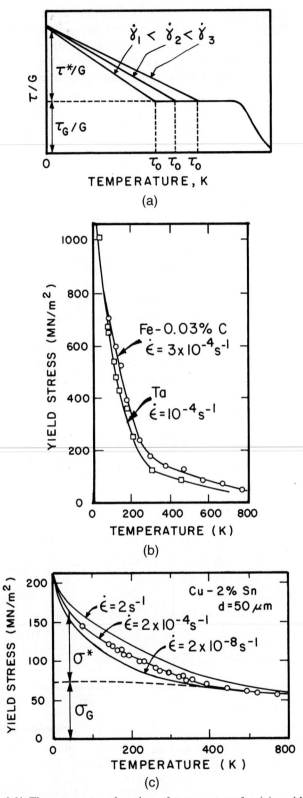

Figure 4.61 Flow stress as a function of temperature for (a) an idealized material, (b) BCC metals, and (c) FCC metals. Notice the greater temperature dependence for Ta and Fe (BCC).

component of stress is $\sigma_G \approx 50$ MPa, whereas the thermal component exceeds 1,000 MPa at 0 K. The increase in flow stress with decrease in temperature is much more gradual for FCC metals, as shown in Figure 4.61 (c). The differences in temperature and strain rate sensitivity are due to different mechanisms controlling the rate of dislocation motion. In BCC metals, Peierls–Nabarro stresses are the major obstacles at low temperatures, and thermal energy can effectively aid the dislocations to overcome these stresses, which constitute a short-range barrier. For FCC metals, dislocations intersecting dislocations ("forest" dislocations) are the main barriers to the motion of dislocations. Thermal energy is less effective in helping dislocations to overcome these barriers.

At temperatures higher than 800 K, there is an additional drop in the flow stress, not shown in Figure 4.61. This drop occurs at $T \approx 0.5T_m$, where T_m is the melting point of the metal (or alloy). The drop is due to creep, which often involves dislocation climb. Creep is treated separately in Chapter 13.

Johnston and Gilman[10] were the first to measure the velocities of dislocations as a function of applied stress. They used LiF crystals for their measurements and observed, as expected, that the distance a dislocation moves increases with the magnitude and duration of the stress pulse. The distance also increases, at a constant stress, with increasing temperature. This relationship is known as the Johnston–Gilman equation and has the form

$$v = A\tau^m e^{-Q/RT},$$

where v is the dislocation velocity, m is a stress dependency that is dependent on v, Q is an activation energy, and A is a preexponential term that depends on the material and the nature of the dislocation (edge or screw). Although this equation predicts an infinite dislocation velocity when the stress is high, it is generally accepted that the limiting dislocation velocity is the velocity of elastic shear waves. Thus, the equation breaks down at velocities close to the shear wave velocity (e.g., $\sim 3,000$ m/s for iron).

SUGGESTED READINGS

Point Defects

C. S. BARRETT and T. B. MASSALSKI. *Structure of Metals,* 3d ed. New York: McGraw-Hill, 1966.

J. H. CRAWFORD JR. and L. M. SLIFKIN (eds.). *Point Defects in Solids.* New York: Plenum Press, 1972.

A. C. DAMASK and G. J. DIENES. *Point Defects in Metals.* New York: Gordon and Breach, 1963.

C. P. FLYNN. *Point Defects and Diffusion.* Oxford: Clarendon Press, 1972.

H. KIMURA and R. MADDIN. *Quench Hardening in Metals,* in the series "Defects in Crystalline Solids, S. Amelinck, R. Gevers, and J. Nihoul (eds.). Amsterdam: North-Holland, 1971.

A. S. NOWICK and B. S. BERRY. *Anelastic Relaxation in Crystalline Solids.* New York: Academic Press, 1972.

H. G. VAN BUEREN. *Imperfections in Crystals.* Amsterdam: North-Holland, 1961.

Line Defects

A. H. COTTRELL. *Dislocations and Plastic Flow in Crystals.* New York: McGraw-Hill, 1953.

J. C. FISHER, W. G. JOHNSTON, R. THOMSON, and T. VREELAND, JR. (eds.). *Dislocations and Mechanical Properties of Crystals.* New York: Wiley, 1957.

J. FRIEDEL. *Dislocations.* Elmsford, NY: Pergamon Press, 1967.

---

[10] W. G. Johnston and J.J. Gilman, *J. Appl. Phys.* 33 (1959) 129.

J. P. Hirth and J. Lothe. *Theory of Dislocations,* 2d ed. New York: J. Wiley, 1981.

D. Hull and D. J. Bacon. *Introduction to Dislocations.* New York: Oxford U. Press, 1989.

A. Kelly and G. W. Groves. *Crystallography and Crystal Defects,* Reading, MA: Addison-Wesley, 1974.

I. Kovacs and L. Zsoldos. *Dislocations and Plastic Deformation.* Elmsford, NY: Pergamon Press, 1973.

D. Kuhlmann-Wilsdorf, in *Physical Metallurgy,* 3d ed. R. W. Cahn and P. Haasen (eds.), Amsterdam: North Holland, 1990, 1983.

F. R. N. Nabarro (ed.). *Dislocations in Solids,* (9 vols.). New York: Elsevier/North-Holland, 1979-1998.

W. T. Read, Jr. *Dislocations in Crystals.* New York: McGraw-Hill, 1953.

J. Weertman and J. R. Weertman. *Elementary Dislocation Theory.* New York: Oxford U. Press, 1992.

EXERCISES

4.1 Calculate the radii of the tetrahedral and octahedral holes in BCC and FCC iron; assume lattice parameters of 0.286 and 0.357 nm, respectively.

4.2 Calculate the concentration of monovacancies in gold at 1,000 K, knowing that $H_f = 1.4 \times 10^{-19}$ J. If the gold is suddenly quenched to ambient temperature, what will be the excess vacancy concentration?

4.3 How many vacancies per cubic centimeter are there in gold, at ambient temperature, assuming a lattice parameter of 0.408 nm?

4.4 What is the effect of vacancies on electrical conductivity?

4.5 What is the effect of vacancies on the amplitude of vibration of the neighboring atoms?

4.6 What stress is required to render operational a Frank–Read source in iron, knowing that the distance between points B and C is 200 Å and that the Goldschmidt radius of the iron atoms is 0.14 nm?

4.7 Make all possible reactions between (perfect) dislocations in $(11\bar{1})$ and $(1\bar{1}\bar{1})$ in an FCC crystal. Among them, which ones are Lomer locks?

4.8 Consider all possible reactions between partial Shockley dislocations (only the front dislocation, from the pair) in (111) and $(11\bar{1})$ in an FCC crystal. Among them, which ones will form a stair-rod dislocation?

4.9 **(a)** Show that the reaction

$$\frac{a}{2}[10\bar{1}] \rightarrow \frac{a}{6}[21\bar{1}] + \frac{a}{6}[11\bar{2}]$$

is either vectorially correct or incorrect?

   (b) Is the reaction energetically favorable?

4.10 10^7 and 10^{11} cm^{-2} are typical values for the dislocation density of annealed and deformed nickel, respectively. Calculate the average space among dislocation lines (assuming a random dislocation distribution), as well as the line energy for edge and screw dislocations, in both cases. In nickel, $E = 210$ GPa, $\nu = 0.3$, and the lowest distance between atom centers is 0.25 nm.

4.11 Calculate the dislocation density for Figure 4.21b; assume a foil thickness of 0.3 μm.

4.12 The concentration of vacancies in aluminum at 600°C is 9.4×10^{-4}; by quenching, this concentration is maintained at ambient temperature. The vacancies tend to form disks, with Frank partials at the edges. Determine the loop concentration and dislocation density, assuming that:

   (a) Disks with a 5-nm radius are formed.

   (b) Disks with a 50-nm radius are formed.

   For aluminum, assume that the radius of the atoms is 0.143 nm. (*Hint:* The length of the Frank dislocation corresponding to a disk is equal to the circumference of the circle.)

4.13 The flow stress of monocrystals is on the order of 10^{-4} G. Using the concept of Frank–Read sources, determine the length of segments required for this stress level. If the length of the segments is determined by dislocations on a second slip plane ("tree" dislocations), obtain an estimate for the dislocation density in annealed monocrystals. Assume that the dislocations are equally distributed on the slip planes of an FCC crystal.

4.14 On what planes of a BCC structure can the $a/2$ [111] move?

4.15 Upon encountering an obstacle, an edge dislocation stops. A second edge dislocation, with identical Burgers vector and moving in the same plane, approaches the first dislocation, driven by a stress equal to 140 MPa.

  (a) What will be the equilibrium separation between the two dislocations? Assume that the metal is nickel ($E = 210$ GPa, $\nu = 0.3, r = 2.49$ Å).

  (b) What would be the equilibrium separation if the dislocations were both screw dislocations?

4.16 LiF is an ionic crystal with a NaCl-type structure (cubic). The Li atoms occupy the vertices and the centers of the faces of the unit cell, while the F atoms occupy the edges, and one F atom is in the body-centered position. There are eight atoms per unit cell. Knowing that the slip plane for LiF is [110], determine the Burgers vector of a perfect dislocation. Remember that one has an ionic crystal and that there is a strong repulsion between ions of the same sign. Explain your results.

4.12 Draw a unit cell for an HCP crystal. Show the perfect dislocations in the base plane. Can they decompose into partials? If so, represent them by the special notation for dislocations.

4.18 Nickel sheet is being rolled at ambient temperature in a rolling mill (roll diameter 50 cm, velocity 200 rpm). The initial thickness is 20 mm and the final thickness is 10 mm (one pass).

  (a) Calculate the average strain rate.

  (b) Calculate the energy that will be stored in the material, assuming that the final density is 10^{11} cm^{-2}.

  (c) Determine the total energy expenditure per unit volume, assuming a flow stress equal to 300 MPa.

  (d) Assuming that all energy not stored as dislocations is converted into heat, calculate the temperature rise if the process is adiabatic ($C_p = 0.49$ J/g°C).

  (e) Why does the energy stored represent only a fraction of the energy expended?

4.19 Calculate the largest atom that would fit interstitially into (a) nickel (FCC; atomic radius = 0.125 nm) and (b) molybdenum (BCC; atomic radius = 1.36 nm).

4.20 Calculate, for tungsten (BCC; atomic radius = 0.1369 nm), the radii of the largest atoms that can fit into (a) a tetrahedral interstitial site (at $0, 1/4, 1/2$) and (b) an octahedral interstitial site (at $0, 1/2, 1/2$).

4.21 If the enthalpy of formation for a vacancy is equal to 80 kJ/mole, what is the fraction of vacant sites at 1,500 K.

4.22 The lattice parameter of a BCC crystal was measured at ambient temperature and at 1,000°C. The parameter showed an increase of 0.5 percent due to thermal expansion. In the same interval of temperature, the density, measured by a separate method, showed a decrease of 2 percent.

  (a) Assuming that, at room temperature, there is one vacancy per 1,000 atoms, what is the vacancy concentration at 1,000°C?

  (b) Calculate the activation energy necessary for the production of vacancies.

4.23 The Burgers vector of a dislocation is 0.25 nm in a crystal. The shear modulus $G = 40$ GPa. Estimate the dislocation energy per unit length in this crystal.

4.24 A dislocation is anchored between two points 10 μm distant. For a metal with $b = 0.35$ nm and $G = 30$ GPa, compute the shear stress necessary to bow the dislocation into a semicircle. Take the dislocation line tension $T \approx (1/2)Gb^2$.

4.25 Consider an aluminum polycrystal with a grain size of 10 μm. If a dislocation source at the center of a grain emits dislocations under an applied shear stress of 50 MPa that pile up at the grain boundaries, what is the stress experienced by a grain boundary? Take $G = 26$ GPa and $b = 0.3$ nm.

4.26 **(a)** Iron ($r = 0.124$ nm, $G = 70$ GPa) is being deformed to a shear strain of 0.3. Assuming a constant dislocation density equal to 10^{10} cm^{-2}, what is the average distance each dislocation has to move? (b) Assuming that the strain rate is 10^{-2} s^{-1}, what is the average dislocation velocity?

4.27 Iron ($r = 0.124$ nm, $G = 70$ GPa) is deformed to a shear strain of 0.3. A dislocation density equal to 10^{10} cm^{-2} results.
   (a) What is the average distance each dislocation had to move?
   (b) If the strain rate were 10^{-2} s^{-1}, what would be the average dislocation velocity?

4.28 Consider the following dislocation reaction in a face-centered cubic material:

$$\frac{a}{2}[1\bar{1}0] \rightarrow \frac{a}{6}[2\bar{1}1] + \frac{a}{6}[1\bar{2}\bar{1}].$$

Show that the reaction will occur.

4.29 Consider dislocations blocked in gold. If the flow stress is controlled by the stress necessary to operate a Frank–Read source, compute the dislocation density ρ in the crystal when it is deformed to a point where the resolved shear stress on the slip plane is 45 MPa. Take $G = 27$ GPa.

5

Imperfections: Interfacial and Volumetric Defects

In Chapter 4, we dealt with point and line defects. There is another class of defects called *interfacial,* or *planar,* defects. These imperfections, as the name signifies, occupy an area or surface and so are two dimensional, as well as being of great importance. Examples of such defects are free surfaces of a material, grain boundaries, twin boundaries, domain boundaries, and antiphase boundaries. Of all these, grain boundaries are the most important from the point of view of the mechanical properties of the material. In what follows, we consider in detail the structure of grain and twin boundaries and their importance in various deformation processes, and, very briefly, the structure of other interfacial defects. Details regarding the strengthening of a material by grain boundaries are given in Section 5.3. Volumetric defects, such as voids, also play a major role in the mechanical properties of materials, affecting the strength and elastic properties of the material significantly. Volumetric defects are briefly described in Section 5.6. In Section 5.7, we present the defects occurring in polymers.

5.1 GRAIN BOUNDARIES

Crystalline solids generally consist of a large number of grains separated by boundaries. Most industrial metals and ceramics are polycrystalline aggregates, and the mechanical properties of these polycrystals can be radically different from those of the monocrystals that form the individual grains. Figure 5.1 illustrates a polycrystalline aggregate, in which each grain has a distinct crystallographic orientation. The sizes of these individual grains vary from submicrometer (for nanocrystalline and microcrystalline structures) to millimeters and even centimeters (for materials especially processed for high-temperature creep resistance). Figure 5.2 shows typical equiaxed grain configurations for polycrystalline tantalum and titanium carbide. Grains often are elongated through plastic deformation. Each

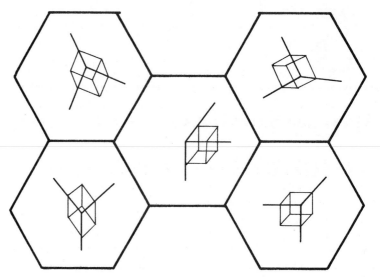

Figure 5.1 Grains in a metal or ceramic; the cube depicted in each grain indicates the crystallographic orientation of the grain in a schematic fashion.

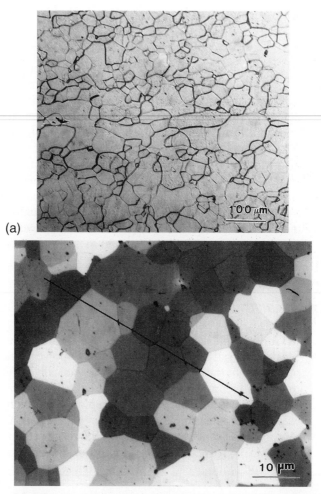

Figure 5.2 Micrographs showing polycrystalline (a) tantalum and (b) TiC.

grain (or subgrain) is a single crystal, and the grain boundaries are thus transition regions between neighboring crystals. These regions may consist of various kinds of dislocations. When the misorientation between two grains is small, the grain boundary can be described by a relatively simple configuration of dislocations (e.g., an edge dislocation wall) and is, fittingly, called a *low-angle boundary*. When the misorientation is large (called, again appropriately, a *high-angle boundary*), more complicated structures are involved (as in a configuration of soap bubbles simulating the atomic planes in crystal lattices). A general grain boundary has five degrees of freedom. Three degrees specify the orientation of one grain with respect to the other, and two degrees specify the orientation of the boundary with respect to one of the grains.

Grain structure is usually specified by giving the average diameter or using a procedure due to ASTM according to which the grain size is specified by the number n in the expression $N = 2^{n-1}$, where N is the number of grains per square inch when the sample is examined at 100 power.

The ASTM procedure is common in engineering applications. In research, it is often preferred to measure the grain size by the lineal intercept technique. In this technique, lines are drawn in the photomicrograph, and the number of grain-boundary intercepts, N_ℓ, along a line is counted. The mean lineal intercept is then

$$\bar{\ell} = \frac{L}{N_\ell M},$$
(5.1)

where L is the length of line and M is the magnification in the photomicrograph of the material. In Figure 5.2, a line is drawn for purposes of illustration. The length of the line is 6.5 cm. The number of intersections, N_ℓ, is equal to 7, and the magnification $M = 1,300$. Thus,

$$\bar{\ell} = \frac{100 \times 10^3}{7 \times 1300} = 11 \ \mu m.$$

Several lines should be drawn to obtain a statistically significant result. The mean lineal intercept $\bar{\ell}$ does not really provide the grain size, but is related to a fundamental size parameter, the grain-boundary area per unit volume, S_v, by the equation

$$\bar{\ell} = \frac{2}{S_v}.$$
(5.2)

The proof of this formula is beyond the scope of this book, but is given by deHoff and Rhines.[†] If we assume, to a first approximation, that the grains are spherical, we have the following relationship between the grain-boundary area and volume:

$$S_v = \frac{1}{2} \frac{4\pi r^2}{\frac{4}{3}\pi r^3}$$

$$= \frac{3}{2r} = \frac{3}{D}.$$
(5.3)

[†]R. T. deHoff and F. N. Rhines (eds.), *Quantitative Microscopy*, New York: McGraw-Hill, 1968.

Here, D is the average grain diameter, and the factor $1/2$ was introduced because each surface is shared between two grains. From Equations 5.2 and 5.3, we get

$$D = \frac{3}{2}\bar{l},$$

which is the most correct way to express the grain size from lineal intercept measurements.

EXAMPLE 5.1

The American Society for Testing and Materials (ASTM) has a simple index, called the ASTM grain size number, n, defined as

$$\boxed{N = 2^{n-1}}$$

where N is the number of grains in an area of 1 in^2 ($= 64.5$ mm^2) in a 100-power micrograph. In one such grain size measurement of an aluminum sample, it was found that there were 56 full grains in the area, and 48 grains were cut by the circumference of the circle of area 1 in^2. (a) Calculate ASTM grain size number n for this sample. (b) Calculate the mean lineal intercept.

Solution: The grains cut by the circumference of the circle are taken as one-half the number. Thus,

$$N = 56 + 48/2$$
$$= 56 + 24 = 80 = 2^{n-1},$$
$$n = \ln N/\ln 2 + 1$$
$$= \ln 80/\ln 2 + 1$$
$$= 4.38/0.69 + 1 = 7.35.$$

(b) For the mean lineal intercept, we use the circle:

$$\pi r^2 = 1 \text{ in}^2,$$
$$r = 0.56 \text{ in},$$
$$\bar{l} = \frac{2\pi r}{N_\ell M} = \frac{2\pi \times 0.56 \times 25.4}{48 \times 100}$$
$$= 0.0186 \text{ mm} = 18.6 \ \mu\text{m}.$$

EXAMPLE 5.2

Determine the grain size for the microstructure shown in Figure E5.2, using both the lineal intercept method and the ASTM method. The straight marks

Figure E5.2

traversing the grains are annealing twins and should be counted in the computation. From the mean lineal intercept, obtain the grain diameter.

Solution: From the ASTM method, $N = 2^{n-1}$, where N is the number of grains per unit
area (in^2) and n is the grain size number.

The number of grains counted is approximately 60, and the area of the
picture is $3.07 \times 4.20 = 12.90$ in^2. So we have

$$N = \frac{60}{12.9} = 4.65.$$

We rewrite N as 2^{n-1}, and taking logarithms, we get

$$\ln N = \ln 2^{n-1}$$
$$= (n-1) \ln 2.$$

So we have

$$1.53 = (n-1) \ln 2,$$
$$n - 1 = 2.24,$$
$$n \approx 3.$$

By the lineal intercept method, $\bar{l} = L/(MN_\ell)$, where $M = 2\,\text{cm}/200\,\text{mm} = 100$ is the magnification, $L = 12$ cm is the straight line drawn, and $N \approx 9$ is the number of intercepts (with grains). Thus, we have

$$\bar{l} = \frac{12}{100 \times 9} = 0.013 \text{ cm}$$

$$= 130 \ \mu\text{m}.$$

5.1.1   Tilt and Twist Boundaries

The simplest grain boundary consists of a configuration of edge dislocations between two grains. The misfit in the orientation of the two grains (one on each side of the boundary) is accommodated by a perturbation of the regular arrangement of atoms in the boundary region. This is very clearly seen in the high-resolution transmission electron micrograph of Figure 5.3. A low-angle grain-boundary with a misorientation $\theta = 10°$ between equivalent (100) planes is shown, and the dislocations are highlighted by circles marking their Burgers vector.

    Figure 5.4 shows some vertical atomic planes terminating in a boundary, and each termination is represented by an edge dislocation. The misorientation at the boundary is related to the spacing between dislocations, D, by the relation (see triangle with dimensions)

$$D = \frac{b/2}{\sin(\theta/2)} \simeq \frac{b}{\theta} \qquad \text{(for very small } \theta\text{),} \qquad (5.4)$$

where b is the Burgers vector.

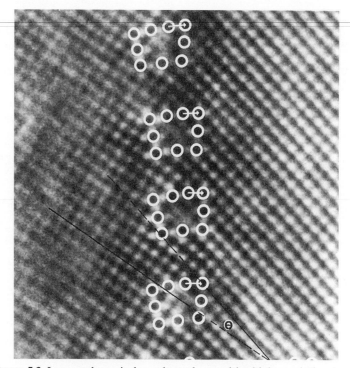

Figure 5.3 Low-angle grain-boundary observed by high-resolution transmission electron microscopy. Positions of individual dislocations are marked by Burgers circuits. (Courtesy of R. Gronsky)

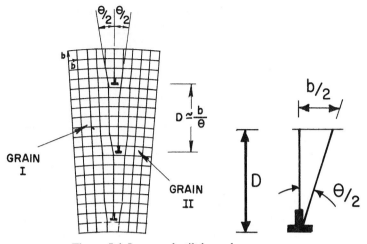

Figure 5.4 Low-angle tilt boundary.

It is instructive to calculate the spacing between dislocations in Figure 5.3 and to compare it with the measured value from the electron micrograph. We will express all values in terms of the lattice spacing along [100] directions. Let us call this value a, so that

$$b = 1.3a.$$

The calculated dislocation spacing (from the measured angle $\theta = 10° \approx (1/6$ rd$)$ is

$$D = \frac{1.3a}{\theta} \approx 7.8a.$$

The measured dislocation spacing in Figure 5.3 is

$$D = 8a.$$

Thus, the agreement with Equation 5.4 and Figure 5.3 is excellent.

As the misorientation θ increases, the spacing between dislocations is reduced, until, at large angles, the description of the boundary in terms of simple dislocation arrangements does not make sense. Theta becomes so large, that the dislocations are separated by one or two atomic spacings; for such small separations, the dislocation core energy becomes important and the linear elasticity does not hold. In these cases, the grain boundary is a region of severe localized disorder.

Boundaries consisting entirely of edge dislocations are called *tilt boundaries,* because the misorientations, as can be seen in Figure 5.4, can be described in terms of a rotation about an axis normal to the plane of the paper and contained in the plane of dislocations. The example shown in that figure is called a *symmetrical tilt wall,* as the two grains are symmetrically located with respect to the boundary. A boundary consisting entirely of screw dislocations is called a *twist boundary,* because the misorientation can be described by a

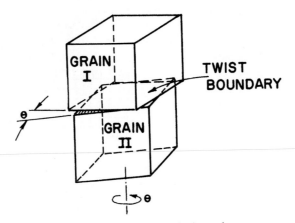

Figure 5.5 Low-angle twist boundary.

relative rotation of two grains about an axis. Figure 5.5 shows a twist boundary consisting of two groups of screw dislocations.

It is possible to produce misorientations between grains by combined tilt and twist boundaries. In such a case, the grain boundary structure will consist of a network of edge and screw dislocations.

5.1.2 Energy of a Grain Boundary

The dislocation model of a grain boundary can be used to compute the energy of low-angle boundaries ($\theta \leq 10°$). For such boundaries, the distance between dislocations in the boundary is more than a few interatomic spaces. We have

$$\frac{b}{D} \simeq \theta \leq 10° \simeq \frac{1}{6} \text{rad} \quad \text{or} \quad D \approx 6b,$$

and

$$E = E_{\perp}\left(\frac{1}{D}\right) = E_{\perp}\left(\frac{\theta}{b}\right). \tag{5.5}$$

Thus, the energy of a tilt boundary is given by (this derivation is not trivial and is not given here)[†]

$$E = E_0\theta(A - \ln \theta), \tag{5.6}$$

where A is a parameter that emerges in the derivation and

$$E_0 = \frac{Gb}{4\pi(1 - \nu)}. \tag{5.7}$$

EXAMPLE 5.3

In a low-angle tilt boundary in an aluminum sample, the misorientation is 5°. Estimate the spacing between dislocations in this boundary, given that $b_{Al} = 0.29$ nm.

Solution: We have

$$b = 0.29 \text{ nm}, \qquad \theta = 5° = 5/57.3 = 0.087 \text{ rad.}$$

[†]M. A. Meyers and K. K. Chawla, *Mechanical Metallurgy,* Prentice Hall (1984), pp. 273–275.

The dislocation spacing is

$$D = b/\theta = 0.29 \text{ nm}/0.087 = 3.33 \text{ nm}.$$

EXAMPLE 5.4

Calculate the energy of a low-angle tilt boundary in nickel as a function of the misorientation θ, for $0 < \theta < 10$. For Ni, $r = 0.125$ nm, $G = 76$ GPa, and $v = 0.31$.

Solution: We have

$$E = \frac{Gb}{4\pi(1 - v)} \theta(A - \ln \theta)$$

We first calculate b; we use $a = 2r\sqrt{2}$, and the magnitude of [100] Burgers vectors is

$$d_{[110]} = \frac{a}{\sqrt{h^2 + k^2 + l^2}} = \frac{a}{\sqrt{2}}.$$

Thus,

$$|\mathbf{b}| = \frac{a}{\sqrt{2}} = 2r = 0.250 \text{ nm}$$

and

$$E = \frac{76 \times 10^9 \times 0.25 \times 10^{-9}}{4\pi(1 - 0.31)} \theta(A - \ln \theta).$$

We can assume that the dislocation energy is equal to the core energy when the separation between them is equal to $10b$. This is twice the core radius used by many scientists. From that value, we obtain the value of the constant of integration, A. The sequence of equations is

$$U = \frac{Gb^2}{10} \quad \text{for} \quad D = 10b = \frac{b}{\theta},$$

$$E = \frac{Gb^2}{10D} = \frac{Gb^2}{100b} = \frac{Gb}{100} \quad (\theta = 0.1),$$

$$\frac{Gb}{100} = \frac{Gb \times 0.1}{4\pi(1 - v)} (A - \ln 0.1),$$

$$A = \frac{4\pi(1 - v)}{10} + \ln 0.1 = 0.866 - 2.30$$

$$= -1.436.$$

So

$$E = 2.2\theta \, (-1.436 - \ln \theta).$$

EXAMPLE 5.5

Calculate the dislocation spacing and energy of a low-angle tilt boundary in copper crystal if $\theta = 0.5°$, $G = 48.3$ GPa, $\nu = 0.343$, and $r_{Cu} = 0.157$ nm.

Solution: The spacing is

$$D = \frac{b}{\theta}.$$

For FCC copper,

$$b = \frac{a}{\sqrt{2}}, \qquad 4r_{Cu} = \sqrt{2}a.$$

$$b = \frac{(4/\sqrt{2})r_{Cu}}{\sqrt{2}} = 2r_{Cu} = 0.314 \text{ nm},$$

$$\theta = 0.5° = \frac{0.5}{180}\pi = 0.009 \text{ rad},$$

$$D = \frac{b}{\theta} = \frac{0.314}{0.009} = 34.9 \text{ nm}.$$

We next assume that $D = 10b$, so

$$D = \frac{b}{\theta} = 10b,$$

$$\theta = 0.1.$$

We thus have

$$E = E_{\perp}\left(\frac{1}{D}\right) = \frac{Gb^2}{10}\left(\frac{1}{D}\right) = \frac{Gb}{100}, \qquad \theta = 0.1.$$

Also,

$$E = \frac{Gb}{4\pi(1-\nu)}\theta(A - \ln\theta).$$

Hence, setting the two equations for E equal to each other, we obtain

$$\frac{Gb}{100} = \frac{Gb}{4\pi(1-\nu)}\theta(A - \ln\theta)$$

and it follows that

$$A = \frac{4\pi(1-\nu)}{100\theta} + \ln\theta$$

$$= \frac{4\pi(1-0.343)}{100 \times 0.1} + \ln 0.1$$

$$= -1.477.$$

Substituting this value of A into the second equation for E yields

$$E = \frac{Gb}{4\pi(1 - \nu)}\, \theta(-1.477 - \ln \theta).$$

Now, given that G = 48.3 GPa, ν = 0.343, and θ = 0.0009 rad, we obtain

$$E = \frac{48.3 \times 10^9 \times 0.314 \times 10^{-9}}{4\pi(1 - 0.343)} \times 0.009 \times (-1.477 - \ln 0.009)$$

$$= 0.053 \text{ J/m}^2.$$

5.1.3 Variation of Grain-Boundary Energy with Misorientation

Consider Equation 5.6. Because of the $(-ln\ \theta)$ term, a merger of two low-angle bound-aries, forming a high-angle boundary, always results in a net decrease in the total energy of the interface. Thus, low-angle boundaries have a tendency to combine and form bound-aries of large misorientation.

A plot of E versus θ gives a curve with a maximum at $\theta_{\max} \approx 0.5$ rad ($\approx 30°$). How-ever, the dislocation model of grain boundaries loses validity at much smaller orientations ($\theta \le 10°$). Some recent studies, using field-ion microscopy, have shown that the high-angle grain boundaries consist of rather large regions of atomic fit separated by regions of misfit, to which are associated the grain-boundary ledges. The boundary thickness is not more than two to three atomic diameters. Low-angle grain boundaries have a dislocation density that increases proportionally to the misorientation angle (see Equations 5.4 and 5.6), and, consequently, the energy of a low-angle boundary increases linearly with θ near $\theta = 0°$. After this, the energy increases slowly as the stress fields of adjacent disloca-tions interact more strongly. This behavior is shown in Figure 5.6. A surface tension, γ_{gb}, can be associated with an ordinary (high-angle) grain boundary, which consists of a mix-ture of various types of dislocations. Because the value of γ_{gb} is relatively high, it is in-structive to determine the stable forms assumed by the grains of a given material. As it happens, there are certain special boundaries for which a particular high angle between two adjacent crystals produces a low value of γ. These special boundaries can be divided

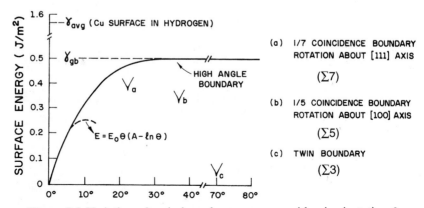

Figure 5.6 Variation of grain-boundary energy γ_{gb} with misorientation θ. (Adapted with permission from A. G. Guy, *Introduction to Materials Science* (New York: McGraw-Hill, 1972), p. 212.

into two categories: coincidence boundaries and coherent twin boundaries. A coincidence boundary (Figure 5.7) is incoherent, as is an ordinary grain boundary; that is, a majority of the atoms of one crystal in the boundary do not correspond to the lattice sites of the other crystal. On an average, however, this noncorrespondence in a coincidence boundary is less as the density of coincidence sites increases. For example, in the figure, one atom in seven in the boundary is in a lattice position for both the crystals. We call this boundary a *one-seventh coincidence boundary,* and the atomic sites (the black atoms in the figure) in question form a coincidence lattice for the two grains. Coincidence lattices occur in all common crystalline structures and have a density of sites varying from $\frac{1}{3}$ to $\frac{1}{9}$ and less.

A twin boundary is frequently a kind of coincidence boundary, but it is convenient to treat it separately. The energy of a twin boundary, γ_{twin}, is generally about $0.1\,\gamma_{gb}$ (see Figure 5.6), whereas the energy of a coincidence boundary is only slightly less than γ_{gb}. The two most common twin orientations are (1) rotation twins (coincidence), produced by a rotation about a direction $[hkl]$ called the twinning axis, and (2) reflection twins, in which the two lattices maintain a mirror symmetry with respect to a plane $[hkl]$ called the twinning plane.

Some of the orientations that give the highest density of coincidence lattice sites in crystals are shown in Table 5.1. These boundaries have lower energies than those of random high-angle boundaries. Contrary to the great majority of low-energy boundaries, coincidence site boundaries have greater mobility than that of random boundaries. Twin boundaries, even with low energies, have lower mobility because they are coherent.

The interfaces between different phases (interphase interfaces) are more complex, since the accommodation of the atoms has to be more drastic. Nevertheless, strong interfaces can be formed, even between different ceramic phases. An illustration of this is provided in Figure 5.8, which shows the interface between alumina (hexagonal) and the spinel structure ($NiAl_2O_4$). In ceramics, the requirement of electrical charges puts additional restrictions on the boundaries. Nevertheless, the boundary shown in Figure 5.8a and the atomic positions clearly marked in Figure 5.8b are of high coherence.

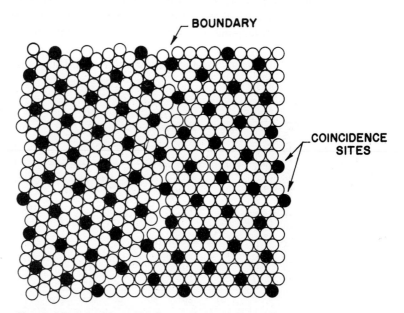

Figure 5.7 Coincidence lattice made by every seventh atom in the two grains, misoriented $22°$ by a rotation around the $\langle 111 \rangle$ axis. (Adapted from M. L. Kronberg and H. F. Wilson, *Trans. AIME,* 85 (1949), 501, p. 506.)

TABLE 5.1 Some Coincidence Site Boundaries in FCC Crystals

Rotation Axis	Rotation Angle (deg)	Density of Coincidence Sites
(111)	38	1 in 7
	22	1 in 7
	32	1 in 13
	47	1 in 19
(110)	39	1 in 9
	$50\frac{1}{2}$	1 in 11
	$26\frac{1}{2}$	1 in 19
(100)	37	1 in 15

Reprinted with permission from J. W. Christian, *The Theory of Transformation in Metals and Alloys* (Elmsford, NY: Pergamon Press, 1965), p. 326.

5.1.4 Coincidence Site Lattice (CSL) Boundaries

It is instructive to consider some other important aspects of coincidence site lattice (CSL) boundaries. As described earlier, we get a CSL boundary when a certain rotation of one grain relative to another grain results in a three-dimensional atomic pattern in which a certain fraction of lattice points coincide in the two grains. The volume of the CSL primitive cell is a small multiple of the volume of the lattice primitive cell. Such a CSL boundary is characterized by a parameter Σ, the reciprocal of the fraction of lattice sites that coincide (in Table 5.1, $\Sigma = 7, 9, 13, 15, 19$). Equivalently, Σ is the ratio of the volume of the CSL primitive cell to that of the lattice primitive cell. A coherent twin boundary is $\Sigma 3$. It has been observed that CSL grain boundaries with relatively low values of Σ can have a significant influence on the mechanical behavior of a polycrystalline material. CSL boundaries with small values of Σ result in short-period ordered structures in the grain boundary. CSL boundaries with Σ less than 29 show the following advantages over random grain boundaries or boundaries with higher Σ values:

- lower grain boundary energy in pure metals
- lower diffusivity
- lower electrical resistivity
- lower susceptibility to solute segregation
- greater resistance to grain boundary sliding, fracture, and cavitation
- greater resistance to initiation of localized corrosion
- greater boundary258 mobility with specific solutes in a specific concentration range.

It would thus appear that control of the character and density of low-Σ boundaries can be a means of producing a superior polycrystalline material.

5.1.5 Grain-Boundary Triple Junctions

Grain-boundary triple junctions are sites where four grains or three grain boundaries meet. Such boundaries are commonly observed in crystalline materials. The number of triple junctions can have a great influence on the mechanical properties of the material. The number of triple junctions in a polycrystalline material will depend on the grain size and

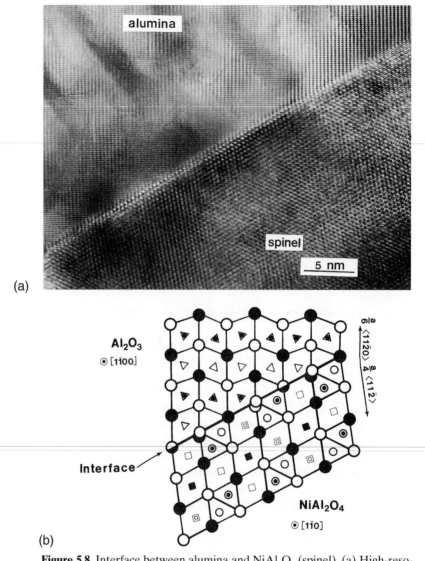

Figure 5.8 Interface between alumina and $NiAl_2O_4$ (spinel). (a) High-resolution TEM. (b) Representation of individual atomic positions. (Courtesy of C. B. Carter)

crystal geometry of the material. Palumbo et al.[1] considered a three-dimensional distribution of tetrakaidecahedral grains and obtained the volume fractions of intercrystalline region (grain-boundary) and triple-boundary junctions. Figure 5.9 shows the effect of grain size on calculated volume fractions of these entities. Note the highly pronounced effect for grain sizes less than 20 nm, i.e., in the nanometer range.

5.1.6 Grain-Boundary Dislocations and Ledges

Various experimental observations of the structure of grain boundaries have demonstrated the existence of grain-boundary dislocations (GBDs) when the orientation rela-

[1]B. Palumbo, S. J. Thorpe, and K. T. Aust, *Scripta Met., 24* (1990) 1347.

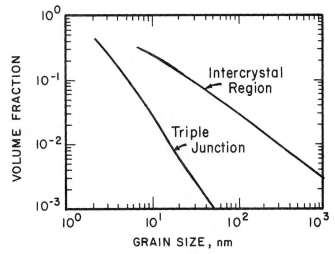

Figure 5.9 The effect of grain size on calculated volume fractions of intercrystal regions and triple junctions, assuming a grain boundary thickness Δ of 1 nm. (Adapted from B. Palumbo, S. J. Thorpe, and K. T. Aust, *Scripta Met.*, 24 (1990) 1347)

tions deviate from the ideal coincidence lattice site orientations. A grain-boundary dislocation belongs to the grain boundary and is not a common lattice dislocation.

Grain-boundary dislocations can acquire the geometry of a grain-boundary ledge by grouping together. This agglomeration, which leads to the formation of a step, is shown in Figure 5.10. Figure 5.10a shows the movement of GBDs along the grain-boundary plane in the direction indicated by the arrow. Figure 5.10b shows the coalescence of GBDs to make a grain-boundary ledge. Another way of ledge formation is shown in Figures 5.10c and d. Under the applied tension, lattice dislocations can move from grain A through the boundary plane to grain B (Figure 5.10c). The passage through the boundary results in heterogeneous shear of the boundary, forming a ledge.

The distinction between a ledge and an intrinsic GBD is one of height; the smallest ledge corresponds to a GBD. Detailed analyses showing how slip can transfer from one grain to another via the formation of intrinsic GBDs have been carried out. Figure 5.11 shows a TEM that reveals ledges and GBDs. The larger steps can be considered ledges, whereas the lines could be GBDs.

In the simplified situation shown in Figure 5.10, the (111) planes of the neighboring grains intersect along the boundaries. Ledges in the grain boundaries constitute an important structural characteristic of the high-angle boundaries. It has been observed that the density of ledges increases with an increase in the boundary misorientation. One of the important aspects of this structure of boundaries is that the ledges can function as effective sources of dislocations, a fact that has important implications for the mechanical properties of polycrystals.

5.1.7 Grain Boundaries as a Packing of Polyhedral Units

The grain-boundary structure can also be described in terms of a packing of polyhedral units. If equal spheres are packed to form a shell such that all spheres touch their neighbors, then the centers of the spheres are at the vertices of a "deltahedron," a polyhedron with equilateral triangles as faces. Ashby et al.[2] regard a crystal as a regular packing of

[2]M. F. Ashby, F. Spaepen, and S. Williams, *Acta Met., 26* (1978) 1053.

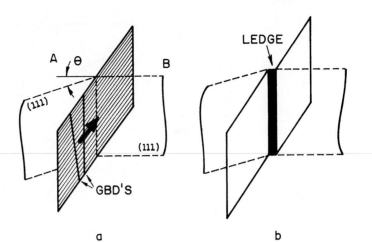

a

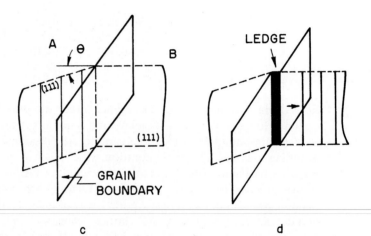

b

c

d

Figure 5.10 Models of ledge formation in a grain boundary. (Reprinted with permission from L. E. Murr, *Interfacial Phenomena in Metals and Alloys,* Reading, MA: Addison Wesley, 1975, p. 255)

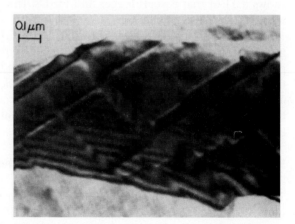

Figure 5.11 Grain boundary ledges observed by TEM. (Courtesy of L. E. Murr)

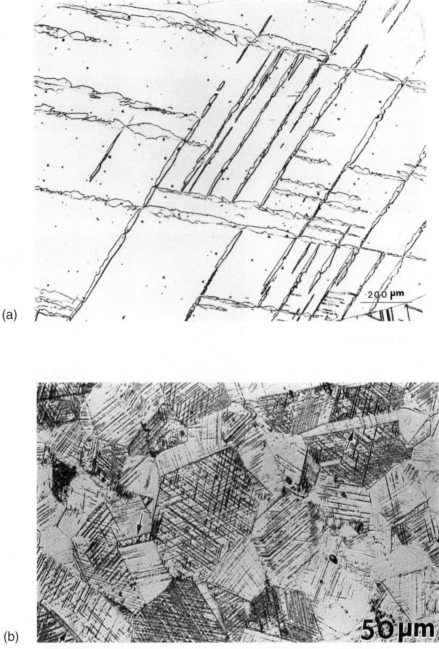

(a)

(b)

Figure 5.14 Deformation twins in (a) iron-silicon (Courtesy of O. Vöhringer) and (b) stainless steel.

requirements. The sagacious reader will note that there is a mixture of annealing and deformation twins in the figure, but this issue will not be discussed. Annealing twins occur during recrystallization and grain growth, and do not contribute to plastic strain. They are parallel faced and not lenticular. In the figure, three annealing twins are designated by arrows.

Figure 5.15 illustrates the formation of deformation twins in ceramics. A grain of silicon nitride subjected to compressive loading is imaged in Figure 5.15a. The diffraction pattern (the spots in Figure 5.15c) reveals more than one crystallographic orientation. It is possible,

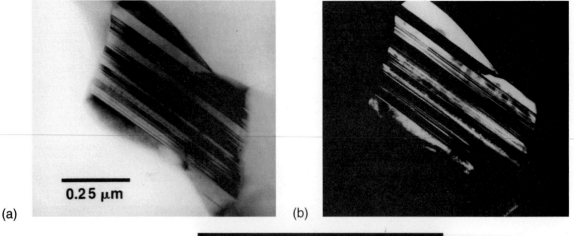

(a)

(b)

(c)

Figure 5.15 Deformation twins in silicon nitride observed by TEM. (a) Bright field. (b) Dark field. (c) Electron diffraction pattern showing spots from two twin variants, *A* and *B*. (Courtesy of K. S. Vecchio)

by focusing on only one family of reflections, to image one family of twins. This is shown in Figure 5.15b. All the bright twins in this dark-field image have the same orientation.

The mechanism of plastic deformation by twinning is very different from that of slip. First, the twinned region of a grain is a mirror image of the original lattice, while the slipped region has the same orientation as that of the original, unslipped grain. Second, slip consists of a shear displacement of an entire block of crystal, while twinning consists of uniform shear strain. Third, the slip direction can be positive or negative (i.e., in tension or compression), while the twinning direction is always polar. Twinning results in a change of shape of a definite type and magnitude, as determined by the crystallographic nature of the twinning elements.

The stress necessary to form twins is, generally, greater, but less sensitive to temperature, than that necessary for slip. This stress required to initiate twinning is much larger than the stress necessary for its propagation. Deformation twinning occurs when the applied stress is high due to work-hardening, low temperatures, or, in the case of HCP metals, when the resolved shear stress on the basal plane is low. Copper and other FCC metals can be made to deform by twinning at very low temperatures or at very high strain rates. Deformation twins, however, play an important role in the straining of HCP metals. The "cry" heard when a polycrystalline sample of tin is bent plastically is caused by the sudden formation of deformation twins. The bursting of twins during straining can lead to a serrated form of stress–strain curve (Figure 5.16). In many HCP metals, the slip is restricted

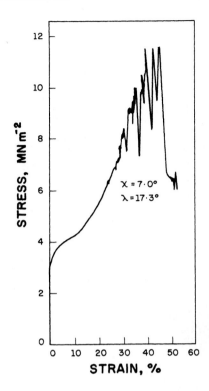

Figure 5.16 Serrated stress–strain curve due to twinning in a Cd single crystal. (Adapted with permission from W. Boas and E. Schmid, *Z. Phys.,* 54 (1929) 16, p. 35)

to basal planes. Thus, twinning can contribute to plastic deformation by the shear that it produces, but this is generally small. (See Table 5.2.) More importantly, the twinning process serves to reorient the crystal lattice to favor further basal slip. In HCP metals, the common twinning elements are the $(10\bar{1}2)$ plane and $[10\bar{1}1]$ direction (see Fig. 5.17). Twinning results in a compression or elongation along the c-axis, depending on the ratio c/a. For $c/a > \sqrt{3}$ (the case of Zn and Cd), twinning occurs on $(10\bar{1}2)$ $[10\bar{1}1]$ when the metal is compressed along the c-axis. When $c/a = \sqrt{3}$, the twinning shear is zero. For $c/a < \sqrt{3}$ (the case of Mg and Be), twinning occurs under tension along the c-axis. Figure 5.17 shows this dependence on the ratio c/a.

5.2.2 Mechanical Effects

One may regard slip and twinning as competing mechanisms; experimentally, it has been found that either an increase in strain rate or a decrease in temperature tends to favor twinning over slip. In this context, the graphical scheme proposed by G. Thomas and presented in Figure 5.18 is helpful. The low temperature dependence of the stress required to initiate twinning is a strong indication that it is not a thermally activated mechanism. Hence, τ/G for twinning is not temperature dependent. On the other hand, the thermally activated dislocation motion becomes very difficult at low temperatures; T_t is the temperature below which the material will yield by twinning in conventional deformation. However, at high strain rates, dislocation generation and dynamics are such that the whole curve is translated upward, while the twinning curve is stationary, for reasons that will be given later. As a consequence, the intersection of the two curves takes place at a higher temperature.

As the stacking-fault energy of an alloy is decreased, the propensity for twinning increases. The addition of zinc to copper decreases the stacking-fault energy dramatically, from 78 mJ/m^2 (for pure Cu) to 7 mJ/m^2 (for 75–25 brass). This leads to a much greater planarity of slip, which eventually results in twinning. Twinning generates internal barriers to slip and breaks down a material's microstructure into progressively smaller domains.

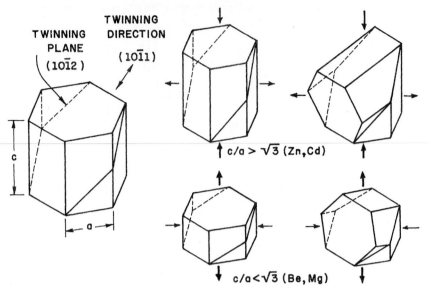

Figure 5.17 Twinning in HCP metals with c/a ratio more than or less than $\sqrt{3}$.

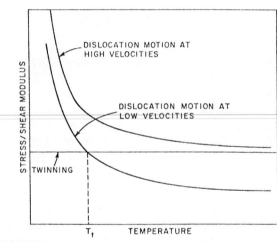

Figure 5.18 Effect of temperature on the stress required for twinning and slip (at low and high strain rates). (Courtesy of G. Thomas)

The result is an increase in work-hardening; that is, the movement of dislocations is hampered. Figure 5.19(a) illustrates this effect. The work-hardening rate of copper decreases with plastic strain, in the expected fashion, while brass, in which twinning is prevalent, shows an almost constant work-hardening, over a significant plastic strain range. The onset of twinning is clearly seen in the plateau of the work-hardening rate, in Fig 5.19(b).

5.3 GRAIN BOUNDARIES IN PLASTIC DEFORMATION (GRAIN-SIZE STRENGTHENING)

Grain boundaries have a very important role in the plastic deformation of polycrystalline materials. The following are among the more important aspects of this role:

1. At low temperatures ($T < 0.5T_m$, where T_m is the melting point in K), the grain boundaries act as strong obstacles to dislocation motion. Mobile dislocations can

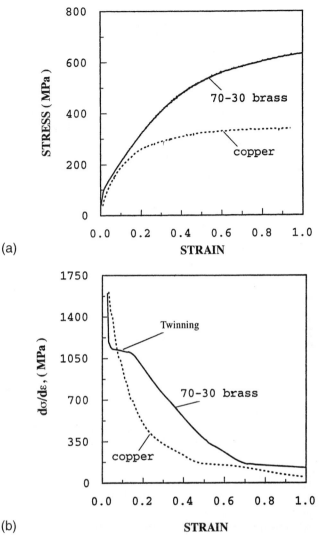

Figure 5.19 (a) Stress–strain curves for copper (which deforms by slip) and 70% Cu–30% Zn brass (which deforms by slip and twinning). (b) Work-hardening slope $d\sigma/d\varepsilon$ as a function of plastic strain; a plateau occurs for brass at the onset of twinning. (Courtesy of A. Staroselsky and L. Anand)

pile up against the boundaries and thus give rise to stress concentrations that can be relaxed by initiating locally multiple slip.

2. There exists a condition of compatibility among the neighboring grains during the deformation of polycrystals that is, if the development of voids or cracks is not permitted, the deformation in each grain must be accommodated by its neighbors.[†] This accommodation is realized by multiple slip in the vicinity of the boundaries, which leads to a high strain-hardening rate. It can be shown, following von Mises, that for each grain to stay in contiguity with others during deformation, at least five independent slip systems must be operating. (See Section 6.2.5.) This condition of strain compatibility leads a polycrystalline sample to have multiple slip in the vicinity of grain boundaries. The smaller the grain size, the larger will be the total boundary surface area per unit volume. In other words, for a given

[†]J. P. Hirth, *Met. Trans.* 3 (1972) 3047.

deformation in the beginning of the stress–strain curve, the total volume occupied by the work-hardened material increases with decreasing grain size. This implies a greater hardening due to dislocation interactions induced by multiple slip.

3. At high temperatures, the grain boundaries function as sites of weakness. Grain boundary sliding may occur, leading to plastic flow or opening up voids along the boundaries. (See Chapter 13.)

4. Grain boundaries can act as sources and sinks for vacancies at high temperatures, leading to diffusion currents, as, for example, in the Nabarro–Herring creep mechanism. (See Chapter 13.)

5. In polycrystalline materials, the individual grains usually have a random orientation with respect to one another. Frequently, however, the grains of a material may be preferentially oriented. For example, an Fe–3% Si solid–solution alloy, used for electrical transformer sheets because of its excellent magnetic properties, has grains with their [110] planes nearly parallel and their ⟨100⟩ direction along the rolling direction of the sheet. This material is said to have a *texture* or *preferred orientation*. A preferred orientation of grains is also frequently observed in drawn wires.

Ever since Hall and Petch[3] introduced their well-known relationship between the lower yield point of low-carbon steels and grain size, a great deal of effort has been devoted to explaining that relationship from a fundamental point of view and applying it to the yield and flow stress of different metals and alloy systems. The Hall–Petch (H–P) equation has the form

$$\sigma_y = \sigma_0 + kD^{-1/2}, \tag{5.8}$$

where σ_y is the yield stress, σ_0 is a frictional stress required to move dislocations, k is the H–P slope, and D is the grain size. This equation has been applied to many systems, with varying degrees of success. It seems to be a satisfactory description of the dependence of yield stress on grain size when a somewhat limited range of grain sizes is being investigated. Figure 5.20 illustrates the Hall–Petch equation for several metals. BCC and FCC metals exhibiting smooth elastic–plastic transitions and yield points are represented. Table 5.3 presents the parameters for a number of metals.

Figure 5.21 shows the yield strength of iron over a much wider range than that presented in Figure 5.20. The plot is of the Hall–Petch line (full line) and the upper bound (theoretical strength, assumed to be $E/30$), as well as the lower bound (single crystal). Substantial deviations from a single Hall–Petch curve that has approximately the slope for ferrovac E steel and 0.05C steel are observed. The very broad range of grain sizes is the reason for the deviation. Thus, the Hall–Petch behavior should be considered not a universal law, but an approximation over a limited range of grain sizes. Since most engineering alloys have grain sizes in the range 10–100 μm, the Hall–Petch equation is indeed very useful.

The principal theories advanced to explain the Hall–Petch relationship are presented next. The first two theories have lost a lot of their credibility, because dislocation pileups are not thought to be as important as they used to be, especially in high-stacking-fault-energy materials.

5.3.1 Hall–Petch Theory

The basic idea behind the separate propositions of Hall and Petch is that a dislocation pileup can "burst" through a grain boundary due to stress concentrations at the head of

---

[3] E. O. Hall, *Proc. Roy. Soc. (London)* B64 (1951) 474; N. J. Petch, *J. Iron Steel Inst.* 174 (1953) 25.

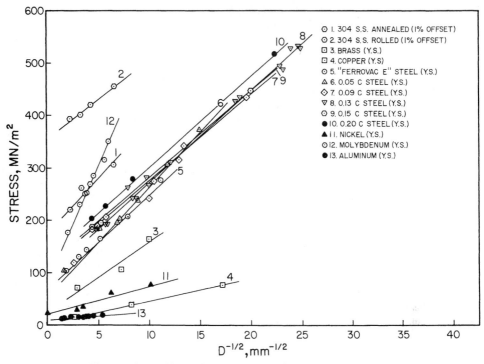

Figure 5.20 Hall–Petch plot for a number of metals and alloys.

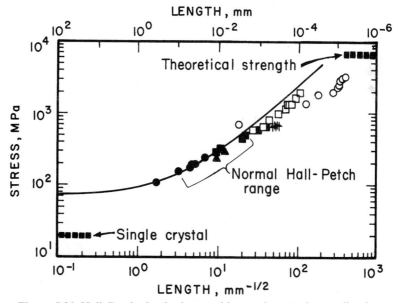

Figure 5.21 Hall–Petch plot for iron and low-carbon steel extending from monocrystal to nanocrystal; notice the change in slope. (After T. R. Smith, R. W. Armstrong, P. M. Hazzledine, R. A. Masumura, and C. S. Pande, *Matls. Res. Soc. Symp. Proc.*, Vol. 362.) 1995, p. 31.

TABLE 5.3 Tabulation of σ_0 and k Values for BCC, FCC, and HCP Structures[a]

Material Specification[b]	σ_0 (MPa)	k MN/m$^{3/2}$
Body-Centered Cubic		
Mild steel, y.p.	70.60	0.74
Mild steel, $\varepsilon = 0.10$	294.18	0.39
Swedish, iron, y.p.	47.07	0.71
Swedish iron, no y.p.	36.28	0.20
Fe–3% Si, y.p.–196°C	505.99	1.54
Fe–3% Si, twinning–196°C	284.37	3.32
Fe–18% Ni, $\varepsilon = 0.002$	650.14	0.22
Fe–18% Ni, twinning–196°C	843.32	1.30
FeCo, ordered, $\varepsilon = 0.004$	50.01	0.90
FeCo, disordered, $\varepsilon = 0.004$	319.68	0.33
Chromium, y.p.	178.47	0.90
Chromium, twinning–196°C	592.52	4.37
Molybdenum, y.p.	107.87	1.77
Molybdenum, $\varepsilon = 0.10$	392.24	0.53
Tungsten, y.p.	640.33	0.79
Vanadium, y.p.	318.70	0.30
Niobium, y.p.	68.64	0.04
Tantalum, with 0_2, y.p. 0°C	186.31	0.64
Face-Centered Cubic		
Copper, $\varepsilon = 0.005$	25.50	0.11
Cu–3.2% Sn, y.p.	111.79	0.19
Cu–30% Zn, y.p.	45.11	0.31
Aluminum, $\varepsilon = 0.005$	15.69	0.07
Aluminum, fracture, 4K	539.33	1.67
Al–3.5% Mg, y.p.	49.03	0.26
Silver, $\varepsilon = 0.005$	37.26	0.07
Silver, $\varepsilon = 0.002$	23.53	0.17
Silver, $\varepsilon = 0.20$	150.03	0.16
Hexagonal Close Packed		
Cadmium, $\varepsilon = 0.001$–196°C	17.65	0.35
Zinc, $\varepsilon = 0.005$, 0°C	32.36	0.22
Zinc, $\varepsilon = 0.175$, 0°C	71.58	0.36
Magnesium, $\varepsilon = 0.002$	6.86	0.28
Magnesium, $\varepsilon = 0.002$–196°C	14.71	0.47
Titanium, y.p.	78.45	0.40
Zirconium, $\varepsilon = 0.002$	29.42	0.25
Beryllium, y.p.	21.57	0.41

[a]Adapted with permission from R. W. Armstrong, in *Advances in Materials Research*, Vol. 5, R. F. Bunshah, ed. (New York: Wiley-Interscience, 1971), p. 101.

[b]y.p.= yield point.

the pileup. If τ_a is the resolved shear stress applied on the slip plane, then the stress acting at the head of a pileup containing n dislocations is $n\tau_a$ (Eq. 4.20). The number of dislocations in a pileup depends on the length of the pileup, which, in turn, is proportional to the grain diameter D. According to Eshelby et al.,[4]

$$n = \frac{\alpha d \tau_a}{Gb/\pi},$$

(5.9)

where d is the length of the pileup—that is, for a pileup with two ends (generated, for instance, by a Frank–Read source in the center of the grain) equal to half the grain diameter D—and α is a geometrical constant equal to unity for the pileup of screw dislocations and equal to $(1 - \nu)$ for edge dislocations. If τ_c is the critical stress required to overcome the grain-boundary obstacles, then the dislocations of the pileup will be able to traverse the grain boundary if

$$n\tau_a \geqslant \tau_c.$$

(5.10)

From Equation 5.9

$$\frac{\alpha D \tau_a}{2Gb/\pi} \tau_a \geqslant \tau_c, \quad \text{or} \quad \frac{\alpha \pi D \tau_a^2}{2Gb} \geqslant \tau_c.$$

In order to take into account the friction stress τ_0 needed to move the dislocations in the absence of any obstacle, we have to add the term τ_0. Thus,

$$\tau_a \geqslant \tau_0 + kD^{-1/2}.$$

(5.11)

Equation 5.11 is essentially identical to Equation 5.8, once the shear stresses are converted into normal stresses. Note that Eshelby's equation is valid only for a large number of dislocations; hence, the equation is not applicable to grain sizes below a few micrometers.

5.3.2 Cottrell's Theory

Cottrell[5] used a somewhat similar approach to that of Hall and Petch; however, he recognized that it is virtually impossible for dislocations to "burst" through boundaries. Instead, he assumed that the stress concentration produced by a pileup in one grain activated dislocation sources in the adjacent grain. Figure 5.22 shows how a Frank–Read source at a distance r from the boundary is activated by the pileup produced by a Frank–Read source in the adjacent grain. The slip band blocked in the boundary was treated by Cottrell as a shear crack. The maximum shear stress at a distance r ahead of a shear crack is given by

$$\tau = (\tau_a - \tau_0)\left(\frac{D}{4r}\right)^{1/2},$$

where τ_0 is the frictional stress required to move dislocations and $r < D/2$. The stress required to activate the Frank–Read source in the neighboring grain is given by

$$\tau_c = (\tau_a - \tau_0)\left(\frac{D}{4r}\right)^{1/2},$$

[4] J. D. Eshelby, F. C. Frank, and F. R. N. Nabarro, *Phil. Mag.* 42 (1951) 351.

[5] A. H. Cottrell, *Trans. TMS-AIME*, 212 (1958) 192.

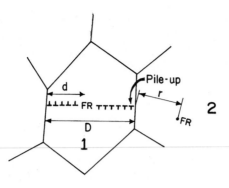

Figure 5.22 Frank–Read source operating in center of grain and producing two pileups at grain boundaries; the Frank–Read source in grain 2 is activated by stress concentration.

or

$$\tau_a = \tau_0 + 4\tau_c\tau^{1/2}D^{-1/2}.$$

This equation is of a Hall–Petch form.

5.3.3 Li's Theory

Li[6] used a different approach to obtain a relationship between the yield stress and grain size. Instead of using pileups, he considered the grain boundary to be a source of dislocations. The concept of grain-boundary dislocation sources is discussed in Section 4.2.6, and it is thought that the onset of yielding in polycrystals is associated with the activation of these sources. Li suggested that the grain-boundary ledges generated dislocations, "pumping" them into the grain. Figure 5.23 shows dislocation activity in stainless steel in the grain-boundary regions. These patterns can be interpreted as being due to dislocation pileups or dislocation emission from grain-boundary ledges. Such dislocations act as Taylor (Section 4.2.8) "forests" in regions close to the boundary. The yield stress is, according to Li, the stress required to move dislocations through these "forests." For many metals, the flow stress is related, under most conditions, to the dislocation density by the relationship (Section 6.3)

$$\tau = \tau_0 + \alpha Gb\sqrt{\rho}, \tag{5.12}$$

where τ_0 is the friction stress, α is a numerical constant, and ρ is the dislocation density. At this point, use was made of the experimental observation: ρ was taken to be inversely proportional to the grain diameter D. Li rationalized this as follows: The ledges "pump" dislocations into the grains. The number of dislocations generated per unit deformation is proportional to the number of ledges, or to the grain-boundary surface per unit volume, assuming the same ledge density per unit area for different grain sizes. That is,

$$\rho \propto S_v. \tag{5.13}$$

Equation 5.3 shows that the grain boundary surface per unit volume, S_v, is inversely proportional to D. *Thus*:

$$\rho \propto \frac{1}{D}. \tag{5.14}$$

[6]J. C. M. Li, *Trans. TMS-AIME*, 227 (1963) 239.

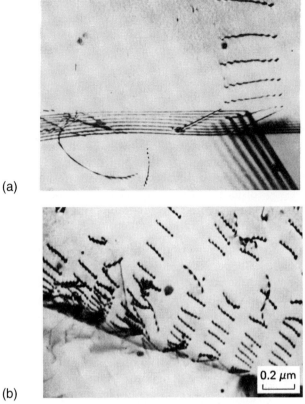

(a)

(b)

Figure 5.23 Dislocation activity at grain boundaries in AISI 304 stainless steel ($\dot{\varepsilon} = 10^{-3}\,\mathrm{s}^{-1}$). (a) Typical dislocation profiles after a strain of 0.15%. (b) Same after a strain of 1.5%. (Courtesy of L. E. Murr)

Substituting Equation 5.14 into Equation 5.12, we obtain

$$\tau = \tau_0 + GbD^{-1/2}.$$

Again, this is a Hall–Petch equation.

5.3.4 Meyers–Ashworth Theory

There have been other proposals, including one by Meyers and Ashworth,[7] who analyzed elastic and plastic incompatibility stress between neighboring grains. Stress concentrations occur at grain boundaries during elastic loading because the strains have to be compatible. For metals having anisotropy ratios different from unity (see Chapter 2, Equation 2-17), the Young's moduli in different directions are different. For example, for nickel,

$$E[100] = 137\ \text{GPa},$$
$$E[110] = 233\ \text{GPa},$$
$$E[111] = 303\ \text{GPa},$$

by finite element analysis. The incompatibility stresses were calculated by Meyers and Ashworth and found to be

$$\tau_{\mathrm{I}} = 1.37\sigma_{\mathrm{AP}},$$

[7]M. A. Meyers and E. Ashworth, *Phil. Mag.* (1982), 737.

where σ_{AP} is the normal stress applied to the specimen. Hence, the interfacial shear stress due to the incompatibility is almost three times higher than the resolved shear stress homogeneously applied on the grain ($\tau_H = \sigma_{AP}/2$). This means that dislocation activity at the grain boundary starts before dislocation activity at the center of the grains.

When the stress reaches the critical level required for emission, localized plastic deformation will start (Figure 5.24b). These dislocations do not propagate throughout the grain, for two reasons:

1. The stress decreases rapidly with distance from the grain boundary.
2. The center of the grains is under homogeneous shear stress control, which is maximum at 45° with the tensile axis. On the other hand, the interfacial and homogeneous shear stresses have different orientations. Figure 5.24 shows how the dislocations emitted from the grain boundaries will undergo cross-slip. Extensive cross-slip and the generation of dislocation locks will result in a localized layer with high dislocation density.

The plastic flow of the grain-boundary region attenuates the stress concentration; geometrically necessary dislocations accommodate these stresses (Figures 5.24b and c). This marks the onset of microyielding. The dislocations do not propagate throughout the whole grain, because of cross-slip induced by the difference in orientation between the maximum shear stress (due to the applied load) and the stress concentration due to elastic incompatibility. The work-hardened grain-boundary layer has a flow stress σ_{fGB}, while the bulk has a flow stress σ_{fB} ($\sigma_{fGB} > \sigma_{fB}$). The material behaves, at increasing applied loads, as a composite made out of a continuous network of grain-boundary film with flow stress σ_{fGB} and of discontinuous "islands" of bulk material with flow stress σ_{fB}. The increasing applied stress σ_{AP} does not produce plastic flow in the bulk in spite of the fact that $\sigma_{AP} > \sigma_{fB}$, because the continuous grain-boundary network provides rigidity to the struc-

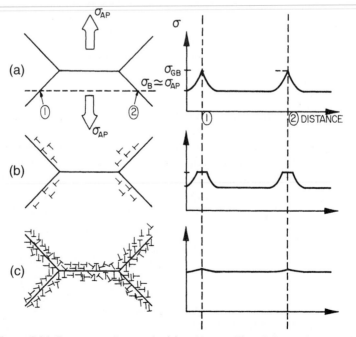

Figure 5.24 Sequence of stages in (a) polycrystalline deformation, starting with (b) localized plastic flow in the grain-boundary regions (microyielding), forming (c) a work-hardened grain-boundary layer that effectively reinforces the microstructure.

ture. The total strain in the continuous grain-boundary network does not exceed 0.005, since it is elastic; hence, plastic deformation in the bulk is inhibited. This situation can be termed "plastic incompatibility."

When the applied load is such that the stress in the grain-boundary region becomes equal to σ_{fGB}, plastic deformation reestablishes itself in this region. The plastic deformation of the continuous matrix results in increases in stress in the bulk with plastic flow (Figure 5.24c). This marks the onset of *macroyielding*. After a certain amount of plastic flow, dislocation densities in the bulk and grain-boundary regions become the same; then, since both regions have the same flow stress, plastic incompatibility disappears, and we have $\sigma_{AP} = \sigma_{GB} = \sigma_B$.

One arrives at a relationship

$$\sigma_y = \sigma_{fB} + 8k(\sigma_{fGB} - \sigma_{fB})D^{-1/2} - 16k^2(\sigma_{fGB} - \sigma_{fB})D^{-1}.$$

The last term becomes important at small grain sizes and decreases the slope.

EXAMPLE 5.6

If you could produce AISI 1020 steel with a grain size of 50 nm, what would be the expected yield stress, assuming a Hall–Petch response? (Use data from Figure 5.20.)

Solution: The Hall–Petch equation for this problem is $\sigma_y = \sigma_0 + kd^{-1/2}$. From Figure 5.20

$$\sigma_0 = 120 \text{ MPa},$$

$$k = 18 \text{ MPa/mm}^{1/2} = 0.56 \text{ MN/m}^{3/2}.$$

Therefore,

$$\sigma_y = (120 \times 10^6) + (0.56 \times 10^6) \times (50 \times 10^{-9})^{-1/2}$$

$$= 2.65 \times 10^9 \text{ Pa}$$

$$= 2.65 \text{ GPa}.$$

5.4 OTHER INTERNAL OBSTACLES

There are other internal obstacles to the motion of dislocations that may have an effect analogous to grain boundaries. Examples are cell walls and deformation twins. These barriers were studied by several investigators, and their effect on flow stress may be represented by the general equation

$$\sigma_f = \sigma_0 + K\Delta^{-m}, \tag{5.15}$$

where the coefficient m has been found to vary between $\frac{1}{2}$ and 1. If we want to include the effects of both grain size and substructure refinement due to the internal barriers, we can use the following overall equation, which describes the response of the material reasonably well:

$$\sigma_f = \sigma_0 + K_1 D^{-1/2} + K_2 \Delta^{-m}. \tag{5.16}$$

Figure 5.25 shows an example of substructural refinement in nickel. The twins were induced by shock loading at 45 GPa and 2 μs. It is easy to understand why these obstacles strengthen the metal. Dislocation movement occurring in subsequent deformation by, say, tensile testing is severely hampered by all such planar obstacles. Internal cells are also very effective barriers.

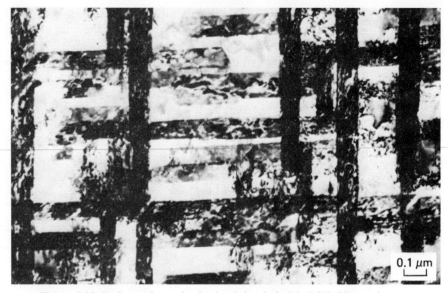

Figure 5.25 Deformation twins in shock-loaded nickel (45 GPa peak pressure; 2 μs pulse duration). Plane of foil (100); twinning planes (111) making 90°. (Courtesy of L. E. Murr)

The effect of the dislocation cell size on the flow stress of highly cold-worked low-carbon steel wire is shown in Figure 5.26. The straining to high levels was accomplished by wire drawing, and the material was recovered and showed thin cell walls and virtually dislocation-free cell interiors. The slope in the log–log plot is −1, and we have, consequently,

$$\log(\sigma_f - \sigma_0) - \log(\sigma_1 - \sigma_0) = -1(\log \overline{d} - \log \overline{d}_1), \tag{5.17}$$

where this equation expresses the straight line passing through $(\sigma_f - \sigma_0, \overline{d})$ and $(\sigma_1 - \sigma_0, \overline{d}_1)$. Notice that the ordinate in Figure 5.26 is $\sigma - \sigma_0$. Manipulation of Equation 5.17 will yield

$$\log \frac{(\sigma_f - \sigma_0)}{(\sigma_1 - \sigma_0)} = \log \left(\frac{\overline{d}}{\overline{d}_1}\right)^{-1}.$$

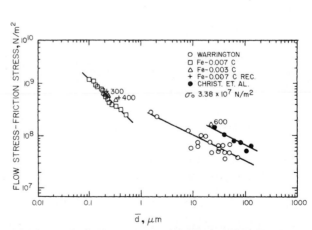

Figure 5.26 Strength of wire-drawn and recovered Fe–0.003% C as a function of transverse linear-intercept cell size. Recovery temperatures (in °C) as indicated. (Adapted with permission from H. J. Rack and M. Cohen, in *Frontiers in Materials Science: Distinguished Lectures,* L. E. Murr, ed., New York: M. Dekker, 1976, p. 365)

Hence,

$$\sigma_f - \sigma_0 = \frac{\sigma_1 - \sigma_0}{\bar{d}_1^{-1}}\bar{d}^{-1} = K\bar{d}^{-1},$$

$$\sigma_f = \sigma_0 + K\bar{d}^{-1}.$$

On the other hand, when the annealings were done at 600°C and above, recrystallization took place, and the group of points on the right side of the plot were found. The slope was decreased to $-\frac{1}{2}$, leading to a regular Hall–Petch relationship.

In low-carbon steels, the yield stress is strongly dependent on grain size; a steel with a grain size of 0.5 mm and σ_y of 104 MN/m^2 has its yield stress increased to approximately 402 MN/m^2 when the grain size is reduced to 0.005 mm. As the carbon content is increased and the steel tends more and more toward eutectoid, other effects, such as the ferrite–pearlite ratio, the spacing of cementite layers in the pearlite, and the size of the pearlite colonies, become important parameters. Gladman, McIvor, and Pickering[8] developed an expression for pearlite–ferrite mixtures, namely,

$$\sigma_y(\text{ksi}) = f_\alpha^{1/3}[2.3 + 3.81(\% \text{ Mn}) + 1.13D^{-1/2}]$$

$$+ (1 - f_\alpha^{1/3})[11.6 + 0.25 S_0^{-1/2}] + 4.1(\% \text{ Si}) + 27.6(\sqrt{\%\text{N}}),$$

where f_α is the ferrite fraction, D is the ferrite grain size (in mm), S is the interlamellar spacing in pearlite (in mm), and % Mn, Si, and N are the weight percentages of manganese, silicon, and nitrogen, respectively.

Hyzak and Bernstein[9] proposed the following equation for fully pearlitic steels:

$$\sigma_y(\text{MPa}) = 2.18 S^{-1/2} - 0.40P^{-1/2} - 2.88 D^{-1/2} + 52.30.$$

Here, S is the pearlite interlamellar spacing, P is the pearlite colony size, and D is the austenite grain size. (The units of S, P, and D are not given by Hyzak and Bernstein, but should be cm.)

5.5 NANOCRYSTALLINE MATERIALS

Since 1985, a great deal of research has been devoted to materials containing grain sizes in the nanometer range. These materials possess mechanical, magnetic, and electronic properties that are quite different from those of conventional crystalline materials (10 μm $\leq$ $d \leq$ 300 μm). It is clear that high strength levels can be achieved through reductions in grain size. Another beneficial effect is an enhanced deformability of ceramics, due to the large grain-boundary interface. A strength level of 4,000 MPa was obtained in a drawn steel that had a grain size of 10 nm (0.01 μm).

Figure 5.27 shows the schematic atomic structure of a nanocrystalline material. The atoms in the centers of the crystals (black circles) have a crystalline periodic arrangement. The configuration was developed by Gleiter, based on a Morse potential fitted to gold. At the boundaries, the spacings are altered. Thus, nanocrystalline materials can be considered a new class of disordered materials created by having a sizeable fraction of the atoms at disordered sites. The boundary region is characterized by a lower atomic density, and this

[8]T. Gladman, I. D. McIvor, and R. E. Pickering, *J. Iron Steel Inst.*, 210 (1972) 916.

[9]J. M. Hyzak and I. M. Bernstein, *Met. Trans.*, 7A (1976), 1217.

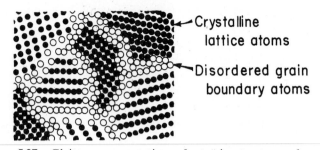

Figure 5.27 Gleiter representation of atomic structure of a nano-crystalline material; white circles indicate grain-boundary regions. (Courtesy of H. Gleiter)

is indeed a characteristic of nanocrystalline materials (between 75 and 90% of the crystalline density). The densities of nanocrystalline materials vary from 83–96% for Pd and 72–97% for Cu. In conjunction with the lower density, the Young's modulus of nanocrystalline materials is also lowered. For Cu and Pd (with theoretical values of Young's modulus E of 120–130 GPa), the reported E value in the nanocrystalline state is 21–66 GPa.

Two principal methods are used to produce these nanocrystalline materials:

1. Evaporation of metal from melt and condensation onto a "cold finger"; this nanosized powder is subsequently densified by pressing.
2. Extreme mechanical deformation of powders in, for instance, a ball-milling machine. Hard spheres impinge upon powders numerous times until a saturation of defects occurs, causing recrystallization.

There are other techniques also: molecular beam epitaxy, rapid solidification from melt, reactive sputtering, sol-gel, electrochemical deposition, and spark erosion.

The mechanical properties of nanocrystalline materials are quite distinct from those of conventional polycrystalline materials. A simple extrapolation using the Hall–Petch equation would predict extraordinarily high values of the yield stress. For example, copper with a grain size of 25 nm should have a yield stress of 720 MPa (data extrapolated from

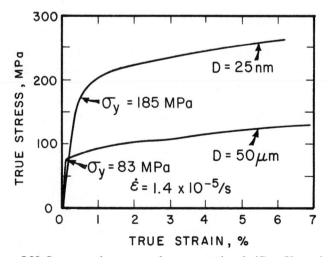

Figure 5.28 Stress–strain curves for conventional ($D = 50$ μm) and nanocrystalline ($D = 25$ μm) copper. (Adapted from G. W. Nieman, J. R. Weertman, and R. W. Siegel, *Nanostructured Materials,* 1 (1992) 185)

Table 5.3). Indeed, experimental results, shown in Figure 5.28, show a very high yield stress (~185 MPa). However, a simple extrapolation using the Hall–Petch equation does not predict quantitatively correct results. The Hall–Petch slope decreases as the grain size is decreased. Figure 5.29 shows the Hall–Petch relationship obtained in the nanocrystalline regime (grain sizes between 10 and 100 nm). The slope k is equal to 470 MPa $\sqrt{nm}$. This can be converted into 0.014 MN/m$^{3/2}$. There has been considerable discussion as to the nature of the strength of nanocrystals. Some of the ideas that have been bandied about are briefly presented in the following list:

1. *Dislocation pileups.* There is a minimum number of dislocations below which the equation for the stress concentration is no longer operative.

2. *Dislocation network models.* Models such as Li's or Meyers and Ashworth's use dislocation networks within the grain-boundary regions as the parameters determining the effects of grain size. Chang and Koch[†] and Scattergood and Koch[††] addressed these phenomena and proposed that, below a critical grain size D_c, a dislocation-network mechanism controlled the flow stress. Meyers and Ashworth's formulation predicted a decrease in the Hall–Petch slope for smaller grain sizes, in line with experimental observations. Their theory is based on the formation of a hardened region along the grain boundaries. (Section 5.3.4)

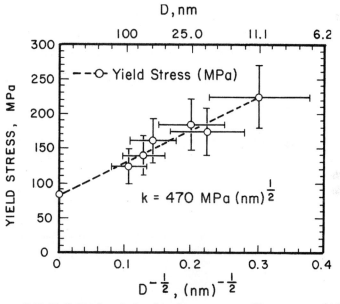

Figure 5.29 Hall–Petch relationship for nanocrystalline copper. (After G. W. Nieman, J. R. Weertman, and R. W. Siegel, *Nanostructured Matls.,* 1 (1992) 185; Figure 2, p. 188)

5.6 VOLUMETRIC OR TRIDIMENSIONAL DEFECTS

Voids and inclusions are among the principal tridimensional defects in materials. Inclusions are often produced in metals by the accidental incorporation of slag or pieces of refractory bricks into the melt or in powder metallurgy processes, from extraneous matter. Inclusions are also often the result of impurities, such as sulphur and phosphor

[†]J. S. C. Chang and C. C. Koch, *Scripta Met. Mat.* 24 (1990) 1599.

[††]R.O. Scattergood and C. C. Koch, *Scripta Met. Mat.* 27 (1992) 1195.

in steel. Vacuum arc remelting and other refining processes lead to alloys in which the inclusion content is minimized. Ceramics and brittle metals and intermetallics are especially sensitive to inclusions and voids. As will be seen in Chapter 8, these are easy sites for the initiation of fracture. Spherical and elongated flaws are the principal failure initiation sites in brittle materials. (Such flaws are activated both in tension and compression, and are responsible for the great differences between compressive and tensile strength (a factor of 5–10).

Ceramics are often produced by sintering or hot pressing of powders. This often leaves a residual porosity, which is a major source of concern. Figure 5.30 shows the microstructure of titanium carbide produced by hot pressing of powders. Residual porosity can be seen, and the voids are indicated by arrows. These voids have diameters of 1–4 μm. It is difficult to completely eliminate porosity in ceramics. Small, intragranular pores that are only visible by TEM, such as the ones in Al_2O_3 in Figure 5.31a, are very difficult to remove, because bulk diffusion is orders of magnitude slower than grain-boundary diffusion. If the voids were at the confluence of grain boundaries, it would be easier to eliminate them by high-temperature sintering. The voids seen in Figure 5.31a are faceted because this shape minimizes the overall surface energy; the surface energy is anisotropic, and the surfaces with the least number of broken bonds per unit area have the least energy. This is evident from the hexagonal voids shown in Figure 5.31a, which all have parallel faces. The TEM of Figure 5.31b also shows dislocations, which are produced during hot pressing of titanium carbide. The difficulty of hot pressing or sintering pure, high-temperature ceramics without voids is often bypassed by using sintering aids, or materials with a lower melting point. These materials—usually glasses—become viscous at high temperatures and fill the existing voids. They also act as a high-temperature lubricant between the ceramic particles and help to densify the ceramic, by capillary action. An illustration of the use of sintering aids to help the consolidation is given in Figure 5.32. Three silicon nitride grains are imaged by the TEM; the interplanar spacing, 0.65 nm, is shown. The three grains surround a glassy material, marked G. If no sintering aid were used, a central void would be formed. Nevertheless, the glassy phase is a volumetric defect and results in a weaker material than a fully dense, pure Si_3N_4.

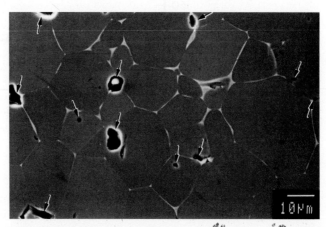

Figure 5.30 Voids (dark spots marked by arrows) in titanium carbide. The intergranular phase (light) is nickel, which was added to increase the toughness of the TiC.

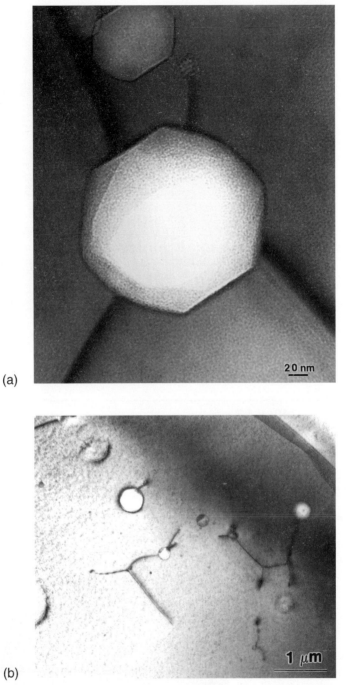

(a)

(b)

Figure 5.31 (a) Transmission electron micrograph illustrating faceted grain-interior voids within alumina and (b) voids in titanium carbide; dislocations are pinned by voids.

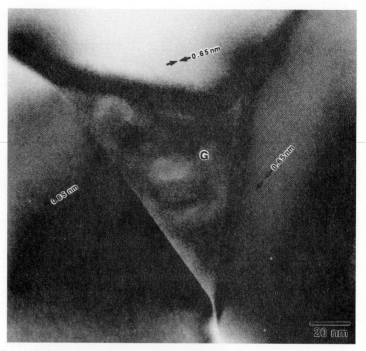

Figure 5.32 Glassy phase at triple point in silicon nitride; notice the individual crystallographic planes in Si_3N_4. (Courtesy of K. S. Vecchio)

EXAMPLE 5.7

(a) Calculate the volume fraction of voids for the micrograph in Figure 5.30.

(b) If Young's modulus for fully dense TiC is 440 GPa, what is Young's modulus for the porous TiC?

Solution: (a) We overlay a grid on the micrograph and count the intersections of lines falling within the voids. (See Figure E5.7.)

Total numbers of intersections in grid $= 72 \times 47 = 3,384$;

Total numbers of intersections inside voids ≈ 66.

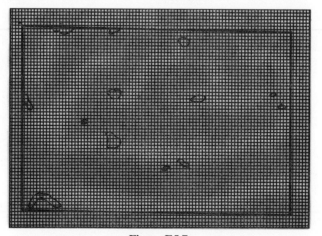

Figure E5.7

Therefore, the porosity is approximately 66/3,384 = 2%.

(b) From Chapter 2, we obtain the equation for Young's modulus (eqn. 2.25):

$$E = E_0(1 - 1.9p)$$
$$= 440 \times (1 - 1.9 \times 0.02)$$
$$= 423 \text{ GPa.}$$

5.7 IMPERFECTIONS IN POLYMERS

Let us consider again the basic "cooked spaghetti" structure of a polymer. In an amorphous polymer, there is no apparent order among the molecules, and the polymeric chains are arranged randomly. As we pointed out in Chapter 1, macromolecules can be made to crystallize. However, unlike metals or ceramics, long-chain polymers or macromolecules (synthetic or natural) do *not* form exact, periodic structures having long-range order in three dimensions. This is because such a highly ordered structure in a polymer, in general, will not be in equilibrium. It is, however, possible to obtain a variety of metastable chain conformations, depending on the route taken to reach a particular state. In any of these metastable states, order may be locally present; that is, we can have crystalline regions interspersed with amorphous regions. Polymers can thus be amorphous or partially crystalline, a 100-percent crystalline polymer being difficult to obtain in practice. In a partially crystalline or semicrystalline polymer, depending on its type, molecular weight, and crystallization temperature, the amount of crystallinity can vary from 30 to 90 percent. The inability to attain a fully crystalline structure is due mainly to the long chain structure of polymers: Some twisted and entangled segments of chains that get trapped between crystalline regions never undergo the conformational reorganization necessary to achieve a fully crystalline state. Molecular architecture also has an important bearing on the polymer crystallization behavior: Linear molecules with small or no side groups crystallize easily; branched chain molecules with bulky side groups do not. For example, linear, high-density polyethylene can be crystallized to more than 90 percent, whereas branched polyethylene can be crystallized only to about 65 percent.

Amorphous polymers can be considered to be fairly homogeneous on a supramolecular scale. Semicrystalline polymers, consisting of tiny crystalline regions randomly distributed in an amorphous material, are heterogeneous, multiphasic, or even *composite* in nature. Lamellar crystals can form when a crystallizable polymer such as a linear polymer is cooled very slowly from its melting point. Small, platelike lamellar single crystals can also be obtained by the precipitation of a polymer from a dilute solution. The long molecular chains in the lamellae are folded in a regular manner. In a lamellar-polymer single crystal, the thickness of a lamella is typically about 10 nm, while the length of the chain is about 100 to 1,000 nm. The extremely long chain is conformed into a narrow lamella by the process of chain folding during crystallization. Figure 1.26c shows this phenomenon of chain folding. Many such lamellar crystallites group together and form spherulites. (See chapter 1.)

Crystalline defects such as those described for metals and ceramics are not at all ubiquitous in polymers. One may define defects in polymers in simple chemical and physical terms. Chemical defects include defects such as a linear polymer branching off into two branches that grow at different rates to give branches of different lengths. One can also have syndiotactic defects, which are stereochemical in nature. For example, an isotactic polymer chain can have syndiotactic defects embedded in it. Physical aspects of defects involve conformational defects in chain coiling. It is easy to see that kinetic and energetic factors will be very important in these type of defects, because such defects involve chain movement. Variables such as temperature, pressure, concentration, molecular weight, chain polarity, etc., are important, and statistical mechanics needs to be used. Thus, point defects in polymers are chain-conformational kinks, jogs, and inclusions. Point defects also include an interstitial or substitutional molecule. For example, if a macromolecular chain consisting of species A has a monomer B trapped inside the polymer crystal, that would be an interstitial point defect. If there is a break in the length of the molecular chain we will have a chain end and a vacancy or a row of vacancies.

As we have seen, in metals dislocations are very important because they are mobile, while in ceramics they are immobile under most conditions. Although dislocations can exist in polymeric crystals also, they do not play such a major role in the deformation of polymers. Direct observations of dislocations have been made in some semicrystalline polymers by transmission electron microscopy, which has been instrumental in elucidating the structural imperfections in metals and ceramics. One of the great limitations to the use of electron microscopy in the study of polymers is the radiation damage produced in the polymers by the electron beam. Images produced by electron diffraction contrast, as well as electron diffraction patterns, depend on the crystallinity of the specimen. A large dose of electrons will tend to destroy the long-range crystalline order, more so in polymers than in metals or ceramics, because nonpolymeric crystalline materials such as metals and ceramics are more resistant to electron irradiation. Thus, only a limited number of scattered electrons can be used to obtain crystallographic information from the sample under study before the diffraction pattern changes from crystalline reflections to broad, amorphous haloes. Radiation damage can establish cross-links and cause strain in the lattice at first. Continued exposure to an electron beam can make the diffraction contrast weaker and eventually disappear. It is therefore necessary to take special precautions before examining the structure of polymers in an electron microscope. Perhaps the most widely studied polymer in this regard is polyethylene, although it is difficult to take high-resolution images of polyethylene at room temperature by TEM because of the sensitivity of the polymer to radiation. By comparison, thermoplastics such as PPS, PEEK, and PEK are fairly resistant to electron irradiation. Experimentally, giant screw dislocations showing growth spirals have been observed in these thermoplastics. Terminating moiré fringes have been used to show the existence of dislocations in a polyethylene crystal. A lattice-imaging technique has been used on poly (para phenylene terephthalamide) PPTA and poly (para-phenylene benzobis thiazole) PBT fibers. In these fibrous materials, one has, relative to polyethylene, rather high radiation stability for electron microscopic observations because of the electronic conjugation of the backbone chain.

In crystalline metals and ceramics, two-dimensional defects such as grain boundaries are thin regions where two grains meet. In polymeric crystals, grain boundaries can be very complex, again because of the chain connectivity. Besides, the electron beam sensitivity of polymers makes TEM observations and their interpretation quite difficult. Planar defects such as stacking faults and twins have been observed in samples of poly (diacetylene) crystals.

SUGGESTED READINGS

Interfacial Defects

H. GLEITER. "On the Structure of Grain Boundaries in Metals." *Mater. Sci. Eng.,* 52 (1982), 91.

H. GLEITER and B. CHALMERS. "High-Angle Grain Boundaries," *Progress in Materials Science,* Vol. 16, B. Chalmers, J. W. Christian, and T. B. Massalski (eds.). Elmsford, NY: Pergamon Press, 1972.

L. E. MURR. *Interfacial Phenomena in Metals and Alloys.* Reading, MA: Addison-Wesley, 1975

A. P. SUTTON and R. W. BALUFFI. *Interfaces in Crystalline Materials.* New York: Oxford University Press, 1994.

Twinning

R. E. REED-HILL, J. P. HIRTH, and H. C. ROGERS (eds.). *Deformation Twinning.* TMS-AIME Conf. Proc. New York: Gordon and Breach, 1965.

J. W. CHRISTIAN and S. MAHAJAN. *Deformation Twinning.* New York: Oxford University Press, 1995.

Grain-Size Effects

R. W. ARMSTRONG. "The influence of Polycrystal Grain Size on Mechanical Properties" in *Advances in Materials Research,* Vol. 4, H. Herman (ed.). John Wiley & Sons Interscience Publishers, 1971, p. 101.

R. W. ARMSTRONG, in *Yield, Flow, and Fracture of Polycrystals,* T. N. Baker (ed.), London: Appl. Sci. Publ., 1983, p. 1.

H. GLEITER, "Nanocrystalline Materials" *Progress in Materials Science,* 33 (1989), 223.

EXERCISES

5.1 Calculate the dislocation spacing in a symmetrical tilt boundary ($\theta = 0.5°$) in a copper crystal.

5.2 Starting from the equation $E = E_0\theta(A - \ln\theta)$ for a low-angle boundary, show how one can obtain graphically the values of E_0 and A.

5.3 Taking $A = 0.3$, compute the value of θ_{max}.

5.4 Show that, for a low-angle boundary, we have

$$\frac{E}{E_{max}} = \frac{\theta}{\theta_{max}}\left(1 - \ln\frac{\theta}{\theta_{max}}\right),$$

where E_{max} and θ_{max} correspond to the maximum in the E-versus-θ curve.

5.5 Consider two parallel tilt boundaries with misorientations θ_1 and θ_2. Show that, thermodynamically, we would expect the two boundaries to join and form one boundary with misorientation $\theta_1 + \theta_2$.

5.6 Can you suggest a quick technique to check whether lines observed in an optical microscope on the surface of a polished sample after deformation are slip lines or twin markings?

5.7 A twin boundary separates two crystals of different orientations; however, we do not necessarily need dislocations to form a twin. Why?

5.8 Let m be the total length of dislocations per unit area of a grain boundary. Assume that at yield, all the dislocations in the grain interiors (ρ) are the ones emitted by the boundaries. Assume also that the grains are spherical (with diameter d). Derive the Hall–Petch relation ($\sigma = \sigma_0 + kd^{-1/2}$) for this case, and give the expression for k.

5.9 Consider a piano wire that has a 100% pearlitic structure. When this wire undergoes a reduction in diameter from D_0 to D_ε, the pearlite interlamellar spacing normal to the wire axis is reduced from d_0 to d_ε, that is,

$$\frac{d_0}{d_\varepsilon} = \frac{D_0}{D_\varepsilon},$$

where the subscript 0 refers to the original dimensions, while the subscript ε refers to the dimensions after a true plastic strain of ε. If the wire obeys a Hall–Petch type of relationship between the flow stress and the pearlite interlamellar spacing, show that the flow stress of the piano wire can be expressed as

$$\sigma = \sigma_i + \frac{k'}{\sqrt{d_0}}\exp\left(\frac{\varepsilon}{4}\right).$$

5.10 **(a)** Determine the mean lineal intercept, the surface area per unit volume, and the estimated grain diameter for the specimen shown in Figure Ex 5.10.

      (b) Estimate the yield stress of the specimen (AISI 304 stainless steel).

      (c) Estimate the parameters of part (a), excluding the annealing twins. By what percentage is the yield stress going to differ?

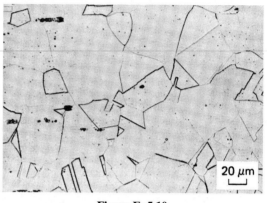

Figure Ex5.10

5.11 Professor M. I. Dum conducted a study on the effect of grain size on the yield stress of a number of metals using thin foil specimens (thickness 0.1 mm and width 6.25 mm). He investigated grain sizes of 5, 25, 45, and 100 μm. Which specimens can be considered truly polycrystalline?

5.12 A. W. Thompson[10] obtained the following results for the yield stress of nickel:

Grain Sizes (μm)	Yield Stress (MN/m²)
0.96	251
2	185
10	86
20	95
95	33
130	25

      (a) Find the parameters in the Hall–Petch equation. Plot the yield stress versus D^{-1}, $D^{-1/2}$, and $D^{-1/3}$. Which plot shows the best linearity?

      (b) Show how you can determine the correct exponent by another plot (not by trial and error).

5.13 If the grain size of a metal is doubled by an appropriate annealing, by what percentage is the surface area per unit volume of the metal changed?

5.14 J. L. Nilles and W. S. Owen[11] found a strong grain-size dependence of the stress required for twinning when deforming an Fe–25% Ni alloy at 4 K. From what you learned in the text, is this behavior expected? Compare the ratio of the Hall–Petch slopes of the twinning and yield stresses for Fe–25% Ni with the ratio found for chromium and Fe–Si.

5.15 Most polycrystalline materials, when etched, form grooves at grain boundaries. When annealed, ceramics form thermal grooves at grain boundaries. A schematic of such a groove is shown in Figure Ex 5.15. If the surface energy per unit area of the material is γ_s, derive an expression for the energy per unit area of the grain boundary between grains 1 and 2.

[10]A. W. Thompson, *Acta Met.,* 25 (1977) 83.

[11]J. L. Nilles and W. S. Owen, *Met. Trans.,* 3 (1972), 1877.

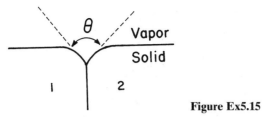

Figure Ex5.15

5.16 Estimate the average grain diameter and the grain-boundary area per unit volume for a material that has isotropic grains (the same dimension in all directions) and ASTM grain size 6.

5.17 How many grains in an area of 5×5 cm would be counted, in a photomicrograph taken at a magnification of $500\times$, for a metal with ASTM grain size 3?

5.18 A graduate student (undergraduates are much brighter!) measured the grain size of a metallic specimen and found that it was equal to ASTM #2. However, he had the wrong magnification in his picture ($400\times$ instead of $100\times$). (a) What is the correct ASTM grain size? (b) What is the approximate grain diameter?

5.19 Nanophase materials show many different characteristics vis-à-vis conventional materials. Discuss the sintering behavior of a nanophase powder in relation to that of a conventional powder.

5.20 Calculate the volume fraction of voids for TiB_2 and TiC specimens shown in Figure 5.31.

5.21 Examine Figure Ex 5.21.
   (a) Using the lineal intercept method, determine the mean lineal intercept and the grain size if the material is TiC.
   (b) Determine the grain size using the ASTM method.

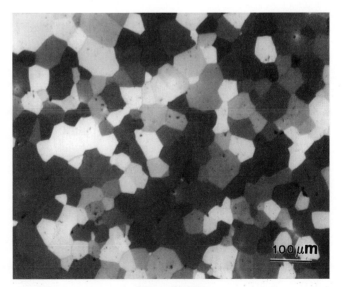

Figure Ex5.21

5.22 **(a)** Using the mean lineal intercept, calculate the grain diameter for tantalum, given the micrograph in Figure 5.2a.
   (b) Calculate the ASTM grain size.
   (c) Estimate the yield stress for this specimen of tantalum, using values from Table 5.3.

5.23 A polycrystalline sample has 16 grains per square inch in a photomicrograph taken at magnification $100\times$. What is the ASTM grain size number?

5.24 A 20-cm line gave seven intersections in a 100× micrograph. Using the lineal intercept method, determine the mean lineal intercept and the grain size.

5.25 How many grains in an area of 5 × 5 cm would be counted in a photomicrograph taken at a magnification of 500× for a metal with an ASTM grain size 3.

5.26 The yield stress of AISI 1020 steel with a grain size of 200 μm is 200 MPa. Estimate the yield stress for a grain size of 10 μm if the Hall–Petch constant $k = 0.8$ MN/m$^{3/2}$.

5.27 A small-angle tilt boundary has a misorientation of 0.1°. What is the spacing between the dislocations in this boundary if the Burgers vector of the dislocation is 0.33 nm?

5.28 Calculate the dislocation spacing of a low-angle tilt boundary in aluminum for $\theta = 0.5°$. Take $G = 26.1$ GPa, $\nu = 0.345$, and $r_{Al} = 0.143$ nm.

6

Geometry of Deformation and Work-Hardening

6.1 INTRODUCTION

The relaxation times for the molecular processes in gases and in a majority of liquids are so short, that molecules/atoms are almost always in a well-defined state of complete equilibrium. Consequently, the structure of a gas or liquid does not depend on its past history. In contrast, the relaxation times for some of the significant atomic processes in crystals are so long, that a state of equilibrium is rarely, if ever, achieved. It is for this reason that metals in general (and ceramics and polymers, under special conditions) show the usually desirable characteristic of work-hardening with straining, or strain-hardening. In other words, plastic deformation distorts the atoms from their equilibrium positions, and this manifests itself subsequently in hardening.

In fact, hardening by plastic deformation (rolling, drawing, etc.) is one of the most important methods of strengthening metals, in general. Figure 6.1 shows a few deformation-processing techniques in which metals are work-hardened. These industrial processes are used in the fabrication of parts and enable the shape of metals to be changed. The figure is self-explanatory. Rolling is used for the production of flat products such as plates, sheets, and also more complicated shapes (with special rolling cylinders). In forging, the top hammer comes down, and the part is pushed into a die (closed-die forging) or is simply compressed. Extrusion uses a principle similar to that in the use of a tube of toothpaste. The material is squeezed through a die, and its diameter is reduced. In stamping, first the ends of a blank are held, and then the upper die comes down, punching the blank into the lower die.

If deformation is carried out at low and moderate temperatures, the metal work-hardens. However, if the temperature is sufficiently high, the dislocations generated in work-hardening are annealed out, and the final metal is in the annealed condition. *Hot working* designates all work done on a metal or alloy above its recrystallization temperature, while *cold working* indicates work done below the recrystallization temperature of the

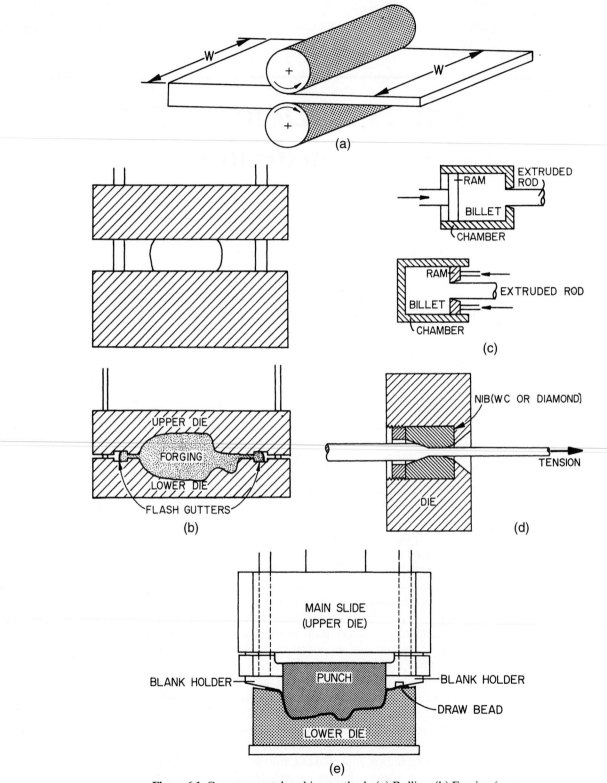

Figure 6.1 Common metalworking methods. (a) Rolling. (b) Forging (open and closed die). (c) Extrusion (direct and indirect). (d) Wire drawing. (e) Stamping.

metal or alloy. Certain metals, in particular (e.g., copper), do not have many precipitation hardening systems, but are ductile and can be appreciably hardened by cold working. If the relaxation times were short, the structure would return almost immediately to its state of equilibrium, and a constant stress for plastic deformation would result, independently of the extent of deformation. This is shown in Figure 6.2 as the elastic, ideally plastic solid. However, when a real crystalline solid is deformed plastically, it turns more resistant to deformation, and a greater stress is required for additional deformation. The phenomenon is called *work-hardening*. If the stress is interrupted, and the material is unloaded after a certain plastic strain, the unloading slope is equal to the Young's modulus. Upon loading, the stress returns to its original value. Thus, for a work-hardening material, the flow stress is increased above σ_0, whereas for an ideally plastic material, the flow stress is constant at σ_0. Under certain conditions, the material can also soften. This is also shown in Figure 6.2 and is discussed in greater detail in Section 6.4.

In Chapter 4, we discussed the various kinds of defects in materials. Of these defects, the primary carriers of plastic deformation in metals and ceramics are dislocations and twins. From the simple motion of dislocations along specific planes, we derived the Orowan Equation 4.23, which relates the global plastic strain to the individual dislocation motion and density. Basically, the hardening in a crystalline structure occurs because these materials deform plastically by the movement of dislocations, which interact directly among themselves and with other imperfections, or indirectly with the internal stress field (short range or long range) of various imperfections and obstacles. All these interactions lead to a reduction in the mean mobility of dislocations, which then require a greater stress for accomplishing further movement (i.e., with continuing plastic deformation, we need to apply an ever greater stress for further plastic deformation); hence the phenomenon of work-hardening.

Figure 6.3 illustrates how a metal (in this case, nickel) work-hardens by cold rolling. As the nickel plate is reduced in thickness (and increased in length), its stress–strain response changes. In the figure, we plot engineering stress versus engineering strain, and all the curves show a softening after hardening. This softening is due not to an inherent structural "softening," but to a localized reduction in cross section, called *necking*. (See Chapter 3.) The yield stress increases from less than 100 MPa (in the annealed condition) to approximately 850 MPa (after 90% reduction in thickness by cold rolling). Concomitantly, the ductility decreases. The sample that received 80% reduction was subjected to one-hour annealings at various temperatures; the resulting mechanical response is shown in Figure 6.3b. After a 700°C annealing, the curve is almost coincident with the original annealed curve, showing that the effects of cold rolling have been eliminated. This occurs because the dislocations produced by plastic deformation have been eliminated by the annealing. The nickel specimens were polycrystalline, with a grain size of 40 μm.

In Chapter 4, we dealt exclusively with monocrystals; we will see, in this chapter, how the plastic deformation in one single crystallographic direction is related to the

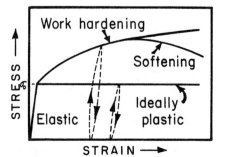

Figure 6.2 Stress–strain curves (schematic) for an elastic, ideally plastic; a work-hardening; and a work-softening material.

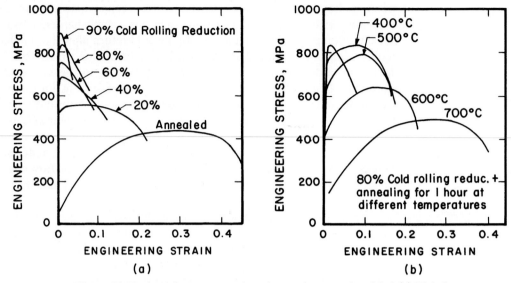

Figure 6.3 Engineering-stress–engineering-strain curves for nickel. (a) Nickel subjected to 0, 20, 40, 60, 80, and 90% cold-rolling reduction. (b) Nickel cold rolled to 80%, followed by annealing at different temperatures. (From D. Jaramillo, V. S. Kuriyama, and M. A. Meyers, *Acta Met.* 34 (1986) 313, Figure 3, p. 316)

overall deformation of a crystal and how different crystals in a polycrystal deform in a "cooperative" manner.

In ceramics, plastic deformation is not so common. At room temperatures many ceramics are brittle, and it was seen in Chapter 4 that the Peierls–Nabarro stress opposing dislocations is much higher and that the mobility of dislocations is much more restricted than for metals. This is illustrated in Figure 6.4 which shows results of compression tests on TiC specimens carried out at different temperatures. Note that the elastic portion of the curves shows a slope that is considerably lower than the prediction from the Young's modulus, because no extensometer was used to measure strain. Thus, the abscissa records both the strain in the specimen and the deflection in the machine; for this reason, the term "apparent strain" is used. The ambient-temperature compressive strength of TiC is approximately 4,000 MPa. As the temperature is increased beyond 950°C, plastic deformation gradually sets in. This is called the *ductile–brittle transition*. As the temperature is increased, the flow stress decreases. In this temperature regime, the material exhibits plasticity. In monocrystalline Al_2O_3 deformed at high temperatures, significant plastic deformation is also observed. Figure 6.5 shows the shear-stress–shear-strain response for Al_2O_3 oriented for prismatic slip. (See Section 4.2.6.)

6.2 GEOMETRY OF DEFORMATION

6.2.1 Stereographic Projections

The mechanical properties of crystals are anisotropic, and slip occurs only in certain planes, along certain directions. For this reason, it is important to define the *orientation* of a crystal. The most common technique for doing so is the stereographic projection, which will be presented here in an abbreviated way; greater details are given in Barrett and Massalski

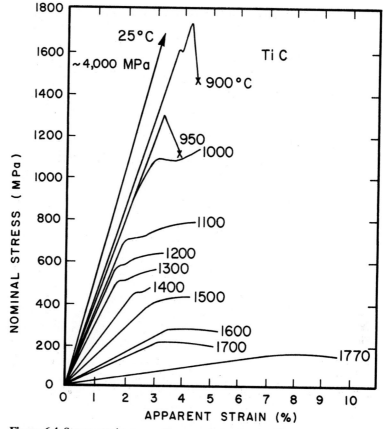

Figure 6.4 Stress–strain curves for annealed polycrystalline TiC deformed in compression at the temperatures indicated ($\varepsilon = 1.7 \times 10^{-4}\,\text{s}^{-1}$). (Adapted from G. Das, K. S. Mazdiyasni, and H. A. Lipsitt, *J. Am. Cer. Soc.*, Feb. 1982, p. 104)

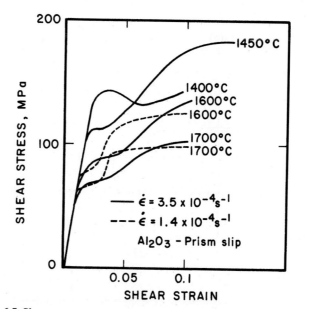

Figure 6.5 Shear stress τ vs. shear strain γ for prism plane slip in Al_2O_3 at various temperatures; $\dot{\varepsilon} = 3.5 \times 10^{-4}\,\text{s}^{-1}$ for the solid curves, $\dot{\varepsilon} = 1.4 \times 10^{-4}\,\text{s}^{-1}$ for the dashed curves. (Courtesy T. E. Mitchell)

(See suggested readings.) The stereographic projection is a geometric representation of the directions and planes of a crystal. From stereographic projections, one can determine the angles between planes, planes and directions, and directions. The stereographic projection is the projection of a sphere on a plane. We imagine a unit cell of a certain crystalline structure at the center of the sphere. (See Figure 6.6a.) The directions and plane poles (normals to the planes passing through the origins) intercept a sphere at points; these points are projected onto a plane. Figure 6.6b shows a standard cubic projection. This projection is known as a [100] standard projection because the [001] direction corresponds to the center. There are a series of other standard projections: [110], [111], [112], and so on. Theoretically, the angles between directions and/or plane poles are measured on the sphere; in practice, however, these angles are measured on the standard projection, making use of a special chart called the Wulff net. This chart is the projection of a plane of a sphere in which all the meridians and parallels are marked at regular degree intervals. The sphere has the same diameter as the standard projection. By inserting a tack at the center and rotating the standard projection around it, we can easily find all desired angles.

An analogy can be made with maps. Imagine that we look at the earth from the "top"; that is, we view the northern hemisphere with the north pole at the center. If we now draw a map on a circle, we have a situation analogous to a stereographic projection. The meridians of the map correspond to great circles on the stereographic projection—that is, circles whose center is coincident with the center of the sphere. The four great circles that are perpendicular to the plane of the paper are projected as straight lines.

In a stereographic projection, the crystalline symmetry can be clearly seen. For instance, the ⟨100⟩ directions form a cross in Figure 6.6, with the crystalline symmetry indicated in Figure 6.6b; two-, three-, four-, and six-field symmetry axes are shown. (The symmetries have been introduced in Section 2.8, and the reader is referred to Table 2.1.) For the ⟨111⟩, ⟨110,⟩, and ⟨100⟩ directions, the symmetry is four-, two-, and threefold, respectively, in the cubic system. Two-, three-, and four-fold symmetries are indicated by lens, triangle, and square, respectively. As a consequence, the standard projection can be divided, by means of great circles, into 24 spherical triangles that are crystallographically equivalent. The vertices of these triangles are ⟨100⟩, ⟨110⟩, and ⟨111⟩, as can be seen in

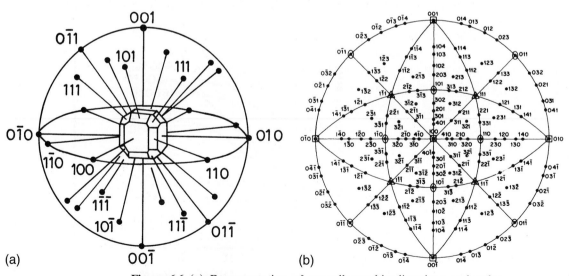

Figure 6.6 (a) Representation of crystallographic directions and poles (normals to planes) for cubic structure. (b) Standard [100] stereographic projection. (Reprinted with permission from C. S. Barrett and F. B. Massalski, *The Structure of Metals,* 3d ed., New York: McGraw-Hill, 1966, p. 39)

Figure 6.7. Comparing this figure with Figure 6.6b, we can see that the directions on the sides and within the spherical triangles are also equivalent. Consequently, one single triangle is sufficient to specify any crystallographic orientation in the cubic system; the [100], [110], [111] triangle is used most commonly. The reader is warned, however, that this simplification is not applicable to the other crystal systems.

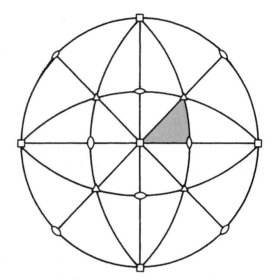

Figure 6.7 Standard [001] stereographic projection divided into 24 triangles.

6.2.2 Stress Required for Slip

The flow stresses of crystals are highly anisotropic. For instance, the yield stress of zinc under uniaxial tension varies by a factor of at least 6. Consequently, it is very important to specify the orientation of the load. In shear or torsion tests, the shear plane and directions are precisely known. Because dislocations can glide only under the effect of shear stresses, these shear stresses have to be determined. In uniaxial tensile and compressive tests (the most common tests), one has to determine mathematically the shear component of the applied stress acting on the plane in which slip is taking place. Figure 6.8 shows a crystal with a normal cross-sectional area A upon which a tensile load acts, generating a uniaxial stress P/A. The slip plane and direction are indicated, respectively, by the angles ϕ and λ that they make with the tensile axis. We take the normal **n** of the plane that makes an angle ϕ with the loading direction ℓ.

The areas A_1 and A are related by the angle ϕ. Area A is the projection of A_1 onto the horizontal plane; thus, we can write

$$A = A_1 \cos \phi.$$

The shear stress τ acting on the slip plane and along the slip direction **S** is obtained by dividing the resolved load along the slip direction ($P \cos \lambda$) by A_1:

$$\tau = \frac{P \cos \lambda}{A_1} = \frac{P}{A} \cos \phi \cos \lambda.$$

But $P/A = \sigma$ is the normal stress applied to the specimen. Hence,

$$\tau = \sigma \cos \phi \cos \lambda.$$

Note that $\cos \phi = \sin \chi$

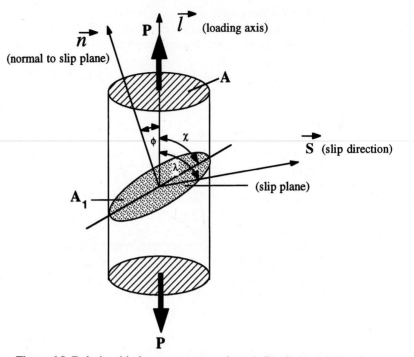

Figure 6.8 Relationship between stress axis and slip plane and direction.

This equation shows that τ will be zero when either λ or ϕ is equal to 90°. On the other hand, the shear component is maximum when both ϕ and λ are equal to 45°. We have, in this case,

$$\tau_{\max} = \sigma \cos 45° \cos 45° = \frac{\sigma}{2}.$$

The angle between any two directions **a** and **b** can be obtained from the scalar product of these vectors:

$$\mathbf{a} \cdot \mathbf{b} = |\mathbf{a}||\mathbf{b}|\cos\theta,$$

or

$$\cos\theta = \frac{\mathbf{a} \cdot \mathbf{b}}{|\mathbf{a}||\mathbf{b}|}.$$

For cubic crystals, planes and directions with the same indices are perpendicular, and the angle is determined from the coefficients, h, k, and l. For two vectors

$$\mathbf{a} = h_1\mathbf{i} + k_1\mathbf{j} + l_1\mathbf{k}$$

and

$$\mathbf{b} = h_2\mathbf{i} + k_2\mathbf{j} + l_2\mathbf{k},$$

the angle θ is given by

$$\cos \theta = \frac{h_1 h_2 + k_1 k_2 + l_1 l_2}{(h_1^2 + k_1^2 + l_1^2)^{1/2}(h_2^2 + k_2^2 + l_2^2)^{1/2}}. \tag{6.1}$$

If two directions are perpendicular, their dot product is zero; and the same is true for a direction that is contained in a plane. From Equation 6.1, it is possible to obtain the $\cos \phi$ and $\cos \lambda$ terms for all desired crystallographic directions of a crystal. For instance, if the loading direction is [123] for an FCC crystal, then the Schmid factors of the various slip systems are found by obtaining the angles of [123] with $\langle 111 \rangle$ (perpendicular to slip planes) and $\langle 110 \rangle$ (in slip directions). Note that each slip plane contains three slip directions and that 12 values (4×3) have to be obtained.

Schmid and coworkers[1] used the variation in the resolved shear stress to explain the great differences in the yield stresses of monocrystals of certain metals. They proposed the following rationalization, known as the *Schmid law*: *Metal flows plastically when the resolved shear stress acting in the plane and along the direction of slip reaches the critical value*

$$\tau_c = \sigma_0 \sin \chi \cos \lambda = M\sigma_0, \tag{6.2}$$

$$M = \sin \chi \cos \lambda = \cos \phi \cos \lambda \tag{6.3}$$

where the factor M is usually known as the *Schmid factor*.

Schmid's law has found experimental confirmation principally in hexagonal crystals. Figure 6.9 shows the experimental results, compared with Schmid's prediction for high-purity zinc. The full line shows the hyperbola obtained by the use of Equation 6.2, assuming a critical resolved shear stress of 184 kPa. It is worth noting that the yield stress is minimum for $M = 0.5$.

For cubic crystals, the correspondence between Schmid's law and experiments is not as good. This is mainly due to the great number of slip systems in these structures. For nickel, the critical resolved shear stress is practically orientation independent. On the other hand, for copper, the critical resolved shear stress is dependent on orientation, being constant in the center of the stereographic triangle and assuming higher values close to the sides. Figure 6.10 shows the inverse of Schmid's factor in the stereographic triangle based on a $\{111\} \langle 110 \rangle$ slip: This is the situation for FCC crystals. The orientation for which FCC crystals are softest is $M = 0.5$, or $M^{-1} = 2$, which occurs approximately at the center of the triangle. The dependence of τ_c on the orientation for cubic systems is thought to be due to the fact that the components of compressive stresses acting normal to the slip planes are different for different orientations at the same applied stress level. These compressive stresses should have an effect on τ_c. Easy glide in FCC crystals is greatest in the center of the stereographic projection, in the region closer to (but not coinciding with) the $\langle 110 \rangle$ corner. It is affected by a number of parameters, the most notable being the following:

1. *Specimen size.* Specimens with a smaller cross-sectional area tend to have a more extended easy-glide region.

[1]E. Schmid and W. Boas, *Kristalplastizitat* (*Plasticity of Crystals*) (Berlin and London: Springer and Hughes, 1950).

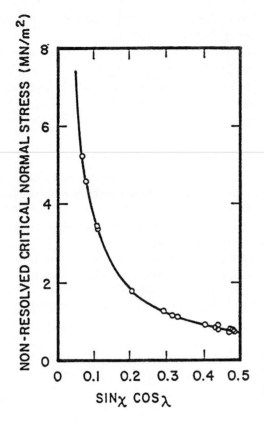

Figure 6.9 Comparison of Schmid law prediction with experimental results. (Adapted with permission from D. C. Jillson, *Trans. AIME,* 188 (1950) 1120)

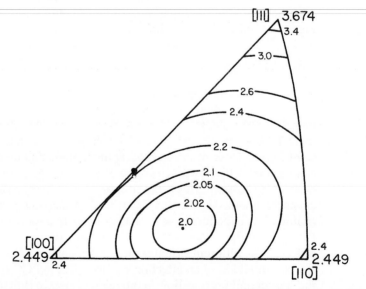

Figure 6.10 Effect of orientation on the inverse of Schmid's factor $(1/M)$ for FCC metals. (Adapted with permission from G. Y. Chin, "Inhomogeneities of Plastic Deformation," in *The Role Preferred Orientation in Plastic Deformation* (Metals Park, OH: ASM, 1973), pp. 83, 85.

2. *Temperature.* Easy glide is more pronounced at lower temperatures and may vanish completely at high temperatures.

3. *Stacking-fault energy.* FCC metals with low stacking-fault energy tend to have a more pronounced easy-glide region. Why?

4. *Solute atoms.* If solute atoms pin the dislocations, they will shorten their mean free path and the extent of stage I. If solute atoms contribute primarily to the lowering of the stacking-fault energy or to ordering, they will increase the easy-glide range.

EXAMPLE 6.1

A single crystal of copper is being deformed in tension. The loading axis is [112].

(a) Calculate the Schmid factors for the different slip systems.

(b) If the critical resolved shear stress is 50 MPa, what is the tensile stress at which the material will start to deform plastically?

Solution: (a) Copper is FCC, which has 12 slip systems of the type $\{111\}\langle 110 \rangle$; thus, we have

$$\cos \phi = \frac{\mathbf{n} \cdot \mathbf{l}}{|\mathbf{n}| \cdot |\mathbf{l}|},$$

$$\cos \lambda = \frac{\mathbf{s} \cdot \mathbf{l}}{|\mathbf{s}| \cdot |\mathbf{l}|},$$

and the following table:

Slip plane (n)	Slip direction (s)	Cos φ	Cos λ	Schmid factor (cos φ cos λ)	σ (MPa)
	[$\bar{1}$10]	$2\sqrt{2}/3$	0	0	Not deformed
(111)	[$\bar{1}$01]	$2\sqrt{2}/3$	$\sqrt{3}/6$	$\sqrt{6}/9$	184
	[0$\bar{1}$1]	$2\sqrt{2}/3$	$\sqrt{3}/6$	$\sqrt{6}/9$	184
	[110]	$\sqrt{2}/3$	$\sqrt{3}/3$	$\sqrt{6}/9$	184
($\bar{1}$11)	[101]	$\sqrt{2}/3$	$\sqrt{3}/2$	$\sqrt{6}/6$	122
	[0$\bar{1}$1]	$\sqrt{2}/3$	$\sqrt{3}/6$	$\sqrt{6}/18$	367
	[110]	$\sqrt{2}/3$	$\sqrt{3}/3$	$\sqrt{6}/9$	184
(1$\bar{1}$1)	[101]	$\sqrt{2}/3$	$\sqrt{3}/6$	$\sqrt{6}/18$	367
	[011]	$\sqrt{2}/3$	$\sqrt{3}/2$	$\sqrt{6}/6$	122
	[$\bar{1}$10]	0	0	0	Not deformed
(11$\bar{1}$)	[101]	0	$\sqrt{3}/2$	0	Not deformed
	[011]	0	$\sqrt{3}/2$	0	Not deformed

A diagram showing the loading axis [112] is given in Figure E6.1.

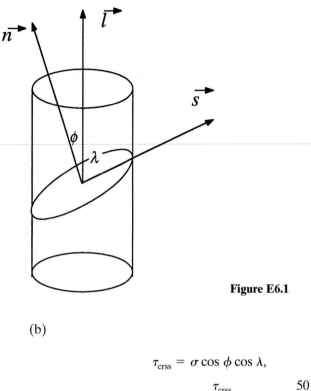

Figure E6.1

(b)

$$\tau_{\text{crss}} = \sigma \cos \phi \cos \lambda,$$

$$\sigma = \frac{\tau_{\text{crss}}}{\cos \phi \cos \lambda} = \frac{50}{\cos \phi \cos \lambda}.$$

Results are shown in the foregoing table.

EXAMPLE 6.2

Calculate the total energy due to dislocations for iron in the annealed condition, after 20% plastic deformation and 100% plastic deformation. Use both the exact (Eqn. 4.14) and the approximate equation (Eqn. 4.15) ($U = Gb^2/2$). Assume that the core has a radius equal to $5b$ and that dislocations are evenly distributed among edge and screw types. Take

$$G = 81.6 \text{ GPa} \qquad \text{(Table 2.5)},$$

$$\nu = 0.293 \qquad\qquad \text{(Table 2.5)},$$

$$r = 0.124 \text{ nm}.$$

The relationship between the stress and dislocation density is (see Section 6.3):

$$\tau = 40 \times 10^6 + 16.67\sqrt{\rho} \quad \text{(in Pa)}.$$

The stress–strain relationship is

$$\tau = \tau_0 + k\gamma^n,$$

where

$$\tau_0 = 50 \times 10^6, \quad k = 100 \times 10^6, \quad \text{and} \quad n = 1/2.$$

Solution: The stress levels for $\gamma = 0, 0.4$, and 2 ($\gamma = 2\varepsilon$) are

$$\tau = 50, 113, \text{ and } 191.4 \times 10^6 \text{ MPa}.$$

The dislocation density is:

$$\rho = (\tau - 40 \times 10^6)^2 \times \frac{1}{16.67^2} = 3.5 \times 10^{-3}(\tau - 40 \times 10^6)^2.$$

Hence, for

$$\varepsilon = 0, \qquad \rho = 3.5 \times 10^9 \text{ m}^{-2},$$
$$\varepsilon = 0.2, \qquad \rho = 3.71 \times 10^{13} \text{ m}^{-2};$$
$$\varepsilon = 1, \qquad \rho = 1.15 \times 10^{14} \text{ m}^{-2}.$$

We now obtain the dislocation spacing. It is known that

$$l \approx \frac{1}{\rho^{1/2}}.$$

So, for

$$\varepsilon = 0, \qquad l \approx 1.69 \times 10^{-5} \text{ m},$$
$$\varepsilon = 0.2, \qquad l \approx 1.64 \times 10^{-7} \text{ m},$$
$$\varepsilon = 1, \qquad l \approx 0.93 \times 10^{-7} \text{ m},$$

and we have

$$U_T = \frac{Gb^2}{10} + \frac{Gb^2}{4\pi(1 - \nu)}(1 - \nu\cos^2 \alpha)\frac{\rho^{-1/2}}{5b}.$$

For 50% edge and 50% screw dislocations, we make $\alpha = 45°$. The Burgers vector can then be calculated from the radius of the atoms. If the lattice parameter is a, the Burgers vector is

$$|\mathbf{b}| = a\sqrt{1^2 + 1^2 + 1^2} = 4r$$
$$= 0.496 \text{ nm}.$$

Thus, for

$$\varepsilon = 0, \qquad U_T = (0.1 + 0.847)Gb^2 = 0.947Gb^2;$$
$$\varepsilon = 0.2, \qquad U_T = (0.1 + 0.402)Gb^2 = 0.502Gb^2;$$
$$\varepsilon = 1, \qquad U_T = (0.1 + 0.34)Gb^2 = 0.44Gb^2.$$

The approximate expression for the dislocation self-energy ($U_T = Gb^2/2$) becomes gradually better as the density is increased. The total energy of dislocations per unit volume is

$$U = U_T\rho,$$

and for

$$\varepsilon = 0, \qquad U = 66.3 \text{ J/m}^3;$$

$$\varepsilon = 0.2, \qquad U = 74.2 \times 10^4 \text{ J/m}^3;$$

$$\varepsilon = 1, \qquad U = 2.3 \times 10^5 \text{ J/m}^3.$$

6.2.3 Shear Deformation

Just as a tensile test does not directly provide the shear stress in the slip plane and along the slip direction, it does not directly provide the corresponding deformation. Accordingly, one must determine shear by taking into account the relative orientations of the tensile axis and the slip system. If a tensile specimen is attached to the grips of a tensile-testing machine by means of universal joints, it can be seen that the slip plane will rotate with respect to the tensile axis as deformation proceeds. Therefore, it is important to know the deformation and, consequently, the change in orientation, along with the attendant alteration in Schmid's factor. In a similar way, it can be shown that the shear strain $d\gamma$ in the slip system is related to the longitudinal strain $d\varepsilon$ by

$$d\gamma = \frac{d\varepsilon}{\sin \chi \cos \lambda} = \frac{d\varepsilon}{M}. \tag{6.4}$$

Therefore, when $M = 0.5$, we have $\tau = 0.5\sigma$ and $\gamma = 2\varepsilon$. (Notice that $\tau = \sigma/2$!).

6.2.4 Slip Systems

Equations 6.3 and 6.4 establish the stress and strain in the plane and in the direction of shear and are therefore important from the point of view of dislocation motion. In HCP structures, the slip is more easily maintained in one plane. However, in BCC and FCC structures, other slip systems are easily activated. The rotation and direction of the slip plane will easily put other systems in a favorable position. This situation is shown in the stereographic projection of Figure 6.11. A certain crystal has its tensile axis within the crosshatched stereographic triangle. The first slip system to be activated will be the one with the highest Schmid factor. (See Equations 6.2 and 6.3). There are eight slip systems around axis P in the figure. There are other ones in the total stereographic projection. By using great circles, the reader can check whether the following systems of directions really belong to the planes:

$$(11\bar{1}) [101], \qquad (11\bar{1}) [1\bar{1}0],$$

$$(111) [1\bar{1}0], \qquad (111) [10\bar{1}],$$

$$(\bar{1}1\bar{1}) [10\bar{1}], \qquad (\bar{1}1\bar{1}) [110],$$

$$(1\bar{1}\bar{1}) [110], \qquad (1\bar{1}\bar{1}) [101].$$

The maximum value of Schmid's factor, $M = 0.5$, is obtained for $\chi = \lambda = 45°$. The angles between P and the $\langle 100 \rangle$ directions are determined by means of a Wulff net, passing a great circle through the two poles. Among the preceding eight systems, the slip system having the highest Schmid factor is $(11\bar{1}) [101]$; slip will initially take place in this system. Plane $(11\bar{1})$ is therefore called the *primary slip plane*. As deformation proceeds, χ and λ will ro-

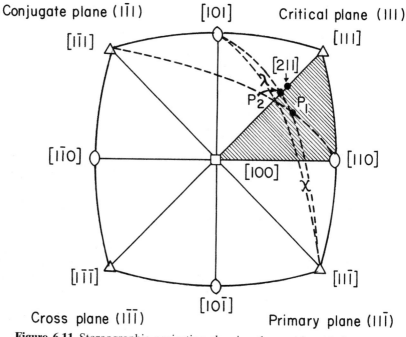

Figure 6.11 Stereographic projection showing the rotation of slip plane during deformation. Direction P, inside stereographic triangle moves towards P_2 on boundary [100]–[111]. Then, P_2 moves toward [211].

tate. In the stereographic projection, this is indicated by rotation of the axis P. Actually, the specimen rotates with respect to the axis. P will tend to align itself with direction [101], decreasing λ in the process; this is shown in Figure 6.11. However, when the great circle passing through [100] and [111] is reached, the primary system and the *conjugate slip system* (111) [110] will have the same Schmid factor. The typical behavior in this case is double slip in both systems: The axis P will tend toward the direction [211], as shown in the figure. In reality, there are deviations from this behavior, and there is a tendency for "overshoot" and subsequent correction. The two other slip systems are called the *cross* system and the *critical* system. This nomenclature, however, is not universal: Often, the term "cross-slip" is used to describe a different situation—small slip segments in a secondary slip system joining slip lines in a primary slip system.

As a conclusion to the foregoing discussion, it can be said that a cubic crystal will initially undergo slip in one system if P is within the stereographic triangle. If P is on the sides of the triangle, two systems have the same Schmid factor. On the other hand, if P coincides with one of the edges, the situation is more complicated: Eight systems will have the same Schmid factor if P coincides with [100], four if it coincides with [110], and six if it coincides with [111]. The term "polyslip" refers to a crystal oriented in such a way that more than one system is activated.

When a cubic monocrystal with an orientation inside the stereographic triangle is deformed, one single slip system is often activated. Such orientations in the center of the stereographic triangle are considered "soft" orientations, and Figure 6.12 illustrates the different stress–strain curves obtained for niobium. Orientations 1 and 2 are close to polyslip, and the stress–strain curves have the characteristic parabolic hardening shape. Several slip systems are activated at the onset of yielding. For orientations 3 through 7, inside of the stereographic triangle, one single slip system is activated first. The onset of conjugate slip requires rotation of the crystal toward an orientation along the sides of the triangle; this occurs only at a certain amount of strain, which depends on the orientation. Single slip

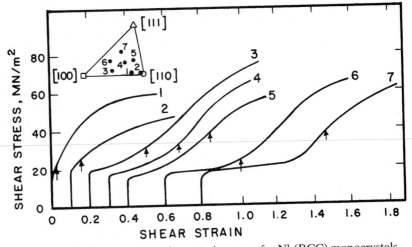

Figure 6.12 Shear-stress vs. shear-strain curves for Nb(BCC) monocrystals at different crystallographic orientations; arrows indicate calculated strain at which conjugate slip is initiated. (From T. E. Mitchell, *Prog. App. Matls. Res.* 6 (1964) 117)

is characterized by a very low work-hardening rate; once the conjugate slip becomes operative, the work-hardening rate increases significantly.

Figure 6.13 shows generic shear-stress–shear-strain curves for FCC single crystals. Any such curve can be divided, conveniently, into three regions: I, II, and III; θ_I, θ_{II}, and θ_{III} are the respective work-hardening slopes ($d\tau/d\gamma$) of the regions. In what follows, we describe the salient points of the various stages.

Stage I starts after elastic deformation at the critical stress τ_0. This stage, called "easy glide," is a linear region of low strain-hardening rate. θ_I is approximately $G/30$. Stage I is characterized by long slip lines (100 to 1,000 μm), straight and uniformly spaced (10 to 100 nm apart). We adopt the nomenclature used by A. Seeger.[2] *Sliplines* are the "elemen-

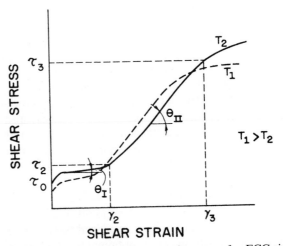

Figure 6.13 Generic shear-stress–shear-strain curves for FCC single crystals for two different temperatures.

[2]Alfred Seeger, "The Mechanism of Glide and Work Hardening in Face-Centered Cubic and Hexagonal Close-Packed Metals" in J. C. Fischer, W. G. Johnston, and T. Vreeland (eds.), *Dislocations and Mechanical Properties of Crystals* (New York: John Wiley, 1957), p. 243.

tary structure" of slip and can be observed only via the electron microscope. With the optical microscope, one observes *slip bands;* they occur at the higher strains and are made up of clusters of slip lines. On the other hand, *slip markings* are observed as steps at the surface of the specimen. Stage I does not exist in polycrystals or in monocrystals oriented for polyslip. The extent of this stage depends strongly on the crystal orientation. The strain at the end of stage I (γ_2) has a maximum value when the crystal orientation is located in the center of the standard stereographic triangle. The end of stage I is considered to be the start of secondary slip (when, in Figure 6.11, point P_1 has moved to P_2).

Stage II, or the linear hardening stage, has the following important characteristics:

1. A linear hardening regime with a high θ_{II}.
2. $\theta_{II}/G \approx 1/300$. This parameter is relatively constant for a great majority of metals. (The maximum variation is a factor of about 2).

θ_{II} is approximately equal to $10\theta_I$ and is relatively independent of temperature, although it has a significant effect on the extent of stage II.

Stage III is characterized by cross-slip. Stage III is difficult to bring about at a low level of stresses, and its operation is aided by high temperatures. Thus, one expects that the stress necessary at the start of stage III, τ_3, would depend on temperature, and such, indeed, is the case in practice: τ_3 increases with a decrease in temperature.

The start of Stage III is also markedly dependent on the stacking-fault energy of the metal. Metals with relatively low stacking-fault energies—for example, brasses, bronzes, and austenitic steels—have a rather wide stacking-fault ribbon and, consequently, need a higher activation energy for cross-slip to occur. (See Figure 6.14.) This is so because, for cross-slip to occur in these metals, it is necessary to form a constriction over a wide ribbon of the stacking fault, in order to have a certain length of perfect dislocation. Thus, in metals and alloys with low stacking-fault energies, cross-slip will be difficult to bring about at normal stress levels. This, in turn, makes it difficult for the screw dislocations to change their slip plane. The dislocation density is high, and the transition from stage II to stage III is retarded. Aluminum, on the other hand, has a higher stacking-fault energy. Thus, the stress necessary for cross-slip to occur in aluminum, at a given temperature, is much lower than in, say, copper or brass.

6.2.5 Independent Slip Systems in Polycrystals

For any FCC crystal whose tensile axis is near the center of the stereographic triangle, deformation should start at the primary system. However, if the crystal is surrounded by other crystals with different crystallographic orientations—as is likely in a polycrystalline aggregate—it may not start deforming in the same manner. The strain then takes place in

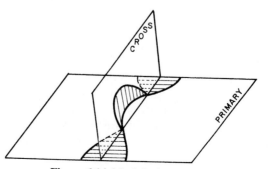

Figure 6.14 Model of cross-slip.

the first crystals (grains). In other words, it is not possible to form discontinuities along the grain boundaries; deformation has to propagate from one grain to another if continuity at the boundary is to be maintained. Five independent slip systems are required to produce a general homogeneous strain in a crystal by slip.

The slip along several parallel systems produces, macroscopically, a translation of one part of the crystal with respect to the other and, consequently, a certain shear. Since the plastic flow generally occurs without any appreciable change in volume, we have $\varepsilon_{11} + \varepsilon_{22} + \varepsilon_{33} = 0$. This relationship reduces the components of strain from six ($\varepsilon_{11}, \varepsilon_{22}, \varepsilon_{33}, \varepsilon_{12}, \varepsilon_{13}, \varepsilon_{23}$) to five; the operation of one slip system produces only one independent component of the strain tensor. Therefore, one can conclude that five independent slip systems are required for the deformation of one grain in a polycrystalline aggregate. Consequently, polycrystals do not exhibit stage I (easy glide) of work-hardening.

6.3 WORK-HARDENING

In preceding sections, work-hardening in single crystals was attributed to the interaction of dislocations with other dislocations and barriers that impede the motion of dislocations through the crystal lattice. In polycrystals, too, this basic idea remains valid. However, due to the mutual interference of neighboring grains and the problem of compatible deformations among adjacent grains, multiple slip occurs rather easily, and, consequently, there is an appreciable work-hardening right at the beginning of straining.

In a manner similar to that in single crystals, primary dislocations interact with secondary dislocations, giving rise to dislocation dipoles and loops which result in local dislocation tangles and, eventually, a three-dimensional network of subboundaries. Generally, the size of these cells decreases with increasing strain. The structural differences between one metal and another are mainly in the sharpness of these cell boundaries. In BCC metals and in FCC metals with high stacking-fault energy, such as Al, the dislocation tangles rearrange into a well-defined cell structure, while in metals or alloys with low stacking-fault energy (e.g., brasses, bronzes, austenitic steels, etc.), where the cross-slip is rather difficult and the dislocations are extended, the sharp subboundaries do not form even at very large strains.

The plastic deformation and the consequent work-hardening results in an increase in the dislocation density. An annealed metal, for example, will have about 10^6 to 10^8 dislocations per cm^2, while a plastically cold-worked metal may contain up to 10^{12} dislocations per cm^2. The relationship between the flow stress and the dislocation density is the same as that observed for single crystals—that is,

$$\tau = \tau_0 + \alpha G b \sqrt{\rho}, \tag{6.5}$$

where α is a constant with a value between 0.3 and 0.6. This relationship has been observed to be valid for a majority of the cases. τ_0 is the stress necessary to move a dislocation in the absence of other dislocations. Figure 6.15 shows that Equation 6.5 is obeyed for copper monocrystals (with one, two, and six slip systems operating), as well as polycrystals. The relationship is very important and serves as a basis for work-hardening theories. In ceramics, only limited observations of such kind have been made. Nevertheless, they show the same trend. Measurements of dislocation densities in sapphire (single-crystal α-alumina) subjected to plastic deformation at high temperatures (1,400–1,720°C), above the ductile-to-brittle transition, are shown in Figure 6.16. These dislocation densities were measured at strains $\gamma < 0.23$, and it was observed that the dislocation density showed a stress dependence analogous to Equation 6.5, with $\tau_0 = 0$. The proportionality coefficient was dependent on temperature and varied in the range 0.2–0.5, which is very similar to the corresponding range for metals.

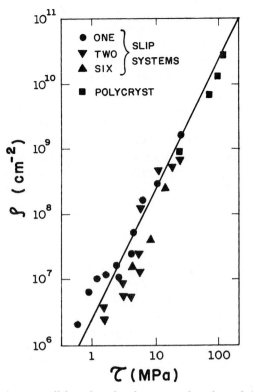

Figure 6.15 Average dislocation density ρ as a function of the resolved shear stress τ for copper. (Adapted with permission from H. Wiederich, *J. Metals,* 16 (1964) p. 425, 427)

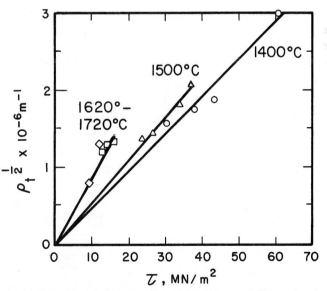

Figure 6.16 Relationship between flow shear stress and dislocation density for monocrystalline sapphire (Al_3O_3) deformed at different temperatures. (From B. J. Pletka, A. H. Heuer, and T. E. Mitchell, *Acta Met.,* 25 (1977) 25, Figure 3, p. 27)

Many theories have been advanced to explain the phenomenon of work-hardening. The most important and difficult part in the attempt to predict work-hardening behavior is to determine how the density and distribution of dislocations vary with the plastic strain. The problem is that stress is a state function in the thermodynamic sense (i.e., it depends only on its position, not on how it got there). Plastic strain, on the other hand, is a path function of its position (i.e., it depends on the actual path traversed in reaching a certain strain value). In other words, plastic strain is dependent on its history. Thus, the presence or absence of dislocations and their distributions can tell us nothing about how a certain amount of strain was accumulated in the crystal, because we do not know the path that dislocations traversed to accumulate that strain. Hence, one constructs models that re-create the processes by means of which the various dislocation configurations emerge; one then tries to correlate the models with the configurations observed experimentally. Both the density and the distribution of dislocations are very sensitive functions of the crystal structure, stacking-fault energy, temperature, and rate of deformation. In view of all this, it is not surprising that a unique theory of work-hardening which would explain all of its aspects does not exist.

In what follows, we briefly review three of the best known theories of work-hardening—those of Taylor, Seeger, and Kuhlmann–Wilsdorf.

6.3.1 Taylor's Theory

Taylor's theory[3] is one of the oldest theories of work-hardening. At the time the theory was postulated (1934), the stress–strain curve for metallic crystals such as aluminum was considered to be parabolic. (The single-crystal stress–strain curve consisting of three stages was unknown; see Figure 6.13.) This being so, Taylor proposed a model that would predict the parabolic curve. The principal idea, which, incidentally, is still used in one form or another by modern theories, was that the dislocations, on moving, elastically interact with other dislocations in the crystal and become trapped. These trapped dislocations give rise to internal stresses that increase the stress necessary for deformation (i.e., the flow stress).

Let l be the average distance that a dislocation moves before it is stopped. The initial and final positions A and B are marked in Figure 6.17. Let ρ be the dislocation density after a certain strain. Then the shear strain is given by (see Eqn. 4.23)

$$\gamma = k\rho bl, \tag{6.6}$$

where k is an orientation-dependent factor and b is the Burgers vector.

Taylor considered only edge dislocations and assumed that the dislocation distribution was uniform; thus, the separation between dislocations, L, will be equal to $\rho^{-1/2}$. (See Figure 6.17.) The effective internal stress τ, caused by these interactions among dislocations, is the stress necessary to force two dislocations past each other. The interactions among dislocations are complex, involving attraction, repulsion, reactions, etc. Taylor considered only a very sim-

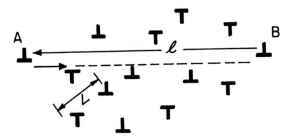

Figure 6.17 Taylor model of interaction among dislocations in a crystal.

[3]G. I. Taylor, *Proc. Roy. Soc. (London)*, A145 (1934) 362.

ple case: As the dislocation moves from A to B, it will approach the other dislocations, with the minimum distance being $L/2$. Taking into account just the repulsion from the dislocations, we can assume that, for an edge dislocation, the shear stress fields given in Chapter 4 are (Eqn. 4.6b)

$$\sigma_{12} = \frac{Gb}{2\pi(1 - \nu)} \frac{x_1(x_1^2 - x_2^2)}{(x_1^2 + x_2^2)^2}.$$

Supposing that $x_1 = L/2$ and $x_2 = 0$, we arrive at

$$\sigma_{12} = \frac{Gb}{\pi(1 - \nu)L} = \frac{KGb}{L},$$

where K is a constant. In order for the moving dislocation to overcome this stress field, a shear stress

$$\tau = \frac{KGb}{L}$$

has to be applied. Or, recalling that $L = \rho^{-1/2}$, we obtain

$$\tau = KGb\sqrt{\rho}. \tag{6.7}$$

From Equations 6.6 and 6.7, we get

$$\tau = KGb\sqrt{\frac{\gamma}{kbl}} = k'G\sqrt{\frac{\gamma}{l}}. \tag{6.8}$$

We could add a frictional term τ_0 that is required to move the dislocation in the absence of other dislocations, and arrive at:

$$\tau = \tau_0 + k''\gamma^{1/2}$$

Equation 6.8 is a parabolic relation between the stress τ and the strain γ. It describes, approximately, the behavior of many materials at large deformations. Among the criticisms of the Taylor theory, one may include the following:

1. Such regular configurations of dislocations are rarely observed in cold-worked crystals.
2. Screw dislocations are not involved, and thus, the cross-slip is excluded; edge dislocations cannot cross-slip.
3. Two dislocations on neighboring planes may be trapped in each other's stress fields and may thus become incapable of moving independently of each other. But the pair of dislocations may be pushed by a third dislocation.
4. We know now that stress–strain curves for hexagonal crystals, as well as those for stage II of cubic crystals, are linear. Taylor's theory does not explain this linear hardening.
5. Taylor's parabolic relation derives from the supposition that there is a uniform distribution of deformed regions inside the crystal. In reality, the distribution is not uniform, and experimentally, we observe slip bands, cells, and other nonuniform arrangements.

6.3.2 Seeger's Theory

Seeger's theory, (see the suggested readings) addresses the three stages of work-hardening of a monocrystal (easy glide, linear hardening, and parabolic hardening) and proposes specific mechanisms for each stage. The values of the slopes for the three stages are obtained from dislocation considerations. In stage I, long-range interactions between well-spaced dislocations are considered. The dislocation loops are blocked by unspecified obstacles, all on the primary system. Slip activity on secondary slip systems begins in stage II of hardening. The secondary activity furnishes barriers such as Lomer-Cottrell barriers. The dislocations pile up against such barriers in Stage II and give rise to long-range internal stresses that control the flow stress. Without going into complex details, we can say that the long-range theory of Seeger et al. does predict that $\theta_{11}/G \approx 1/300$ for FCC metals.

6.3.3   Kuhlmann–Wilsdorf's Theory

The substructures developed during metal deformation processes resemble the idealized models only in the first stages. As the imposed deformation increases, dislocation cells start to form in alloys with medium and high stacking-fault energies. With increasing deformation, the cell diameters decrease, and the cells become elongated in the general direction of the deformation. The cell walls tend to become progressively sharper as the misorientation between two adjacent cells increases. A cell wall is essentially a low-angle grain boundary, but when the misorientation between adjacent cell walls reaches a certain critical value, we can no longer refer to the boundary in these terms. The boundary between two cells becomes freer of dislocations, and subgrains are formed in a process called *polygonization*. This transition from cells to subgrains occurs at different effective strains for different materials: 0.80 for 99.97% pure Al and 1 to 1.20 for copper. A detailed treatment of the work-hardening and formation of texture at large imposed plastic strains is given by Gil Sevillano et al.[4] For metals with low stacking-fault energies, the development of a fine lamellar substructure consisting of microtwins, twin bundles, shear bands, and stacking faults is the characteristic feature of high-strain deformation.

   Figure 6.18 shows the changes in substructure observed in nickel rolled at room temperature. At reductions up to 40%, we clearly have a cellular structure. We can see that at 40% (Figure 6.18b) we already have a large dislocation density. At 80% reduction, we can clearly see that many of the cell walls have disappeared and are replaced by well-defined boundaries. The observation is made more difficult because of the large density of dislocations. The electron diffraction patterns (right-hand corner of photomicrographs) show the effect very well. Up to 40% reductions, the diffraction spots are fairly clear, with little asterism (elliptical distortion). At 80% (Figure 6.18c), the asterism is very pronounced, and elongated spots break down into smaller spots, indicating that a distorted grain has broken down into subgrains, which have relatively little distortion. Based on observations of dislocation cells in plastically deformed metals with medium and high stacking-fault energies, Kuhlmann–Wilsdorf[5] proposed the so-called mesh-length theory, which is based on the stress necessary for dislocation bowing. In stage I, the dislocations multiply into certain restricted regions and penetrate into regions as yet substantially free of mobile dislocations, until a quasiuniform distribution of dislocations is obtained. The only resistance to deformation is the dislocation line tension. Thus, hardening occurs due to the fact that free segments of dislocations become ever smaller. Stage II starts when there are no more "virgin" areas left for penetration by new dislocations. The stress required to bow segments of dislocation is responsible for a great part of stage II hardening: Dislocation segments can bow out inside the cells. Figure 6.19 shows, in a schematic manner, dislocation cells of size L in which the cell walls occupy a fraction f of the total crystal. Dislocation sources with mean width l are activated and form loops, shown in the figure. As these loops are formed, the dislocation density increases and the cell size decreases. Kuhlmann–Wilsdorf was able to explain, in quantitative manner, the three stages of work-hardening.

[4]J. Gil Sevillano, P. van Houtte, and E. Aernoudt, *Prog. Mater. Sci.,* 25 (1981) 69.

[5]D. Kuhlmann–Wilsdorf, *Met. Trans.,* 16A (1985) 2091.

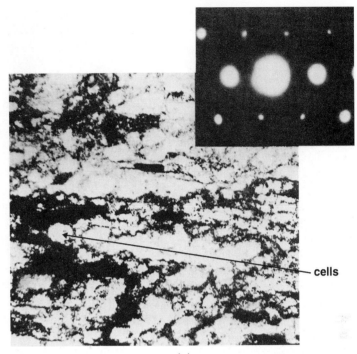

(a)

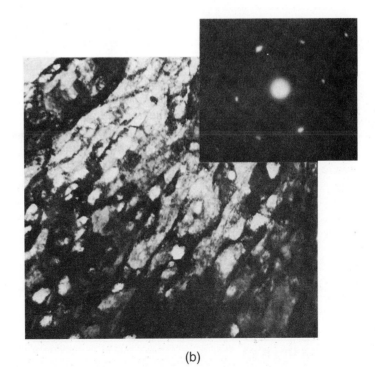

(b)

Figure 6.18 Development of substructure of Nickel-200 as a function of plastic deformation by cold rolling. (a) 20% reduction. (b) 40% reduction. (c) 80% reduction.

(c)

Figure 6.18 (Continued)

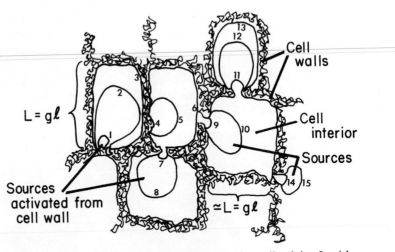

Figure 6.19 Schematic representation of dislocation cells of size L, with activation of dislocation sources from the cell walls and bowing out of loops into the cell interior. (Courtesy of D. Kuhlmann–Wilsdorf)

EXAMPLE 6.3

Consider dislocations blocked in a copper crystal. If the flow stress is controlled by the stress necessary to operate a Frank–Read source, compute the dislocation density ρ in this crystal when it is deformed to a point where the resolved shear stress in the slip plane is 42 MPa. Take $G = 50$ GPa.

Solution: The dislocation line length is related to the dislocation density by

$$\ell = \rho^{-1/2}.$$

The flow stress is the shear stress necessary to operate a Frank–Read source. Hence (from Eqn. 4.17d),

$$\tau = Gb/\ell = Gb\sqrt{\rho}.$$

For copper, $b = 3.6 \times 10^{-10}(\sqrt{2}/2)$ m $= 2.55 \times 10^{-10}$ m, where 3.6×10^{-10} m is the Cu lattice parameter. Rearranging the preceding expression, we obtain the dislocation density

$$\rho = \tau^2/G^2b^2 = (42 \times 10^6)^2/(50 \times 10^9)^2 \times (2.55 \times 10^{-1})^2,$$

or

$$\rho = 1.09 \times 10^{13} \text{ m}^{-2}.$$

EXAMPLE 6.4

For the single crystal of an FCC metal, the work-hardening rate in shear is $d\tau/d\gamma = 0.3$ GPa. Compute the work-hardening rate in tension, $d\sigma/d\varepsilon$, for a polycrystal of this metal. Take the Schmid factor M_p to be 1/3.1.

Solution: The tensile stress is related to the shear stress by the Schmid factor

$$\sigma = M_p^{-1}\tau.$$

Thus,

$$d\sigma = M_p^{-1}d\tau. \tag{1}$$

Also, the tensile strain ε is related to the shear strain γ by

$$\varepsilon = M_p\gamma.$$

Thus,

$$d\varepsilon = M_p d\gamma. \tag{2}$$

Dividing Equation 1 by Equation 2, we have

$$d\sigma/d\varepsilon = M_p^{-2}(d\tau/d\gamma) = (d\tau/d\gamma)(3.1)^2,$$

or

$$d\sigma/d\varepsilon = 9.61(d\tau/d\gamma) = 9.61 \times .3 = 2.88 \text{ GPa}.$$

6.4 SOFTENING MECHANISMS

Under special circumstances, materials can undergo softening during plastic deformation. This degradation of a material's strength can be caused by a number of mechanisms. *Damage accumulation* is the most prevalent mechanism in ceramics and composites. Damage can be of many types: microcracks forming in the material, a breakup of the matrix–materials reinforcement interface, cracking of second phase, etc. Figure 6.20 shows softening observed in concrete. The compression was halted at several points, and the specimen was unloaded and subsequently reloaded. The damage consists of microcracks, which are expressed in the reduction in the Young's modulus of concrete as the compression evolves ($E_1 > E_2 > E_3$). In Chapter 2, we saw how microcracks affect the Young's modulus of brittle materials. A discussion of damage accumulation in composites is given in Chapter 15.

Softening of radiation-hardened materials occurs when the sweeping of radiation-induced defects (point defects) by dislocations leads to the formation of "soft" channels.

In *geometric softening*, during plastic deformation, individual grains rotate toward crystallographic orientations for which the Schmid factor is increased. This rotation can lead to global softening in spite of the hardening along the individual slip systems.

We describe the last of the major softening mechanisms, *thermal softening*, in detail. The plastic deformation of a metal is an irreversible process, and most of the work of deformation is converted into heat. At most, only 10% of plastic deformation is stored as defects (primarily dislocations). Let us calculate the figure for a crystal with a dislocation density of $\sim 10^{11}$ cm^{-2}, characteristic of highly deformed metal. We first find the total energy in the crystals which is equal to

$$U = \rho \frac{Gb^2}{2}.$$

For copper, $G = 48.3$ GPa and $b = 0.25$ nm. Thus, the total deformation energy is (10^{11} cm^{-2} = 10^{15} m^{-2}):

$$U_d = \frac{1}{2} \times 10^{15} \times 48.3 \times 10^9 \times 0.0625 \times 10^{-18}$$

$$= 1.5 \times 10^6 \text{ J/m}^3.$$

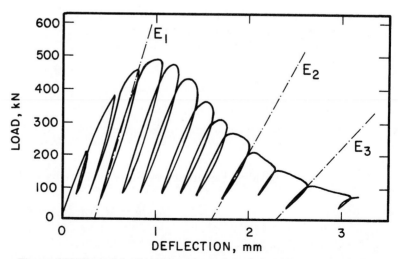

Figure 6.20 Typical load deformation curve for concrete under uniaxial compression; the specimen was unloaded and reloaded at different stages of deformation. (From G. A. Hegemier and H. E. Reed, *Mech. of Matls.*, 4 (1985) 215, Fig. 1, p. 217; data originally from A. Anvar)

Assuming that this sample of copper exhibits work-hardening and that the constitutive equation is (see Eqn. 3.11)

$$\sigma = \sigma_0 + K\varepsilon^n,$$

where

$$\sigma_0 = 50 \text{ MPa},$$

$$n = 0.5,$$

$$K = 500 \text{ MPa}.$$

We can calculate the total deformation energy per unit volume at a strain of 0.5:

$$U = \int_0^\varepsilon \sigma d\varepsilon = \int_0^{\varepsilon_1} (\sigma_0 + K\varepsilon^n) d\varepsilon$$

$$= \sigma_0\varepsilon_1 + K\frac{\varepsilon_1^{n+1}}{2} = 50 \times 10^6 \times 0.5 + 500 \times 10^6 \times \frac{0.35}{2}$$

$$= (25 + 87.5) \times 10^6$$

$$= 1.12 \times 10^8 \text{ J/m}^3.$$

Thus, the dislocation energy represents 1.4 percent of the total work of deformation. The latter leads to a rise in the temperature of the specimen. If there is insufficient time for the heat to escape from the specimen during deformation, the material cannot be considered isothermal any longer, and the loss of strength due to the increase in temperature will, at a certain point, exceed the increase in strength due to work-hardening. At this point, the stress–strain curve starts to go down, and thermal softening sets in. This is shown in Figure 6.21a. At lower strain rates (2×10^{-4} s^{-1}, 10^{-3} s^{-1}, and 10^{-2} s^{-1}), the curves show the normal work-hardening behavior up to high strains. However, for the strain rates of 1.44 s^{-1} and 3.9 s^{-1}, the stress–strain curves show maxima beyond which softening sets in. It is easy to understand and to predict this softening. Figure 6.21b shows shear-stress–shear-strain curves for titanium at different temperatures. For simplicity, linear work hardening was assumed. These curves are all isothermal. If we now compute the temperature elevation produced by plastic deformation, we have to apply the equation

$$dT = \frac{\beta}{\rho C_p} \sigma d\varepsilon,$$

where β is the conversion of mechanical energy into heat, C_p is the heat capacity and ρ the density of the material. By taking small increments of strain, we obtain

$$\Delta T = \frac{\beta}{\rho C_p} \sigma \Delta\varepsilon.$$

In Figure 6.21b, an adiabatic curve was built in such a fashion. The work-to-heat conversion factor β_1 is usually taken to be in the range 0.9–1.0. (Most of the work is converted to heat.) The adiabatic curve shows a maximum at γ approximately equal to 1; this marks the shear strain at which softening starts.

The softening of the material will lead to the phenomenon of adiabatic shear localization. Adiabatic shear bands are narrow regions where softening occurs and where

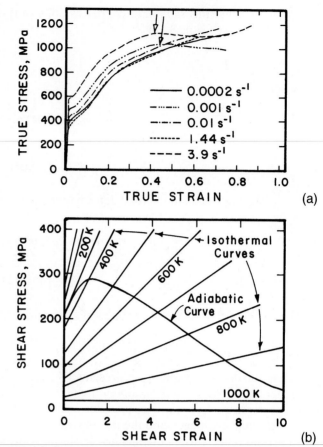

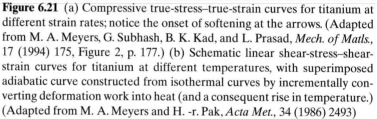

Figure 6.21 (a) Compressive true-stress–true-strain curves for titanium at different strain rates; notice the onset of softening at the arrows. (Adapted from M. A. Meyers, G. Subhash, B. K. Kad, and L. Prasad, *Mech. of Matls.*, 17 (1994) 175, Figure 2, p. 177.) (b) Schematic linear shear-stress–shear-strain curves for titanium at different temperatures, with superimposed adiabatic curve constructed from isothermal curves by incrementally converting deformation work into heat (and a consequent rise in temperature.) (Adapted from M. A. Meyers and H. -r. Pak, *Acta Met.*, 34 (1986) 2493)

concentrated plastic deformation takes place. Steels, titanium alloys, and aluminum alloys are quite prone to shear-band formation, which occurs in machining and which is responsible for the breakup of the machining chips. Shear-band formation also occurs in high-strain-rate operations, such as forging and shearing, as well as in ballistic impact.

Shear bands incurred during forging operations are highly undesirable, because they can lead to subsequent fracture of the specimen. The microstructure within shear bands is quite different from the surrounding material. The shear bands often undergo dynamic recrystallization, due to the high local temperature.

In the ballistic impact of projectiles against armor, shear bands play a major role both in the defeat of the armor and in the breakup of the projectiles. Since recrystallization occurs very rapidly, the resultant grain size is very small, typically 0.1 μm. Figure 6.22a shows a shear band in titanium with a width of approximately 10 μm. The microcrystalline structure inside of the shear band is seen in the photomicrograph of Figure 6.22b; the initial grain size of the material was 50 μm.

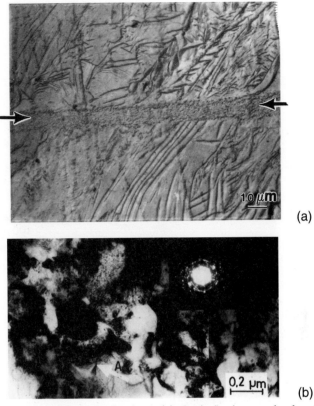

Figure 6.22 Shear bands in titanium. (a) Optical micrograph, showing band. (b) Transmission electron micrograph, showing microcrystalline structure, with grain size approximately equal to 0.2 μm. The original grain size of the specimen was 50 μm.

6.5 TEXTURE STRENGTHENING

A single crystal rotates when it deforms plastically in a particular slip system. (See Section 6.2.4.) When a polycrystal is deformed in rolling, forging, drawing, and so on, the randomly oriented grains will slip on their appropriate glide systems and rotate from their initial conditions, but this time under a constraint from the neighboring grains. Consequently, a strong preferred orientation or texture develops after large strains; that is, certain slip planes tend to align parallel to the rolling plane, while certain slip directions tend to align in the direction of rolling or wire drawing. In metals, annealing can also result in a texture generally different from that obtained by mechanical working, but still dependent on the history of the mechanical working. As an illustration, Figure 6.23 shows the microstructures along three perpendicular planes for nickel cold rolled to a reduction in thickness of 60%. The highly elongated grains along the rolling direction are readily seen.

A strongly textured material can exhibit highly anisotropic properties. This is not intrinsically bad; in fact, controlled anisotropy in sheet metals can be exploited to obtain an improved final product. The Young's modulus E of steel can, theoretically, have a value between the extreme values of the iron monocrystal (i.e., between Fe[111] and Fe[100]), as shown in Figure 6.24. The Young's modulus cannot be changed much by alloying, but texture can—again, theoretically—have some influence. We caution the reader that the effect on E, for all practical purposes, is rather small. This is not the case, however, for many

Figure 6.23 Perspective view of microstructure of Nickel-200 cold rolled to a reduction in thickness of 60%.

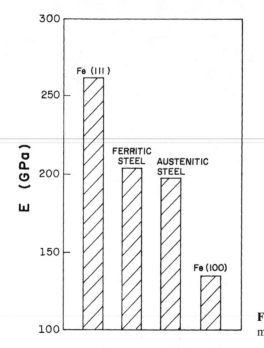

Figure 6.24 Theoretical bounds on the Young's modulus E of steel.

other properties. For example, Figure 6.25 shows the rather marked orientation dependence of the yield strength σ_y and the strain to fracture, ε_f, of a rolled copper sheet. Clearly, cups made out of this material by deep drawing would show "earing" at 90° intervals due to this texture (see Figure 3.41 for illustration of "earing"). Use is made of such texture development in Fe–3% Si. Sheets of this material are used to make transformer cores, wherein thermomechanical treatments are given to develop a desirable magnetic anisotropy that improves electrical performance.

Crystallographic texture is commonly represented in the form of normal-pole or inverse-pole figures. A normal-pole figure is a stereographic projection showing the intensity of normals to a specific plane in all directions, while an inverse-pole figure is a sterographic projection showing the intensities of all planes in a specific direction. The ex-

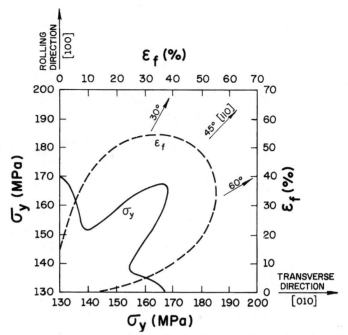

Figure 6.25 Orientation dependence of yield strength σ_y and strain to fracture, ε_f, of a rolled copper sheet.

perimental procedure involves measuring relative intensities of X-ray reflections from the polycrystalline material at different angular settings. Details of the experimental determination of pole figures can be found in standard texts on the subject.

Figure 6.26 shows the [111] pole figure of a heavily deformed α-brass (70% Cu–30% Zn) sheet. This texture, called brass-type texture, is a (110)[1$\bar{1}$2] texture, i.e., with (110) planes parallel to the rolling plane and [1$\bar{1}$2] directions parallel to the rolling direction. The double texture indicated for FCC structures on Table 6.1 is not obtained in α-brass, but single (110) [1$\bar{1}$2] texture develops, due to the material's low stacking-fault energy or (probably) to mechanical twinning.

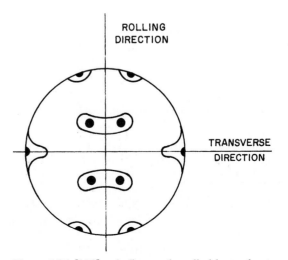

Figure 6.26 [111] pole figure of a rolled-brass sheet.

TABLE 6.1 Some Common Wire and Sheet Textures

	Wire (Fiber Texture)	Sheet (Rolling Texture)
FCC	[111] + [100]	$(110)[1\bar{1}2] + (112)[11\bar{1}]$
BCC	110	(100)[011]
HCP	$[10\bar{1}0]$	$(0001)[11\bar{2}0]$

SUGGESTED READINGS

Geometry of Deformation

J. GILSEVILLANO, P. VAN HOUTTE, and E. AERNOUDT, *Large Strain Work Hardening and Textures,* in Progress in Materials Science, Vol. 25, J. W. Christian, P. Haasen, and T. B. Massalski (eds.). Elmsford, NY: Pergamon Press, 1981, p. 69.

R. W. K. HONEYCOMBE and H. K. D. H. BHADESHIA. *The Plastic Deformation of Metals.* New York: St. Martin's Press, 1995.

W. F. HOSFORD, *The Mechanics of Crystals and Textured Polycrystals,* New York; Oxford U. Press: 1993.

Inhomogeneities of Plastic Deformation. Metals Park, Ohio: ASM, 1970.

Work Hardening

L. M. CLAREBROUGH and M. E. HARGREAVES. "Work Hardening of Metals," in *Progress in Metal Physics,* Vol. 8, B. Chalmers and W. Hume-Rothery (eds.). New York: Pergamon Press, 1959, p. 1.

A. H. COTTRELL. *Dislocations and Plastic Flow in Crystals.* London: Oxford University Press, 1953.

P. B. HIRSCH (ed.). *The Physics of Metals, Vol. 2: Defects.* Cambridge, U.K.: Cambridge University Press, 1975.

J. P. HIRTH and J. LOTHE. *Theory of Dislocations,* 2d. ed. New York: J. Wiley, 1982.

D. Kuhlmann-Wilsdorf, *Met-Trans.* 11A (1985) 2091.

A. SEEGER, in *Work Hardening,* TMS-AIME Conf., Vol. 46, 1966, p. 27.

A. W. THOMPSON (ed.). *Work Hardening in Tension and Fatigue,* New York: TMS-AIME, 1977.

EXERCISES

6.1 Discuss the merits and demerits of the use of transmission electron microscopy techniques to study the dislocation behavior in crystalline materials.

6.2 Explain why a metal like lead does not work-harden when deformed at room temperature, whereas a metal such as iron does.

6.3 What is the effect of cold work and annealing on the Young's modulus of a metal?

6.4 If we strain an FCC and an HCP single crystal, which of the two will have a larger amount of easy glide, and why?

6.5 In a cold-worked metal, a dislocation density of 1×10^{16} m^{-2} was measured after a shear strain of 10%. Assuming that the dislocations are uniformly distributed, estimate the flow stress of this metal. Take $G = 25$ GPa.

6.6 Consider dislocation blocked with an average spacing of 1—in a copper crystal. If the flow stress is controlled by the stress necessary to operate a Frank–Read source, compute the dislocation density ρ in this crystal when it is deformed to a point where the resolved shear stress in the slip plane is 42 MPa. Take $G = 50$ GPa.

6.7 Make a schematic plot showing the variation in the following parameters with percent cold work:
    (a) ultimate tensile strength
    (b) yield strength in tension

Liberty

    (c)   strain to failure

    (d)   reduction in area

6.8 The stress axis in an FCC crystal makes angles of 31° and 62° with the normal to the slip plane and with the slip direction, respectively. The applied stress is 10 MN/m².

    (a)   Determine the resolved stress in the shear plane.

    (b)   Is the resolved stress larger when the angles are 45° and 32°, respectively?

    (c)   Using a stereographic projection, determine the resolved stresses on the other slip systems.

6.9 Magnesium oxide is cubic (having the same structure as NaC1). The slip planes and directions are [110] and ⟨110⟩, respectively. Along which directions, if any, can a tensile (or compressive) stress be applied without producing slip?

6.10 A Cu monocrystal (FCC) of 10 cm length is pulled in tension. The stress axis is [$\bar{1}23$].

    (a)   Which is the stress system with the highest resolved shear stress?

    (b)   If the extension of the crystal continues until a second slip system becomes operational, what will this system be?

    (c)   What rotation will be required to activate the second system?

    (d)   How much longitudinal strain is required to activate the second system?

6.11 Flow stress varies with strain rate; one equation that has been used to express this dependence is

$$\sigma = c\dot{\varepsilon}^{m'}f(\varepsilon, T),$$

where m' is the strain-rate sensitivity, which is generally lower than 0.1. Some metals, called superplastic, can undergo elongations of up to 1,000% in uniaxial tension. Assuming that these tests are performed at a uniform velocity of the crosshead, will the metals have a very high or a very low value of m'? Explain, in terms of the formation and inhibition of the neck.

6.12 W. G. Johnston and J. J. Gilman[6] experimentally determined the relationship between dislocation velocity and applied stress

$$\nu = A\sigma^m,$$

where A is the constant of proportionality. Assuming that the mobile dislocation density does not depend on the velocity of the dislocations, obtain a relationship between m and m' (from Exercise 6.11).

6.13 The following results were obtained in an ambient-temperature tensile test, for an aluminum monocrystal having a cross-sectional area of 9 mm² and a stress axis making angles of 27° with [100], 245° with [110], 24.5° with [110], and 29.5° with [111]:

Load (N)	Length (cm)
0	10.000
12.40	10.005
14.30	10.040
16.34	10.100
18.15	10.150
21.10	10.180
23.60	10.200
26.65	10.220

    (a)   Plot the results in terms of true stress versus true strain.

    (b)   Determine the resolved shear stress on the system that will slip first.

    (c)   Determine the longitudinal strain at the end of the easy-glide stage (when a second slip system become operative).

---

[6]W. G. Johnston and J. J. Gilman, *J. Appl. Phys.*, 30 (1959) 129.

6.14 Take a stereographic triangle for a cubic metal. If the FCC slip systems are operative, indicate the number of slip systems having the same Schmid factor if the stress axis is
   (a) [111]
   (b) [110]
   (c) [100]
   (d) [123]
   Use the stereographic projections to show your results.

6.15 A copper bicrystal is composed of two monocrystals separated by a coherent twin boundary (111). The bicrystal is being compressed in a homogeneous upset test in such a way that the twin boundary is perpendicular to machine plates. The compression direction is the same for both crystals, namely, [134].
   (a) Is this crystal isoaxial?
   (b) Is deformation in the two crystals compatible or incompatible?

6.16 The flow stress σ is related to the dislocation density ρ by the relationship

$$\sigma_1 = \sigma_i + \alpha Gb\sqrt{\rho},$$

where the symbols have their usual significance. If the dislocation density is inversely related to the grain size d, show that a Hall–Petch type of dependence of flow stress on grain size is obtained.

6.17 For an FCC polycrystalline metal, TEM analysis showed that the dislocation density after cold working was 5×10^{10} m^{-2}. If the friction stress is 100 MPa, $G = 40$ GPa, and $b = 0.3$ nm, compute the flow stress of this metal.

6.18 The stress–strain curve of a polycrystalline aluminum sample can be represented by

$$\sigma = 25 + 200\varepsilon^{0.5}.$$

Calculate the energy of deformation per unit volume corresponding to uniform strain (i.e., just prior to the onset of necking) in this material.

6.19 An FCC crystal is pulled in tension along the [100] direction.
   (a) Determine the Schmid factor for all slip systems.
   (b) Identify the slip system(s) that will be activated first.
   (c) What is the tensile stress at which this crystal will flow plastically? ($\tau = 50$ MPa.)

6.20 Calculate the total energy due to dislocations for copper that underwent 20% plastic deformation, resulting in a dislocation density of 10^{14} m^{-2}. Assume that $b = 0.3$ nm.

6.21 Using data from Figure Ex 6.21 for (Ni–22%Cr–12%Co–9%Mo), obtain appropriate parameters for the Johnson–Cook equation (see Chapter 3). Assume $\dot{\varepsilon}_0 = 3 \times 10^{-4}$ s^{-1} and $T_m = 1,600$ K. Using the Johnson–Cook equation plot stress–strain curves for temperatures of 77, 177, 477, and 1,477 K.

6.22 A monocrystal (diameter 4 mm, length 100 mm) is being pulled in tension.
   (a) What is the elongation undergone by the specimen if 1,000 dislocations on slip planes making 45° with the tension axis cross the specimen completely? Take $b = 0.25$ nm.
   (b) What would the elongation be if all dislocations that exist in the crystal (10^6 cm^{-2}) were ejected by the applied stress? Assume a homogeneous distribution of dislocations. Assume that the crystal is FCC and all the dislocations are in the same slip system.

6.23 A long crystal with a square cross section (1×1 cm) is bent to form a semicircle with radius $R = 25$ cm.
   (a) Determine the total number of dislocations generated if all bending is accommodated by edge dislocations.
   (b) Determine the dislocation density ($b = 0.3$ nm).

6.24 The response of copper to plastic deformation can be described by Hollomon's equation $\sigma = \kappa\varepsilon^{0.7}$.

It is known that for $\varepsilon = 0.25$, $\sigma = 120$ MPa. The dislocation density varies with flow stress according to the well-known relationship.

$$\sigma = K'\rho^{1/2}.$$

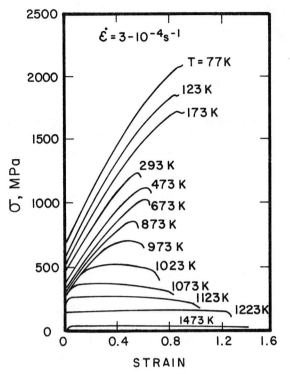

Figure E6.21 (After D Viereck, G. Merckling, K. H. Lang, D. Eifler, and
D. Löhe, in *Festigkeit und Verformung bei Höher Temperatur,* K. Schneider,
ed. (Oberursel: Informationsgesellshaft, pp. 102–208)

(a) If the dislocation density at a plastic strain of 0.4 is equal to 10^{11} cm^{-2}, plot the dislocation density versus strain.

(b) Calculate the work performed to deform the specimen.

(c) Calculate the total energy stored in the metal as dislocations after a plastic deformation of 0.4, and compare this value with the one obtained in part (b). Explain the difference.

6.25 A single crystal of silver is pulled in tension along the [100] direction. Determine the Schmid factor for all slip systems. What is the tensile stress at which this crystal will flow plastically? ($\tau = 100$ MPa.)

6.26 Determine the area of the slip plane in Ni deformed parallel to [100] and under a load $P = 150 \times 10^3$ N. The shear stress is 600 MPa.

6.27 Compute the dislocation density in tungsten if the flow stress is controlled by the stress necessary to operate a Frank–Read source. The shear stress in the slip plane is 50 MPa. Take $G = 166$ GPa.

7

Fracture: Macroscopic Aspects

7.1 INTRODUCTION

The separation or the fragmentation of a solid body into two or more parts, under the action of stresses, is called *fracture*. The subject of fracture is vast and involves disciplines as diverse as solid-state physics, materials science, and continuum mechanics. Fracture of a material by cracking can occur in many ways, principally the following:

1. Slow application of external loads.
2. Rapid application of external loads (impact).
3. Cyclic or repeated loading (fatigue).
4. Time-dependent deformation (creep).
5. Internal stresses, such as thermal stresses caused by anistropy of the thermal expansion coefficient or temperature differences in a body.
6. Environmental effects (stress corrosion cracking, hydrogen embrittlement, liquid metal embrittlement, etc.)

The process of fracture can, in most cases, be subdivided into the following categories:

1. Damage accumulation.
2. Nucleation of one or more cracks or voids.
3. Growth of cracks or voids. (This may involve a coalescence of the cracks or voids.)

Damage accumulation is associated with the properties of a material, such as its atomic structure, crystal lattice, grain boundaries, and prior loading history. When the local strength or ductility is exceeded, a crack (two free surfaces) is formed. On continued loading, the crack propagates through the section until complete rupture occurs. Linear elastic fracture mechanics (LEFM) applies the theory of linear elasticity to the phenomenon of fracture—mainly, the propagation of cracks. Defining the fracture toughness of a material as its resistance to crack propagation, LEFM provides us with a quantitative measure of fracture toughness. Various standardization bodies, including the American Society for Testing and Materials (ASTM), British Standards Institution (BSI), and Japan Institute of Standards (JIS), have standards for fracture toughness tests.

In this chapter, we will develop a quantitative understanding of cracks. It is very important to calculate the stresses at the tip (or in the vicinity of the tip) of a crack, because these calculations help us answer a very important practical question: At what value of the external load will a crack start to grow?

Figure 7.1 shows a simple analog that will assist the student in the visualization of different types of crack. In Figure 7.1a, "goofy duck" has its beak initially closed. Let us consider the spacing between the upper and lower beaks as a crack. Depending on how the goofy duck moves its beak, different modes of crack loading are generated:

- The opening mode, shown in Figure 7.1(b) is caused by loading that is perpendicular to the crack plane.
- The sliding mode, shown in Figure 7.1(c) is produced by forces parallel to the crack plane and perpendicular to the crack "line" (crack extremity).
- The tearing mode (Figure 7.1d) is produced by forces parallel to the crack surface and to the crack "line."

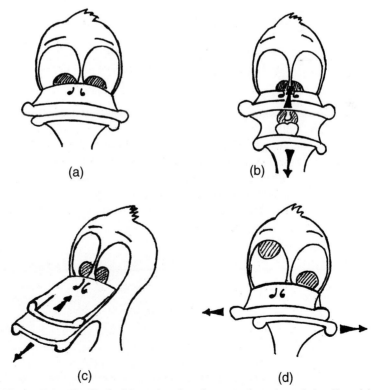

(a) (b)

(c) (d)

Figure 7.1 "Goofy duck" analog for three modes of crack loading. (a) Crack/beak closed. (b) Opening mode. (c) Sliding mode. (d) Tearing mode. (Courtesy of M. H. Meyers)

Among the parameters and tests that have been developed, mostly during the last quarter of the 20th century, to describe the resistance to fracture of a material in a quantitative and reproducible manner, is the *plane strain fracture toughness,* defined as the critical stress intensity factor under plane strain conditions and mode I loading. This is the stress intensity factor at which a crack of a given size starts to grow in an unstable manner. The fracture toughness is related to the applied stress by the equation

$$K_{Ic} = Y\sigma\sqrt{\pi a},$$

where K_{Ic} is the fracture toughness, a is the characteristic dimension (semilength) of the crack and Y is a factor that depends on the geometry of the specimen, the location of the crack, and the loading arrangement of the material. One can see that the stress which can be safely applied decreases with the square root of the size of the crack. Also, note that K_{Ic} is a parameter of the material in the same manner as are hardness and yield strength. We will explain this in detail in Section 7.5.

7.2 STRESS CONCENTRATION AND GRIFFITH CRITERION OF FRACTURE

The most fundamental requisite for the propagation of a crack is that the stress at the tip of the crack must exceed the theoretical cohesive strength of the material. This is indeed the fundamental criterion, but it is not very useful, because it is almost impossible to measure the stress at the tip of the crack. An equivalent criterion, called the Griffith criterion, is more useful and predicts the force that must be applied to a body containing a crack for the propagation of the crack. The Griffith criterion is based on an energy balance and is described in Section 7.3. Let us first grasp the basic idea of stress concentration in a solid.

7.2.1 Stress Concentrations

The failure of a material is associated with the presence of high local stresses and strains in the vicinity of defects. Thus, it is important to know the magnitude and distribution of these stresses and strains around cracklike defects.

Consider a plate having a through-the-thickness notch and subjected to a uniform tensile stress away from the notch (Figure 7.2). We can imagine the applied external force being transmitted from one end of the plate to the other by means of lines of force (simi-

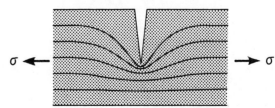

Figure 7.2 "Lines of force" in a bar with a side notch. The direction and density of the lines indicate the direction and magnitude of stress in the bar under a uniform stress σ away from the notch. There is a concentration of the lines of force at the tip of the notch.

lar to the well-known magnetic lines of force). At the ends of the plate, which is being uniformly stretched, the spacing between the lines is uniform. The lines of force in the central region of the plate are severely distorted by the presence of the notch (i.e., the stress field is perturbed). The lines of force, acting as elastic strings, tend to minimize their lengths and thus group together near the ends of the elliptic hole. This grouping together of lines causes a decrease in the line spacing locally and, consequently, an increase in the local stress (a stress concentration), there being more lines of force in the same area.

7.2.2 Stress Concentration Factor

The theoretical fracture stress of a solid is on the order $E/10$ (see Chapter 1), but the strength of solids (crystalline or otherwise) in practice is orders of magnitude less than this value. The first attempt at giving a rational explanation of this discrepancy was due to Griffith. His analytical model was based on the elastic solution of a cavity elongated in the form of an ellipse.

Figure 7.3 shows an elliptical cavity in a plate under a uniform stress σ away from the cavity. The maximum stress occurs at the ends of the major axis of the cavity and is given by Inglis's formula,[1]

$$\sigma_{\max} = \sigma\left(1 + 2\,\frac{a}{b}\right), \tag{7.1}$$

where $2a$ and $2b$ are the major and minor axes of the ellipse, respectively.[2] The value of the stress at the leading edge of the cavity becomes extremely large as the ellipse is flattened.

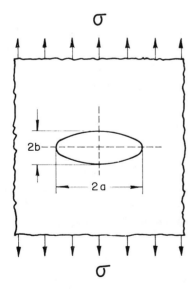

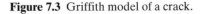

Figure 7.3 Griffith model of a crack.

[1]C. E. Inglis, Proc. Inst. Naval Arch., 55 (1913) 163, 219.

[2]The derivation of this equation, which can be found in more advanced tests [e.g., J. F. Knott, *Fundamentals of Fracture Mechanics*, (London: Butterworths, 1973), p. 51], involves the solution of the biharmonic equation, the choice of an appropriate Airy stress function, and complex variables.

In the case of an extremely flat ellipse or a very narrow crack of length $2a$ and having a radius of curvature $\rho = b^2/a$, Equation 7.1 can be written as

$$\sigma_{max} = \sigma\left(1 + 2\sqrt{\frac{a}{\rho}}\right) \simeq 2\sigma\sqrt{\frac{a}{\rho}} \qquad \text{for } \rho \ll a \qquad (7.2)$$

We note that as ρ becomes very small, σ_{max} becomes very large, and in the limit, as $\rho \to 0$, $\sigma_{max} \to \infty$. We define the term $2\sqrt{a/\rho}$ as the stress concentration factor K_t (i.e., $K_t = \sigma_{max}/\sigma$). K_t simply describes the geometric effect of the crack on the local stress (i.e., at the tip of the crack). Note that K_t depends more on the *form* of the cavity than on its size. A number of texts and handbooks give a compilation of stress concentration factors K_t for components containing cracks or notches of various configurations.

As an example of the importance of stress concentration, we point out the use of square windows in the COMET commercial jet aircraft. Fatigue cracks, initiated at the corners of the windows, caused catastrophic failures of several of these aircraft.

In addition to producing a stress concentration, a notch produces a local situation of biaxial or triaxial stress. For example, in the case of a plate containing a circular hole and subject to an axial force, there exist radial as well as tangential stresses. The stresses in a large plate containing a circular hole (with diameter $2a$) and axially loaded (Figure 7.4a) can be expressed as[3]

$$\sigma_{rr} = \frac{\sigma}{2}\left(1 - \frac{a^2}{r^2}\right) + \frac{\sigma}{2}\left(1 + 3\frac{a^4}{r^4} - 4\frac{a^2}{r^2}\right)\cos 2\theta,$$

$$\sigma_{\theta\theta} = \frac{\sigma}{2}\left(1 + \frac{a^2}{r^2}\right) - \frac{\sigma}{2}\left(1 + 3\frac{a^4}{r^4}\right)\cos 2\theta,$$

$$\sigma_{r\theta} = -\frac{\sigma}{2}\left(1 - \frac{3a^4}{r^4} + \frac{2a^2}{r^2}\right)\sin 2\theta. \qquad (7.3)$$

In many modern materials that have a high strength and limited ductility, the toughness determines the load-bearing ability. In design, one has to assume a maximum flaw size (determined by the quality control during processing) and determine the maximum stress from that. Section 7.5 introduces linear elastic fracture mechanics and the concepts leading to K_{Ic}. Materials exhibiting a large ductility require different parameters to describe this resistance to a crack's propagation; the J integral has been successfully used for these materials. There are comprehensive tables of fracture toughness data for different materials. (See Tables 7.1 and 8.4). The use of K_{Ic} is widespread in the description of the resistance of materials to crack propagation. In this chapter, we describe the macroscopic phenomena associated with the fracture process. The maximum stress occurs at point A in Figure 7.4a, where $\theta = \pi/2$ and $r = a$. In this case,

$$\sigma_{\theta\theta} = 3\sigma = \sigma_{max},$$

where σ is the uniform stress applied at the ends of the plate. We first define the stress concentration $K_t = \sigma_{max}/\sigma$. K_t is 3. Figure 7.4b shows the stress concentration for a circular hole in a plate of finite lateral dimensions. When D, the lateral dimension, decreases, the stress concentration K_t drops from 3 to 2.2.

[3]See, for example, S. Timoshenko and J. N. Goodier, *Theory of Elasticity,* 2d ed. (New York: McGraw-Hill, 1951), p. 78.

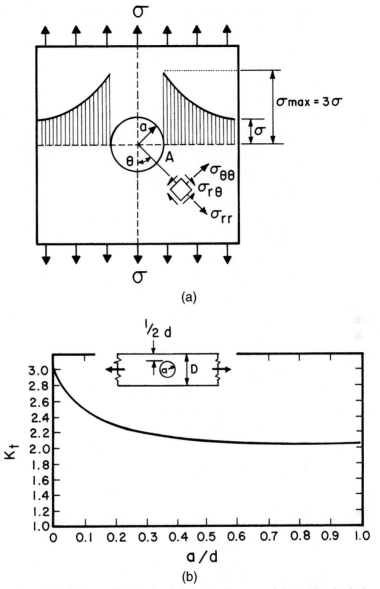

Figure 7.4 (a) Stress distribution in a large plate containing a circular hole. (b) Stress concentration factor K_t as a function of the radius of a circular hole in a large plate in tension.

Goodier[4] calculated the stresses around spherical voids in perfectly elastic materials. Although his solution was obtained when the applied stress was tensile, it can be extended to compressive stress by changing the signs. The stresses given by Timoshenko and Goodier can be determined from the methods of elasticity theory. For the equatorial plane ($\theta = \pi/2$), the stress σ_θ is equal to

$$\sigma_{\theta\theta} = \left[1 + \frac{4 - 5\nu}{2(7 - 5\nu)} \frac{a^3}{r^3} + \frac{9}{2(7 - 5\nu)} \frac{a^5}{r^5} \right] \sigma, \qquad (7.4)$$

[4]J. N. Goodier, *App. Mech.* 1 (1933) 39; see also Timoshenko and Goodier, *op. cit.*

where a is the radius of the hole, r is the radial coordinate, and ν is the Poisson's ratio. For $r = a$, $\nu = 0.3$, and we have

$$(\sigma_{\theta\theta})_{max} = \frac{45}{22} \sigma \approx 2\sigma.$$

For $\nu = 0.2$, $(\sigma_{\theta\theta})_{max} = 2\sigma$; thus, as expected, the stress concentration for a spherical void is approximately 2. The stress $\sigma_{\theta\theta}$ decays quite rapidly with r, as can be seen from Equation 7.4; the decay is given by r^{-3}. For $r = 2a$, we have $\sigma_\theta = 1.054$. This decay is faster than for the circular hole, where it goes with r^{-2} (Equation 7.3). For $\theta = 0$ (north and south poles), Timoshenko and Goodier have the equation

$$(\sigma_{rr})_\theta = (\sigma_{\theta\theta})_{\theta=0} = -\frac{3 + 15\nu}{2(7 - 5\nu)} \sigma.$$

Hence, a compressive stress generates a tensile stress at $\theta = 0$. This result is very important and shows that compressive stress can generate cracks at spherical flaws such as voids. Taking $\nu = 0.2$–0.3 (typical of ceramics), one arrives at the following values:

$$\frac{1}{2} \leq (\sigma_{\theta\theta})_{\theta=0} \leq \frac{7.5}{11}.$$

Thus, the tensile stress is 50–80% of the applied compressive stress. If failure is determined by cracking at spherical voids, cracking should start at a compressive stress level equal to $-4\sigma_t$ (depending on ν; in this case, for $\nu = 0.2$), where σ_t is the tensile strength of the material. This value represents, to a first approximation, the marked differences between the tensile and compressive strengths of cast irons, intermetallic compounds, and ceramics. The result is fairly close to the stress generated around a circular hole, given in Equation 7.3. In that case, for $r = a$, we find that

$$\sigma_{\theta\theta} = -\sigma.$$

In tensile loading, the stress $\sigma_{\theta\theta} = 3\sigma$, which would predict a threefold difference in tensile and compressive strengths. More general (elliptical) flaws can be assumed, and their response under compressive loading provides a better understanding of the compressive strength of brittle materials. The generation and growth of cracks from these flaws also needs to be analyzed, for more realistic predictions. This will be carried out in Section 8.3.4.

In the case of an elliptical hole, for $a = 3b$, Figure 7.5 shows that σ_{22} falls from its maximum value and attains σ asymptotically. σ_{11} increases to a peak and then falls to zero with the same tendency as σ_{22}. The general result is that the major perturbation in the applied stress state occurs over a distance approximately equal to a from the boundaries of the cavity, with the major stress gradients being confined to a region of dimensions roughly equal to ρ surrounding the maximum concentration position.

Although the exact formulas vary according to the form of the crack, in all cases K_t increases with an increase in the crack length a and a decrease in the root radius at the crack tip, ρ.

Despite the fact that the analysis of Inglis represented a great advance, the fundamental nature of the fracture mechanism remained obscure. If the Inglis analysis was applicable to a body containing a crack, how does one explain that, in practice, larger cracks propagate more easily than smaller cracks? What is the physical significance of the root radius at the tip of the crack?

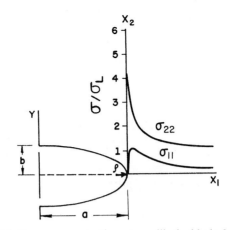

Figure 7.5 Stress concentration at an elliptical hole for $a = 3b$.

EXAMPLE 7.1

Although the elastic modulus of silica-based glass is rather low ($E = 70$ GPa), the theoretical strength of a defect-free glass can be as high as 3 GPa. Generally, such high strength values are not measured in practice. Why?

Solution: Extremely minute cracklike defects form rather easily on the glass surface. Such imperfections can lead to a drastic reduction in the strength of glass. This is the reason that, in the making of a glass fiber, a protective coating called a *size* is applied to the fiber immediately as it comes out of the spinneret. Just to get an estimate of the reduction in strength caused by a tiny imperfection—say, a 1-μm-long, atomically sharp scratch—we can use the Inglis expression (Equation 7.2),

$$\sigma_{th} = 2\sigma(a/\rho)^{0.5}, \quad \text{or} \quad \sigma = 0.5\,\sigma_{th}(\rho/a)^{-0.5},$$

where σ_{th} is the theoretical strength (3 GPa), a is the crack length (1 μm), and ρ is the root radius at the crack tip, which, since the tip is atomically sharp, can be taken to be 0.25 nm. Plugging these values into the preceding expression, we find that the real strength of such a glass is only 24 MPa! Note that in this problem we made an estimate of the notch root radius. In practice, this is very difficult to measure. That is why the concept of stress intensity factor, involving the far-field stress and the square root of the crack length, is much more convenient to deal with in fracture toughness problems, as we shall see later in this chapter (Section 7.5).

EXAMPLE 7.2

Determine the stresses at distances equal to 0, $a/2$, a, $(3a)/2$, and $2a$ from the surface of a spherical hole and for $\theta = 0$ and π/a.

Solution: We use Equation 7.3. By setting $\theta = 0$, we have

$$\sigma_{rr} = \frac{\sigma}{2}\left(2 - \frac{5a^2}{r^2} + \frac{3a^4}{r^4}\right),$$

$$\sigma_{\theta\theta} = \frac{\sigma}{2}\left(\frac{a^2}{r^2} - \frac{3a^4}{r^4}\right),$$

$$\tau_{r\theta} = 0.$$

For $\theta = \pi/2$,

$$\sigma_{rr} = \frac{\sigma}{2}\left(\frac{3a^2}{r^2} - \frac{3a^4}{r^4}\right),$$

$$\sigma_{\theta\theta} = \frac{\sigma}{2}\left(2 + \frac{a^2}{r^2} + \frac{3a^4}{r^4}\right),$$

$$\tau_{r\theta} = 0.$$

We calculate the stresses for $r = 0, a/2, a, (3a)/2,$ and $2a$ and plot them as shown in Figure E7.2 in terms of a dimensionless parameter r/a.

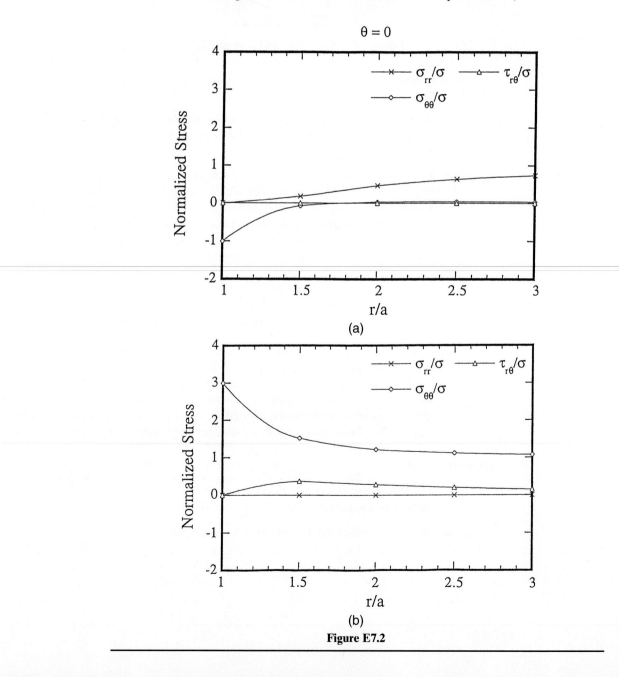

Figure E7.2

EXAMPLE 7.3

Two flat plates are being pulled in tension. (See Figure E7.3.) The flow stress of the materials is 150 MPa.
(a) Calculate the maximum stresses inside the plate.
(b) Will the material flow plastically?
(c) For which configuration is the stress higher?

Solution:

(a) Normal stress: $\sigma = \dfrac{P}{A} = \dfrac{100 \text{ kN}}{10 \text{ cm} \times 1 \text{ cm}}$

$$= 100 \text{ MPa},$$

$$\sigma_{max} = \sigma\left(1 + 2\frac{a}{b}\right).$$

Circular hole: $a = b = 3/2 \text{ cm} = 1.5 \text{ cm},$

$$\sigma_{max} = 100 \times \left(1 + 2 \times \frac{1.5}{1.5}\right) = 300 \text{ MPa}.$$

Elliptical hole: $a = 3/2 \text{ cm} = 1.5 \text{ cm}, b = 1/2 \text{ cm} = 0.5 \text{ cm},$

$$\sigma_{max} = 100 \times \left(1 + 2 \times \frac{1.5}{0.5}\right) = 700 \text{ MPa}.$$

(b) Yes, because in both cases, the stress is greater than the flow stress (150 MPa).
(c) The elliptical hole has higher stress than the circular one.

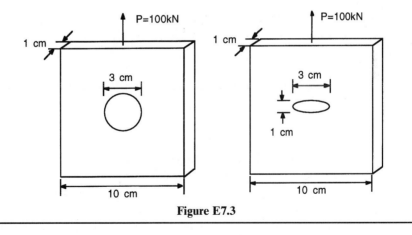

Figure E7.3

7.3 *GRIFFITH CRITERION*

Griffith proposed a criterion based on a thermodynamic energy balance. He pointed out that two things happen when a crack propagates: Elastic strain energy is released in a volume of material, and two new crack surfaces are created, which represent a surface-energy term. Thus, according to Griffith, an existing crack will propagate if the elastic strain energy released by doing so is greater than the surface energy created by the two new crack surfaces. Figure 7.6a shows an infinite plate of thickness t that contains a crack of length $2a$ under plane stress. As the stress is applied, the crack opens up. The shaded region

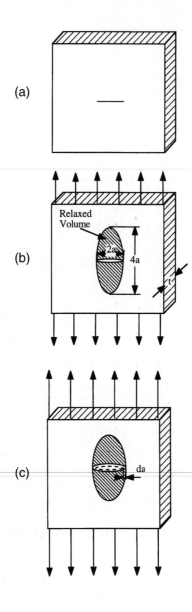

Figure 7.6 A plate of thickness t containing a crack of length $2a$. (a) Unloaded condition. (b) and (c) Loaded condition.

denotes the approximate volume of material in which the stored elastic strain energy is released (Figure 7.6b). When the crack extends a distance da on the extremities, the volume over which elastic energy is released increases, as shown in Figure 7.6c. The elastic energy per unit volume in a solid under stress is given by $\sigma^2/2E$. (See Chapter 2.) To get the total strain energy released, we need to multiply this quantity by the volume of the material in which this energy is released. In the present case, this volume is the area of the ellipse times the plate thickness. The area of the shaded ellipse is $\pi(2a)a = 2\pi a^2$; therefore, the volume in which the strain energy is relaxed is $2\pi a^2 t$. The total strain energy released is thus

$$\left(\frac{\sigma^2}{2E}\right)(2\pi a^2 t) = \frac{\pi\sigma^2 a^2 t}{E},$$

or, in terms of the per-unit thickness of the plate under plane stress, the energy released is

$$\pi\sigma^2 a^2/E.$$

This decrease in strain energy when a crack propagates is balanced by an increase in the surface energy produced by the creation of the two new crack surfaces. The increase in energy due to the latter equals $(2at)(2\gamma_s)$, where γ_s is the specific surface energy, i.e., the energy per unit area. Once again, in terms of the per-unit thickness of the plate, the increase in surface energy is $4a\gamma_s$. Now, when an elliptical crack is introduced into the plate, we can write, for the change in potential energy of the plate,

$$\Delta U = 4a\gamma_s - \frac{\pi\sigma^2 a^2}{E},$$

where ΔU is the potential energy per unit thickness of the plate in the presence of the crack, less is the potential energy per unit thickness of the plate in the absence of the crack, σ is the applied stress, a is half the crack length, E is the modulus of elasticity of the plate, and γ_s is the specific surface energy (i.e., the surface energy per unit area) of the plate.

As the crack grows, strain energy is released, but additional surfaces are created. The crack becomes stable when these energy components balance each other. If they are not in balance, we have an unstable crack (i.e., the crack will grow). We can obtain the equilibrium condition by equating to zero the first derivative of the potential energy ΔU with respect to the crack length. Thus,

$$\frac{\delta\Delta U}{\delta a} = 4\gamma_s - \frac{2\pi\sigma^2 a}{E} = 0, \tag{7.5a}$$

or

$$2\gamma_s = \frac{\pi\sigma^2 a}{E}. \tag{7.5b}$$

The reader can check the nature of this equilibrium further by taking the second derivative of U with respect to a. A negative second derivative would imply that Equations 7.5 represent an unstable equilibrium condition and that the crack will advance.

Rearranging Equation 7.5b, we may write, for the critical stress required for the crack to propagate in the plane-stress situation,

$$\sigma_c = \sqrt{\frac{2E\gamma_s}{\pi a}} \quad \text{(plane stress).} \tag{7.6a}$$

For the plane-strain situation, we will have the factor $(1 - \nu^2)$ in the denominator because of the confinement in the direction of thickness. The expression for the critical stress for crack propagation then becomes

$$\sigma_c = \sqrt{\frac{2E\gamma_s}{\pi a(1 - \nu^2)}} \quad \text{(plane strain).} \tag{7.6b}$$

The distinction between plane stress and plane strain is shown in Figure 7.7. Normal and shear stresses at free surfaces are zero; hence, for a thin plate, $\sigma_{33} = \sigma_{23} = \sigma_{13} = 0$. This is a plane-stress state (Figure 7.7a). In very thick planes ($t_2 > t_1$), the flow of material in the x_3 direction is restricted. Therefore, $\varepsilon_{33} = 0$, and consequently, $\varepsilon_{23} = \varepsilon_{13} = 0$. This is a plane-strain condition (Figure 7.7b). Note that the factor $(1 - \nu^2)$ is less than unity and is in the denominator. Therefore, the critical stress in a plane-strain situation will be higher than that in the plane-stress state. This is as expected, because of the confinement in the direction of thickness in the case of plane strain. For many metals, $\nu \approx 0.3$, and $(1 - \nu^2) \approx 0.91$. Thus, the difference is not very large.

The importance of the length of the crack is implicit in Griffith's analysis. In modern fracture mechanics, as we shall see later, the crack length enters as a square-root term in the product $\sigma\sqrt{a}$. According to Griffith's thermodynamic analysis, a necessary condition for crack propagation is

$$-\frac{\delta U_e}{\delta a} \geq \frac{\delta U_\gamma}{\delta a},$$

where U_e is the elastic energy of the system (i.e., the machine plus the test piece) and U_γ is equal to 2γ, in the simplest case. This is a necessary condition for fracture by rapid crack propagation. But it may not always be sufficient: If the local stress at the crack tip is not sufficiently large to break the atomic bonds, the energy criterion of Griffith will be inadequate.

Let us consider Equation 7.6a or 7.6b again. Note that the fracture stress, or critical stress required for crack propagation, σ_c, is inversely proportional to $\sqrt{a}$. More importantly, the quantity $\sigma_c\sqrt{a}$ depends only on material constants. It is instructive, then, to examine the Inglis result, Equation 7.2, and the Griffith result, Equation 7.6a, or 7.6b in the form

$$\sigma_c\sqrt{a} = \frac{1}{2}(\sigma_{max})_c\sqrt{\rho} = \text{constant}.$$

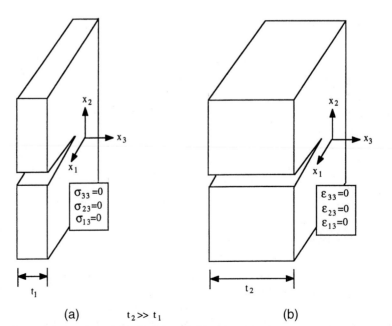

Figure 7.7 Crack in (a) thin (t_1) and (b) thick (t_2) plates. Note the plane-stress state in (a) and the plane-strain state in (b).

Here, σ_c is the critical far-field or uniform stress (i.e., the stress at fracture), a is the crack length corresponding to σ_c, $(\sigma_{max})_c$ is the stress at the crack tip at fracture, and ρ is the root radius at the tip of the crack.

Both analyses, Inglis's and Griffith's lead to the same result, viz., that a crack will propagate when an appropriate quantity with dimensions of stress times the square root of length reaches a critical value, a material constant. It is easy to see that the parameters in the Inglis analysis, $(\sigma_{max})_c$ and ρ, are local parameters and very difficult to measure, while the Griffith analysis allows us to use the far-field applied stress and crack length, which are easy to measure. It is this quantity, $\sigma_c\sqrt{a}$, that is called the fracture toughness and is denoted by K_{Ic}. We treat fracture toughness in detail in Section 7.5.

EXAMPLE 7.4

Consider a brittle material with $\gamma_s = 1$ J/m² and $E = 100$ GPa. (a) What is the breaking strength of this material if it contains cracklike defects as long as 1 mm? (b) Should it be possible to increase γ_s to 3,000 J/m², what would be the breaking strength for a 1-mm-long crack?

Solution: (a) We have

$$\gamma_s = 1 \text{ J/m}^2 \quad \text{and} \quad E = 100 \text{ GPa},$$

and

$$2a = 1 \text{ mm} \quad \text{and} \quad a = 0.5 \text{ mm}.$$

Thus,

$$\sigma_c = \sqrt{\frac{2E\gamma_s}{\pi a}} = \sqrt{\frac{2 \times 100 \times 10^9 \times 1}{\pi \times (0.5 \times 10^{-3})}}$$

$$= 11.3 \text{ MPa}.$$

(b) if γ_s increases to 3,000 J/m²,

$$2a = 1 \text{ mm} \quad \text{and} \quad a = 0.5 \text{ mm},$$

so that

$$\sigma_c = \sqrt{\frac{2E\gamma_s}{\pi a}} = \sqrt{\frac{2 \times 100 \times 10^9 \times 3,000}{\pi \times (0.5 \times 10^{-3})}}$$

$$= 618 \text{ MPa}.$$

7.4 CRACK PROPAGATION WITH PLASTICITY

If the material in which a crack is propagating can deform plastically, the form of the crack tip changes because of plastic strain. A sharp crack tip will be blunted. Another important factor is time: Because plastic deformation requires time, the amount of plastic deformation that can occur at the crack tip will depend on how fast the crack is moving. Figure 7.8 shows dislocations that were generated at a crack tip and that propagated along crystallographic planes. The crack is at the left-hand side, and the plane of the copper foil is (123).

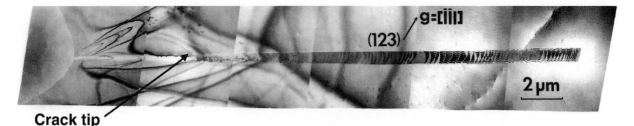

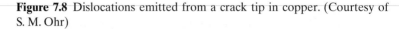

Crack tip

Figure 7.8 Dislocations emitted from a crack tip in copper. (Courtesy of S. M. Ohr)

In a great majority of materials, localized plastic deformation at and around the crack tip is produced because of the stress concentrations there. In such a case, a certain amount of plastic work is done during crack propagation, in addition to the elastic work done in the creation of two fracture surfaces. The mechanics of fracture will, then, depend on the magnitude of γ_p, the plastic work done, which in its turn depends on the crack speed, temperature, and the nature of the material. For an inherently brittle material, at low temperatures and at high crack velocities γ_p is relatively small ($\gamma_p < 0.1\gamma_s$). In such a case, the crack propagation would be continuous and elastic. These cases are usefully treated by means of linear elastic fracture mechanics, which is dealt with in Section 7.5. In any event, in the case of plastic deformation, the work done in the propagation of a crack per unit area of the fracture surface is increased from γ_s to $(\gamma_s + \gamma_p)$. Consequently, the Griffith criterion (Equation 7.6a or 7.6b) is modified to

$$\sigma_c = \sqrt{\frac{2E}{\pi a}(\gamma_s + \gamma_p)} \qquad \text{(plane stress)} \qquad (7.7a)$$

and

$$\sigma_c = \sqrt{\frac{2E}{\pi a(1 - \nu^2)}(\gamma_s + \gamma_p)} \qquad \text{(plane strain).} \qquad (7.7b)$$

Rearranging Equation 7.7a, we get

$$\sigma_c = \sqrt{\frac{2E\gamma_s}{\pi a}\left(1 + \frac{\gamma_p}{\gamma_s}\right)}.$$

For $\gamma_p/\gamma_s \gg 1$,

$$\sigma_c \simeq \sqrt{\frac{2E\gamma_p}{\pi a}}.$$

Thus, the plastic deformation around the crack tip makes it blunt and serves to relax the stress concentration by increasing the radius of curvature of the crack at its tip. Localized plastic deformation at the crack tip therefore improves the fracture toughness of the material.

    This is the conventional treatment of the plastic work contribution to the fracture process, wherein γ_p is considered to be a constant. However, the reader should be warned that this is not strictly true. As a matter of fact, the value of γ_p increases with the stress intensity factor K ($= Y\sigma\sqrt{a}$). Consider Equation 7.7a. As was pointed out, in the conventional approach γ_p will be very much larger than γ_s for a ductile material such as polycrystalline copper. Thus, according to this conventional treatment, the fracture stress σ_c should be relatively insensitive to changes in γ_s. However, in the embrittlement of copper with beryllium, all we change is the γ_s part of Equation 7.7a (along the grain boundaries where the fracture proceeds). The γ_p part in that equation (i.e., the plastic behavior of copper) does not change appreciably by the addition of beryllium to copper.

As pointed out earlier, equations of the type 7.6 or 7.7 are difficult to use in practice. It is not a trivial matter to measure quantities such as surface energy and the energy of plastic deformation. In a manner similar to that of Griffith, Irwin made a fundamental contribution to the mechanics of fracture when he proposed that fracture occurs at a stress that corresponds to a critical value of the crack extension force

$$G = \frac{1}{2}\frac{\delta U}{\delta a} = \text{rate of change of energy with crack length.}$$

G is sometimes called the *strain energy release rate.*

Now, $U = \pi a^2\sigma^2/E$, the energy released by the advancing crack per unit of plate thickness. This is for plane stress. For plane strain, a factor of $(1 - \nu^2)$ is introduced in the denominator. Thus,

$$G = \frac{\pi a\sigma^2}{E}$$

At fracture, $G = G_c$, and

$$\sigma_c = \sqrt{\frac{EG_c}{\pi a}} \qquad \text{(plane stress)} \tag{7.8a}$$

or

$$\sigma_c = \sqrt{\frac{EG_c}{\pi a(1 - \nu^2)}} \qquad \text{(plane strain).} \tag{7.8b}$$

From Equations 7.7 and 7.8, we see that

$$G_c = 2(\gamma_s + \gamma_p).$$

We shall come back to this idea of crack extension force later in the chapter.

7.5 LINEAR ELASTIC FRACTURE MECHANICS

A nonductile material has a very low capacity to deform plastically; that is, it is not capable of relaxing peak stresses at cracklike defects. In such a material, a crack will propagate very rapidly with little plastic deformation around the crack tip, resulting in what is called a brittle fracture. Typically, such a fracture is also characterized by a crack propagation that is sudden, rapid, and unstable. In practical terms, this definition of brittleness, which refers to the onset of instability under an applied stress smaller than the stress corresponding to plastic yielding of the material, is much more useful. Numerous brittle fractures have occurred in service, and there are abundant examples of them in a great variety of structural and mechanical engineering fields involving ships, bridges, pressure vessels, oil ducts, turbines, and so on. In view of the great importance of brittle fracture in real life, the discipline called linear elastic fracture mechanics (LEFM) has emerged, enabling us to obtain a quantitative measure of the resistance of a brittle material to unstable or catastrophic crack propagation. Extension of these efforts into nonlinear elastic and plastic regimes has led to the development of elasto-plastic fracture mechanics (EPFM), also called postyield fracture mechanics (see Section 7.8).

7.5.1 Fracture Toughness

Fracture mechanics gives us a quantitative handle on the process of fracture in materials. Its approach is based on the concept that the relevant material property, the fracture toughness, is the force necessary to extend a crack through a structural member. Under certain circumstances, this crack extension force (or an equivalent parameter) becomes independent of the dimensions of the specimen. The parameter can then be used as a quantitative measure of the fracture toughness of the material.

Fracture mechanics adopts an entirely new approach to designing against fracture. Admittedly defects will always be present in a structural component. But consider a structure or a component with a cracklike defect. We can simulate this with single edge notch of length a in a plate. (See Figure 7.9.) Alternatively, we can say that we are increasing the applied stress intensity factor K at the crack tip. The material at the tip, however, presents resistance to crack growth. We denote this inherent material resistance by K_R (sometimes the symbol R alone is used in place of K_R.) The discipline of fracture mechanics can then be represented by a triangle as shown in Figure 7.9; that is, we have an interplay among the following three quantities:

1. The far-field stress, σ.
2. The characteristic crack length, a.
3. The inherent material resistance to cracking, K_R.

Various parameters can represent K_R. Their equivalence is discussed in Section 7.6.5.

We now seek an answer to the question: Given a certain applied stress, what is the largest size defect (crack) that can be tolerated without the failure of the member? Once we know the answer to this question, it remains only to use appropriate inspection tech-

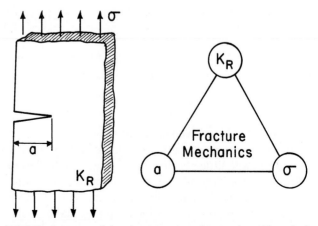

Figure 7.9 Inherent material resistance to crack growth and its relationship to the applied stress σ and crack size a.

niques to select a material that does not possess defects larger than the critical size for the given design stress.

7.5.2   Hypotheses of LEFM

The basic hypotheses of LEFM are as follows:

1. Cracks are inherently present in a material, because there is a limit to the sensibility of crack-detecting equipment.
2. A crack is a free, internal, plane surface in a linear elastic stress field. With this hypothesis, linear elasticity furnishes us stresses near the crack tip as

$$\sigma_{r\theta} = \frac{K}{\sqrt{2\pi r}} f(\theta), \tag{7.9}$$

   where r and θ are polar coordinates and K is a constant called the *stress intensity factor* (SIF).
3. The growth of the crack leading to the failure of the structural member is then predicted in terms of the tensile stress acting at the crack tip. In other words, the stress situation at the crack tip is characterized by the value of K. It can be shown by elasticity theory that $K = Y\sigma\sqrt{\pi a}$, where σ is the applied stress, a is half the crack length, and Y is a constant that depends on the crack opening mode and the geometry of the specimen.

7.5.3   Crack-Tip Separation Modes

The three modes of fracture are shown in Figure 7.10. Mode I (Figure 7.10a), called the opening mode, has tensile stress normal to the crack faces. Mode II (Figure 7.10b) is called the sliding mode or the forward shear mode. In this mode, the shear stress is normal to the advancing crack front. Mode III (Figure 7.10c) is called the tearing mode or transverse shear mode, with the shear stress parallel to the advancing crack front. The "goofy duck" analog of Fig. 7.1 shows this in a more illustrative fashion.

7.5.4   Stress Field in an Isotropic Material in the Vicinity of a Crack Tip

The stress components for the three fracture modes in an isotropic material are given next. In the case of anisotropic materials, these relations must be modified to permit the

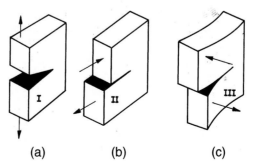

Figure 7.10 The three modes of fracture. (a) Mode I: opening mode. (b) Mode II: sliding mode. (c) Mode III: tearing mode (see Figure 7.1).

asymmetry of stress at the crack tip. K_I, K_{II}, and K_{III} indicate modes I, II, and III, respectively. We have (the derivation of these expressions is to Westergaard*)

Mode I:

$$\begin{bmatrix} \sigma_{11} \\ \sigma_{22} \\ \sigma_{12} \end{bmatrix} = \frac{K_I}{\sqrt{2\pi r}} \cos\frac{\theta}{2} \begin{bmatrix} 1 - \sin\frac{\theta}{2} \ \sin\frac{3\theta}{2} \\ 1 + \sin\frac{\theta}{2} \ \sin\frac{3\theta}{2} \\ \sin\frac{\theta}{2} \ \cos\frac{3\theta}{2} \end{bmatrix},$$

$$\sigma_{13} = \sigma_{23} = 0,$$
$$\sigma_{33} = 0 \qquad \text{(plane stress)},$$
$$\sigma_{33} = \nu(\sigma_{11} + \sigma_{22}) \quad \text{(plane strain)}. \tag{7.10}$$

Mode II:

$$\begin{bmatrix} \sigma_{11} \\ \sigma_{22} \\ \sigma_{12} \end{bmatrix} = \frac{K_{II}}{\sqrt{2\pi r}} \cdot \begin{bmatrix} -\sin\frac{\theta}{2} \ \left(2\cos\frac{\theta}{2} \ \cos\frac{3\theta}{2}\right) \\ \sin\frac{\theta}{2} \ \cos\frac{\theta}{2} \ \cos\frac{3\theta}{2} \\ \cos\frac{\theta}{2} \ \left(1 - \sin\frac{\theta}{2} \ \sin\frac{3\theta}{2}\right) \end{bmatrix},$$

$$\sigma_{13} = \sigma_{23} = 0,$$
$$\sigma_{33} = 0 \qquad \text{(plane stress)},$$
$$\sigma_{33} = \nu(\sigma_{11} + \sigma_{22}) \quad \text{(plane strain)}. \tag{7.11}$$

Mode III:

$$\begin{bmatrix} \sigma_{13} \\ \sigma_{23} \end{bmatrix} = \frac{K_{III}}{\sqrt{2\pi r}} \begin{bmatrix} -\sin\frac{\theta}{2} \\ \cos\frac{\theta}{2} \end{bmatrix},$$

$$\sigma_{11} = \sigma_{22} = \sigma_{33} = \sigma_{12} = 0. \tag{7.12}$$

7.5.5 Details of the Crack-Tip Stress Field in Mode I

Consider an infinite, homogeneous, elastic plate containing a crack of length $2a$ (Figure 7.11). The plate is subjected to a tensile stress σ far away from and normal to the crack. The stresses at a point (r, θ) near the tip of the crack are given by Equation 7.10. Ignoring the subscript of K, we may write the stress components in expanded form as

$$\sigma_{11} = \frac{K}{\sqrt{2\pi r}} \cos\frac{\theta}{2}\left(1 - \sin\frac{\theta}{2}\sin\frac{3\theta}{2}\right),$$

$$\sigma_{22} = \frac{K}{\sqrt{2\pi r}} \cos\frac{\theta}{2}\left(1 + \sin\frac{\theta}{2}\sin\frac{3\theta}{2}\right),$$

$$\sigma_{12} = \frac{K}{\sqrt{2\pi r}} \cos\frac{\theta}{2}\sin\frac{\theta}{2}\cos\frac{3\theta}{2},$$

*H. M. Westergaard, *J. Appl. Mechan.*, *5A* (1939) 49.

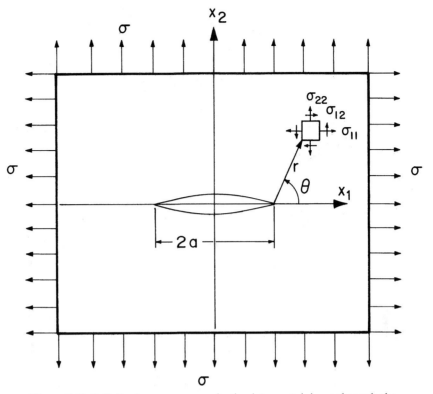

Figure 7.11 Infinite, homogeneous, elastic plate containing a through-the-thickness central crack of length $2a$, subjected to a tensile stress σ.

$$\sigma_{13} = \sigma_{23} = 0,$$
$$\sigma_{33} = 0 \quad \text{(plane stress)},$$
$$\sigma_{33} = \nu(\sigma_{11} + \sigma_{22}) \quad \text{(plane strain)}, \tag{7.13}$$

where

$$K = \sigma\sqrt{\pi a} \tag{7.14}$$

is the stress intensity factor for the plate and has the units $(\text{N/m}^2)\sqrt{m}$, or $\text{Pa}\sqrt{m}$, or $\text{Nm}^{-3/2}$. Note that Equation 7.13 is applicable in the region $r \ll a$ (i.e., in the vicinity of the crack tip). For $r \approx a$, higher order terms must be included.

For a thin plate, one has plane-stress conditions, and $\sigma_{33} = \sigma_{13} = \sigma_{23} = 0$. For a thick plate (infinite in the direction of thickness), there exist plane-strain conditions [i.e., $\sigma_{33} = \nu(\sigma_{11} + \sigma_{22})$ and $\sigma_{13} = \sigma_{23} = 0$].

Consider again Equation 7.13. The right-hand side has three quantities: K, r, and $f(\theta)$. The terms r and $f(\theta)$ describe the stress distribution around the crack tip. These two characteristics [i.e., dependence on $\sqrt{r}$ and $f(\theta)$] are identical for all cracks in two- or three-dimensional elastic solids. The stress intensity factor K includes the influence of the applied stress σ and the appropriate crack dimensions, in this case half the crack length a. Thus, K will characterize the external conditions (i.e., the nominal applied stress σ and half the crack length a) that correspond to fracture when stresses and strains at the crack tip reach a critical value. This critical value of K is designated as K_c. It turns out, as we shall see later, that K_c depends on the dimensions of the specimen. In the case of a thin sample (plane-stress conditions), K_c depends on the thickness of the sample, whereas in the case of a sufficiently thick sample (plane-strain conditions), K is independent of the thickness of the specimen and is designated as K_{Ic}.

The stress intensity factor K measures the amplitude of the stress field around the crack tip and should not be confused with the stress concentration factor K_t discussed in Section 7.2.2. It is also important to distinguish between K and K_c or K_{Ic}. The stress intensity factor K is a quantity, determined analytically or not, that varies as a function of configuration (i.e., the geometry of the crack and the manner of application of the external load). Thus, the analytical expression for K varies from one system to another. However, once K attains its critical value, K_{Ic}, in plane strain for a given system and material, it is essentially a constant for all the systems made of this material. The difference between K_c and K_{Ic} is that K_c depends on the thickness of the specimen, whereas K_{Ic} is independent of the thickness. The forms of K for various load and crack configurations have been calculated and are available in various handbooks. Some of the more common configurations and the corresponding expressions for K are presented in Figure 7.12.

For samples of finite dimensions, the general practice is to consider the solution for an infinite plate and modify it by an algebraic or trigonometric function that would make the surface tractions vanish. Thus, for a central through-the-thickness crack of length $2a$, in a plate of width W, we have

$$K = \sigma\left(W \tan \frac{\pi a}{W}\right)^{1/2}. \tag{7.15}$$

For the same crack in an infinite plate, we have

$$K = \sigma\sqrt{\pi a}.$$

If we expand Equation 7.15, we get

$$K = \sigma W^{1/2}\left(\frac{\pi a}{W} + \frac{\pi^3 a^3}{3W^3} + \cdots\right)^{1/2}$$

$$= \sigma\sqrt{\pi a}\left(1 + \frac{\pi^2 a^2}{3W^2} + \cdots\right)^{1/2}.$$

Thus, for an infinite solid, $a/W = 0$, and we have $K = \sigma\sqrt{\pi a}$, as expected. For an edge crack in a semi-infinite plate, we have $K = 1.12\,\sigma\sqrt{\pi a}$. The factor 1.12 here takes care of the fact that stresses normal to the free surface must be zero.

At this point, it is appropriate to make some comments on the limitations of LEFM. It was pointed out earlier that the expressions for stress components (Equations 7.10–7.12) are valid only in the neighborhood of the crack tip. The reader will have noticed that these stress components tend to infinity as we approach the tip (i.e., as r goes to zero). Now, there does not exist a material in real life that can resist an infinite stress. The material in the neighborhood of the crack tip, in fact, would inevitably deform plastically. Thus, these expressions for stress components based on linear elasticity theory are not valid in the plastic zone at the crack tip. The deformation process in a plastic zone, as is well known, will be a sensitive function of the microstructure, among other things. However, in spite of ignorance of the exact nature of the plastic zone, the LEFM treatment is valid for low-enough stresses such that the size of the plastic zone at the crack tip is small with respect

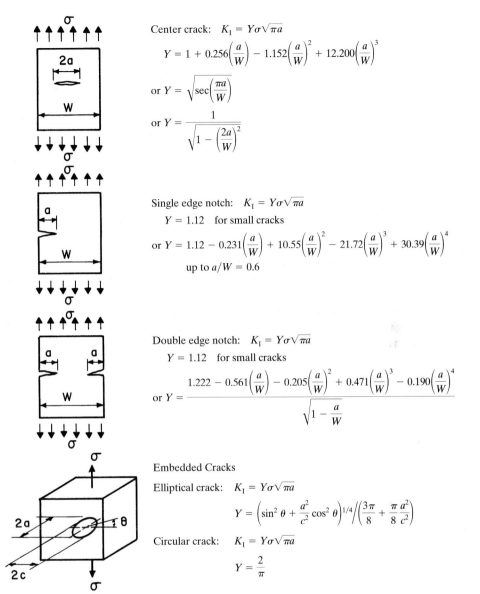

Center crack: $K_I = Y\sigma\sqrt{\pi a}$

$$Y = 1 + 0.256\left(\frac{a}{W}\right) - 1.152\left(\frac{a}{W}\right)^2 + 12.200\left(\frac{a}{W}\right)^3$$

or $Y = \sqrt{\sec\left(\dfrac{\pi a}{W}\right)}$

or $Y = \dfrac{1}{\sqrt{1 - \left(\dfrac{2a}{W}\right)^2}}$

Single edge notch: $K_I = Y\sigma\sqrt{\pi a}$

$Y = 1.12$ for small cracks

or $Y = 1.12 - 0.231\left(\dfrac{a}{W}\right) + 10.55\left(\dfrac{a}{W}\right)^2 - 21.72\left(\dfrac{a}{W}\right)^3 + 30.39\left(\dfrac{a}{W}\right)^4$

up to $a/W = 0.6$

Double edge notch: $K_I = Y\sigma\sqrt{\pi a}$

$Y = 1.12$ for small cracks

or $Y = \dfrac{1.222 - 0.561\left(\dfrac{a}{W}\right) - 0.205\left(\dfrac{a}{W}\right)^2 + 0.471\left(\dfrac{a}{W}\right)^3 - 0.190\left(\dfrac{a}{W}\right)^4}{\sqrt{1 - \dfrac{a}{W}}}$

Embedded Cracks

Elliptical crack: $K_I = Y\sigma\sqrt{\pi a}$

$$Y = \left(\sin^2\theta + \frac{a^2}{c^2}\cos^2\theta\right)^{1/4} \bigg/ \left(\frac{3\pi}{8} + \frac{\pi}{8}\frac{a^2}{c^2}\right)$$

Circular crack: $K_I = Y\sigma\sqrt{\pi a}$

$$Y = \frac{2}{\pi}$$

Figure 7.12 Some common load and crack configurations and the corresponding expressions for the stress intensity factor, K.

to the crack length and the dimensions of the sample. We shall see in the next section how to incorporate a correction term for the presence of a plastic zone at the crack tip.

7.5.6 Plastic-Zone Size Correction

Equations 7.10–7.12 show a $\sqrt{r}$ singularity; that is, σ_{11}, σ_{22}, and σ_{12} go to infinity when $\sqrt{r}$ goes to zero. For a great majority of materials, local yielding will occur at the crack tip, which would relax the peak stresses. As we shall see shortly, the utility of the elastic stress field equations is not affected by the presence of this plastic zone as long as the nominal stress in the material is below the general yielding stress of the material.

When yielding occurs at the crack tip, it becomes blunted; that is, the crack surfaces separate without any crack extension. (See Figure 7.13.) The plastic zone (radius r_y) will

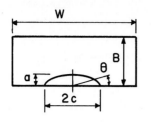

Semi-elliptical surface flaw in tension:

$$K_I = Y \frac{\sigma \sqrt{\pi a}}{\dfrac{3\pi}{8} + \dfrac{\pi}{8}\dfrac{a^2}{c^2}} \left(\sin^2 \theta + \frac{a^2}{c^2}\cos^2 \theta\right)^{1/4}$$

$$Y = 1.12$$

or: $K_I = Y\sigma \dfrac{\sqrt{\pi a}}{\Phi}$

$$Y = \left(\frac{a}{B}, \frac{a}{c}, \frac{c}{W}, \theta\right)$$

$$Y = \left[Y_1 + Y_2\left(\frac{a}{B}\right)^2 + Y_3\left(\frac{a}{B}\right)^4\right]Y_4 g(\theta)g(W)$$

$$Y_1 = 1.13 - 0.09\left(\frac{a}{c}\right)$$

$$Y_2 = -0.54 + \frac{0.89}{0.2 + \left(\dfrac{a}{c}\right)}$$

$$Y_3 = 0.5 - \frac{1.0}{0.65 + \left(\dfrac{a}{c}\right)} + 14\left(1.0 - \frac{a}{c}\right)^{24}$$

$$Y_4 = 1 + \left[0.1 + 0.35\left(\frac{a}{B}\right)^2\right](1 - \sin\theta)^2$$

$$g(\theta) = \left[\sin^2 \theta + \left(\frac{a}{c}\right)^2 \cos^2 \theta\right]^{1/8}$$

$$g(W) = \left[\sec\frac{\pi c}{W}\sqrt{\frac{a}{B}}\right]^{1/2}$$

for: $0 < \dfrac{a}{c} < 1 \qquad 0 < \dfrac{a}{B} < 1$

$\qquad 0 < \dfrac{c}{B} < 0.5 \qquad 0 < \theta < \pi$

Quarter elliptical corner crack in tension:

$$K_I = Y \frac{\sigma \sqrt{\pi a}}{\dfrac{3\pi}{8} + \dfrac{\pi}{8}\dfrac{a^2}{c^2}} \left(\sin^2 \theta + \frac{a^2}{c^2}\cos^2 \theta\right)^{1/4}$$

$$Y = 1.2$$

Figure 7.12b (Continued)

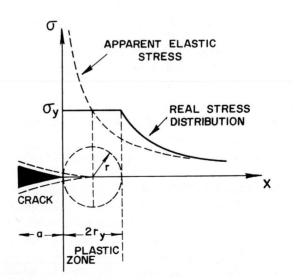

Figure 7.13 Plastic-zone correction. The effective crack length is $(a + r_v)$.

then be embedded in an elastic stress field. Outside an far away from the plastic zone, the elastic stress field "sees" the crack and the perturbation due to the plastic zone, as if there were present a crack in an elastic material with the leading edge of the crack situated inside the plastic zone. A crack of length $2(a + r_y)$ in an ideal elastic material produces stresses almost identical to elastic stresses in a locally yielded member outside the plastic zone. If the stress applied is too large, the plastic zone increases in size in relation to the crack length, and the elastic stress field equations lose precision. When the whole of the reduced section yields, the plastic zone spreads to the edges of the sample, and K does not have any validity as a parameter defining the stress field.

When the plastic zone is small in relation to the crack length, it can be visualized as a cylinder (Figure 7.13) of radius r_y at the crack tip. From Equation 7.13, for $\theta = 0, r = r_y$, and $\sigma_{22} = \sigma_y$, the yield stress, we can write

$$\sigma_y = \frac{K}{\sqrt{2\pi r_y}},$$

and, to a first approximation, the plastic-zone radius will be

$$r_y = \frac{1}{2\pi}\left(\frac{K}{\sigma_y}\right)^2. \tag{7.16}$$

In fact, the plastic-zone radius is a little bigger than $(1/2\pi)(K/\sigma_y)^2$, due to redistribution of load in the vicinity of the crack tip. Irwin,[5] taking into account the plastic constraint factor in the case of plane strain, gave the following expressions for the size of the plastic zone:

$$r_y \approx \frac{1}{2\pi}\left(\frac{K}{\sigma_y}\right)^2 \quad \text{(plane stress),}$$

$$r_y \approx \frac{1}{6\pi}\left(\frac{K}{\sigma_y}\right)^2 \quad \text{(plane strain).}$$

Thus, the center of perturbation, the apparent crack tip, is located a distance r_y from the real crack tip. The effective crack length is, then,

$$(2a)_{\text{eff}} = 2(a + r_y).$$

Substituting $(a + r_y)$ for a in the elastic stress field equations gives an adequate adjustment for the crack-tip plasticity under conditions of small-scale yielding. With this adjustment, the stress intensity factor K is useful for characterization of the fracture conditions.

There is another model for the plastic zone at the crack tip for the plane-stress case, called the Dugdale–BCS model.[6] In this model, the plasticity spreads out at the two ends of a crack in the form of narrow strips of length R (Figure 7.14). These narrow plastic strips

[5]G. R. Irwin, in *Encyclopaedia of Physics*, Vol. VI (Heidelberg: Springer-Verlag, 1958); see also *J. Basic Eng., Trans. ASME*, 82 (1960) 417.

[6]B. A. Bilby, A. H. Cottrell, and K. H. Swinden, *Proc. Roy. Soc.*, A272 (1963) 304; D. S. Dugdale, *J. Mech. Phys. Solids*, 8 (1960) 100.

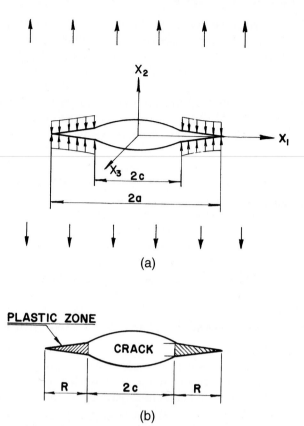

Figure 7.14 Dugdale–Bilby–Cottrell–Swinden model of a crack.

in front of the actual crack tips are under the yield stress σ_y that tends to close the crack. Mathematically, the internal crack of length $2c$ is allowed to extend elastically a distance $2a$, and then internal stress is applied to reclose the crack in this region. Combining the internal stress field surrounding the plastic enclaves with the external stress field associated with the applied stress σ acting on the crack, Dugdale showed that

$$\frac{c}{a} = \cos \frac{\pi\sigma}{2\sigma_y}.$$

From this relation, one notes that as $\sigma \to \sigma_y, c/a \to 0, a \to \infty$ (i.e., general yielding occurs). On the other hand, as σ/σ_y decreases, we can write (using the series expansion for cosine),

$$\frac{c}{a} = 1 - \frac{\pi^2\sigma^2}{8\sigma_y^2} + \cdots.$$

Noting that $a = c + R$ and using the binomial expansion, we have

$$\frac{c}{a} = \frac{c}{c + R} = \left(1 + \frac{R}{c}\right)^{-1} = 1 - \frac{R}{c} + \cdots.$$

Thus, for $\sigma <\!\!< \sigma_y$,

$$\frac{R}{c} \approx \frac{\pi^2}{8}\left(\frac{\sigma}{\sigma_y}\right)^2,$$

or

$$R \approx \frac{\pi}{8} \left(\frac{K}{\sigma_y} \right)^2.$$ (7.17)

Comparing Equation 7.17 with Equation 7.16, we see that there is good agreement between the two ($\pi/8 \approx 1/\pi$). In fact, the size of the plastic zone varies with θ also. A formal representation of the plastic zone at the crack front through the plate thickness is shown in Figure 7.15. The reader is also advised to see Exercises 7.16 and 7.17 at the end of the chapter.

7.5.7 Variation in Fracture Toughness with Thickness

The elastic stress state is markedly influenced by the plate thickness, as indicated by Equation 7.13. The material in the plastic zone deforms in such a way that its volume is kept constant. Thus, the large deformations in the x_1 and x_2 directions tend to induce a contraction in the x_3 direction (parallel to the direction of the crack front or the plate thickness), which is resisted by the surrounding elastic material (Figure 7.13). We next perform a dimensional analysis. Since the elastic material surrounding the plastic zone is the primary source of constraint, the size of the plastic zone, $2r_y$, will be the significant dimension to be compared with the plate thickness B. The ratio of the plate thickness B to the size of the plastic zone, $2r_y$, is given by

$$\frac{B}{2r_y} = \pi \frac{B}{(K_c/\sigma_y)^2},$$

and this would be a convenient parameter to characterize the variation of fracture toughness, K_c, with thickness. Data for Al 7075-T6 and H-11 steel are plotted in Figure 7.16a in the form[7] of K_c/σ_y versus $B/(K_c/\sigma_y)^2$. Observe that when $B/(K_c/\sigma_y)^2$ is greater than $1/\pi$ (i.e., $B >> 2r_y$), the fracture toughness value K_c does not change with B. Apparently,

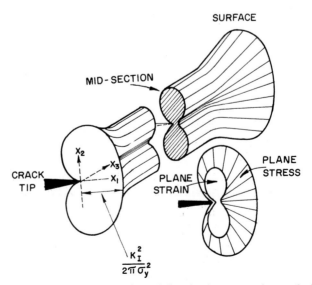

Figure 7.15 Formal representation of the plastic zone at the crack tip in a through-the-thickness crack in a plate.

[7]J. E. Srawley and W. F. Brown, ASTM STP 381, ASTM, Philadelphia, p. 133; W. F. Brown and J. E. Srawley, ASTM STP 410, ASTM, Philadelphia, 1966, p. 1.

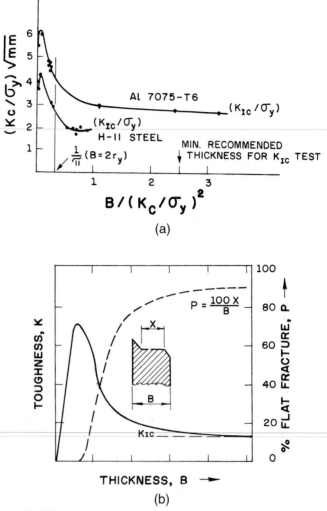

Figure 7.16 (a) Variation in fracture toughness (K_c) with plate thickness (B) for Al 7075-T6 and H-11 Steel. (Reprinted with permission from J. E. Srawley and W. F. Brown, ASTM STP 381, ASTM, Philadelphia, pp. 133, and G. R. Irwin, in *Encyclopaedia of Physics,* Vol. VI (Heidelberg: Springer Verlag, 1958); see also *J. Basic Eng., Trans. ASME, 82* (1960) 417. (b) Schematic variation of fracture toughness K_c and percentage of flat fracture P with the plate thickness B.

beyond a thickness $B >> 2r_y$, the constraint in the thickness direction (x_3) is completely effective, and additional plate thickness does not change K_c. This particular value of K_c that is independent of the thickness of the specimen is labeled the fracture toughness of the material, and the symbol K_{Ic} is used to denote it.

On the other extreme, when the ratio $B/(K_c/\sigma_y)^2$ is much smaller than $1/\pi$ (i.e., $B << 2r_y$), we expect the fracture toughness to increase linearly with the plate thickness. In the region of $B/(K_c/\sigma_y)^2 = 1/\pi$ corresponding to $B = 2r_y$, the data for both materials show a rapid fall to a constant level of K_{Ic}. This decrease in the peak value of K_c (Figure 7.16b) to the K_{Ic} level represents a change in the fracture mode from a plane-stress type to a plane-strain condition. The fracture in a relatively thin plate (plane stress) usually consists of a certain fraction of slant fracture (high energy) and another fraction of flat fracture (low energy). In general, with increasing thickness of the specimen, the percentage of slant fracture decreases, and the energy necessary for crack propagation also decreases—hence the fall in the K_c value. At a certain critical thickness, the crack propagates under plane-strain conditions, and the stress intensity factor reaches the minimum value designated as K_{Ic}. Figure 7.16b shows schematically the variation of K_c and the percentage of flat fracture P with the plate thickness B. K_{Ic} is especially relevant in the evaluation of the material, as it is a constant that is essentially independent of the dimensions of the specimen. One may say that the relation between K_{Ic} and K_c is the same as that existing between the yield stress σ_y and the strength σ.

EXAMPLE 7.5

Establish the maximum load that the component shown in Figure E7.5, made of Ti-6A1-4V alloy, can withstand ($\sigma_y = 900$ MPa, $K_{Ic} = 100$ MPa $\sqrt{m}$).

Solution:

$$a = 1 \text{ cm},$$

$$W = 10 \text{ cm}.$$

$$K_{Ic} = Y\sigma\sqrt{\pi a},$$

$$Y = 1.12 - 0.231\left(\frac{a}{w}\right) + 10.55\left(\frac{a}{w}\right)^2$$

$$= 1.12 - 0.231\left(\frac{1}{10}\right) + 10.55\left(\frac{1}{10}\right)^2$$

$$= 1.20.$$

We rewrite Equation 1 as

$$\sigma = \frac{K_{Ic}}{Y\sqrt{\pi a}}$$

to get

$$\sigma = \frac{100}{1.20\sqrt{\pi \times 10^{-2}}} = 470 \text{ MPa} < \sigma_y.$$

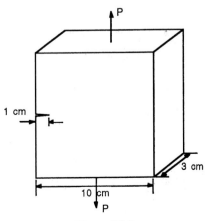

Figure E7.5

Therefore,

$$\frac{P}{A} = \sigma \quad \text{and} \quad P = \sigma A = (470 \times 10^6) \times (10 \times 10^{-2} \times 3 \times 10^{-2})$$

$$= 1,410 \text{ kN}.$$

Hence, the existing flaw, and not the yield stress, limits the maximum load.

7.6 FRACTURE TOUGHNESS PARAMETERS

In this section, we describe the variety of fracture toughness parameters that have come into being.

7.6.1 Crack Extension Force *G*

The concept of the crack extension force G, due to Irwin, can be interpreted as a generalized force. One can say that fracture mechanics is the study of the response of a crack (measured in terms of its velocity) to the application of various magnitudes of the crack extension force. Let us consider an elastic body of uniform thickness B, containing a through-the-thickness crack of length $2a$. Let the body be loaded as shown in Figure 7.17a. With increasing load P, the displacement e of the loading point increases. The load–displacement diagram is shown in Figure 7.17b. At point 1, we have the load as P_0 and displacement as e_0. Now let us consider a "*gedanken*" experiment in which the crack extends by a small increment, δa. Due to this small increment in crack extension, the loading point is displaced by δe, while the load falls by δP. Now, before the crack extension, the potential energy stored in body was

$$U_1 = \frac{1}{2} Pe,$$

represented by the area of the triangle through point 1 in the figure. After the crack extension, the potential energy stored in the body is

$$U_2 = \frac{1}{2} (P - \delta P)(e + \delta e),$$

represented by the area of the triangle passing through point 2 in the figure. In this process of crack extension, the change in potential energy, $U_2 - U_1$ is given by the difference in the areas of the two crosshatched regions in the figure. Considering the small increment δa in crack length, we can write an equation for G, the crack extension force per unit length, as

$$GB \, \delta a = U_2 - U_1 = \delta U.$$

The change in elastic strain energy with respect to the crack area, in the limit of the area going to zero, equals the crack extension force; that is,

$$G = \lim_{\delta A \to 0} \frac{\delta U}{\delta A},$$

where $\delta A = B \, \delta a$.

It is convenient to evaluate G in terms of the compliance c of the sample, defined as

$$e = cP. \tag{7.18}$$

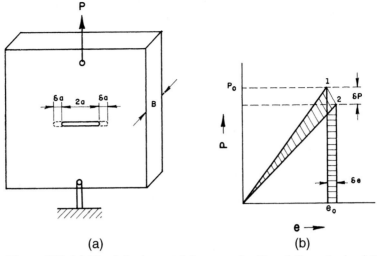

Figure 7.17 (a) Elastic body containing a crack of length $2a$ under load P. (b) Diagram of load P versus displacement e.

Now,

$$\delta U = U_2 - U_1 = \frac{1}{2}(P - \delta P)(e + \delta e) - \frac{1}{2}Pe,$$

or

$$\delta U = \frac{1}{2}P\,\delta e - \frac{1}{2}e\,\delta P - \frac{1}{2}\delta P\,\delta e. \qquad (7.19)$$

Differentiating Equation 7.18, we have

$$\delta e = c\,\delta P + P\,\delta c. \qquad (7.20)$$

Substituting Equation 7.20 in Equation 7.19, we obtain

$$\delta U = \frac{1}{2}Pc\,\delta P + \frac{1}{2}P^2\,\delta c - \frac{1}{2}e\,\delta P - \frac{1}{2}e(\delta P)^2 - \frac{1}{2}P\,\delta P\,\delta c. \qquad (7.21)$$

Remembering that $e = cP$ and ignoring the higher order product terms, we can write

$$\delta U = \frac{1}{2}Pc\,\delta P + \frac{1}{2}P^2\,\delta c - \frac{1}{2}Pc\,\delta P,$$

or

$$\delta U = \frac{1}{2}P^2\,\delta c. \qquad (7.22)$$

Then

$$G = \lim_{\delta A \to 0} \frac{\delta U}{\delta A} = \lim_{\delta A \to 0} \frac{\frac{1}{2}P^2\,\delta c}{\delta A},$$

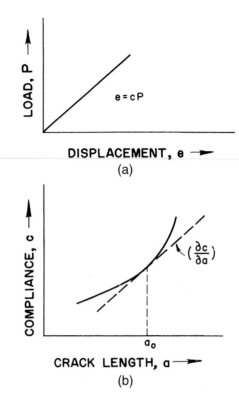

Figure 7.18 (a) Load P versus displacement e. Compliance c is the inverse of the slope of this curve. (b) Compliance c versus crack length a. a_0 is the initial crack length.

or

$$G = \frac{1}{2} \frac{P^2}{B} \frac{\delta c}{\delta a}.$$ (7.23)

From Equation 7.23, we see that G is independent of the rigidity of the surrounding structure and the test machine. In fact, G depends only on the change in compliance of the cracked member due to crack extension. Thus, to obtain G for a specimen, all we need to do is to determine the compliance of the specimen as a function of crack length and measure the gradient of the resultant curve, dc/da, at the appropriate initial crack length (Figure 7.18).

This method is more useful for relatively small test samples, on which exact measurements can be made in the laboratory. One of the important uses of Equation 7.23 is that it provides a value of G (or K) for complex structures that have not been (or cannot be) treated analytically. An experimental determination of G_c, the critical crack extension force, using this equation requires the value of fracture load (measured experimentally) and the value of dc/da. The compliance can be measured by calibrating a series of samples with different crack lengths. We obtain a diagram of c versus a, and dc/da is evaluated as the slope at the appropriate initial crack length.

EXAMPLE 7.6

A titanium alloy (Ti–6% Al–4% V) is used for aircraft applications. The NDE methods used cannot detect flaws whose size is smaller than 1 mm. You are asked, as the design engineer, to specify the maximum tensile stress that the part can bear in plane-stress and plane-strain situations. The yield stress of the alloy is 1,450 MPa.

$$E = 115\,\text{GPa},$$

$$\nu = 0.321,$$

$$G_c = 23.6\,\text{kN/m},$$

Solution: We have

$$2a = 1\,\text{mm},$$

so that

$$a = 0.5 \times 10^{-3}\,\text{m}$$

The critical stress in plane stress is

$$\sigma_c = \sqrt{\frac{EG_c}{\pi a}}$$

$$= \left(\frac{115 \times 10^9 \times 23.6 \times 10^3}{\pi \times 0.5 \times 10^{-3}}\right)^{1/2}$$

$$= 1.31 \times 10^9\,\text{Pa}.$$

The critical stress in plane strain is

$$\sigma_c = \sqrt{\frac{EG_c}{\pi a(1 - \nu^2)}}$$

$$= 1.385 \times 10^9\,\text{Pa}.$$

Thus, the maximum stresses are 1.31 GPa (plane stress) and 1.385 GPa (plane strain).

From consideration of fracture toughness, the maximum stress is lower than the yield stress; hence, the former is the limiting stress.

7.6.2 Crack Opening Displacement

The development of a plastic zone at the tip of the crack results in a displacement of the faces without crack extension. This relative displacement of opposite crack edges is called the crack opening displacement (COD) (Figure 7.19). Wells[8] suggested that when this displacement at the crack tip reaches a critical value δ_c, fracture would ensue.

LEFM is applicable only when the plastic zone is small in relation to the crack length (i.e., well below the yield stress and in plane strain). Consider a small crack in a brittle material. We have

$$\sigma_c = K_{Ic}(\sqrt{\pi a})^{-1}, \qquad \text{as } a \to 0, \qquad \sigma_c \to \infty.$$

[8]A. A. Wells, *Brit. Weld. J., 13* (1965) 2.

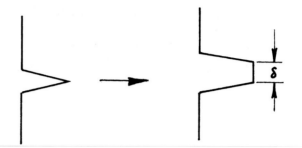

Figure 7.19 Crack opening displacement.

But this, as we very well know, does not occur. Instead, a plastic zone develops and may extend through the section such that

$$\sigma_{net} = \sigma \frac{W}{W - a} \geq \sigma_y,$$

where W is the width of sample and σ_y is the yield stress. In practice, $\sigma_c \leq 0.66\sigma_y$ for the K_{Ic} validity.

In more ductile materials, the critical stress predicted by LEFM will be higher than σ_y. One can use the concept of COD in such cases. In the elastic case (Figure 7.19),

$$\text{COD} = \Delta = \frac{4\sigma}{E} \sqrt{(a^2 - x^2)}. \tag{7.24}$$

At the center of the crack ($x = 0$), the maximum opening is

$$\Delta_{max} = \frac{4\sigma a}{E}.$$

Applying the plastic zone correction, we have, from Equation 7.24,

$$\Delta = \frac{4\sigma}{E} \sqrt{(a + r_y)^2 - x^2},$$

where $(a + r_y)$ is the effective crack length.

The crack-tip opening displacement (CTOD), δ, is given for $x = a$ and $r_y \ll a$ as

$$\delta = \frac{4\sigma}{E} \sqrt{2ar_y}. \tag{7.25}$$

A displacement of the origin to the crack tip gives a general expression for the crack opening:

$$\Delta = \frac{4\sigma}{E} \sqrt{2a_{eff}r_y}.$$

Substituting $r_y = \sigma^2 a / 2\sigma_y^2$ (see Equation 7.16) in Equation 7.25 gives

$$\delta = \frac{4}{\pi} \frac{K_I^2}{E\sigma_y}. \tag{7.26}$$

Equation 7.26 is valid in the LEFM regime, and fracture occurs when $K_I = K_{Ic}$, which corresponds to $\delta = \delta_{Ic}$, a material constant.

The use of the COD criterion demands the measurement of δ_c. Direct measurement of δ_c is not easy. An indirect way is the following. We have

$$\Delta = \frac{4\sigma}{E} \sqrt{(a + r_y)^2 - x^2}$$

$$= \frac{4\sigma}{E} \sqrt{a^2 + 2ar_y + r_y^2 - x^2}.$$

Ignoring the r_y^2 term and using the relationship of Equation 7.25, we can write

$$\Delta = \frac{4\sigma}{E} \left(a^2 - x^2 + \frac{E^2}{16\sigma^2} \delta^2 \right)^{1/2}. \tag{7.27}$$

According to this equation, δ can be measured indirectly from a COD measurement (e.g., at $x = 0$, at the center of the crack) without making any simplifications about the plastic-zone size correction. Δ can be measured by means of a clip gage.

Another way of obtaining δ is to use the equations of Dugdale–BCS model of the crack. (See Section 7.5.6.) According to Dugdale–BCS model (Bilby, Cotrell, Swinden, op. cit.; Dugdale, op. cit.)

$$\delta = \frac{8\sigma_y a}{\pi E} \log \sec \frac{\pi \sigma}{2\sigma_y}.$$

Expanding the log sec function in series, we get

$$\delta = \frac{8\sigma_y a}{\pi E} \left[\frac{1}{2} \left(\frac{\pi \sigma}{2\sigma_y} \right)^2 + \frac{1}{12} \left(\frac{\pi \sigma}{2\sigma_y} \right)^4 + \cdots \right].$$

For $\sigma \ll \sigma_y$, we can write (neglecting fourth and higher order terms)

$$\delta = \frac{\pi \sigma^2}{E \sigma_y} = \frac{G_I}{\sigma_y}. \tag{7.28}$$

Comparing Equation 7.28 with Equation 7.26, we note that the difference is in the factor $4/\pi$, which comes from the plastic-zone correction. In general,

$$\boxed{\delta = \frac{G_I}{\lambda \sigma_y} = \frac{K_I^2 (1 - \nu^2)}{E \lambda \sigma_y} \qquad \text{(for plane strain)}. \tag{7.29}}$$

The factor $(1 - \nu^2)$ should be ignored in the case of plane stress.

In the literature, we encounter various values of λ. These depend on the exact location where CTOD is determined (i.e., the exact location of the crack tip). Wells[9] suggested that, experimentally, $\lambda \approx 2.1$ for compatibility with LEFM (i.e., limited plasticity). For cases involving extensive plasticity, the engineering design application approach is to take $\lambda \approx 1$.

[9] A. A. Wells, *Eng. Fract. Mech., 1* (1970) 399.

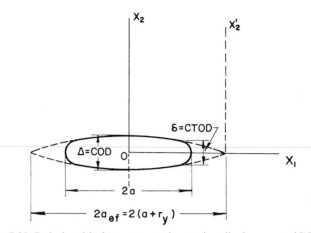

Figure 7.20 Relationship between crack opening displacement (COD, Δ), crack-tip opening displacement (CTOD, δ), crack length ($2a$), and size of plastic zone (r_y).

Thus, at unstable fracture, $G_{Ic} = \lambda \sigma_y \, \delta c$. The important point about COD is that, theoretically, δ_c can be computed for both elastic and plastic materials, whereas G_{Ic} is restricted only to the elastic regime. The COD thus allows one treat fracture under plastic conditions. A word of caution is in order, however. Figure 7.20 presents a comparison between COD and CTOD. We should realize that the strain fields and crack opening displacements associated with a crack tip will be different for different specimen configurations. Thus, we cannot define a single critical COD value for a given material in a manner equivalent to that of K_{Ic}, as the COD value will be affected by the geometry of the test specimen.

EXAMPLE 7.7

If the toughness of a thermoplastic polymer is $G_c = 10^3$ J m^{-2}, what would be the critical crack length under an applied stress of 200 MPa? Take Young's modulus of the polymer to be 70 GPa.

Solution:

We have

$$G_c = 10^3 \text{ J m}^{-2}, \qquad E = 70 \text{ GPa}, \qquad \sigma = 200 \text{ MPa}.$$

Thus, the critical crack length $a_c = EG_c/\pi\sigma^2 = 70 \times 10^9 \times 10^3/\pi(200 \times 10^6)^2 = 0.56$ mm.

7.6.3 *J* Integral

The *J* integral provides a value of energy required to propagate a crack in an elastic–plastic material. The mathematical foundation for the *J* integral was laid by Eshelby,[10] who applied it to dislocations. Cherepanov[11] and Rice[12] applied it, independently, to cracks.

[10]J. D. Eshelby, *Phil. Trans. Roy. Soc London,* A244 (1951) 87.

[11]G. P. Cherepanov, *Appl. Math. Mech.* (*Prinkl. Mat. Mekh.*), 31, no. 3 (1967) 503.

[12]J. R. Rice, *J. Appl. Mech.,* 35 (1968) 379.

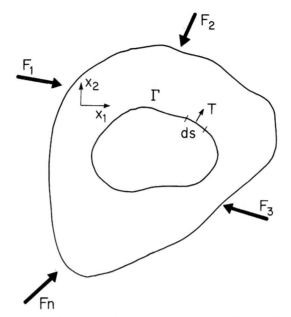

Figure 7.21 A body subjected to external forces $F_1, F_2, \ldots, F_n$ and with a closed contour Γ.

Figure 7.21 shows a closed contour Γ in a two-dimensional body. When such a body is subjected to external forces, internal stresses arise in it. On the basis of the theory of conservation of energy, Eshelby showed that the integral J is equal to zero for a closed contour; that is,

$$J = \int_\Gamma \left(W \, dx_2 - T \, \frac{\partial u}{\partial x_1} \, ds \right) = 0, \tag{7.30}$$

where

$$W = \int_0^{\Sigma_{ij}} \sigma_{ij} \, d\varepsilon_{ij}$$

is the strain energy per unit volume (see Chapter 2), T is the tension vector (traction) perpendicular to Γ and pointing to the outside of the contour, ds is an element of length along the contour, and u is the displacement in the x_1 direction. The J integral is an energy-related quantity; similar to the crack extension force G, J has the units of energy per unit area (J/m²) or force per unit length (N/m).

Figure 7.22 shows a crack, around which a contour $ABCDEFA$ is made. The total J must be zero, i.e.,

$$J = J_{\Gamma_1 + \Gamma_2} = 0.$$

Along AF and CD (crack surfaces), the tractions T are equal to zero. The same is true for the normal and shear stresses. Thus, $J_{AF} = J_{CD} = 0$. It can therefore be concluded that

$$J_{\Gamma_1 + \Gamma_2} = J_{\Gamma_1} + J_{\Gamma_2} + J_{AF} + J_{CD} = 0, \qquad J_{\Gamma_1} = -J_{\Gamma_2}.$$

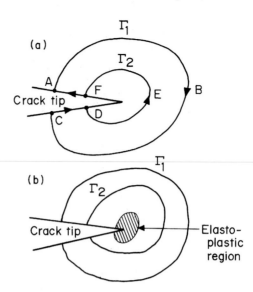

Figure 7.22 Eshelby contours around cracks.

Hence, the J integral along two different paths around a crack has the same value. That is, in general, the J integral around a crack is path independent.

From a physical point of view, the J integral represents the difference in the potential energies of identical bodies containing cracks of length a and $a + da$; in other words, the J integral around a crack is equal to the change in potential energy for a crack extension da. For a body of thickness B, this can be written as

$$J = -\frac{1}{B}\frac{\delta U}{\delta a},\tag{7.31}$$

where U is the potential energy, a the crack length, and B the plate thickness. U is equal to the area under the curve of load versus displacement. Figure 7.23 shows this interpretation, where the shaded area is $\delta U = JB\,\delta a$. Like G_{Ic}, J_{Ic} measures the critical energy associated with the initiation of crack growth, but in this case accompanied by substantial plastic deformation. In fact, Begley and Landes[13] showed the formal equivalence of J_{Ic} and G_{Ic} by measuring the former from small fully plastic specimens and the latter from large elastic specimens satisfying the plane-strain conditions for the LEFM test.

The path independence of the J integral, together with this interpretation in terms of energy, makes it a powerful analytical tool. The J integral is path independent in the case of either linear or nonlinear materials behaving elastically. When extensive plastic deformation occurs, the practice is to assume that the plastic yielding can be described by the deformation theory of plasticity. According to this theory, stresses and strains are functions only of the point of measurement and not of the path taken to get to that point. As in the case of slow, stable crack growth, there will be a relaxation of stresses at the crack tip, so there will be a violation of this postulate. Thus, the use of the J integral should be limited to the initiation of crack propagation, by stable or unstable processes. Studies using incremental plasticity or flow theories with finite elements indicate the path independence of the J integral.

[13]J. A. Begley and J. D. Landes, ASTM STP 514, ASTM, Philadelphia, 1972, p. 1.

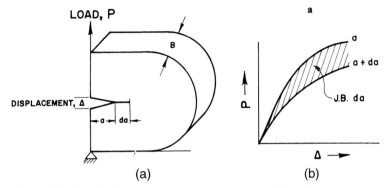

Figure 7.23 Physical interpretation of the J integral. The J integral represents the difference in potential energy (shaded area) of identical bodies containing cracks of length a and $a + da$.

7.6.4 *R* Curve

The R curve characterizes the resistance of a material to fracture during slow and stable propagation of a crack. An R curve graphically represents this resistance to crack propagation of the material as a function of crack growth. With increasing load in a cracked structure, the crack extension force G at the crack tip also increases. (See Equation 7.23.) However, the material at the tip presents a resistance R (sometimes, the symbol K_R is used) to crack growth. According to Irwin, failure will occur when the rate of change of the crack extension force $(\delta G/\delta a)$ equals the rate of change of this resistance to crack growth in the material $(\delta R/\delta a)$. The resistance of the material to crack growth, R, increases with an increase in the size of the plastic zone. Since the plastic zone size increases nonlinearly with a, R will also be expected to increase nonlinearly with a. G increases linearly with a. Figure 7.24 shows the instability criterion: the point of tangency between the curves of G versus a and R versus a. Figure 7.24a shows the R curve for a brittle material, and Figure 7.24b shows the R curve for a ductile material. Crack extension occurs for $G > R$. Consider the G line for a stress σ', shown in Figure 7.24b. At the stress σ', the crack in the material will grow only from a_0 to a', since $G > R$ for $a < a'$. $G < R$ for $a > a'$, and the crack does not extend beyond a'. As the load is increased, the position of the G line changes, as indicated in the figure. When G becomes tangent to R, unstable fracture ensues. The R curve for a brittle material (Figure 7.24a) is a "square" curve, and the crack does not extend at all until the contact is reached, at which point $G = G_c$ and the unstable fracture follows.

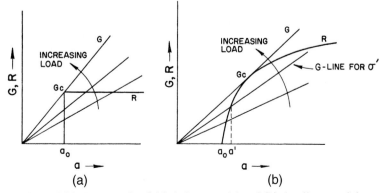

Figure 7.24 R curves for (a) brittle material and (b) ductile material.

The *R*-curve method is another version of the Griffith energy balance. One can conveniently make this kind of analysis if an analytical expression for the *R* curve is available. Experimental determination of *R* curves, however, is complicated and time consuming.

7.6.5 Relationships among Different Fracture Toughness Parameters

So far, we have seen that, in our effort to develop a quantitative description of fracture toughness, various parameters, such as K, G, J, δ, R, etc., have been developed. Since all these parameters define the same physical quantity, it is not unexpected that they are interrelated. And we have mentioned in different sections the relationships among the parameters. Figure 7.25 summarizes these relationships. It would, however, be helpful to the reader to recapitulate these relationships, even at the risk of repeating. That is what we will do in this section.

If we take into account the stress distribution around the tip of a crack, we get the stress-intensity-factor (K) approach. The magnitude or the intensity of the local stresses is determined by K, because the form of the local crack-tip stress field is the same for all situations involving a remote stress σ. Thus, K, and not σ, is the local characterizing parameter. The fracture then occurs when the applied K attains the critical value K_c. In particular, when the specimen's dimensions satisfy the plane-strain condition, we call this value the plane-strain fracture toughness and denote it by K_{Ic}. The stress and the crack length corresponding to K_{Ic} are the fracture stress σ_c and the fracture crack length a_c. Note that the elastic constants of the material are not involved. The energy-release-rate approach gives us the crack extension force G, which is related to the parameters K by the equation

$$K_I^2 = E'G,$$

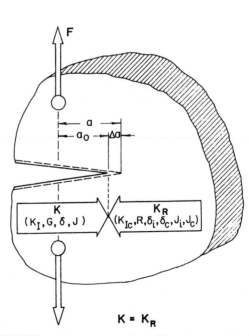

Figure 7.25 Different parameters describing the growth of a crack.

where $E' = E$, Young's modulus, in the case of plane stress and $E' = E/(1 - \nu^2)$ in the case of plane strain. Note that, in characterizing the fracture behavior in terms of G, we need to know the elastic constants of the material. Because in the case of polymers E is time dependent and very precise modulus data are not available, there is some advantage to using the K approach.

The critical crack opening displacement δ_c is another useful parameter. It is related to K by the equation

$$\delta_c = K_{Ic}^2/\lambda E \sigma_y,$$

where λ is a dimensionless constant that depends on the geometry of the specimen, its state of stress, and the work-hardening capacity of the material. λ has a value between 1 and 2. In particular, for the strip-yielding model of Dugdale–BCS, $\lambda = 1$.

The J integral provides yet another measure of fracture toughness. And, for small-scale yielding, we have

$$J = \lambda \delta \sigma_y.$$

In short, for small-scale yielding, we can sum up the relationships among the different fracture toughness parameters as

$$J = G = K^2/E' = \lambda \sigma_y \delta,$$

where the symbols have the usual meaning.

7.7 IMPORTANCE OF K_{Ic} IN PRACTICE

K_{Ic} is the critical stress intensity factor under conditions of plane strain ($\varepsilon_{33} = 0$), which is characterized by small-scale plasticity at the crack tip. The material is fully constrained in the direction of thickness. When determined under these rigorous conditions, K_{Ic} will be a material constant. Thus, when one needs to characterize materials by their toughness (in the same way that one characterizes materials by their ultimate tensile strength or tensile yield strength), only valid K_{Ic} data should be considered. This will be explained in Chapter 9.

K_c is the critical stress intensity factor under conditions of plane stress ($\sigma_{33} = 0$), which is characterized by large plasticity at the crack tip. In this case, the through-thickness constraint is negligible. K_c values can be up to two times greater than the K_{Ic} values of the same material. K_{Ic} depends on the temperature T, on the strain rate $\dot{\varepsilon}$, and on microstructural variables.

In general, K_c or K_{Ic} decreases as the (yield or ultimate) strength of a material increases. This inverse relationship between fracture toughness and strength is shown schematically in Figure 7.26. With concurrent improvement in the material's strength and toughness, this curve shifts in the direction of the arrow. The dependence of K_{Ic} on tensile strength and on sulfur level in a steel is shown in Figure 7.27. As expected, K_{Ic} decreases monotonically with increases in tensile strength or sulfur content. (Sulfur is well known to embrittle steels.) Figure 7.28 shows that the same holds for K_{Ic} as a function of the yield strength. K_c also depends on these variables.

Table 7.1 shows representative fracture toughnesses for selected materials. Metals have the highest toughness. For most ceramics, K_{Ic} does not exceed 5 MPa$\sqrt{}$m. The

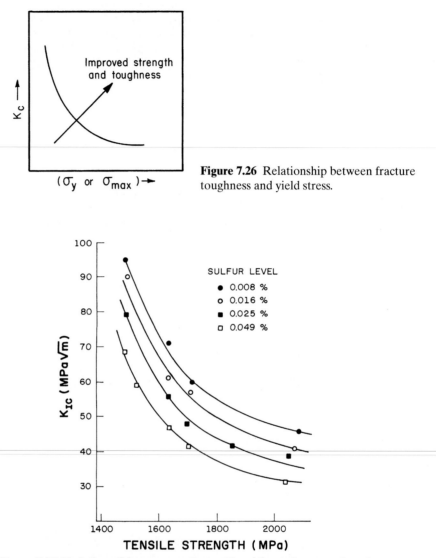

Figure 7.26 Relationship between fracture toughness and yield stress.

Figure 7.27 Variation of fracture toughness K_{Ic} with tensile strength and sulfur content in a steel. (Adapted with permission from A. J. Birkle, R. P. Wei, and G. E. Pellissier, *Trans. ASM,* 59 (1966) 981, p. 982)

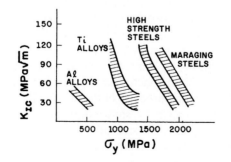

Figure 7.28 Variation of fracture toughness K_{Ic} with yield strength σ_y for a series of alloys. (Adapted with permission from D. Broek, *Elementary Engineering Fracture Mechanics,* 3d ed. (Amsterdam: Martinus Nijhoff, 1978), p. 270).

TABLE 7.1 Plane-Strain Fracture Toughnesses for Representative Materials

MATERIAL	K_{Ic} (MPa$\sqrt{m}$)
(a) *Metals*	
300M steel 300°C temper	65
300M steel 650°C temper	152
18-Ni maraging steel, vacuum melted	176
18-Ni maraging steel, air melted	123
AISI 4130 steel	110
2024-T651 aluminum	24
2024-T351 aluminum	34
6061-T651 aluminum	34
7075-T651 aluminum	29
Ti-6Al-4V, mill annealed	106–123
Ti-6Al-4V, recrystallized, annealed	77–116
(b) *Ceramics*	
Cement/concrete	0.2
Soda-lime glass	0.7–0.9
MgO	3
Al_2O_3	3–5
Al_2O_3 + 15% ZrO_2	10
SiC	3–4
Si_3N_4	4–5
(c) *Polymers*	
Epoxy	0.3–0.6
Polyethylene, high density	2
Polyethylene, low density	1
Polypropylene	3
ABS	3–4
Polycarbonate	1–2.6
PVC	2.4
PVC (rubber modified)	3.4
Acrylic	1.8

addition of partially stabilized zirconia to alumina increases K_{Ic} to 10 MPa$\sqrt{m}$ and even higher. The reason for this is a martensitic transformation that is described in greater detail in Chapter 11. Plastics have low K_{Ic}; however, we should remember that their density is only a small fraction of that of metals.

7.8 POST-YIELD FRACTURE MECHANICS

The concepts of crack opening displacement and the J integral are complementary. The crack tip opening displacement (CTOD), δ, is the parameter that controls crack extension. But the notion of CTOD is not problem free. For example, there exists a considerable amount of diversity in its very definition. Figure 7.29 shows some ways of measuring δ. The experimental determination of δ and the calculation of the relevant value for a cracked structure also involve uncertainties. We can split the CTOD value into an elastic and a plastic component, to wit:

$$\delta_t = \delta_{el} + \delta_{pl}.$$

The elastic portion is, of course, related to K or G, as indicated earlier. In particular, K_{Ic} or G_{Ic} correspond to δ_{Ic}, the CTOD value at the initiation of unstable fracture. The

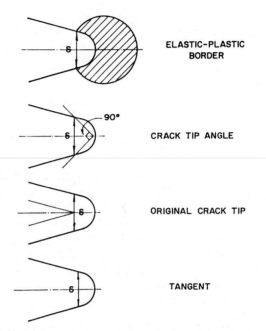

Figure 7.29 Different measures of crack-tip opening displacement.

plastic portion is not strictly a material property, inasmuch as it depends on the specimen's dimensions, constraints, etc.

The J integral is, mathematically, a path-independent integral. From a practical engineering point of view, the J integral represents, similarly to G, a strain energy release rate and is related to the area under the curve load, P, and the *load line displacement*.

Just as we did for the concept of COD, we can write, for J,

$$J_{\text{total}} = J_{\text{el}} + J_{\text{pl}},$$

where J_{el} is the elastic portion, equal to K_I^2/E'. Here again, E' equals E for plane stress and $E/(1 - \nu^2)$ for plane strain. J_{pl} is a function of the geometry of the component and the crack load corresponding to extensive plastic deformation, and the material characteristics such as the yield strength, ultimate tensile strength, etc.

This division of the crack driving force into elastic and plastic parts is conceptually very convenient. Tests for the J integral, as well as for COD, are based on the fact that a ductile structure containing a crack is characterized by three successive stages:

1. Crack blunting and the initiation of propagation.
2. Slow and stable crack growth under increasing load.
3. Unstable crack growth, i.e., the instability.

A curve showing these stages is called a *resistance curve* (δ–R or J–R). It describes the material resistance as a function of stable crack growth a (see Sec. 9.6).

7.9 STATISTICAL ANALYSIS OF FAILURE STRENGTH

As we have repeatedly pointed out, materials in real life are never perfect. No matter how carefully processed a material is, it will always contain a distributions of flaws. The presence of flaws in ductile metals is not very serious, because these metals have the ability to deform plastically and thus attenuate, at least to some extent, the insidious effect of flaws on strength. The same cannot be said of brittle materials. Such preexisting flaws are responsible for the phenomenon of catastrophic fracture in these materials. In general, flaws vary in size, shape, and orientation; consequently, the strength of a material will vary from specimen to specimen. When we test a brittle material, one or several of the larger flaws propagate. In the case of a ductile material such as aluminum, most of the flaws get blunted because of plastic deformation, and only after considerable plastic deformation do microvoids form and coalesce, leading to an eventual fracture. (See Chapter 8.) If we were to test a large number of identical samples and plot the strength distribution of a brittle and a ductile solid, we would get the curves shown in Figure 7.30. The strength distribution curve for the ductile solid is very narrow and close to a Gaussian or normal distribution, while that for the brittle solid is very broad with a large tail on the high-strength side—that is, a non-Gaussian distribution. It turns out that the strength distribution of a brittle solid can be explained by a statistical distribution called the *Weibull distribution*, named after the Swedish engineer who first proposed it.[14] We next describe this distribution and its application to the analysis of the strength of brittle solids.

The basic assumption in Weibull distribution is that a body of material with volume V has a statistical distribution of noninteracting flaws. Thus, the body of volume V can be considered to be made up of n volume elements, each of unit volume V_0 and having the same flaw distribution. Now, if we subject such a solid to an applied stress σ, the probability that the solid will survive can be written as

$$P(V) = P(V_0)^n, \tag{7.32}$$

where V_0 is the volume of an element and n is the number of volume elements. Taking logarithms, we have

$$\ln P(V) = n \ln P(V_0),$$

or

$$P(V) = \exp[n \ln P(V_0)]. \tag{7.33}$$

Weibull defined a risk-of-rupture parameter

$$R = -[\ln P(V_0)], \tag{7.34}$$

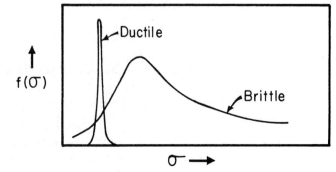

$f(\sigma)$

Ductile

Brittle

$\sigma \longrightarrow$

Figure 7.30 Strength distribution of a brittle and a ductile solid.

[14]W. Weibull, *J. App. Mech.*, 18 (1951), 293.

or

$$\exp(-R) = P(V),$$

or, alternatively,

$$P(V_0) = \exp(-R). \tag{7.35}$$

He then postulated that this parameter is given by

$$R = [(\sigma - \sigma_u)/\sigma_0]^m, \tag{7.36}$$

where σ is the applied stress and m, σ_0, and σ_u are material constants for a constant-flaw population, i.e., the flaw population does not change from element to element. σ_u is the stress below which the probability of failure is zero. If we assume that any tensile stress will cause failure in a brittle solid, then we can take σ_u to be zero. For such a material, σ_0 is a characteristic strength—often taken to be approximately the mean strength—of the material, and m, called the *Weibull modulus,* is a measure of the variability of the strength of the material; the higher the value of m, the less is the material's variability in strength. m can have any value between 0 and ∞, i.e., $0 < m < \infty$. As $m \rightarrow 0$, $R \rightarrow 1$, and the material will fail at any stress. Also, when $m \rightarrow \infty$, the material will not fracture at any stress below σ_0. Table 7.2 gives some typical values of m for some materials.

   From Equations (7.35) and (7.36), we can write, for the survival probability of a brittle material

$$P(V_0) = \exp\left[-\left(\frac{\sigma - \sigma_u}{\sigma_0}\right)^m\right]. \tag{7.37}$$

We can write the failure probability as

$$F(V_0) = 1 - P(V_0) = 1 - \exp\left[-\left(\frac{\sigma - \sigma_u}{\sigma_0}\right)^m\right]. \tag{7.38}$$

As explained in the preceding paragraph, we can take $\sigma_u = 0$ for a brittle material. This will make Equation 7.37 become

$$P(V_0) = \exp\left[-\left(\frac{\sigma}{\sigma_0}\right)^m\right]. \tag{7.39}$$

TABLE 7.2 Typical values of the Weibull modulus m for some materials

Material	m
Traditional ceramics: Brick, Pottery, Chalk	<3
Engineered Ceramics: SiC, Al_2O_3, Si_3N_4	5–10
Metals: Aluminum, Steel	90–100

Equation 7.39 says that when the applied stress $\sigma = 0$, the survival probability $P(V_0) = 1$, and all samples of the material tested survive. As the applied stress increases, more samples fail, and the survival probability decreases. Eventually, as $\sigma \to \infty, P(V_0) \to 0$; that is, all samples fail at very high stresses. We can arrive at a value of σ_0 by noting that, when $\sigma = \sigma_0$,

$$P(V_0) = \frac{1}{e} = 0.37.$$

Thus, σ_0 is the stress corresponding to a survival probability of 37%. Taking logarithms of Equation 7.39, we get

$$\ln\left[\frac{1}{P(V_0)}\right] = \left(\frac{\sigma}{\sigma_0}\right)^m. \tag{7.40}$$

Thus, a double-logarithmic plot of Equation 7.40 will give a straight line with slope m. This yields a convenient way of obtaining a Weibull analysis of the strength of a given material. If N samples are tested, we rank their strengths in ascending order and obtain the probability of survival for the ith strength value as

$$P_i(V_0) = (N + 1 - i)/(N + 1).$$

Note that there will be $N + 1$ strength intervals for N tests. Alternatively, we can use the failure probability $F_i(V_0) = 1 - P_i(V_0) = i/(N + 1)$.

We can incorporate a volume dependence into Equation 7.39. Let V_0 be a reference volume of a material with a survival probability of $P(V_0)$, i.e., fraction of samples, each of volume V_0, that survive when loaded to a stress, σ. Now consider a volume V of this material such that $V = nV_0$. Then, from Equation 7.32, we can write

$$P(V) = P(V_0)^n = [P(V_0)]^{V/V_0}.$$

Taking logarithms, we get

$$\ln P(V) = \frac{V}{V_0} \ln P(V_0),$$

or

$$P(V) = \exp\left[\frac{V}{V_0} \ln P(V_0)\right]. \tag{7.41}$$

From Equations 7.39 and 7.41, we have

$$P(V) = \exp\left[-\frac{V}{V_0}\left(\frac{\sigma}{\sigma_0}\right)^m\right], \tag{7.41a}$$

or

$$\ln P(V) = -\frac{V}{V_0}\left(\frac{\sigma}{\sigma_0}\right)^m. \tag{7.42}$$

We can convert Equation 7.42 to the following form by taking logarithms again.

$$\ln \ln \left[\frac{1}{P(V)}\right] = \ln \frac{V}{V_0} + m \ln \frac{\sigma}{\sigma_0}. \qquad (7.43)$$

Equation 7.42 tells us that, for a given probability of survival and for two volumes V_1 and V_2 of a material,

$$\ln P(V) = -\frac{V_1}{V_0}\left[\frac{\sigma_1}{\sigma_0}\right]^m = -\frac{V_2}{V_0}\left[\frac{\sigma_2}{\sigma_0}\right]^m,$$

where σ_1 and σ_2 are the strengths of the material in volumes V_1 and V_2, respectively. Hence,

$$V_1 \sigma_1^m = V_2 \sigma_2^m,$$

or

$$\frac{\sigma_1}{\sigma_2} = \left(\frac{V_2}{V_1}\right)^{1/m}. \qquad (7.44)$$

Thus, we see that, for an equal probability of survival, the larger the volume $(V_2 > V_1)$, the smaller must be the fracture strength $(\sigma_1 < \sigma_2)$.

An interesting application of the Weibull distribution is illustrated in Figure 7.31, which shows a double-logarithmic plot as per Equation 7.40. Note that the failure proba-

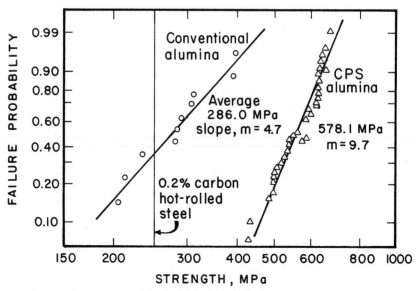

Figure 7.31 A Weibull plot for a steel, a conventional alumina, and a controlled-particle-size (CPS) alumina. Note that the slope (Weibull modulus m) $\rightarrow \infty$ for steel. For CPS alumina, m is double that of conventional alumina. (After E. J. Kubel, *Adv. Mater. Proc.* (Aug 1988) 25)

bility $F(V) = 1 - P(V)$, rather than the survival probability $P(V)$ is plotted. The figure shows the following items:

1. The Weibull modulus m of steel $\rightarrow \infty$. (Note the vertical line.)
2. The Weibull modulus m of conventionally processed alumina is 4.7.
3. If we process alumina carefully—say, by using a controlled particle size—the value of m is doubled, to 9.7. By a controlled particle size, we mean a monosize powder that enhances packing, less use of a binder material (which produces flaws after sintering), more uniform shrinkage, etc.

Figure 7.32 shows the probability of failure as a function of stress for three important engineering ceramics: AlN, SiC, and Si_3N_4. As the Weibull modulus increases, the slope of the curve becomes steeper. When we plot the curves on logarithmic abscissa and ordinate axes, a straight line is obtained that can be used to obtain m as shown in Figure 7.31.

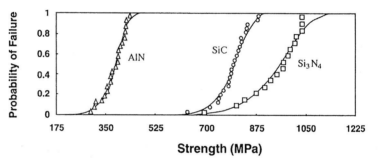

Figure 7.32 Flexural strengths (4-point bend test with inner and outer spans 20 and 40 mm, respectively, and cross section of 3 × 4 mm) for three ceramics. (Courtesy of C. J. Shih)

Some words of caution regarding the use of Weibull probability plots are in order. The tail of the distribution (see Figure 7.30) must be included in the analysis. In practical terms, this means that the statistical sample size should be sufficiently large. Typically, for an allowable failure rate $P = 0.01$, the sample size would be greater than 100. Also, the preceding analysis assumes a "well-behaved flaw population." Bimodal flaw populations can result in two linear parts on the Weibull plot, indicating two values of the Weibull modulus.

EXAMPLE 7.8

The data obtained in four-point bend (or flexure) tests on SiC specimens processed in three different ways are reported in Table E7.8.1. Calculate the Weibull modulus m and the characteristic strength σ_0, and make the Weibull plot, for each specimen. Each specimen had outer and inner spans of 40 and 20 mm, respectively. The height and width of the specimens are 3 mm and 4 mm, respectively.

Solution: We first obtain the stresses from the loads in Table E7.8.1. The moment is

$$M = \frac{P}{2} \times \frac{L}{4}.$$

Table E7.8.1 Fracture load (N) of three hot-pressed SiC specimens

Test No.	SiC-A	SiC-B	SiC-N
1	497	421	466
2	291	690	618
3	493	556	529
4	605	573	627
5	511	618	564
6	524	609	564
7	327	690	573
8	484	654	394
9	394	618	618
10	448	645	493
11	511	591	511
12	497	739	475
13	426	739	618
14	345	703	493
15	358	569	591
16	287	685	627
17	412	708	618
18	466	573	600
19	493	717	645
20	591	676	614

(See Figure E7.8.1.) The maximum tensile stress is

$$\sigma = \frac{Mc}{I},$$

where

$$c = \frac{h}{2},$$

$$I = \frac{bh^3}{12}.$$

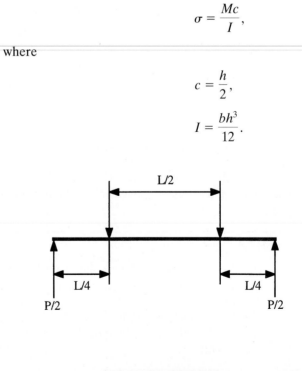

Figure E7.8.1

Hence,

$$\sigma = \frac{PLh \times 12}{8 \times bh^3} = \frac{3}{4} \frac{PL}{bh^2}$$

The calculated stresses are shown in Figure E7.8.2.

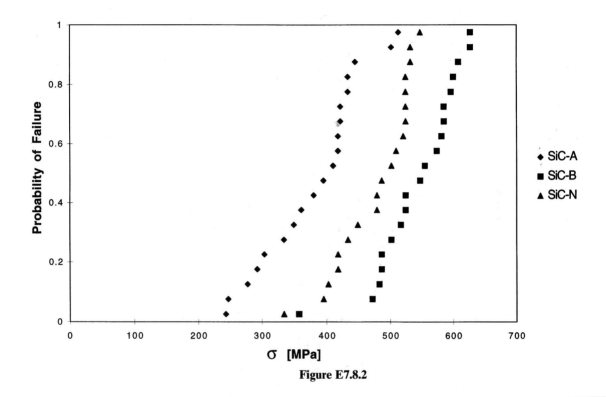

Figure E7.8.2

To obtain the Weibull parameters, we use Equation 7.39:

$$P(V) = \exp\left[-\left(\frac{\sigma}{\sigma_0}\right)^m\right],$$

or

$$1 - F(V) = \exp\left[-\left(\frac{\sigma}{\sigma_0}\right)^m\right].$$

Taking logarithms yields

$$\ln[1 - F(V)] = -\left(\frac{\sigma}{\sigma_0}\right)^m.$$

Taking logarithms again results in

$$\ln \ln[1 - F(V)] = -m(\ln \sigma - \ln \sigma_0),$$

or

$$\ln \ln\left[\frac{1}{1 - F(V)}\right] = m(\ln \sigma - \ln \sigma_0).$$

To obtain $F(V)$ for each point, we use the following Equation

$$1 - P_i(V) = F_i(V) = \frac{i}{N + 1}.$$

In the present case, $N = 20$. Hence, $F_1(V) = 1/21$, $F_2(V) = 2/21$, $F_3(V) = 3/21, \ldots$. These results are plotted in Figure E7.8.3. We use a double logarithm for $1/[1 - F(V)]$ and the logarithm for σ. The slope of this plot provides m. The horizontal line passing through zero gives the values of the characteristic strengths. We summarize our results in Table E7.8.2. Figure E7.8.4 shows the Weibull curves with the preceding parameters superimposed on the data points of Figure E7.8.3.

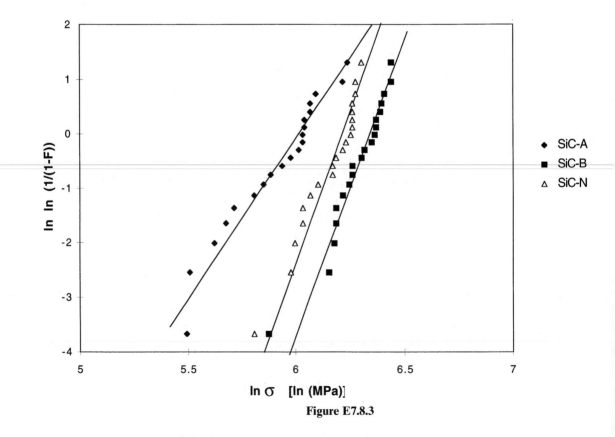

Figure E7.8.3

TABLE E7.8.2

Specimen	m	σ_0 (MPa)	Average Stress $\pm$ S.D.
SiC-A	5.61	411.3	380.7 $\pm$ 63.1
SiC-B	9.10	572.1	542.0 $\pm$ 52.6
SiC-N	9.22	502.9	476.8 $\pm$ 48.7

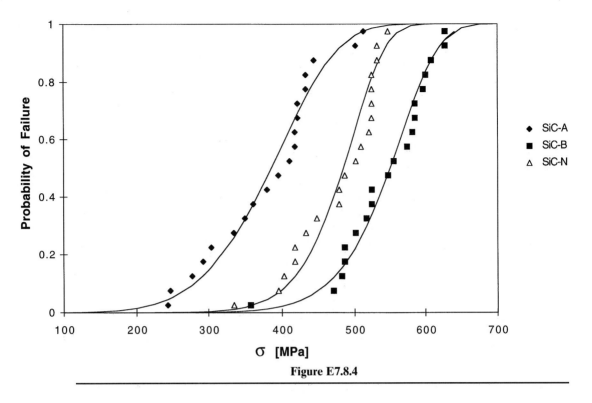

Figure E7.8.4

SUGGESTED READINGS

T. L. ANDERSON. *Fracture Mechanics,* 2d ed. Boca Raton, Fl: CRC Press, 1995.

J. M. BARSOM AND S. T. ROFFE. *Fracture and Fatigue Control in Structures,* 2d ed. Englewood Cliffs, NJ: Prentice Hall, 1987.

D. BROEK. *Elementary Engineering Fracture Mechanics,* 3d ed. The Hague: Sijthoff and Noordhoff, 1978.

H. L. EWALDS AND R. J. H. WANHILL. *Fracture Mechanics.* London: Arnold, 1984.

R. W. HERTZBERG. *Deformation and Fracture Mechanics of Engineering Materials,* 4th ed. New York: John Wiley, 1994.

M. F. Kanninen and C. H. Popelar. *Advanced Fracture Mechanics,* New York: Oxford University Press, 1985.

J. F. KNOTT. *Fundamentals of Fracture Mechanics,* 3d ed. London: Butterworths, 1993.

A. J. KINLOCH AND R. J. YOUNG. *Fracture Behavior of Polymers.* London: App. Sci., 1983.

EXERCISES

7.1 In a polyvinyl chloride (PVC) plate, there is an elliptical, through-the-thickness cavity. The dimensions of the cavity are:

major axis = 1 mm,
minor axis = 0.1 mm.

Compute the stress concentration factor K_t at the extremities of the cavity.

7.2 Calculate the maximum tensile stress at the surfaces of a circular hole (in the case of a thin sheet) and of a spherical hole (in the case of a thick specimen) subjected to a tensile stress of 200 MPa. The material is Al_2O_3 with $\nu = 0.2$.

7.3 Calculate the maximum tensile stress if the applied stress is compressive for a circular hole for which $\sigma_c = 200$ MPa and $\nu = 0.2$.

7.4 The strength of alumina is approximately $E/15$, where E is the Young's modulus of alumina, equal to 380 GPa. Use the Griffith equation in the plane-strain form to estimate the critical size of defect corresponding to fracture of alumina.

7.5 Compute the ratio of stress required to propagate a crack in a brittle material under plane-stress and plane-strain conditions. Take Poisson's ratio ν of the material to be 0.3.

7.6 An Al_2O_3 specimen is being pulled in tension. The specimen contains flaws having a size of 100 μm. If the surface energy of SiC is 0.8 J/m², what is the fracture stress? Use Griffith's criterion. $E = 380$ GPa.

7.7 A thin plate is rigidly fixed at its edges (see Figure Ex. 7.7). The plate has a height L and thickness t (normal to the plane of the figure). A crack moves from left to right through the plate. Every time the crack moves a distance Δx, two things happen:

1. Two new surfaces (with specific surface energy) are created.

2. The stress falls to zero behind the advancing crack front in a certain volume of the material.

Obtain an expression for the critical stress necessary for crack propagation in this case. Explain the physical significance of this expression.

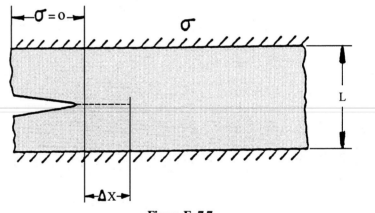

Figure Ex7.7

7.8 A central through-the-thickness crack, 50 mm long, propagates in a thermoset polymer in an unstable manner at an applied stress of 5 MPa. Find K_c.

7.9 Machining of SiC produced surface flaws of a semielliptical geometry. The flaws that were generated have dimensions $a = 1$ mm, width $w = 100$ mm, and $c = 5$ mm, and the thickness of the specimen is $B = 2$ mm. Calculate the maximum stress that the specimen can withstand in tension. $K_{IC} = 4$ MPa m$^{1/2}$.

7.10 An AISI 4340 steel plate has a width W of 30 cm and has a central crack $2a$ of 3 mm. The plate is under a uniform stress σ. This steel has a K_{Ic} value of 50 MPa $\sqrt{m}$ and a services stress of 1,500 MPa. Compute the maximum crack size that the steel may have without failure.

7.11 A microalloyed steel, quenched and tempered at 250°C, has a yield strength (σ_y) of 1,750 MPa and a plane-strain fracture toughness K_{Ic} of 43.50 MPa $\sqrt{m}$. What is the largest disk-type inclusion, oriented most unfavorably, that can be tolerated in this steel at an applied stress of $0.5\sigma_y$?

7.12 A 25-mm² bar of cast iron contains a crack 5 mm long and normal to one face. What is the load required to break this bar if it is subjected to three-point bending with the crack toward the tensile side and the supports 250 mm apart?

7.13 Consider a maraging steel plate of thickness (B) 3 mm. Two specimens of width (W) equal to 50 mm and 5 mm were taken out of this plate. What is the largest through-the-thickness crack that can be tolerated in the two cases at an applied stress of $\sigma = 0.6\sigma_y$, where σ_y (yield stress) = 2.5 GPa? The plane-strain fracture toughness K_{Ic} of the steel is 70 MPa $\sqrt{\text{m}}$. What are the critical dimensions in the case of a single-edge notch specimen?

7.14 An infinitely large plate containing a central crack of length $2a = 50/\pi$ mm is subjected to a nominal stress of 300 MPa. The material yields at 500 MPa. Compute:

   (a) The stress intensity factor at the crack tip.
   (b) The size of the plastic zone at the crack tip.

   Comment on the validity of Irwin's correction for the size of the plastic zone in this case.

7.15 A steel plate containing a through-the-thickness central crack of length 15 mm is subjected to a stress of 350 MPa normal to the crack plane. The yield stress of the steel is 1,500 MPa. Compute the size of the plastic zone and the effective stress intensity factor.

7.16 The size of the plastic zone at the crack tip in the general plane-stress case is given by

$$r_y = \frac{K_I^2}{2\pi\sigma_y^2} \cos^2 \frac{\theta}{2}\left(4 - 3\cos^2 \frac{\theta}{2}\right).$$

   (a) Determine the radius of the plastic zone in the direction of the crack.
   (b) Determine the angle θ at which the plastic zone is the largest.

7.17 For the plane-strain case, the expression for the size of the plastic zone is

$$r_y = \frac{K_I^2}{2\pi\sigma_y^2} \cos^2 \frac{\theta}{2}\left\{4\left[1 - \nu(1 - \nu) - 3\cos^2 \frac{\theta}{2}\right]\right\}.$$

   (a) Show that this expression reduces to the one for plane stress.
   (b) Make plots of the size of the plastic zone as a function of θ for $\nu = 0$, $\nu = \frac{1}{3}$, and $\nu = \frac{1}{2}$. Comment on the size and form of the zone in the three cases.

7.18 A sheet of polystyrene has a thin central crack with $2a = 50$ mm. The crack propagates catastrophically at an applied stress of 10 MPa. The Young's modulus polystyrene is 3.8 GPa, and the Poisson's ratio is 0.4. Find G_{Ic}.

7.19 Compute the approximate size of the plastic zone, r_p, for an alloy that has a Young's modulus $E = 70$ GPa, yield strength $\sigma_v = 500$ MPa, and toughness $G_c = 20$ kJ/m^2.

7.20 300-M steel, commonly used for airplane landing gears, has a G_c value of 10 kN/m. A nondestructive examination technique capable of detecting cracks that are 1 mm long is available. Compute the stress level that the landing gear can support without failure.

7.21 A thermoplastic material has a yield stress of 75 MPa and a G_{IC} value of 300 J/m^2. What would be the corresponding critical crack opening displacement? Take $\lambda = 1$. Also, compute J_{Ic}.

7.22 A line pipe with overall diameter of 1 m and 25-mm thickness is constructed from a microalloyed steel ($K_{Ic} = 60$ MPa m$^{1/2}$; $\sigma_y = 600$ MPa.). Calculate the maximum pressure for which the leak-before-break criterion will be obeyed.[†]

7.23 Al_2O_3 has a fracture toughness of approximately 3 MPa m$^{1/2}$. Suppose you carried out a characterization of the surface of the specimen and detected surface flaws with a radius $a = 50$ μm. Estimate the tensile and compressive strengths of this specimen; show by sketches, how flaws will be activated in compression and tension.

7.24 Using the Weibull equation, establish the tensile strength, with a 50% survival probability, of specimens with a length of 60 mm and a diameter of 5 mm. Uniaxial tensile tests carried out on specimens with a length of 20 mm and the same diameter yielded the following results in MPa (10 tests were carried out): 321, 389, 411, 423, 438, 454, 475, 489, 497, 501.

7.25 An engineering ceramic has a flexure strength that obeys Weibull statistics with $m = 10$. If the flexure strength is equal to 200 MPa at 50% survival probability, what is the flexure strength level at which the survival probability is 90%?

7.26 What would be the flexure strength, at 90% survival probability, if the ceramic in the preceding problem is subjected to a hot isostatic processing (HIP) treatment that greatly reduces the

---

[†]The leak-before-break criterion states that a through-the-thickness crack $(a = t)$ will not propagate catastrophically.

population of flaws and increases m to 60. Assume that the flexure strength at 50% survival probability is unchanged.

7.27 Ten rectangular bars of Al_2O_3 (10 mm wide and 5 mm in height) were tested in three-point bending, the span being 50 mm. The failure loads were 1,040, 1,092, 1,120, 1,210, 1,320, 1,381, 1,410, 1,470, 1,490, and 1,540 N. Determine the characteristic flexure strength and Weibull's modulus for the specimens. (See Section 9.6.1 for the flexure formula.)

7.28 Verify the values of m in Figure 7.32, and obtain the characteristic strengths σ_0 for the three materials. If the fracture toughness of SiC, Si_3N_4, and AlN are equal to 5.2, 5.7, and 2.4 $MNm^{3/2}$, respectively, what are the largest flaws that can be tolerated in these specimens?

7.29 Aluminum has a surface energy of 0.5 Jm^{-2} and a Young's modulus of 70 GPa. Compute the stress at the crack tip for two different crack lengths: 1 mm and 1 cm.

7.30 Determine the stress required for crack propagation under plane strain for a crack of length equal to 2 mm in aluminum. Take the surface energy equal to 0.018 J/m^2, Poisson's ratio to 0.345, and the modulus of $E = 70.3$ GPa.

7.31 Calculate the maximum load that a 2024-T851 aluminum alloy (10 cm $\times$ 2 cm) with a central through-the-thickness crack (length 0.1 mm) can withstand without yielding. Given: $\sigma_y = 500$ MPa and $K_{Ic} = 30$ MPa $\sqrt{m}$.

7.32 An infinitely large sheet is subjected to a far-field stress of 300 MPa. The material has a yield strength of 600 MPa, and there is a central crack $7/\pi$ cm long.

    (a) Calculate the stress intensity factor at the tip of the crack.

    (b) Estimate the size of the plastic zone size at the tip of the crack.

7.33 What is the maximum allowable crack size for a material that has $K_{Ic} = 55$ MPa $\sqrt{m}$ and $\sigma_y = 1,380$ MPa. Assume a plane-strain condition and a central crack.

7.34 Two specimens of concrete were tested in compression. One was wrapped with a very strong tape. They exhibited substantial differences in strength, shown in Figure Ex.7.34. Explain, in terms of microstructural behavior, the reason for the difference in response. Use sketches.

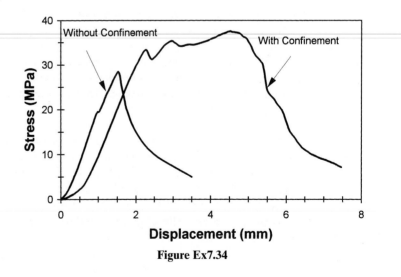

Figure Ex7.34

8

Fracture: Microscopic Aspects

8.1 INTRODUCTION

In Chapter 7, we described the macroscopic aspects of the fracture behavior of materials. As with other characteristics, the microstructure of a material has a great influence on its fracture behavior. In what follows, we present a brief description of the microstructural aspects of crack nucleation and propagation, as well as the effect of the environment on the fracture behavior of different materials. Figure 8.1 shows, schematically, some important fracture modes in a variety of materials. These different modes will be analyzed in some detail in this chapter. Metals fail by two broad classes of mechanisms: *ductile and brittle* failure.

Ductile failure occurs by (a) the nucleation, growth, and coalescence of voids, (b) continuous reduction in the metal's cross-sectional area until it is equal to zero, or (c) shearing along a plane of maximum shear. Ductile failure by void nucleation and growth usually starts at second-phase particles. If these particles are spread throughout the interiors of the grains, the fracture will be transgranular (or transcrystalline). If these voids are located preferentially at grain boundaries, fracture will occur in an intergranular (or intercrystalline) mode. The appearance of a ductile fracture, at high magnification ($500\times$ or higher) is of a surface with indentations, as if marked by an ice-cream scooper. This surface morphology is appropriately called *dimpled*. Rupture by total necking is very rare, because most metals contain second-phase particles that act as initiation sites for voids. However, high-purity metals, such as copper, nickel, gold, and other very ductile materials, fail with very high reductions in their areas.

Brittle fracture is characterized by the propagation of one or more cracks through the structure. While totally elastic fracture describes the behavior of most ceramics fairly well, metals and some polymers undergo irreversible deformation at the tip of the crack, which affects its propagation. Figure 8.1 shows the variety of morphologies and processes

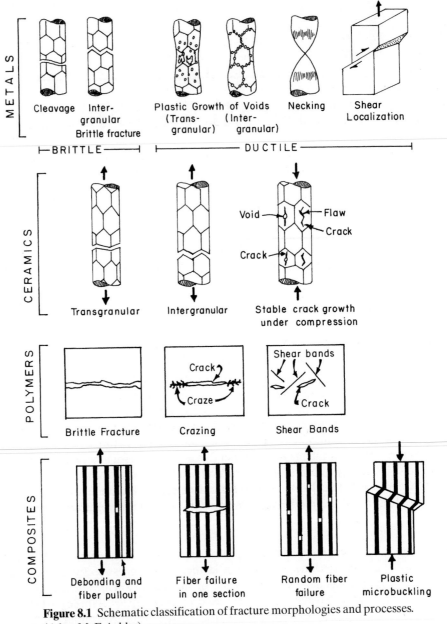

Figure 8.1 Schematic classification of fracture morphologies and processes. (After M. F. Ashby)

occurring during fracturing of materials. For metals and ceramics, two modes of crack propagation: transgranular fracture (or cleavage) and intergranular fracture are observed. For energy-related reasons, a crack will tend to take the path of least resistance. If this path lies along the grain boundaries, the fracture will be intergranular.

Often, a crack also tends to run along specific crystallographic planes, as is the case for brittle fracture in steel. Upon observation at high magnification, transgranular brittle fracture is characterized by clear, smooth facets that have the size of the grains. In steel, brittle fracture has the typical shiny appearance, while ductile fracture has a dull, grayish aspect. In addition to brittle fracture, polymers undergo a mode of fracture called *crazing*, in which the polymer chains ahead of a crack align themselves along the tensile axis, so that the stress concentration is released.

Another mode of deformation that is a precursor to fracture is the phenomenon of *shear banding* in a polymer. If one stretches the polymer material, one observes the formation of a band of material with a much higher flow stress than exists in the unstretched state. Shear banding (or localization) is also prevalent in metals.

Composites—especially fibrous ones—can exhibit a range of failure modes that is dependent on the components of the material (matrix and reinforcement) and on bonding. If the bond strength is higher than the strength of the matrix and reinforcement, the fracture will propagate through the latter (Figure 8.1). If the bonding is weak, one has debonding and fiber pullout. In compression, composites can fail by a kinking mechanism, also shown in the figure; the fibers break, and the entire structure rotates along a band, resulting in a shortening of the composite. This mechanism is known as *plastic microbuckling*.

8.2 FRACTURE IN METALS

Metals are characterized by a highly mobile dislocation density, and they generally show a ductile fracture. In this section, we discuss the various aspects of void and crack nucleation and propagation in metals.

8.2.1 Crack Nucleation

Nucleation of a crack in a perfect crystal essentially involves the rupture of interatomic bonds. The stress necessary to do this is the theoretical cohesive stress, which was dealt with in Chapter 1, starting from an expression for interatomic forces. From this expression, we see that ordinary materials break at much lower stresses than do perfect crystals—on the order of $E/10^4$, where E is Young's modulus of the material. The explanation of this behavior lies in the existence of surface and internal defects that act as preexisting cracks and in the plastic deformation that precedes fracture. When both plastic deformation and fracture are eliminated—for example, in "whiskers"—stresses on the order of the theoretical cohesive stresses are obtained.

Crack nucleation mechanisms vary according to the type of material: brittle, semibrittle, or ductile. The brittleness of a material has to do with the behavior of dislocations in the region of crack nucleation. In highly brittle materials the dislocations are practically immobile, in semibrittle materials dislocations are mobile, but only on a restricted number of slip planes, and in ductile materials there are no restrictions on the movement of dislocations other than those inherent in the crystalline structure of the material. Table 8.1 presents various materials classified according to this criterion regarding the mobility of dislocations.

The exposed surface of a brittle material can suffer damage by mechanical contact with even microscopic dust particles. If a glass fiber without surface treatment were rolled over a tabletop, it would be seriously damaged mechanically.

Any heterogeneity in a material that produces a stress concentration can nucleate cracks. For example, steps, striations, depressions, holes, and so on act as stress raisers on apparently perfect surfaces. In the interior of the material, there can exist voids, air bubbles, second-phase particles, etc. Crack nucleation will occur at the weakest of these defects, where the conditions would be most favorable. We generally assume that the sizes as well as the locations of defects are distributed in the material according to some function of standard distribution whose parameters are adjusted to conform to experimental data. In this assumption, there is no explicit consideration of the nature or origin of the defects.

In semibrittle materials, there is a tendency for slip initially, followed by fracture on well-defined crystallographic planes. That is, there exists a certain inflexibility in the deformation process, and the material, not being able to accommodate localized plastic strains, initiates a crack to relax stresses.

TABLE 8.1 Materials of Various Degrees of Brittleness[a]

Type	Principal Factors	Materials
Brittle	Bond rupture	Structures of type diamond, ZnS, silicates, alumina, mica, boron, carbides, and nitrides
Semibrittle	Bond rupture, dislocation mobility	Structures of type NaCl, ionic crystals, hexagonal compact metals, majority of body-centered cubic metals, glassy polymers
Ductile	Dislocation mobility	Face-centered cubic metals, nonvitreous polymers, some body-centered cubic metals

[a]Adapted with permission from B. R. Lawn and T. R. Wilshaw, *Fracture of Brittle Solids* (Cambridge: Cambridge University Press, 1975), p. 17.)

Various models are based on the idea of crack nucleation at an obstruction site. For example, the intersection of a slip band with a grain boundary, another slip band, and so on, would be an obstruction site.

8.2.2 Ductile Fracture

In ductile materials, the role of plastic deformation is very important. The important feature is the flexibility of slip. Dislocations can move on a large number of slip systems and even cross from one plane to another (in cross-slip). Consider the deformation of a single crystal of copper, a ductile metal, under uniaxial tension. The single crystal undergoes slip throughout its section. There is no nucleation of cracks, and the crystal deforms plastically until the start of plastic instability, called necking. From this point onward, the deformation is concentrated in the region of plastic instability until the crystal separates along a line or a point. (See Figure 8.2a). In the case of a cylindrical sample, a soft single crystal of a metal such as copper will reduce to a point fracture. Figure 8.2b shows an example of such a fracture in a single crystal of copper. However, if, in a ductile material, there are microstructural elements such as particles of a second phase, internal interfaces, and so on, then microcavities may be nucleated in regions of high stress concentration in a manner similar to that of semibrittle materials, except that, due to the ductile material's large plasticity, cracks generally do not propagate from these cavities. The regions between the cavities, though, behave as small test samples that elongate and break by plastic instability, as described for the single crystal.

In crystalline solids, cracks can be nucleated by the grouping of dislocations piled up against a barrier. Such cracks are called *Zener–Stroh* cracks.[1] High stresses at the head of a pileup are relaxed by crack nucleation, as shown in Figure 8.3, but this would occur only in the case where there is no relaxation of stresses by the movement of dislocations on the other side of the barrier. Depending on the slip geometry in the two parts and the kinetics of the motion and multiplication of dislocations, such a combination of events could

[1]See C. Zener, *The Fracturing of Metals* (Metals Park, OH: ASM, 1948).

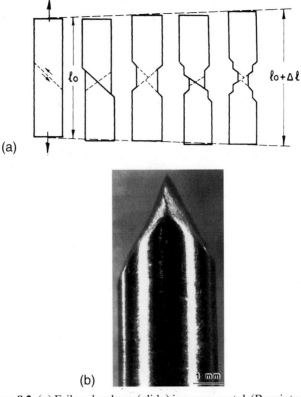

Figure 8.2 (a) Failure by shear (glide) in a pure metal. (Reprinted with permission from D. Broek, *Elementary Engineering Fracture Mechanics,* 3d ed. The Hague, Netherlands: Martinus Nijhoff, (1982), p. 33.) (b) A point fracture in a soft single-crystal sample of copper. (Courtesy of J. D. Embury)

occur. (See Table 8.1.) Figure 8.4(a) shows a bicrystal that has a slip band in grain I. The stress concentration at the barrier due to the slip band is completely relaxed by slip on two systems in grain II. Figure 8.4(b) shows the case of only a partial relaxation and the resulting appearance of a crack at the barrier. Lattice rotation associated with the bend planes and deformation twins can also nucleate cracks. Figure 8.5 shows crack nucleation in zinc as per the model shown in Figure 8.5a. Cracks can also begin at the intersections of various boundaries in a metal, which represent sites at which there is a concentration of

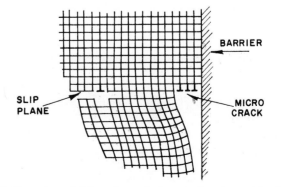

Figure 8.3 Grouping of dislocations piled up at a barrier and leading to the formation of a microcrack (Zener–Stroh crack).

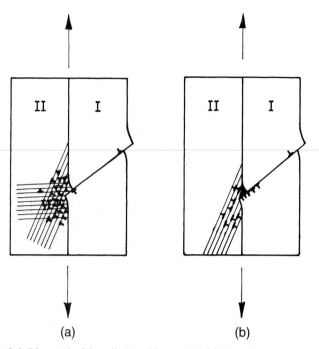

(a)   (b)

Figure 8.4 Bicrystal with a slip band in grain I. (a) The stress concentration at the boundary of the barrier due to slip band is fully relaxed by multiple slip. (b) The stress concentration is only partially relaxed, resulting in a crack at the boundary.

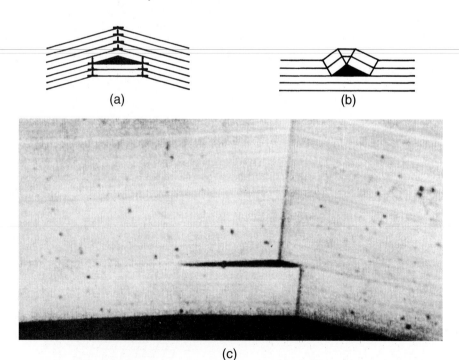

Figure 8.5 Crack nucleation by (a) lattice rotation due to bend planes and (b) deformation twins. (c) Crack nucleation in zinc due to lattice rotation associated with bend planes. (Reprinted with permission from [J. J. Gilman, *Physical Nature of Plastic Flow and Fracture*, General Electric Report No. 60-RL-2410M, April, 1960.]. p. 83.)

stress. Figure 8.6 presents examples of crack nucleation at the intersection of twin boundaries and at the intersection of twin steps and boundaries.

Fracture at high temperature can occur by a variety of other modes as well. For example, grain-boundary sliding occurs rather easily at high temperatures. Grain-boundary sliding can lead to the development of stress concentrations at grain-boundary triple points (where three grain boundaries meet). Cracks nucleate at such triple points as shown schematically in Figure 8.7. Figure 8.8 shows a micrograph of copper in which such a crack nucleation has occurred. This type of crack is called *w-type cavitation* or *w-type cracking*. Yet another type of cracking occurs, characteristically, under conditions of low stresses and high temperature. Small cavities form at grain boundaries that are predominantly at approximately 90° to the stress axis, as shown in Figure 8.9. This is called *r-type cavitation* or *r-type cracking*. Figure 8.10 shows intergranular voids in copper.

The most familiar example of ductile fracture is that in uniaxial tension, giving the classic "cup and cone" fracture. When the maximum load is reached, the plastic deformation in a cylindrical tensile test piece becomes macroscopically heterogeneous and is concentrated in a small region. This phenomenon is called necking (see Section 3.2.2). The

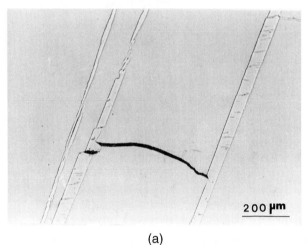

(a)

(b)

Figure 8.6 Initiation of failure by microcrack formation in tungsten deformed at approximately 10^4 s^{-1} at room temperature. (a) Twin steps. (b) Twin steps and twin–twin intersection. (From T. Dümmer, J. C. LaSalvia, M. A. Meyers, and G. Ravichandran, *Acta Mat.,* (1998))

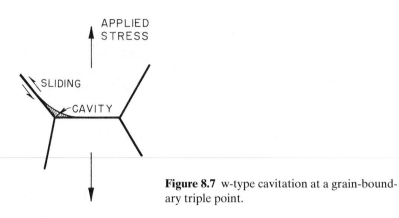

Figure 8.7 w-type cavitation at a grain-boundary triple point.

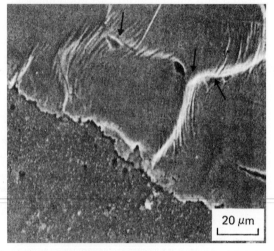

Figure 8.8 w-type cavities nucleated at grain boundaries in copper, seen through a scanning electron microscope.

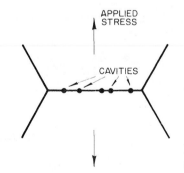

Figure 8.9 r-type cavitation at a grain boundary normal to the stress axis.

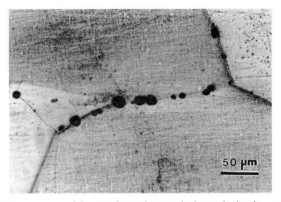

Figure 8.10 r-type cavities nucleated at grain boundaries in copper, seen through an optical microscope.

final fracture occurs in this necked region and has the characteristic appearance of a conical region on the periphery resulting from shear and a central flat region resulting from the voids created there. In extremely pure metal single crystals (e.g., those free of inclusions, etc.), plastic deformation continues until the sample section is reduced to a point, a geometric consequence of slip, as shown in Figure 8.2.

In practice, materials generally contain a large quantity of dispersed phases. These can be very small particles (1 to 20 nm) such as carbides of alloy elements, particles of intermediate size (50 to 500 nm) such as alloy element compounds (carbides, nitrides, carbonitrides) in steels, or dispersions such as Al_2O_3 in aluminum and ThO_2 in nickel. Precipitate particles obtained by appropriate heat treatment also form part of this class (e.g., an Al–Cu–Mg system), as do inclusions of large size (on the order of millimeters)—for example, oxides and sulfides.

If the second-phase particles are brittle and the matrix is ductile, the former will not be able to accommodate the large plastic strains of the matrix, and consequently, these brittle particles will break in the very beginning of plastic deformation. In case the particle/matrix interface is very weak, interfacial separation will occur. In both cases, microcavities are nucleated at these sites (Figure 8.11). Generally, the voids nucleate after a few percent of plastic deformation, while the final separation may occur around 25%. The microcavities grow with slip, and the material between the cavities can be visualized as a small tensile test piece. The material between the voids undergoes necking on a microscopic scale, and the voids join together. However, these microscopic necks do not contribute significantly to the total elongation of the material. This mechanism of initiation, growth, and coalescence of microcavities gives the fracture surface a characteristic appearance. When

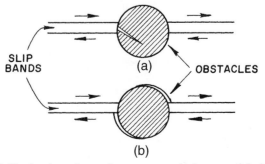

Figure 8.11 Nucleation of a cavity at a second-phase particle in a ductile material. (Adapted with permission from B. R. Lawn and T. R. Wilshaw, *Fracture of Brittle Solids* (Cambridge: Cambridge University Press, 1975), p. 40)

viewed in the scanning electron microscope, such a fracture appears to consist of small dimples, which represent the microcavities after coalescence (Figure 8.12a). In many of these dimples, one can see the inclusions that were responsible for the void nucleation. At times, due to unequal triaxial stresses, these voids are elongated in one or the other direction. Figure 8.12b shows a fractured carbide that contributed to the overall fracture. We describe the process of fracture by void nucleation, growth, and coalescence in some detail because of its great importance in metals.

Fracture by Void Nucleation, Growth, and Coalescence.   Figure 8.13 shows the classic cup-and-cone fracture observed in many tensile specimens with a cylindrical cross section. The configuration is typical of ductile fracture, and upon observation at a higher magnification ($1,000\times$ or higher, best done in a scanning electron microscope), one sees the typical "dimple" features. The dimples are equiaxal in the central portion of the fracture and tend to be inclined in the sidewalls of the "cup." The top two pictures show scanning electron micrographs of these two areas. In the central region fracture is essentially tensile, with the surface perpendicular to the tensile axis. On the sides, the fracture has a strong shear character, and the dimples show the typical "inclined" morphology, i.e.,

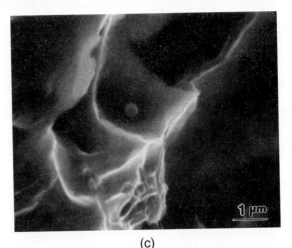

(a)   (b)

(c)

Figure 8.12 (a) Scanning electron micrograph of dimple fracture resulting from the nucleation, growth, and coalescence of microcavities. Note the inclusion particles, which served as the microcavity nucleation sites. (b) Fractured metal carbide particle (see the center of the picture) in Inconel 718 superalloy. (c) Inclusion in steel.

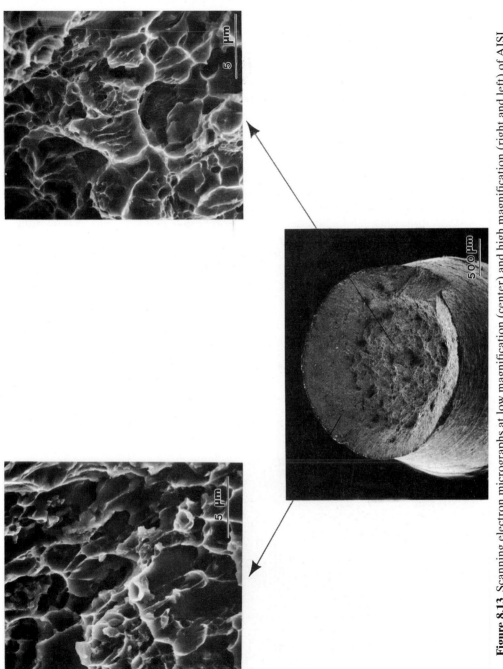

Figure 8.13 Scanning electron micrographs at low magnification (center) and high magnification (right and left) of AISI 1008 steel specimen ruptured in tension. Notice the equiaxal dimples in the central region and elongated dimples on the shear walls (the sides of the cup).

they appear to be elliptical with one side missing. Figure 8.14 shows, in a very schematic fashion, what is thought to occur in the specimen that leads to failure. Voids nucleate and grow in the interior of the specimen when the overall plastic strain reaches a critical level. The voids grow until they coalesce. Initially equiaxial, their shape changes in accordance with the overall stress field. As the voids coalesce, they expand into adjoining areas, due to the stress concentration effect. When the center of the specimen is essentially separated, this failure will grow toward the outside. Since the elastic–plastic constraints change, the plane of maximum shear (approximately 45° to the tensile axis) is favored, and further growth will take place along these planes, which form the sides of the cup. Although it is easy to describe this process in a qualitative way, an analytical derivation is very complex and involves plasticity theory, which is beyond the scope of the text. Figure 8.15 shows the sequence of ductile fracture propagation, with the formation of dimples. The dimples are produced by voids nucleating ahead of the principal crack (Figures 8.15a and b), which has a blunted tip because of the plasticity of the material. The void ahead of the crack grows (Figure 8.15c) and eventually coalesces with the main crack (Figure 8.15d). New voids nucleate ahead of the growing crack, and the process repeats itself. Figure 8.16 shows the propagation of ductile fracture in a specimen of AISI 304 stainless steel undergoing extension, as seen in a high-voltage transmission electron microscope. A referential fixed to the material was added to help visualize the progression of the crack. Figure 8.16a shows the growth of a void ahead of the tip of the crack, while Figure 8.16b shows new voids being nucleated. In Figure 8.16(c), the crack has advanced by joining with these growing voids. New voids have nucleated.

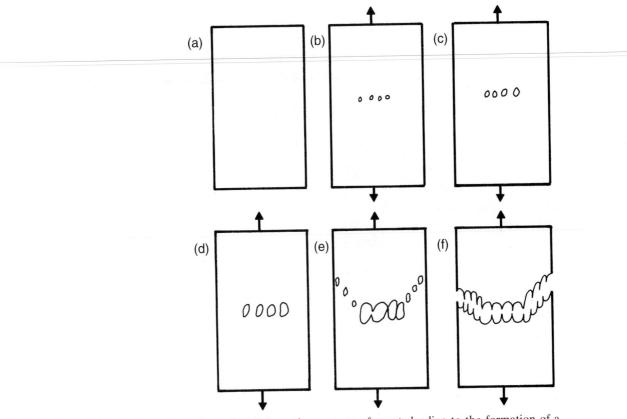

Figure 8.14 Schematic sequence of events leading to the formation of a cup-and-cone fracture.

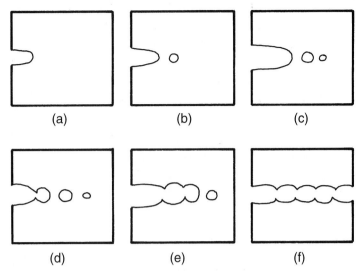

Figure 8.15 Sequence of events in the propagation of ductile fracture by nucleation, growth, and coalescence of voids.

The nucleation and growth of voids is of great importance in determining the fracture characteristics of ductile materials. Many researchers have identified second-phase particles and inclusions as the main sources of voids.[2] Indeed, Figure 8.12 shows dimples, at the bottoms of which second-phase particles can be seen. The size, separation, and interfacial bonding of these particles determine the overall propagation characteristics of ductile cracks and, therefore, the ductility of the material. The role of second-phase particles is illustrated in Figure 8.17. Copper-based alloys with different amounts of

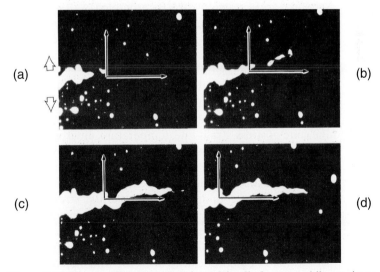

Figure 8.16 Observation of progression of ductile fracture while specimen is stressed in high-voltage transmission electron microscope. Referential is fixed to material (Courtesy of L. E. Murr).

[2]See, for example, H. C. Rogers, in *Ductility* (Metals Park, OH: ASM, 1967), p. 31; and L. M. Brown and J. D. Embury, in *Microstructure in Design of Alloys,* Vol. 1 (*London:* Institute of Metals/Iron and Steel Institute, 1973), p. 164.

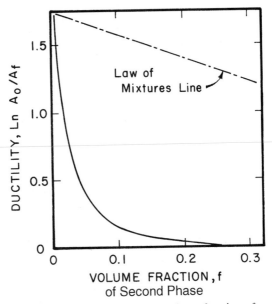

Figure 8.17 Combined plot of ductility vs. volume fraction of second phase, f, for copper specimens containing various second phases and for notched copper specimens. The dashed line represents the prediction from the law of mixtures, assuming zero ductility for the second-phase particles. (From B. I. Edelson and W. J. Baldwin, Jr., *Trans. ASM*, 55, p. 230 (1962))

second-phase particles (fractions from 0 to 0.24) were tested in tension, and the ductility of the material was measured. The ductility is given by the reduction in area of the specimens at the fracture point (ln A_0/A_f where A_0 and A_f are the initial and final cross-sectional areas, respectively). By a simple law of mixtures, assuming that the second-place particles have zero ductility, one obtains the straight line shown in the figure. However, the effect of second-phase particles is much more drastic, and ductility is reduced to zero at $f = 0.24$. This is a clear indication that second-phase particles play a key role in the propagation of ductile fracture.

Before we close this section, it is worth pointing out here that the term "ductility" signifies a material's capacity to undergo plastic deformation. Ductility is not a fundamental property of the material, because the plastic strain before fracture is a function of the state of stress, strain rate, temperature, environment, and prior history of the material. The state of stress is defined by the three-dimensional distribution of normal and shear stresses at a point or by the three principal stresses at a point. (See Chapter 2.) The multi-axial stresses may be obtained by external multiaxial loading, by geometry of the structure or microstructure under load, by thermal stresses, or by volumetric microstructural changes. One can define a simple "triaxiality" factor by the ratio σ_3/σ_1, where $\sigma_1 > \sigma_2 > \sigma_3$ are the principal stresses. If ε_0 is the plastic strain at fracture in uniaxial tension and ε_1 is the maximum principal plastic strain, one can define a ductility ratio as $\varepsilon_1/\varepsilon_0$. This ductility ratio shows, theoretically, a decrease with increasing triaxiality; that is, $\varepsilon_1/\varepsilon_0$ goes to zero as σ_3/σ_1 goes to unity (Figure 8.18). Thus, an increase in the degree of stress triaxiality results in a decrease in the ductility of the material.

The temperature and the strain rate have contrary effects. A high temperature (or a low strain rate) leads to high ductility, whereas a low temperature (or a high strain rate) leads to low ductility.

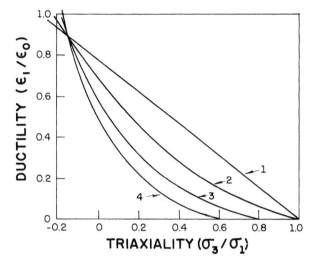

Figure 8.18 Variation of maximum plastic strain (ductility) with the degree of triaxiality, according to (1) theory of maximum tensile stress failure, (2) plane-strain conditions, (3) von Mises criterion, and (4) power law of plastic strain. (Adapted with permission from M. J. Manjoine, in *Fracture: An Advanced Treatise,* Vol. 3, H. Liebowitz (ed.) (New York: Academic Press, 1971), p. 265.

8.2.3 Brittle, or Cleavage, Fracture

The most brittle form of fracture is cleavage fracture. The tendency for a cleavage fracture increases with an increase in the strain rate or a decrease in the test temperature of a material. This is shown, typically, by a ductile–brittle transition in steel (Figure 8.19). The ductile–brittle transition temperature (DBTT) increases with an increase in the strain rate. Above the DBTT the steel shows a ductile fracture, while below the DBTT it shows a brittle fracture. The ductile fracture needs a lot more energy than the brittle fracture. We deal with these aspects of DBTT in more detail in Chapter 9.

Cleavage occurs by direct separation along specific crystallographic planes by means of a simple rupturing of atomic bonds (Figure 8.20). Iron, for example, undergoes cleavage along its cubic planes (100). This gives the characteristic flat surface appearance within a grain on the fracture surface.

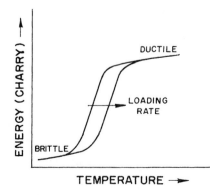

Figure 8.19 Ductile–brittle transition in steel and the effect of loading rate (schematic).

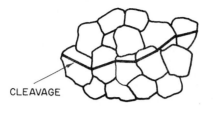

Figure 8.20 Propagation of transgranular cleavage.

There is evidence that some kind of plastic yielding and dislocation interaction is responsible for cleavage fracture. Low studied the fracture behavior of a low-carbon steel at 77 K, comparing the yield stress in compression (in which case fracture does not occur) with the stress for cleavage in tension. He did this for a number of samples with different grain sizes and obtained the plot shown in Figure 8.21. The variation in grain size in both cases followed a Hall–Petch type of relationship, which showed that the controlling mechanism in yielding was also the controlling mechanism for initiating fractures. At 77 K, yielding is closely associated with mechanical twinning (see Section 5.2).

Earlier, we mentioned that cleavage occurs along specific crystallographic planes. As in a polycrystalline material, the adjacent grains have different orientations; the cleavage crack changes direction at the grain boundary in order to continue along the given crystallographic planes. The cleavage facets seen through the grains have a high reflectivity, which gives the fracture surface a shiny appearance (Figure 8.22a). Sometimes the cleavage fracture surface shows some small irregularities—for example, the river markings in Figure 8.22b. What happens is that, within a grain, cracks may grow simultaneously on two parallel crystallographic planes (Figure 8.23a). The two parallel cracks can then join together, by secondary cleavage or by shear, to form a step. Cleavage steps can be initiated by the passage of a screw dislocation, as shown in Figure 8.23b. In general, the cleavage step will be parallel to the crack's direction of propagation and perpendicular to the plane containing the crack, as this configuration would minimize the energy for the step formation by creating a minimum of additional surface. A large number of cleavage steps can join and form a multiple step. On the other hand, steps of opposite signs can join and disappear. The junction of cleavage steps results in a figure of a river and its tributaries. River markings can appear by the passage of a grain boundary, as shown in Figure 8.23c. We know that cleavage crack tends to propagate along a specific crystallographic plane. This being so, when a crack passes through a grain boundary, it has to propagate in a grain with

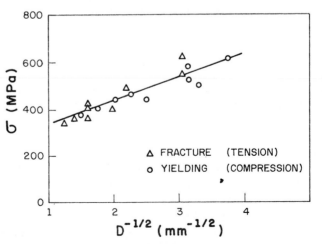

Figure 8.21 Effect of grain size on fracture and yield stress of a carbon steel at 77 K. (Adapted from J. R. Low, in *Madrid Colloquium on Deformation and Flow of Solids* (Berlin: Springer-Verlag, 1956), p. 60)

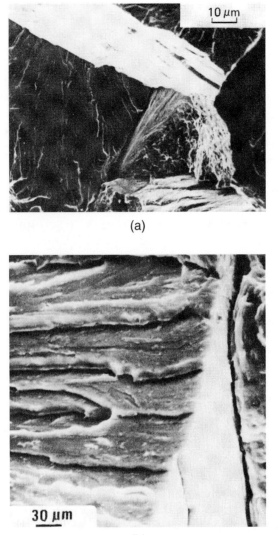

Figure 8.22 (a) Cleavage facets in 300-M steel (scanning electron micrograph). (b) River markings on a cleavage facet in 300-M steel (scanning electron micrograph).

a different orientation. Figure 8.23c shows the encounter of a cleavage crack with a grain boundary. After they meet, the crack should propagate on a cleavage plane that is oriented in a different manner. The crack can do this at various points and spread into the new grain. Such a process gives rise to the formation of a number of steps that can group together, generating a river marking (Figure 8.23c). The convergence of tributaries is always in the direction of flow of the river (i.e., "downstream"). This fact furnishes the possibility of determining the local direction of propagation of crack in a micrograph.

Under normal circumstances, face-centered cubic (FCC) metals do not show cleavage. In these metals, a large amount of plastic deformation will occur before the stress necessary for cleavage is reached. Cleavage is common in body-centered cubic (BCC) and hexagonal close-packed (HCP) structures, particularly in iron and low-carbon steels (BCC). Tungsten, molybdenum, and chromium (all BCC) and zinc, beryllium, and magnesium (all HCP) are other examples of metals that commonly show cleavage.

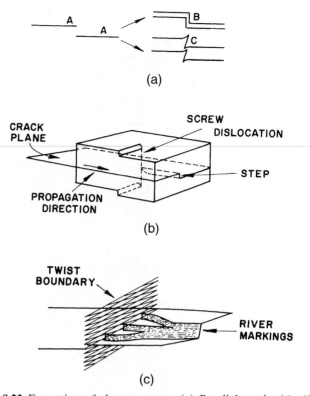

Figure 8.23 Formation of cleavage steps. (a) Parallel cracks (*A, A*) join together by cleavage (*B*) or shear (*C*). (b) Cleavage step initiation by the passage of a screw dislocation. (c) Formation of river markings after the passage of a grain boundary. (Adapted from D. Broek, *Elementary Engineering Fracture Mechanics,* 3d ed. (The Hague, Netherlands: Martinus Nijhoff, 1982), p. 33)

Quasi cleavage is a type of fracture that is formed when cleavage occurs on a very fine scale and on cleavage planes that are not very well defined. Typically, one sees this type of fracture in quenched and tempered steels. These steels contain tempered martensite and a network of carbide particles whose size and distribution can lead to a poor definition of cleavage planes in the austenite grain. Thus, the real cleavage planes are exchanged for small and ill-defined cleavage facets that initiate at the carbide particles. Such small facets can give the appearance of a much more ductile fracture than that of normal cleavage, and generally, river markings are not observed.

Intergranular fracture is a low-energy fracture mode. The crack follows the grain boundaries, as shown schematically in Figure 8.24, giving the fracture a bright and reflective appearance on a macroscopic scale. On a microscopic scale, the crack may deviate around a particle and make some microcavities locally. Figure 8.24b shows an example of this deviation in a micrograph of an intergranular fracture in steel. Intergranular fractures tend to occur when the grain boundaries are more brittle than the crystal lattice. This occurs, for example, in stainless steel when it is accidentally sensitized. This accident in the heat treatment produces a film of brittle carbides along the grain boundaries. The film is then the preferred trajectory of the crack tip. The segregation of phosphorus or sulfur to grain boundaries can also lead to intergranular fracture. In many cases, fracture at high temperatures and in creep tends to be intergranular.

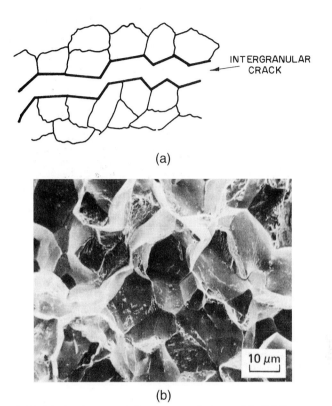

(a)

(b)

Figure 8.24 (a) An intergranular fracture (schematic). (b) Intergranular fracture in steel (scanning electron micrograph).

The ductile–brittle transition temperature of steels and other BCC metals and alloys is significantly affected by grain size. Failure by cleavage (or quasi-brittle crack propagation) and by ductile means are competing mechanisms. When cleavage cracks form and propagate at a greater rate than plastic deformation, the material fails in a brittle manner. It is well known that a reduction in grain size causes a reduction in the ductile-to-brittle transition temperature in steels. Indeed, a reduction in grain size is a very effective means of producing steels that are ductile at low temperature. The explanation of this effect is known as the *Armstrong criterion*[3] and is discussed briefly next.

The yield stress is well represented by the Hall–Petch Equation (see Section 5.3), namely,

$$\sigma_y = \sigma_0 + k_y D^{-1/2}.$$

The temperature effect can be expressed by

$$\sigma_0 = B \exp(-\beta T),$$

where B and β are thermal softening parameters. As T increases, σ_0 decreases. The cleavage stress, on the other hand, is also represented by a Hall–Petch relationship:

$$\sigma_c = \sigma_{0c} + k_c D^{-1/2}.$$

[3]See R. W. Armstrong, *Phil. Mag.,* 9 (1964) 1063.

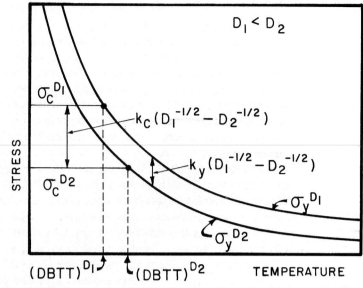

Figure 8.25 Armstrong criterion showing effect of grain size on ductile-to-brittle transition temperature.

Note that σ_c is not dependent on temperature. Note also that $k_c > k_y$.* By setting σ_c equal to σ_y, we can obtain the ductile-to-brittle transition temperature:

$$\sigma_y = \sigma_c,$$

$$T_c = \frac{1}{\beta}\left[\ln B - \ln\{(k_c - k_y) + \sigma_{0c}D^{1/2}\} - \ln D^{-1/2}\right].$$

Figure 8.25 shows, in a schematic fashion, how the yield stress of a steel with two grain sizes ($D_1 < D_2$) varies with temperature. The ductile-to-brittle transition temperatures (DBTT) for the two grain sizes are also marked in the figure. The Armstrong criterion applied to the two grain sizes leads to the prediction:

$$(\text{DBTT})^{D_2} > (\text{DBTT})^{D_1}.$$

Thus, the steel with the smaller grain size (D_1) has the lower DBTT. One can see that grain-size reduction is important in increasing both the strength and the range of temperatures over which it is ductile and tough.

8.3 FRACTURE IN CERAMICS

8.3.1 Microstructural Aspects

Ceramics are characterized by high strength and very low ductility. Among the approaches developed to enhance the ductility (and, consequently, the fracture toughness) of ceramics are:

1. The addition of fibers to the ceramic to form a composite, making crack propagation more difficult because of crack bridging, crack deflection, fiber pullout, etc.
2. The addition of a second phase that transforms at the crack tip with a shear and dilational component, thus reducing the stress concentration at the tip of the crack.

---

*In Figure 8.21 yielding (by twinning) and fracture (by cleavage) have the same k. This is becuase the Hall–Petch slope for twinning is much higher than the one for slip.

3. The production of microcracks ahead of the crack, causing crack branching and distributing the strain energy over a larger area.

4. Careful processing in such a manner that all flaws of a size greater than the grain size are eliminated.

Figure 8.26 shows three toughening mechanisms for ceramics. The addition of fibers renders the propagation of a crack more difficult by one or more of the mechanisms to be explained in Chapter 15, on composites. The addition of a phase that undergoes a transformation is an ingenious strengthening method with great potential. It is described in detail in Chapter 11. Partially stabilized zirconia (zirconia with small additions of yttria is the phase most commonly added. This phase has a tetragonal structure. At the crack tip, the stress field is such that the transformation from a tetragonal to a monoclinic structure takes place. This transformation produces a volume expansion and a shear. The dilation (volumetric strain) is approximately 4%, and the shear strain is approximately 0.16. The regions ahead of the crack tip (Figure 8.26b) that have the right stress state will undergo the transformation, which has the effect of adding a compressive stress at the crack tip that will tend

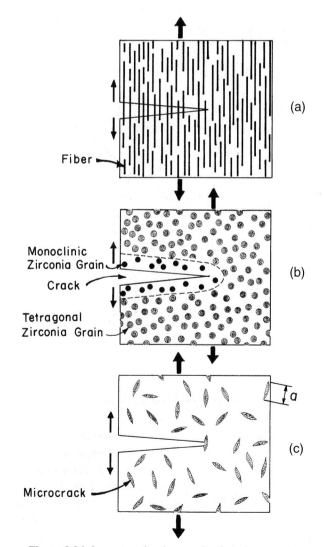

Figure 8.26 Some toughening mechanisms in ceramics.

to close it. Thus, further progression of crack is more difficult. Also, if microcracks are generated around the second-phase particles, they will decrease the stress concentration ahead of the crack tip. The fracture toughness of alumina with partially stabilized zirconia can be much higher than that of alumina alone.

A third mechanism for toughening ceramics is to form microcracks ahead of the main crack. This is shown in Figure 8.26(c). The microcracks have the effect of decreasing the stress intensity factor at the root of the principal crack. An additional effect is that they can lead to crack branching. One single crack branches into several cracks, and the stress required to drive a number of cracks is higher than that required to drive a single crack.

A fourth mechanism for strengthening ceramics is careful processing so as to eliminate, as much as possible, flaws in the material. Figure 8.27 shows identical materials with two flaw-size distributions. Application of the simple equation from fracture mechanics, $K_{Ic} = \sigma\sqrt{\pi a}$, tells us that, for a common ceramic having a fracture toughness of 4 MPa $\sqrt{m}$, a reduction in flaw size from 1 to 0.1 mm has the effect of increasing the maximum tensile stress that the ceramic will withstand from 16 to 56 MPa.

In spite of the processes just described to increase the ductility of ceramics, as a rule, ceramics are not very ductile. Their low ductility and relatively low resistance to crack propagation are responsible for the great differential between the compressive and tensile strength of ceramics. In metals, the difference is relatively small, because failure is often initiated only after considerable plastic deformation. The compressive strength of ceramics is close to 10 times their tensile stress. This same proportion is also observed in rocks. Table 8.2 shows the compressive and tensile strengths of a number of ceramics.

Figure 8.27 Ceramic with two flaw-size distributions.

TABLE 8.2 Compressive, Tensile, and Flexural Strengths of Ceramics (Adapted with permission from *Guide to Engineered Materials* (Metals Park, OH: ASM International, 1985), p. 16)

		Compressive Strength, MPa	Tensile Strength, MPa	Flexural Strength, MPa
Alumina	85	1,620	120	290
(different	90	2,415	140	320
purities)		2,067	170	310
	95	2,411	190	340
	99	2,583	210	340
Alumina silicate		275	17	62
ZrO_2–Al_2O_3		2,411		
3% 1/2 03 PSZ*		2,962		1,170
Transformation Toughened Zirconia		1,757	350	630
9% MgO Partially Stabilized Zirconia*		1,860		690
Cast Si_3N_4		138	24	69
Reaction-bonded SiC		689	140	255
Pressureless sintered SiC		3,858	170	550
Sintered SiC with free silicon		1,030	165	320
Sintered SiC with graphite		410	35	55
Reaction-bonded Si_3N_4		770		210
Hot-pressed Si_3N_4		3,445		860

*Data are from a variety of commercial sources.

It is the *inability* of ceramics to undergo plastic deformation that is responsible for the drastic difference in mechanical performance between metals and ceramics. This inability renders ceramics much stronger, but their ability to resist the propagation of cracks is decreased drastically.

The surface morphology of fractures in ceramics tends to present some markedly different features from those appearing in metals. Usually, failure begins at a flaw and propagates slowly. As it accelerates, its energy release rate increases, and there is a tendency for branching; Figure 8.28a shows a crack schematically. The origin of the crack is shown by the leftmost arrow. At O and O' in a brittle material, branching starts, and the crack becomes a multitude of cracks. This is seen most clearly in glass, but is also observed in crystalline ceramics. Figure 8.28b shows a sequence of photographs of crack branching (or bifurcation) in glass. A sharp hammer impacted the left-hand size of the glass at different velocities in Figure 8.28 (b). As the velocity (and force) of the blow increase, the extent of bifurcation of the cracks increases. Hence, we can understand how shattering of brittle materials occurs. The student is well aware that a glass or a coffee mug (a ceramic!) will break into more parts if the fall is from a greater height. In Figure 8.28 (b), $V_4 > V_3 > V_2 > V_1$. If one looks at the fracture surface, one can often identify the origin of the failure by a smooth area, called the *mirror* region. At the center of this smooth area, the vestiges of the initial flaw can be seen. This mirror area becomes more irregular as the crack propagates from the initial flaw. This is called the *mist* region. As branching becomes prevalent, the flat, smooth surface becomes markedly irregular, and this region is called the *hackle* region (These are similar to the ones observed in polymers and shown in Figure 8.41).

When crack branching (bifurcation) starts, the fracture surface becomes increasingly irregular, because, on separation, different fracture planes become interconnected. In ceramics, the flaws are extremely important, and their concentration and size determine the

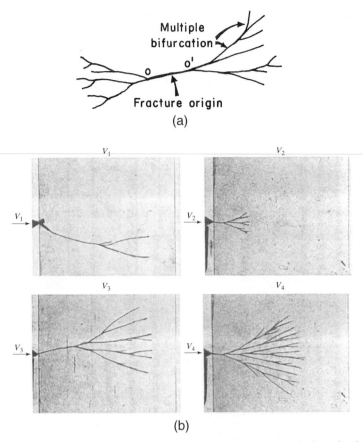

Figure 8.28 (a) Schematic illustrating a typical crack morphology in the vicinity of the origin, and (b) crack bifurcation in glass from an edge initiated failure, caused by sharp instrument blow on left-hand side; blow velocity $V_1 < V_2 < V_3 < V_4$. (Adapted from H. Schardin, in "Fracture," eds. B. L. Averbach, D. K. Felbeck, G. T. Hahn, and D. A. Thomas, MIT Press, Cambridge, Mass., 1959, p. 297, Chapter 16).

strength of the ceramic. These flaws can be classified into three groups: flaws produced during processing, flaws induced by improper design, and flaws introduced during service. Table 8.3 gives examples of the various flaws. Since the strength of ceramics is determined by the basic equation of fracture mechanics, that is,

$$K_{Ic} = \sigma \sqrt{\pi a},$$

it is essential to take care to eliminate flaws (or reduce their number as much as possible).

Flaws induced during processing can be of various kinds. Thermal stresses are an important source of cracks. Thermal stresses are caused by the anisotropy of thermal expansion coefficients in noncubic ceramics. If the ceramic is rapidly cooled, intense internal stresses are set up, leading to microcracks and even total fragmentation of the object. Figure 8.29 shows an intergranular crack produced by thermal shock in alumina. This piece of alumina was cooled fairly rapidly by quenching it in water, from 1,000°C. The intergranular crack can be clearly seen. Figure 8.30 shows voids in AD85 alumina (85% Al_2O_3, plus a glassy phase). These voids account for approximately 12% of the volume of

TABLE 8.3 Sources of Flaws in Ceramics

Processing	Thermal stresses
	Machining
	Large pores
	Isolated large grains
	Cracked grains
	Inclusions
	Laminations during pressing
Design	Stress concentrations due to sharp corners, holes, improper design, etc.
Service	Impact
	Environmental degradation

the material and are a commonly encountered feature of crystalline ceramics. The voids are larger than the individual grains of the material; the micrograph in Figure 8.30b reveals the inside of a void, which clearly shows the individual grains. Defects introduced by machining are also elusive, but dangerous, flaws. These flaws are close to the surface. Isolated large grains, cracked grains, and inclusions are other sources of flaws. Inclusions are often the result of contamination during processing. During pressing or subsequent drying, laminations are often formed. During drying, the laminations can become separated from each other.

The second category of sources of flaws listed in Table 8.3 consists of flaws introduced during the design of the material. Ceramic components have to be designed in such a manner as to avoid sharp corners that are stress raisers. Rather large stress concentrations can be generated by notches, holes, etc. (See Chapter 7.) Regions of components under tension should be minimized because of the great differences between the compressive and tensile strength of ceramics. Design should maximize compression for this reason. It is also known that a component under compression and having a hole will be subjected to some tension (see Section 7.2.2).

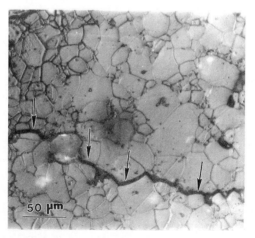

Figure 8.29 Intergranular crack produced by thermal shock (rapid cooling) of alumina. (See arrows.)

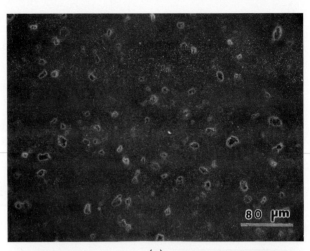

(a)

(b)

Figure 8.30 Voids in AD85 alumina. (a) Scanning electron micrograph of sectioned surface at low magnification. (b) Enlarged view of one void. These voids are larger than the grains.

The last category of sources of flaws given in Table 8.3 involves those flaws introduced during service of the material, as a product. The classic example of the flaw introduced by a rock hitting the glass windshield of an automobile is the most common. The small flaw thus created will eventually generate a crack, which, in most cases, will make the windshield unusable after a certain time. Ceramics are subjected to similar effects, by impact or other sources. Another source of flaws in ceramics is the environment, coupled with stresses. Environmentally assisted cracking in ceramics has many similarities with that in metals.

The fracture of ceramics under tensile loading is essentially dictated by linear elastic fracture mechanics. Thus, the concepts K_{Ic}, J_{Ic}, and R curve are all applicable to ceramics. Table 8.4 presents fracture toughness values for a number of ceramics; they are much lower than those of metals, shown in Table 7.1. The second-highest toughness listed in the table is that for zirconia, approximately 8 MPa $\sqrt{m}$. The highest listed is that of WC, which is bonded by Co and therefore is a metal–ceramic composite (CERMET); this is the reason for the high value of 13 MPa $\sqrt{m}$. Zirconia undergoes a tetragonal monoclinic transfor-

TABLE 8.4 Toughness Values for Ceramics[1] (Adapted with permission from *Guide to Engineered Materials,* Metals Park, OH: ASM International, p. 16).

Material	Comments	Toughness, K_c (MPa m$^{1/2}$)[1]
NaCl	Monocrystal	0.4
Soda–lime glass[2]	Amorphous	0.74 DCB
Aluminosilicate glass	Amorphous	0.91 DCB
WC	Co bonded	13.0
ZnS	Vapor deposited	1.0
Si_3N_4	Hot pressed	5.0
Al_2O_3	MgO doped	4.0
Al_2O_3 (sapphire)	Monocrystal	2.1
SiC	Hot pressed	4.0
$SiC–ZrO_2$	Hot pressed[3]	5.0
MgF	Hot pressed	0.9
MgO[2]	Hot pressed	1.2
B_4C	Hot pressed	6.0
Si	Monocrystal	0.6
ZrO_2	Ca stabilized	7.6 DCB

[1]Obtained by double torsion measurement technique, except where double cantilever beam test (DCB) is indicated.

[2]Commercial sheet glass.

[3]20% ZrO_2 14% mullite by weight. ZrO_2 present in monoclinic form; no transformation toughening.

mation at the tip of the crack, decreasing the stress concentration there. Ceramics with tetragonal zirconia particles can benefit from this transformation (see Chapter 11).

EXAMPLE 8.1

Consider polycrystalline alumina samples with two grain sizes: 0.5 and 50 μm. During cooling, the thermal expansion mismatch produces cracks that have approximate dimensions equal to the grain-boundary facets. If $K_{Ic} = 4$ MNm$^{1/2}$, determine the tensile strength of each sample.

Solution: We assume that the flaw size, i.e., $2a$, is equal to the grain size. Then

$$a_1 = \frac{0.5}{2} \times 10^{-6} \text{ m},$$

$$a_2 = 25 \times 10^{-6} \text{ m},$$

$$K_{Ic} = Y\sigma\sqrt{\pi a}.$$

We take $Y = 1.12$:

$$\sigma_1 = \frac{K_{Ic}}{Y\sqrt{\pi a}} = \frac{4 \times 10^6}{9.9 \times 10^{-4}} \approx 4 \text{ GPa},$$

$$\sigma_2 = \frac{4 \times 10^6}{9.9 \times 10^{-3}} = 400 \text{ MPa}.$$

8.3.2 Effect of Grain Size on Strength of Ceramics

Mechanical properties of ceramics are affected by grain size in several ways. The most important effect is the reduction in the sizes of inherent flaws, as the grain size is reduced. One often finds flaws in a ceramic, caused by processing, that have a characteristic size of the same order of the grain size. The fracture toughness K_{Ic} of a ceramic being an intrinsic property, the tensile stress at which a flaw will be activated is dictated by the equation

$$\sigma = K_{Ic}/\sqrt{\pi a}. \tag{8.1}$$

Since the flaw size is often established by the grain size ($2a = D$), one has

$$\sigma = \frac{K_{Ic}}{\sqrt{\pi D/2}}. \tag{8.2}$$

This factor is important in the tensile strength of ceramics.

The microindentation hardness of ceramics has also been found to be somewhat sensitive to grain size. Figure 8.31a shows microhardnesses for hot pressed (HP) and sintered (S) silicon nitride, as well as for sialon (a silicon–aluminum–oxygen–nitrogen compound). The hardness increases with a decrease in grain size (D), and the results are plotted in a Hall–Petch fashion i.e., hardness vs. $D^{-1/2}$. However, this effect is not as important as in metals.

Figure 8.31b shows the effect of grain size of the strength of alumina. The solid line represents the application of Equation 8.2. For smaller grain sizes, there are deviations (dashed lines), and other factors enter into consideration as well. Nanocrystalline ceramics possess a property of considerable technological significance: superplasticity. This property enables ceramics to undergo plastic deformation in tension and compression. Tensile elongations as high as 800% have been obtained at moderate temperatures (half the melting point of the material). Nanocrystalline TiO_2 deforms superplasticity at temperatures as low as 600°C, around 300°C lower than the submicrometer-size oxide. Nanocrystalline zirconia

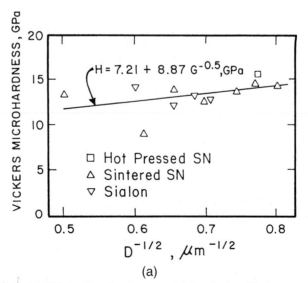

(a)

Figure 8.31 (a) Effect of grain size on microhardness of hot pressed and sintered Si_3N_4 and sialon. (From A. K. Mukhopadhyay, S. K. Datta, and D. Chakraborty, *J. European Cer. Soc.* 6 (1990) 303, Figure 3, p. 308)

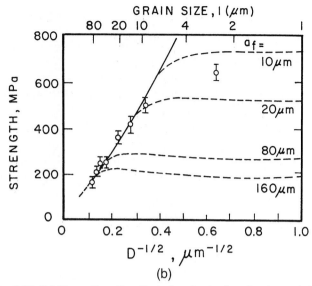

Figure 8.31 (b) Strength as function of grain size for alumina; a_f is the flaw size. The solid curve refers to flaws having size $a_f = 0.5D$ (facet flaws) and the dashed lines to flaws of fixed size. (Adapted from P. Chantikul, S. J. Bennison, and B. R. Lawn, *J. Am Cer. Soc.* 73 (1990.) p. 2419)

has been shown to exhibit superplastic strain rates 34 times faster than 0.3-μm zirconia. These results can be rationalized in terms of the decreasing distance between the grain boundaries, helping plastic deformation by both Coble (grain-boundary sliding) or Nabarro–Herring creep, each of which is described in detail in Chapter 15. Nabarro–Herring creep predicts a strain rate that is a function of D^{-3}, whereas grain-boundary creep predicts a strain rate that varies with D^{-2}. Clearly, the strain rate in creep is a strong function of the grain size D.

8.3.3 Fracture of Ceramics in Tension

Most often, tensile stresses produce mode I fracture in ceramics. Such tensile stresses can be generated by actual tensile testing or by flexural testing. Flexural testing produces a tensile stress in the outer layers of the specimen. The crack propagation path is the one that requires the least energy, and intergranular fracture is often observed in ceramics. Figure 8.32 shows an intergranular fracture in alumina produced by bending. The fracture follows, for the most part, the grain boundaries, although transgranular fracture is also observed in some places. Figure 8.33 shows a primarily intergranular fracture in TiB$_2$.

In single-crystal alumina (sapphire), there are no grain boundaries; therefore, the fracture cannot be intergranular, but will instead propagate through the crystal. In such a monocrystal, different crystallographic planes have different surface energies, and fracture will occur on those planes with the least energy. For polycrystalline alumina, the tensile strength is approximately 0.20 GPa, while the tensile strength of single-crystal alumina is 7–15 GPa. This is fairly close to the theoretical strength. The fracture of sapphire usually does not occur along the basal plane, because the surface energy of the (0001) plane is very high. Separation along this plane is difficult, as it is not electrostatically neutral. The basal plane can be visualized as consisting of oxygen atoms (see Fig. 4.42). Thus, basal plane (0001) fracture would necessitate the separation of oppositely charged ions between planes and would require great energy. Table 8.5 shows the surface energies for sapphire along different planes. From these values, one can see that the {1010} and {1012} planes would be the preferred fracture planes. These high values of surface energy also explain why

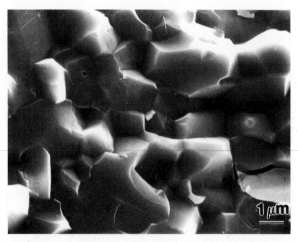

Figure 8.32 Scanning electron micrograph of fracture surface in 99.4% pure alumina. Fracture is primarily intergranular.

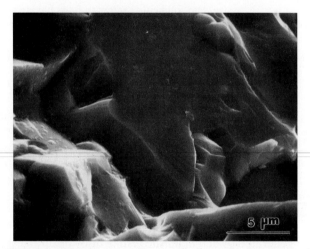

Figure 8.33 Scanning electron micrograph of tensile fracture surface in TiB$_2$. Fracture is primarily transgranular.

TABLE 8.5 Fracture Surface Energy of Sapphire at Room Temperature[a]

Fracture Plane	Fracture Surface Energy (J/m²)
{0001}	>40
{1010}	7.3
{1012}	6
{1126}	24.4

[a]From S. M. Wiederhorn, *J. Am. Cer. Soc.* 50 (1967) 407, Table I, p. 486.

fracture in polycrystals tends to be intergranular. Another reason is that, in anisotropic materials (materials that are anisotropic in their elastic constants or thermal expansion coefficients), the grain boundaries are regions of stress concentration in which the initiation of a fracture is more likely to occur than in other regions. Figure 8.34 shows a scanning electron micrograph of a fracture surface in a single crystal of sapphire. The flat surfaces are the planes where the surface energy is the lowest.

According to fracture mechanics, internal flaws intensify the externally applied forces; furthermore, this intensification factor depends on the size of the flaw. Thus, specimens with different flaw size distributions will have different strengths. It is well known that the tensile strength of ceramics shows a much greater variability than that of metals. While the yield stress of most metals shows a standard deviation of 5% or less, the tensile strength of ceramics often shows a standard deviation of 25%. The great variation in results from test to test necessitates the use of statistics. In this regard, Weibull's contribution is universally recognized.[4] The Weibull analysis is described in Chapter 7; it is sufficient here to give the basic equation for the probability that a specimen of volume V will not fail at an applied tensile stress σ. This equation is (see Eqns. 7.38 and 7.41a)

$$P(V) = \exp\left[-\frac{V}{V_0}\left(\frac{\sigma - \sigma_u}{\sigma_0}\right)^m\right]$$

where σ_0 and V_0 are normalizing parameters, σ_u is the stress below which fracture is assumed to have zero probability, and m is called the *Weibull modulus*, a measure of the variability of the strength of the material: The greater m, the more variation there is in the strength. The Weibull modulus for ceramics is usually between 5 and 20. The equation also shows that the strength of a ceramic decreases as its volume increases. This is due to a greater probability of finding large flaws in a large specimen than in a small one. The important conclusion that can be drawn is that *it is the largest flaw that determines failure.*

The fracture toughness of monolithic ceramics varies between 1 and 5 MPa $\sqrt{m}$. This toughness is dictated by the strength of the material's interatomic bonds, since little plastic deformation is involved in propagating a crack. Many methods can be used to enhance fracture toughness. However, ceramics retain sharp cracks and low ductility. Note that,

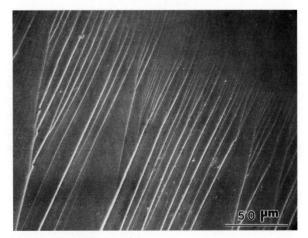

Figure 8.34 Scanning electron micrograph of fracture surface in sapphire (monocrystalline alumina).

[4]W. Weibull, *J. Appl. Mech.*, 18 (1951) 293.

although one idealizes the propagation of a crack as an isolated event in a perfectly elastic material, it has been found that the stresses set up at the tip of the crack tend to generate microcracks. These microcracks change the stress field ahead of the major crack, altering its response to the applied load.

8.3.4 Fracture in Ceramics Under Compression

Fracture under tension is easy to understand, since it involves the formation of cracks (mainly at imperfections in the material) and their propagation. When a brittle metal, an intermetallic compound, or a ceramic is subjected to compression, it will eventually fail, although at stresses much higher than the tensile strength. How does it fail, since we know that cracks propagate (in mode I) under tension only? Several mechanisms have been proposed, and they all involve the formation of localized regions of tension in the material, caused by the interaction of the externally applied compressive stresses with microstructural defects. Figure 8.35a shows a specimen of grout (cement and sand) that failed in compression. The cracks are aligned primarily with the compression axis. The student can reproduce this type of failure by taking a piece of chalk and compressing it in a clamp or vise. The same pattern of cracks will form. This failure mode is called *axial splitting* and is very prominent for unconfined brittle materials. The sequence of events leading from the activation of existing flaws to the growth of cracks, their coalescence, and the formation of slender columns under compression is shown in Figure 8.35b. The columns become unstable and buckle under the applied compressive loads, ejecting fragments, increasing the load on the remaining specimen, and leading to complete failure.

Griffith was the first to propose a mechanism for the compressive fracture of brittle materials.[5] The mechanism is shown in Figure 8.36. It is based on a preexisting crack of length $2a$ oriented at an angle Ψ to the highest compressive stress (σ_c). This compressive stress will cause a shear stress acting on the opposite faces of the preexisting flaw. Thus, sliding of the two surfaces will take place. At the ends of the flaw, this sliding is prevented. This will lead to a localized tensile stress ahead of flaw (marked by a plus sign in Figure 8.36a) that will, eventually, nucleate two cracks (Figure 8.36b). Initially, the cracks will grow at an angle of 70° to the face of the flaw and will then align themselves with the direction of the maximum compressive stress (Figure 8.36b). The equations developed by Griffith, called the Griffith criterion, are given in Section 3.7.5. They predict a compressive strength for brittle material eight times larger than the tensile strength. The mathematical analysis of the stresses created at the end of the flaw is based on the scheme shown in Figure 8.36c. Normal and shear stresses σ'_{22} and σ'_{12} are determined in the plane of the flaw. A frictional resistance μ can be assumed at the flaw surfaces. The wing cracks have length l in Figure 8.36c. A simpler situation is when the flaw is spherical. In this case, (tangential) tensile stresses generated at the north and south pole of the flaw can generate cracks. (See Figure 8.36d.) The introduction of lateral stresses σ_l (also called lateral confinement) alters the propagation of wing cracks and their interaction and has a profound effect on final failure.

Failure of brittle materials under compression is activated by existing flaws. Brittle metals (cast iron, intermetallics), ceramics, ceramic composites, concrete, and rock are subjected to these mechanisms. Spherical voids and sharp (cracklike) flaws are often produced during the processing of brittle materials. For example, spherical voids are generated during sintering and hot pressing of ceramic powders and are the remnants of the material's initial porosity. Microcracks are created by thermal expansion mismatch (especially in noncubic materials). Frequently, the scale of the microcracks is that of the grain size; they tend to extend from boundary to boundary. The compressive failure of brittle materials is

[5]A. A. Griffith, *Proc. First Int. Cong. App. Mech.*, 1 (1924) 55.

(a)

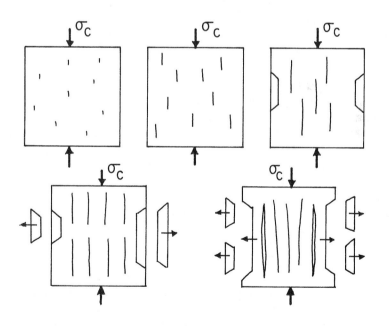

(b)

Figure 8.35 (a) Compressive failure of brittle material by axial splitting.
(b) Schematic representation of growth of critical cracks, producing axial
splitting and spalling of fragments; separate columns under compression
will collapse.

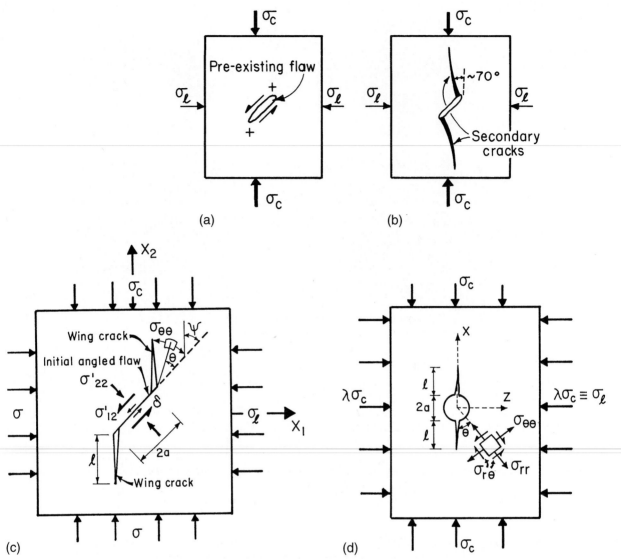

Figure 8.36 (a) Schematic representation of elliptical flaw subjected to compressive stress σ_c; σ_1 is lateral stress. (b) Formation of "wing" cracks from ends of flaw. (c) Stresses generated by flaw of orientation Ψ with compressive axis. (Adapted from M. F. Ashby and S. D. Hallam, *Acta. Met.*, 34 (1986) 497) (d) Circular flaw generating crack. (Adapted from C. G. Sammis and M. F. Ashby, *Acta Met.*, 34 (1986) 511)

strongly affected by lateral confinement (stresses transverse to the loading direction). Figure 8.37 shows how cracks aligned with the principal direction of loading are generated at spherical and sharp flaws and how they lead to failure. The stress–strain curves and the interactions between the cracks are dependent on the lateral confinement of the material, which is increased from left to right. In the absence of confinement ($\sigma_\ell = 0$), the cracks generated at flaws can grow indefinitely under increasing compressive stress σ_c. They split the specimen vertically, and the segments become unstable and crumble—for instance, from Euler instability. As confinement is increased, the growth of cracks is hindered, and

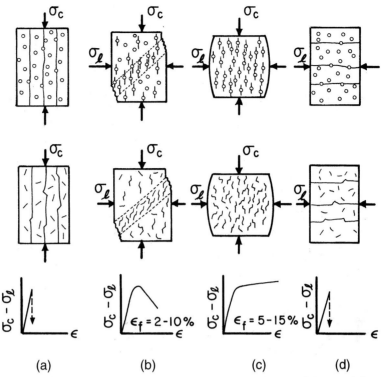

Figure 8.37 Failure modes in compression for brittle materials containing spherical and flat flaws, as a function of increasing confinement (σ_ℓ/σ_c). (a) Simple compression, giving failure by "axial splitting," or "slabbing". (b) Small confining stress, resulting in shear failure. (c) Large confining stresses σ_ℓ, providing homogeneous microcracking and a "pseudoplastic" response. (d) σ_c becomes equal to zero; the situation is identical to (a), but rotated by 90°. (Adapted from C. G. Sammis and M. F. Ashby, *Acta Met.*, 34 (1986) 511 (Figure 1, p. 512) and M. F. Ashby and S. D. Hallam, *Acta Met.*, 34 (1986) 497 (Figure 1, p. 498)).

failure occurs along a band of shear localization, where a larger number of cracks is formed. At a still larger confinement (Figure 8.37c) the brittle material exhibits a "pseudo-plastic" response, with numerous flaws activating cracks. Finally, in Figure 8.37d, axial splitting (also called "slabbing") occurs at 90° to the first case (Figure 8.37a).

Calculations analogous to those of Griffith were carried out for elliptical flaws by Ashby and Hallam[6] and Horii and Nemat–Nasser.[7] The Ashby–Hallam equations are given here.

When the friction coefficient μ is zero, the angle ψ for which K_I is maximum is, as expected, 45°. By making $\sigma_\ell/\sigma_c = \lambda$, the following equation is obtained:

$$K_I = -\frac{\sigma_c\sqrt{\pi a}}{\sqrt{3}}[(1 - \lambda)(1 + \mu^2)^{1/2} - (1 + \lambda)\mu].$$

[6]See M. E. Ashby and S. D. Hallam, *Acta Met.*, 34 (1986), 497.

[7]See H. Horii and S. Nemat–Nasser, *J. Geophys. Res.*, 90 (1985) 3105; and *Phil. Trans. Roy Soc.* (London), 319 (1986), 337.

The critical value of the stress intensity factor, K_{Ic}, is reached at the stress level at which wing crack growth starts. For a crack making an angle $\psi = (1/2)\tan^{-1}(1/\mu)$ with the principal loading axis, we have

$$\frac{\sigma_c\sqrt{\pi a}}{K_{Ic}} = \frac{-\sqrt{3}}{[(1-\lambda)(1+\mu^2)^{1/2} - (1+\lambda)\mu]}.$$

Ashby and Hallam also obtained an expression for the increase in length of the winged cracks, $\ell/a = L$, as a function of normalized stress:

$$\frac{\sigma_c\sqrt{\pi a}}{K_{Ic}} = \frac{-(1+L)^{3/2}}{[1-\lambda - \mu(1+\lambda) - 4.3\lambda L]\left[0.23L + \dfrac{1}{\sqrt{3}(1+L)^{1/2}}\right]}.$$

For spherical and circular flaws, the equations given in Section 7.1 can be applied. The crack grows when $K_I = K_{Ic}$, i.e., when the critical stress intensity factor reaches its critical level. This is the Sammis–Ashby equation:

$$\frac{\sigma_c\sqrt{\pi a}}{K_{Ic}} = -\frac{1}{L^{1/2}\left[\dfrac{1.1(1-2.1\lambda)}{(1+L)^{3.3}} - \lambda\right]}.$$

Under simple compression ($\lambda = 0$, $\sigma_\ell = 0$), the crack grows in a stable fashion from an initial value of normalized stress equal to 4. In an initial stage, from $L = 0$ to $L = 0.2$, the stress actually drops with increasing length. This corresponds to the initial "pop-in" stage of crack formation. Since K_{Ic} is a material constant, the compressive stress at which a crack grows decreases with increasing void size. Hence, *larger voids are more effective crack starters*. For lateral tension ($\lambda < 0$), the crack grows in a stable fashion to a certain size and then grows unstably (in the region where σ_c decreases with increasing L). The equations also show the total suppression of crack growth when $\lambda \geq \frac{1}{3}$.

Additional mechanisms involving dislocations, anisotropy of the elastic properties of adjacent grains, and dislocation–grain-boundary interactions were proposed by Lankford,[8] who studied the behavior of alumina under compression and found localized plasticity (caused by either twinning or dislocations) at stresses below the compressive failure stress. The interaction of deformation bands with grain boundaries caused microcracks to begin forming. Figure 8.38 shows a schematic indicating how microstructural anisotropy can lead to stress concentrations at the grain boundaries. If two adjacent grains have different elastic moduli along the axis of compression (because of differences in crystallographic orientation), they will tend to deform differently. This will impose additional stresses on the grain boundaries because of compatibility requirements (Fig. 8.38a). In a similar way, deformation bands (whether they be dislocations or twins) will create stress concentrations at the grain boundaries (Fig. 8.38b). Figure 8.38c shows examples of a crack produced by different interactions with a grain boundary. Thus, failure of a ceramic under compression is a gradual process, although the actual fracture often occurs in an "explosive" manner, as the ceramic fragments into many pieces due to the coalescence of microcracks.

8.3.5 Thermally Induced Fracture in Ceramics

Thermal stresses induced during cooling can have a profound effect on the mechanical strength of the ceramics. This can be explained in a qualitative manner by Figure 8.39. The polycrystalline aggregate is schematically represented by an array of hexagons. When the

[8]See J. Lankford, *J. Mater. Sci.*, 12 (1977) 791.

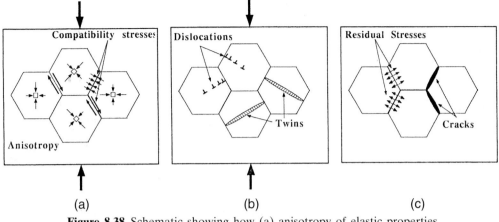

Figure 8.38 Schematic showing how (a) anisotropy of elastic properties and (b) localized plastic deformation can lead to stress concentrations and (c) cracking at grain boundaries during unloading. (After M. A. Meyers, *Dynamic Behavior of Materials* (New York: J. Wiley, 1994), p. 559)

temperature is reduced from T_2 to T_1, the hexagonal grains contract. The noncubic structure of alumina and many other ceramics results in different contractions along different crystallographic orientations. The same effect manifests itself in noncubic metals. In some metals, substantial plastic deformation is observed after thermal cycling (numerous heating and cooling cycles). The problem is especially crucial in composites, where the different components often have quite different thermal expansion coefficients. The thermal expansion coefficient along the c-axis of Al_2O_3 is about 10% higher than perpendicular to it. The stresses set up by these differences in thermal expansion are sufficient to introduce microcracks into the material after cooling. In Figure 8.39, we would have $\Delta L_1 \neq \Delta L_2$ if the grains were free. However, each grain is constrained by its neighbors, and stresses therefore arise. These stresses are given by

$$\sigma = \frac{2}{3(1 - \nu)} \int_{T_1}^{T_2} E(\alpha_c - \alpha_a)dT \,,$$

where T_1 and T_2 are the extreme temperatures of the thermal cycle, and α_c and α_a are the thermal expansion coefficients perpendicular and parallel to the c-axis, respectively. For constant expansion coefficients, and assuming a constant E, we get

$$\sigma = \frac{2E}{3(1 - \nu)}\Delta\alpha\Delta T \,.$$

Cooling a polycrystalline alumina sample from 1,020°C to 20°C would generate stresses on the order of

$$\sigma = \frac{2 \times 400}{3(1 - 0.31)}(0.7 \times 10^{-6}) \times 1,000 = 0.27 \text{ GPa} = 270 \text{ MPa} \,.$$

between two grains of orientations a and c. This is approximately 1½ times the tensile strength of alumina, as can be seen from Table 8.2. Thus, microcracks can be generated by anisotropy of an expansion coefficient. Even in the case where no microcracks are generated, internal stresses remain within the grains. When a ceramic is subjected to external

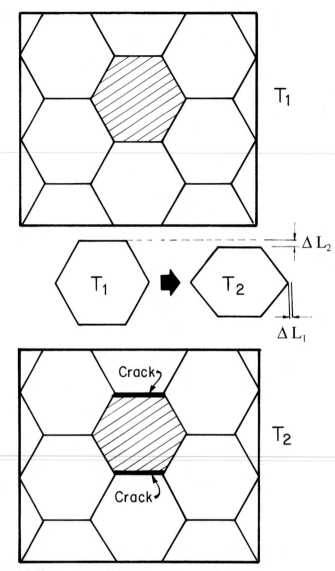

Figure 8.39 Thermally induced cracks created when grains contract in an anisotropic fashion during cooling from T_1 to T_2.

loading, the internal stresses due to thermal differences interact with the externally applied loads and can considerably reduce the stresses required for fracture.

The anisotropic effect of expansion on microcracking affects the strength of ceramics in a manner that is dependent on grain size. This effect, is illustrated in Figure 8.40. Here we assume that microcracks are generated by thermal anisotropy in the two specimens. The microcracks will extend over one grain face. The sizes of the two microcracks are ℓ_1 and ℓ_2 for the small and large grain-sized specimens, respectively. If the grain sizes are D_1 and D_2, we can say that

$$\frac{\ell_1}{D_1} = \frac{\ell_2}{D_2}.$$

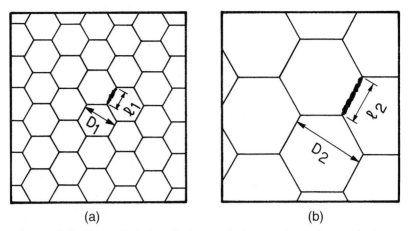

Figure 8.40 Thermally induced microcracks in ceramic specimens with two grain sizes.

The fracture mechanics equation

$$K_{Ic} = \sigma\sqrt{\pi a}$$

can then be applied to determine the tensile strength of the ceramics.

Thus, the tensile strength can be written as

$$\sigma = \frac{K_{Ic}}{\sqrt{\pi a}} = \frac{K_{Ic}}{\sqrt{\dfrac{\pi\ell}{2}}}.$$

By substituting D for ℓ and combining all constants into one, we obtain

$$\sigma = K_{Ic}\, k\, D^{-1/2}.$$

k is a parameter. This simple equation expresses the experimentally observed fact that thermal anisotropy is much more effective in weakening specimens with a large grain size than specimens with a small grain size.

Another serious problem of a thermal nature affecting ceramics is cracking, because of temperature differentials within one component. We all know that china will fracture if rapidly cooled. Ceramics are subject to very intense stress concentrations if temperature differentials are set up within them. This is so because plastic deformation, which serves to accommodate stresses due to severe temperature gradients in metals, is mostly absent in ceramics. Thus, there are limits to the rates at which components can be cooled or heated. If these rates are exceeded, the components fail. A simple example is a furnace tube that is heated to a high temperature. If the resistance wire that heats the furnace touches the ceramic, a significant temperature gradient is established over a small distance. This temperature gradient creates stresses that lead to fracture if the tensile strength of the ceramic is exceeded. It is very common for ceramic bricks (refractory bricks) to break during cooling. In ceramics thermal shock or rapid cooling can have catastrophic effects, and the superb high-temperature properties of ceramics are of no advantage if the ceramic fails during cooling. When ceramics are used in conjunction with metals in machines, the difference between the thermal expansion coefficient of the metal and that of the ceramic can lead to failure. These aspects must be considered in the design of ceramic components, and heat transfer equations should be used to estimate the temperature differentials and the associated stresses within the ceramic.

8.4 FRACTURE IN POLYMERS

The fracture process in polymers involves the breaking of inter- and intramolecular bonds. Recall that amorphous or glassy polymers have a glass transition temperature T_g, but no melting point T_m. These glassy polymers are rigid below T_g and less viscous above that temperature. Semicrystalline polymers have both a melting point and a glass transition temperature, the former referring to the crystalline phase, the latter to the amorphous phase surrounding the crystalline phase. More information about the structure of polymers is given in Chapter 1.

8.4.1 Brittle Fracture

Many polymers fracture in a brittle manner below their glass transition temperature. This is particularly true of polymers having large, bulky side groups or a high density of cross-links. Under either of these circumstances, the molecular chain structure of the polymer becomes so rigid, that chain disentanglement and/or slipping becomes very difficult. Examples of such polymers are thermosets, such as epoxy, polyester, and polystyrene. The stress–strain curve of these polymers is quite linear to fracture, and the strain to failure is typically less than 1%. Figure 8.41a, a scanning electron micrograph, shows an example of

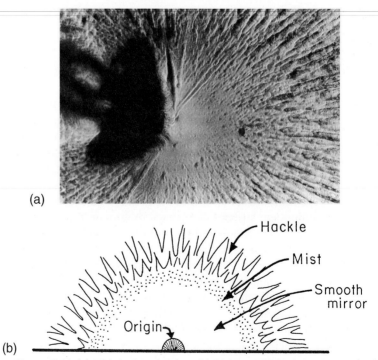

(a)

(b)

Figure 8.41 (a) Brittle fracture in a highly cross-linked thermoset (polyester). (b) The three different regions that compose the brittle fracture surface in (a).

a brittle fracture surface in a cross-linked polyester; Figure 8.41b shows schematically the different regions that compose such a surface. There are three regions:

1. A mirrorlike, or specular, region adjoining the crack nucleation site, indicating slow crack growth.
2. A coarse and flat region indicative of fast crack growth; sometimes this region is called the region of *hackle,* and one can see that the crack has propagated on different levels over small areas. When hackle is elongated in the direction of crack propagation, the pattern is called "river markings."
3. A transitional region between the preceding two that has a *misty* appearance and no resolvable features.

Similar brittle fracture surface features are observed in ceramics. (See Figure 8.28.) In highly cross-linked thermosets, such as epoxies and polyesters, the plastic deformation before fracture is negligible. Consequently, manifestations of plastic deformation, such as crazing and shear yielding, are generally not observed.

8.4.2 Crazing and Shear Yielding

Frequently, the phenomena of crazing and shear yielding precede actual fracture in a polymer. Both these phenomena involve a localization of the plastic deformation in the material. The major difference between the two is that crazing occurs with an increase in volume, whereas shear yielding occurs at constant volume.

In glassy polymers, one can regard crazing and shear yielding as competing processes. In brittle glassy polymers, such as polymethyl methacrylate (PMMA) or polystyrene (PS), crazing precedes the final brittle fracture. In comparatively more ductile polymers (for example, polycarbonate or oriented polyethylene), which have flexible main-chain linkages, shear yielding is the dominant mode of deformation, and the final fracture is ductile. In particular, if an oriented high-density polyethylene sheet is deformed in a direction oblique to the initial draw direction, it will show a shear deformation band in which highly localized plastic deformation occurs.

It is thought that molecular entanglements control the geometry of crazes and shear yield zones. A craze is a region of a polymer in which the normal "cooked-spaghettilike" chain arrangement characteristic of the amorphous state has been transformed into drawn-out molecular chains interspersed by voids. The crazed region is a very small percentage of the total region of the polymer (a few nanometers to a few micrometers). Because of the presence of voids in a craze, the plastic deformation of the small volume of material in the craze occurs without an accompanying lateral contraction; that is, the constancy-of-volume condition which holds in the regular bulk polymer does not hold in the crazed material.

A craze is neither a void nor a crack. Detailed optical- and electron-microscopic observations of crazed regions show that crazes are not voids and that they are capable of transmitting load. The refractive index of a craze in a polymer such as polycarbonate, in the dry state and after immersion in ethanol, would be different. From such measurements, it was concluded that crazes contain about 50–55% by volume of free space; that is, the density of the material in the crazed region is lower than that of bulk polymer. The lower density of the crazed region reduces the refractive index of the region and causes its characteristic reflectivity. Figure 8.42 shows a series of crazes reproduced in a tensile specimen of polycarbonate. Note that several crazes have run through the entire cross section without failure of the specimen, indicating the load-bearing nature of the crazes. The volume fraction of the polymer in the craze is inversely proportional to λ, the draw ratio (final length $\div$ original length) of the craze.

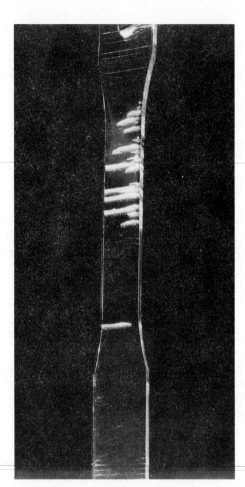

Figure 8.42 A series of crazes produced in a tensile specimen of polycarbonate. (Used with permission from R. P. Kambour, *Polymer,* 4 (1963) 143)

Although crazes are not cracks, cracks leading to final fracture may indeed start at a craze. The polymeric chains in the crazed region get highly oriented in the direction of the applied stress. The void content, as previously mentioned, can be as high as 50 to 60%. Molecular chain entanglements play an important role in controlling craze geometry. Figure 8.43 shows, schematically, craze formation at a crack tip. Crazes are usually nucleated either at surface flaws (scratches, gouge marks, and cracks) or at internal flaws (dust particles and pores). In polymers, microvoids, which are an integral part of crazes, can form at various inhomogeneities in the microstructure, such as random density fluctuations in amorphous polymers, ordered regions in semicrystalline polymers, and particulate matter or inclusions such as fillers, flame retardants, or stabilizers in either kind of polymer. Craze formation is a process of dilatation and is aided by hydrostatic tension and retarded by hydrostatic compression.

The competition between shear yielding and crazing and the importance of the microstructure are shown in Figure 8.44. Polystyrene and polyphenylene oxide (PPO) are completely miscible at all concentrations. Atactic polystyrene (APS) shows the phenomenon of crazing preceding brittle fracture. By mixing the APS and PPO, we can suppress this embrittling tendency. In fact, near 50–50 concentration, crazing in APS is completely suppressed. Instead, extensive shear yielding occurs. The figure shows this phenomenon of transition between shear yielding and crazing in 300-nm films made of blends of APS and PPO and deformed 10% at room temperature. The lower left-hand corners show the weight percentages of APS in the mixture. The letters *C, D,* and *S* indicate crazing, diffuse shear

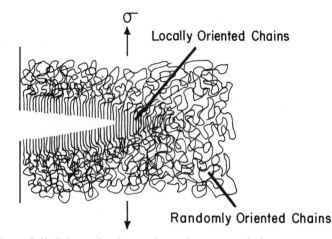

Figure 8.43 Schematic of craze formation at a crack tip.

banding, and sharp shear banding, respectively. The upper left-hand corners indicate the direction of deformation. Note that, as the amount of PPO increases, more and more crazes are blunted by shear bands. At 70% APS (or 30% PPO), only diffuse shear bands appear.

8.4.3 Fracture in Semicrystalline and Crystalline Polymers

The crystalline regions in a semicrystalline polymer have a folded chain structure; that is, the molecular chains fold back upon themselves to form thin platelets called lamellae. (See Chapter 1.) Amorphous material, containing chain ends, tie molecules, and other

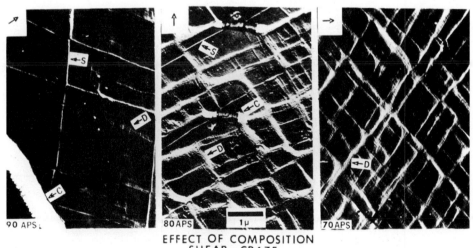

Figure 8.44 A transition between shear yielding in film blends of polypropylene oxide (PPO) and atactic polystyrene (APS) deformed 10% at room temperature (Used with permission from E. Baer, A. Hiltner, and H. D. Keith, *Science,* 235 (1987) 1015.). The APS weight percentages are shown in the lower left-hand corners. *C, D,* and *S* indicate crazing, diffuse shear, and sharp shear banding, respectively. The arrows indicate the direction of deformation.

material that is difficult to crystallize, separates the different lamellae. The properties of such semicrystalline polymers can be highly anisotropic—very strong and stiff in the main chain direction and weak in the transverse direction. Parameters such as the degree of crystallinity, molecular weight, orientation of the crystals, etc., affect the mechanical behavior in general and the fracture behavior in particular. Because the polymers show a significant amount of viscoelastic behavior at their service temperature, the strain rate has a profound effect on their fracture behavior. Figure 8.45 shows schematically the effect of strain rate on the fracture path through a spherulitic polypropylene. At low strain rates the fracture follows an interspherulitic path, while at high strain rates the fracture becomes transspherulitic.

As described in Chapter 1, polymers are generally amorphous or semicrystalline; it is almost impossible to get a 100% crystalline polymer. Invariably, there is some amorphous material in between crystalline regions, because defects such as chain ends, loops, chain folds, and entanglement are almost impossible to eliminate completely. Single crystals of *monomeric* polymers are prepared from dilute solutions or vapor phase deposition. These are transformed into polymers by means of a solid-state reaction. The technique has been successful with only certain substituted diacetylines, and that, too, in an essentially one-dimensional form, i.e., short fibers. Nevertheless, these can be used to study the behavior of single-crystal polymers. Specifically, in terms of their fracture behavior, it has been observed that single-crystal polymers cleave parallel to the chain direction because of rather weak van der Waals bonding normal to the chain and strong covalent bonding in the direction of the chain. In polydiacetylene single-crystal fibers, the fracture strength σ_f shows the following dependence on fiber diameter, d: $\sigma \propto d^{-1/2}$.[9] This is similar to the size

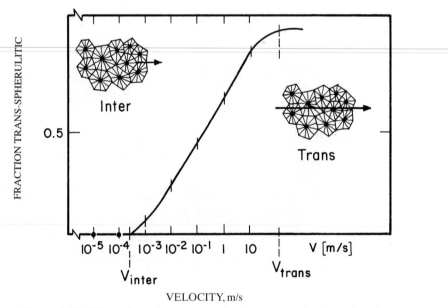

VELOCITY, m/s

Figure 8.45 Effect of strain rate on the fracture path through polypropylene. At low strain rates the fracture is interspherulitic, while at high strain rates it is transspherulitic (after J. M. Schultz, *Polym. Sci. & Eng.,* 24 (1984) 770).

---

[9]See R. J. Young, in *Developments in Oriented Polymers,* Vol. 2, I. M. Ward (ed.) (Essex, U.K.: Elsevier Applied Science, 1987), p. 1.

effect seen in other fibers; that is, preexisting defects lead to fracture, and the size of these defects is proportional to the fiber diameter.

8.4.4 Toughness of Polymers

Thermosetting polymers such as polyesters, epoxies, and polyimides are highly cross-linked and provide adequate modulus, strength, and creep resistance. But the same cross-linking of molecular chains causes extreme brittleness, i.e., very low fracture toughness. Table 8.6 gives the plane-strain fracture toughness values of some common polymers at room temperature and in air. Figure 8.46 compares some common materials in terms of their fracture toughness, as measured by the fracture energy (G_{Ic}) in J/m². Note that thermosetting resins have values only slightly higher than those of inorganic glasses. Thermoplastic resins, such as polymethyl methacrylate, have fracture energies of about 1 kJ/m², while polysulfone thermoplastics have fracture energies of several kJ/m², almost approaching those of the 7075-T6 aluminum alloy. One reason that amorphous thermoplastic polymers have higher fracture energy values is that they have free volume available, which absorbs the energy associated with crack propagation.

Many approaches have been used to improve the toughness of polymers. Alloying or blending a given polymer with a polymer of higher toughness improves the toughness

TABLE 8.6 Plane–Strain Fracture Toughness (K_{Ic}) of Some Polymers in Air at 20°C.

Polymer	K_{Ic} (MPa $\sqrt{m}$)
Epoxy, unsaturated polyester	0.6
Polycarbonate	2.2
Polystyrene	1.0
Polymethylmethacrylate (PMMA)	1.7
Polyethylene	
High density	2.1
Medium density	5.0
Nylon	2.8
Polyvinyl chloride (PVC)	2.5

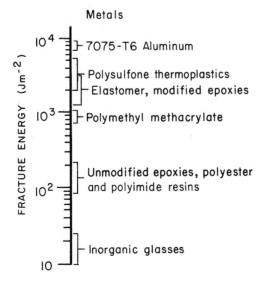

Figure 8.46 Fracture energy (G_{Ic}) of some common materials. (After R. Y. Ting, in *The Role of Polymeric Matrix in the Processing and Structural Properties of Composites* (New York: Plenum Press, 1983), p. 171)

of the polymer. Among the well-known modified thermoplastics are acrylonitrile–butadiene–styrene (ABS) copolymer, high-impact polystyrene (HIPS), and nylon containing a polyolefin. Copolymerization can also lead to improved toughness levels. Generally, thermoplastics are tougher than thermosets, but there are ways to raise the toughness level of thermosets to that of thermoplastics or even higher. One such approach involves the addition of rubbery, soft particles to a brittle thermoset. For example, a class of thermosetting resins that comes close to polysulfones, insofar as toughness is concerned, is the elastomer-modified epoxies. Elastomer- or rubber-modified thermosetting epoxies make multiphase systems, i.e., a kind of composite. Small (a few micrometers or less), soft, rubbery inclusions distributed in a hard, brittle epoxy enhance its toughness by several orders of magnitude. The methods of incorporation of elastomeric particles can be simple mechanical blending of the soft, rubbery particles and the resin or copolymerization of a mixture of the two. Mechanical blending allows only a small amount (less than 10%) of rubber to be added, whereas larger amounts can be added during polymerization. Figure 8.47 shows toughness as a function of temperature for an unmodified epoxy and a rubber-modified epoxy. Note the higher toughness and enhanced temperature dependence of the rubber-modified epoxy. Epoxy and polyester resins can also be modified by introducing carboxyl-terminated butadiene–acrylonitrile copolymers (CTBNs). Figure 8.48 shows the

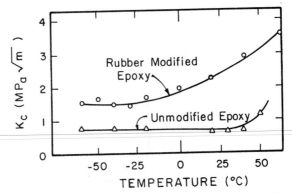

Figure 8.47 Fracture toughness as a function of temperature of unmodified epoxy and rubber-modified epoxy (After J. N. Sultan and F. J. McGarry, *Polymer Eng. Sci.*, 13 (1973) 29)

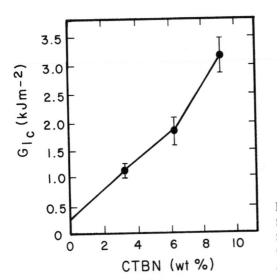

Figure 8.48 Increase in fracture energy as a function of percent weight of carboxyl-terminated butadiene–acrylonitrile (After A. K. St. Clair and T. L. St. Clair, *Int. J. Adhesion and Adhesives*, 1 (1981) 249)

increase in fracture surface energy of an epoxy as a function of the percent weight of CTBN elastomer. Toughening of glassy polymers by elastomeric additions involves different mechanisms in different polymers. Among the proposed mechanisms for this enhanced toughness are triaxial dilatation of the rubber particles at the crack tip, particle elongation, craze initiation, and shear yielding of the polymer.

Like the fracture toughness of a metal or a ceramic, the fracture toughness of a polymer is a sensitive function of its microstructure and test temperature. Most polymers, however, are viscoelastic, and this time-dependent property can influence their fracture toughness as well. The data in Table 8.6 were obtained at ambient temperature, and we see that the toughness range for polymers is 1–5 MPa $\sqrt{m}$, compared to 10–100 MPa $\sqrt{m}$ for metals and 1–10 MPa $\sqrt{m}$ for ceramics. In an elastic or time-independent material, fracture toughness is independent of the crack velocity; in a viscoelastic or time-dependent material, steady-state crack growth can occur at an applied stress intensity that is less than the critical value. Figure 8.49a shows this schematically, while Figure 8.49b shows an actual curve of stress intensity vs. crack velocity for PMMA. Note that the data are plotted on a log–log scale. A semi-log plot of the same curve for PMMA to much higher crack velocities is shown in Figure 8.50. The same trend is observed in metals, where the yield stress increases with strain rate.[†]

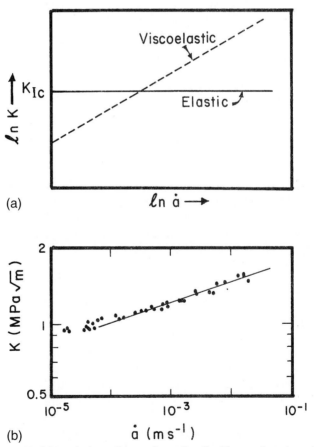

Figure 8.49 (a) Variation of stress intensity (ln K) as a function of crack velocity (ln $\dot{a}$) for an elastic and viscoelastic material. (b) Stress intensity (K) vs. crack velocity (ln $\dot{a}$) for PMMA. K_{Ic} corresponds to a crack velocity of several hundred ms⁻¹. (After G. P. Marshall, L. H. Coutts, and J. G. Williams, *J. Mater. Sci.*, 9 (1974) 1409)

[†]e.g., M. A. Meyers, *Dynamic Behavior of Material*, New York: J. Wiley, 1994.

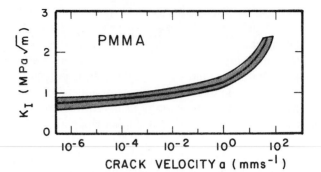

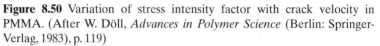

Figure 8.50 Variation of stress intensity factor with crack velocity in PMMA. (After W. Döll, *Advances in Polymer Science* (Berlin: Springer-Verlag, 1983), p. 119)

EXAMPLE 8.2

Describe how the phenomenon of crazing can be exploited to improve the toughness of a polymer.

Answer: Craze formation requires energy. Thus, if we increase the number of crazes nucleated, but do not allow them to grow to fracture, we can improve the toughness of a polymer. Such a mechanism is made use of in acrylonitrile butadiene styrene (ABS), which has a much higher toughness than polystyrene (PS). The acrylonitrile and styrene form a single-phase copolymer. Butadiene is dispersed in this copolymer matrix as elastomeric particles. These particles have a layer of styrene–acrylonitric grafted onto them. Thus, ABS has a two-phase structure. When ABS is stressed, crazes nucleate at rather low strains at the elastomer–styrene interface. However, the high extensibility of elastomeric particles inhibits the growth of these crazes. A small and uniform particle size aids in producing a high density and an even distribution of crazes in ABS. The reader can easily verify this phenomenon by bending a thin strip of ABS. It will become white, called *stress whitening,* due to the formation of a large number of crazes.

8.5 FRACTURE MECHANISM MAPS

Data presented in the form of mechanism maps can be very useful, inasmuch as such maps organize information that is widely scattered. The idea of mechanism maps is just an extension of the concept of phase diagrams in alloy chemistry, in which different phases coexisting in multicomponent systems are represented as a function of composition and temperature.

Fracture mechanism maps provide information about mechanical properties in a compact form. With these maps, one can plot normalized tensile strength σ_{UTS}/E against the homologous temperature T/T_m. Regions of different types of fracture are classified on the basis of fractography, or fracture–time or fracture–strain, *studies.* Figure 8.51 shows examples of fracture mechanism maps. Such maps can be developed for metals (see Figure 8.51a for nickel) as well as ceramics (see Figure 8.51b for alumina). One can also plot stress intensity factor against temperature and obtain information about crack growth during the fracture process.

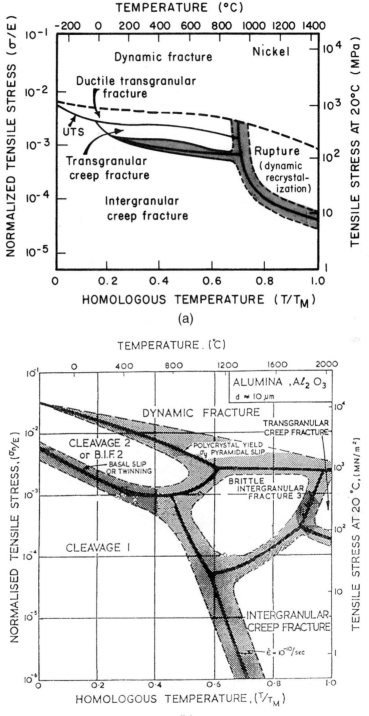

Figure 8.51 (a) Fracture mechanism map for nickel tested in tension; shading indicates a mixed mode of fracture. (Adapted from M. F. Ashby, G. Gandhi, and D. M. R. Taplin, *Acta Met.,* 27 (1979), 699, Figure 7, p. 707) (b) Fracture mechanism map for alumina with grain size of 10 μm. (Adapted from C. Gandhi and M. F. Ashby, *Acta Met.,* 27 (1979), 1565, Figure 21, p. 1591)

EXERCISES

8.1   In Figure 8.6, mechanical twinning has generated microcracks that, in subsequent tensile tests, weakened the specimen. The ultimate tensile strength of tungsten is 1.2 GPa, and its fracture toughness is approximately 70 MPa m$^{1/2}$. By how much is the fracture stress decreased due to the presence of the microcracks?

8.2   Explain why FCC metals show a ductile fracture even at low temperatures, while BCC metals do not.

8.3   Show, by a sequence of sketches, how the neck in pure copper and in copper with 15% volume fraction of a second phase will develop. Using values from Figure 8.17, show the approximate configuration of the final neck.

8.4   Alumina specimens contain flaws introduced during processing; these flaws are, approximately, the grain size. Plot the fracture stress vs. grain size (for grains below 200 μm), knowing that the fracture toughness for alumina is equal to 4 MPa m$^{1/2}$.

8.5   Calculate the theoretical cleavage stress for sapphire (monocrystalline Al_2O_3) along its four crystallographic orientations. (See Table 8.5.)

8.6   **(a)**   Calculate the compressive strength for a ceramic containing crack of size 100 μm. Let μ be the coefficient of friction between the flaw walls and l be the length of the cracks generated at the ends of an existing flaw. (See Figure 8.36.) Assume that the onset of failure corresponds to a value of $l = 2a$ and that $\mu = 1$, and that $K_{Ic} = 4$ MNm$^{1/2}$.

   (b)   Compare the tensile strength with the compressive strength that you obtained in part (a).

8.7   A ceramic with $K_{Ic} = 4$ MPam$^{1/2}$ contains pores with radii $a = 5$ μm due to incomplete sintering. These pores lead to a decrease in the failure stress of the material in both tension and compression. One in every ten grain-boundary junctions contains a void; the grain size of the ceramic is 50 μm. The ceramic fails in compression when the length l of each crack generated at the voids equals one-half of the spacing between the voids.

   (a)   Determine the compression strength of the ceramic, using the equation from Sammis and Ashby.

   (b)   Determine the tensile strength of the ceramic, assuming flaws with size a.

8.8   Using the micrographs of Figure 8.30, establish, for Al_2O_3 ($K_{Ic} = 2.5$ MPam$^{1/2}$), (a) the strength in compression, using the Sammis–Ashby equation from the previous problem, and (b) the tensile strength.

8.9   Tempering is the treatment given to flat glass (e.g., the glass window in the oven in your kitchen) by quenching the glass in a suitable liquid. Draw schematically the stress distribution in such a glass as a function of the thickness of a glass sheet. Discuss the significance of the stress distribution obtained in tempered glass.

8.10   Estimate the internal thermal stress generated in a polycrystalline sample of titanium dioxide for $\Delta T = 1,000°C$. Young's modulus for $TiO_2 = 290$ GPa, and the expansion coefficients along the direction a and c are:

$$\alpha_a = 6.8 \times 10^{-6}\ K^{-1},$$

$$\alpha_c = 8.3 \times 10^{-6}\ K^{-1}.$$

8.11 Si_3N_4 has a surface energy equal to 30 J/m^2 and an atomic spacing $a_0 \approx 0.2$ nm. Calculate the theoretical strength of this material (see Chapter 1), and compare the value you get with the one experimentally observed in tension testing ($\sigma = 550$ MPa). Calculate the flaw size that would cause this failure stress.

8.12   The theoretical density of a polymer is 1.21 g cm^{-3}. By an optical technique, it was determined that the crazed region in this polymer had 40% porosity. What is the density of the crazed region? Can you estimate the elastic modulus of the crazed region as a percentage of the modulus of the normal polymer?

8.13   A polycarbonate sample showed a craze growth length and time relationship of

$$l = k \log(t/t_0),$$

where l is the craze length at time t, t_0 is the time crazing is initiated after the application of the load, and k is a constant. For a given temperature and stress, find the rate of craze growth. Comment on the implications of the relationship that you obtain.

8.14 Craze formation is a plastic deformation mechanism that occurs without lateral contraction. What can you say about the Poisson ratio of the crazed material?

8.15 The velocity of a crack in a material submerged in an aggressive medium such as humid air can be represented by

$$V = \frac{da}{dt} = 0.5K_I^{20}.$$

Using the relationship $K_I = \sigma\sqrt{\pi a}$, compute the time to failure for this material. K_{IC} for the material is 5 MPa $\sqrt{m}$.

8.16 For a silica-based glass, the following data are available for a $V = AK_I^n$ type of relationship:

Relative Humidity	Preexponential constant A	Crack velocity exponent n
10%	2.8	25
100%	4.0	22

Take $K_{Ic} = 1$ MPa $\sqrt{m}$. For a crack length $a = 1$ nm, compute the fracture strength σ_c in an inert atmosphere. Then compute the lifetime of the material under $0.3\sigma_c$ in 10% and 100% relative humidity.

8.17 The stable, slow crack growth in a polymer in an aggressive environment can be represented by

$$\frac{da}{dt} = 0.03K_I^2,$$

where a is the crack length in meters, t is the time in seconds, and K_I is the stress intensity factor in MPa $\sqrt{m}$. K_{Ic} for this polymer is 5 MPa $\sqrt{m}$. Calculate the time to failure under a constant applied stress of 50 MPa. Use $K_I = \sigma\sqrt{\pi a}$.

8.18 It has been observed experimentally that, in cold-worked brass under stress-corrosion conditions, crack propagation is adequately described by

$$\frac{da}{dt} = AK^2,$$

where A is a constant and the other symbols have their normal significance. Derive an expression for the time to failure of the material, t_f, in terms of A, the applied stress σ, the initial crack length a_0, and the critical stress intensity corresponding to a_f (i.e., K_{Ic}).

9

Fracture Testing

9.1 INTRODUCTION

Fracture of any material (be it a recently acquired child's toy or a nuclear pressure vessel) is generally an undesirable happening, resulting in economic loss, an interruption in the availability of a desired service, and, possibly, damage to human beings. Besides, one has good, technical reasons to do fracture testing: to compare and select the toughest (and most economical material) for given service conditions; to compare a particular material's fracture characteristics against a specified standard; to predict the effects of service conditions (e.g., corrosion, fatigue, stress corrosion) on the material toughness; and to study the effects of microstructural changes on material toughness. One or more of these reasons for fracture testing may apply during the design, selection, construction, and/or operation of material structures. There are two broad categories of fracture tests; qualitative and quantitative. The Charpy impact test exemplifies the former, and the plane-strain fracture toughness (K_{Ic}) test illustrates the latter. We describe briefly important tests in both of these categories.

9.2 IMPACT TESTING

We saw in Chapter 7 that stress concentrations, like cracks and notches, are sites where failure of a material starts. It has been long appreciated that the failure of a given material in the presence of a notch is controlled by the material's fracture toughness. Many tests have been developed and standardized to measure this "notch toughness" of a material. Almost all are qualitative and comparative in nature. As pointed out in Chapter 7, a triaxial stress state, high strain rate, and low temperature all contribute to a brittle failure of the material. Thus, in order to simulate most severe service conditions, almost all of these tests involve a notched sample, to be broken by impact over a range of temperatures.

9.2.1 Charpy Impact Test

The Charpy V-notch impact test is an ASTM standard. The notch is located in the center of the test specimen, which is supported horizontally at two points. The specimen receives an impact from a pendulum of a specific weight on the side opposite that of the notch (Figure 9.1). The specimen fails in flexure under impact.

The energy absorbed by the specimen when it receives the impact from the hammer is equal to the difference between the potential energies of the hammer before and after impact. If the hammer has mass m, then

$$E_f = mg(h_0 - h_1),$$

where E_f is the sum of the energy of plastic deformation, the energy of the new surfaces generated, and the vibrational energy of the entire system; h_0 is the initial height of the hammer; h_1 is the hammer's final height; and g is the acceleration due to gravity. Of these, the first is the most significant term, and it can be assumed that the Charpy energy is

$$CV \approx mg(h_0 - h_1). \tag{9.1}$$

At impact with the specimen, the hammer has a velocity (the student should consult his or her physics textbook)

$$v = (2gh_0)^{1/2}.$$

For a difference in height of 1 m,

$$v = 4.5 \text{m/s}.$$

If we assume that the average length over which plastic deformation takes place is 5 mm, we have

$$\dot{\varepsilon} = \frac{v}{L} \approx 10^3 \, s^{-1}.$$

We see, then, that the strain rate in a Charpy test is very high.

In the region around the notch in the test piece, there exists a triaxial stress state due to a plastic yielding constraint there. This triaxial stress state and the high strain rates enhance the tendency toward brittle failure. Generally, we present the results of a Charpy test as the energy absorbed in fracturing the test piece. An indication of the tenacity of the material can be obtained by an examination of the fracture surface. Ductile materials show a fibrous aspect, whereas brittle materials show a flat fracture.

A Charpy test at only one temperature is not sufficient, however, because the energy absorbed in fracture drops with decreasing test temperature. Figure 9.2 shows this variation in the energy absorbed as a function of temperature for a steel in the annealed and in the quenched and tempered states. The temperature at which a change occurs from a high-energy fracture to a low-energy one is called the *ductile–brittle transition temperature* (DBTT). However, since, in practice, there occurs not a sharp change in energy, but instead,

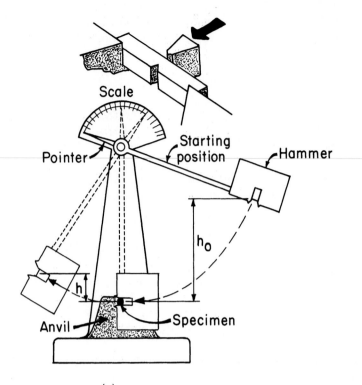

(a)

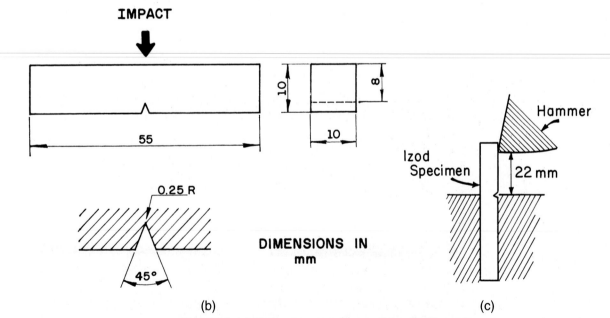

(b)   (c)

Figure 9.1 (a) Charpy impact testing machine. (b) Charpy impact test specimen. (c) Izod impact test specimen.

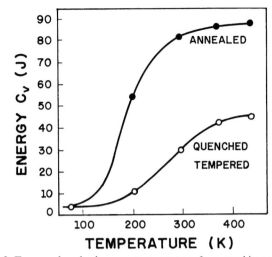

Figure 9.2 Energy absorbed versus temperature for a steel in annealed and in quenched and tempered states. (Adapted with permission from J. C. Miguez Suarez and K. K. Chawla, *Metalurgia-ABM, 34* (1978) pp. 825–829.)

a transition zone, it becomes difficult to obtain this DBTT with precision. Figure 9.3 shows how the morphology of the fracture surface changes in the transition region. The greater the fraction of fibrous fracture, the greater is the energy absorbed by the specimen. A brittle fracture has a typical cleavage appearance and does not require as much energy as a fibrous fracture. BCC and HCP metals or alloys show a ductile–brittle transition, whereas FCC structures do not. Thus, generally a series of tests at different temperatures is conducted that permits us to determine a transition temperature. This temperature,

Figure 9.3 Effect of temperature on the morphology of fracture surface of Charpy steel specimen. Test temperatures $T_a < T_b < T_c < T_d$. (a) Fully brittle fracture. (b, c) Mixed-mode fractures. (d) Fully ductile (fibrous) fracture.

however arbitrary, is an important parameter in the selection of materials, from the point of view of tenacity, or the tendency of occurrence of brittle fracture. Because the transition temperature is, generally, not very well defined, there exist a number of empirical ways of determining it, based on a certain absorbed energy (e.g., 15 J), change in aspect of the fracture (e.g., the temperature corresponding to 50% fibrous fracture), lateral contraction (e.g., 1%) that occurs at the notch root, or lateral expansion of the specimen. The transition temperature depends on the chemical composition, heat treatment, processing, and microstructure of the material. Among these variables, grain refinement is the only method that results in both an increase in strength of the material in accordance with the Hall–Petch relation and, at the same time, a reduction in the transition temperature (see Section 8.2.2). Heslop and Petch[1] showed that the transition temperature T_c depended on the grain size D according to the formula

$$\frac{dT_c}{d \ln D^{1/2}} = -\frac{1}{\beta},$$

where β is a constant. Thus, a graph of T_c against $\ln D^{1/2}$ will be a straight line with slope $-1/\beta$.

In Figure 9.4, the fraction of the fracture area that is cleavage and the lateral expansion of the Charpy specimen are plotted, in addition to the energy absorbed by the hammer. The excellent correlation among the three curves is plain, and this test simulates the dynamic response of a metal.

Figure 9.1(c) shows a second specimen geometry also commonly used (especially for plastics) in the same experimental configuration as the Charpy test. It is called the "Izod" specimen. The cross section (10 × 10 mm) and V-notch geometry of the specimen are identical, but one of the sides is longer. The specimen is held up vertically, and the notch is, in this case, on the same side as the impact.

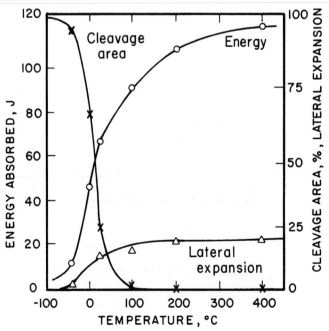

Figure 9.4 Results of Charpy tests for AISI 1018 steel (cold drawn).

[1]J. Heslop and N. J. Petch, Phil. Mag., 3 (1958) 1128.

9.2.2 Drop-Weight Test

The drop-weight test is used to determine a reproducible and well-defined ductile–brittle transition in steels. The specimen consists of a steel plate containing a brittle weld on one surface. A cut is made in the weld with a saw to localize the fracture (Figure 9.5). The specimen is treated as a "simple edge-supported beam" with a stop placed below the center to limit the deformation to a small amount (three percent) and prevent general yielding in different steels. The load is applied by means of a freely falling weight striking the side of the specimen opposite to the crack starter. Tests are conducted at 5-K intervals, and a break/no break temperature, called the *nil ductility transition (NDT) temperature,* is determined. The NDT temperature is thus the temperature below which a fast, unstable fracture (i.e., brittle fracture) is highly probable. Above that temperature, the toughness of the steel increases rapidly with temperature. This transition temperature is more precise than the Charpy-based transition temperature. The drop-weight test uses a sharp crack that moves rapidly from a notch in a brittle weld material, and thus, the NDT temperature correlates well with the information from a K_{Ic} test, described in Section 9.3. The drop-weight test provides a useful link between the qualitative "transition temperature" approach and the quantitative "K_{Ic}" approach to fracture.

The test affords a simple means of quality control through the NDT temperature, which can be used to group and classify various steels. For some steels, identification of the NDT temperature indicates safe minimum operating temperatures for a given stress. That the drop-weight NDT test is more reliable than a Charpy V-notch value of the transition temperature is illustrated in Figure 9.6 for a pressure-vessel steel. The vessel fractured in an almost brittle manner near its NDT temperature, although, according to the Charpy curve, it was still very tough.

The drop-weight test is applicable primarily to steels in the thickness range 18 to 50 mm. The NDT temperature is unaffected by section sizes above about 12 mm; because of the small notch and the limited deformation due to brittle weld bead material, sufficient notch-tip restraint is ensured.

9.2.3 Instrumented Charpy Impact Test

The Charpy impact test described in Section 9.2.1 is one of the most common tests for characterizing the mechanical behavior of materials. The principal advantages of the test are the ease of preparation of the specimen, the execution of the test proper, speed, and low cost. However, one must recognize that the common Charpy test basically furnishes information of only a comparative character. The transition temperature, for example, depends on the thickness of the specimen (hence, the need to use standard samples); that is, this transition temperature can be used to compare, say, two steels, but it is not an absolute material property. Besides, the common Charpy test measures the total energy absorbed (E_T), which is the sum of the energies spent in initiation (E_i) and in propagation (E_p) of the crack

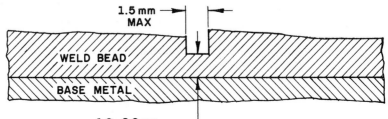

Figure 9.5 Drop-weight test specimen.

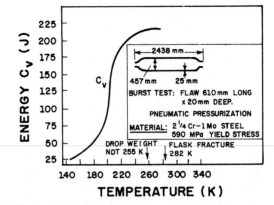

Figure 9.6 Charpy V-notch curve for a pressure-vessel steel. Note that the NDT temperature determined by the drop-weight test corresponds to the high-tough region of the Charpy curve. (after W. J. Langford, *Can. Met. Quart.*, 19 (1980) 13)

(i.e., $E_T = E_i + E_p$). In view of this problem, a test called the *instrumented Charpy impact test* has been developed. This test furnishes, besides the absorbed energy, the variation in the applied load with time. The instrumentation involves the recording of the signal from a load cell on the pendulum by means of an oscilloscope in the form of a load–time curve of the test sample. Figure 9.7a shows a typical oscilloscope record, and Figure 9.7b shows a schematic representation of that record. This type of curve can provide information about the load at general yield, maximum load, load at fracture, and so on. The energy spent in impact can also be obtained by integration of the load–time curve. From this curve, one can obtain the energy of fracture if the velocity of the pendulum is known. Assuming this velocity to be constant during the test, we can write the energy of fracture as

$$E' = V_0 \int_0^t P \, dt, \tag{9.2}$$

where E' is the total fracture energy, based on the constant velocity of the pendulum, V_0 is the initial velocity of the pendulum, P is the instantaneous load, and t is the time.

In fact, the assumption that the velocity of the pendulum is constant is not valid. According to Augland,[2]

$$E_t = E'(1 - \alpha), \tag{9.3}$$

where E_t is the total fracture energy, $E' = V_0 \int_0^t P \, dt$, $\alpha = E'/4E_0$, and E_0 is the initial energy of the pendulum. The values of total energy absorbed in fracture computed this way from the load–time curves show a one-to-one correspondence with the values determined in a conventional Charpy test. Based on this correspondence, we can use Equation 9.3 for computing the initiation and propagation energies at a given temperature. This information, together with the load at yielding, maximum load, and load at fracture, can allow us to identify the various stages of the fracture process.

It is well known (see Section 9.3) that the plane-strain fracture toughness (K_{Ic}) test gives a much better and precise idea of a material's tenacity than the instrumented Charpy test does. Also, K_{Ic} is a material property. However, as will be seen shortly, the K_{Ic} test

---

[2]See B. Augland, *Brit. Weld. J.*, 9 (1962) 434.

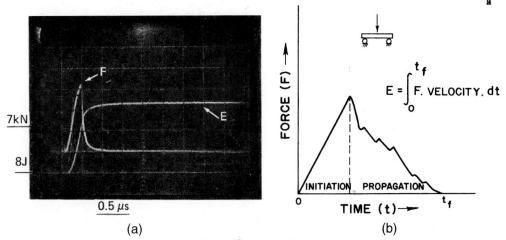

Figure 9.7 (a) Typical oscilloscope record of an instrumented Charpy impact test. (b) Schematic representation of (a).

possesses certain disadvantages: The preparation of equipment and the specimen is rather expensive, the test is relatively slow and not simple to execute, and so on. Consequently, there have been attempts at developing empirical correlations between the energy absorbed in a conventional Charpy test (C_v) and the plane-strain fracture toughness (K_{Ic}). The reader is warned that such correlations are completely empirical and are valid only for the specific metals tested. The instrumented Charpy test, with samples precracked and containing side grooves in order to assure a plane-strain condition, can be used to determine the dynamic fracture toughness K_{ID}. For ultrahigh-strength metals (σ_y very large), $K_{ID} \approx K_{Ic}$. Thus, we may use the instrumented Charpy test to determine K_{Ic} or I_{ID} for very high-strength steels. But we must check the results obtained with those obtained from a standard ASTM K_{Ic} test, as described in the next section.

9.3 PLANE-STRAIN FRACTURE TOUGHNESS TEST

The fracture toughness K_{Ic} of a material may be determined by means of a number of standards, e.g., ASTM E399 or BS 544. The essential steps in fracture toughness tests involve the measurement of crack extension and load at the sudden failure of the sample. Because it is difficult to measure crack extension directly, one measures the relative displacement of two points on opposite sides of the crack plane. This displacement can be calibrated and related to the real crack front extension.

The typical test samples used in fracture toughness tests carried out in accordance with the ASTM standard are shown in Figure 9.8. Figure 9.8(c) shows the size of the specimens. (Tensile and Charpy specimens are also shown for comparison.) The relation between the applied load and the crack opening displacement depends on the size of the crack and the thickness of the sample in relation to the extent of the plastic zones. When the crack length and the sample thickness are very large in relation to the quantity $(K_{Ic}/\sigma_y)^2$, the load–displacement curve is of the type shown in Figure 9.9(a). The load at the brittle fracture that corresponds to K_{Ic} is then well defined. When the specimen is of reduced thickness, a step called "pop-in" occurs in the curve, indicating an increase in the crack opening displacement without an increase in the load (Figure 9.9b). This phenomenon is attributed to the fact that the crack front advances only in the center of the plate thickness, where the material is constrained under plane-strain condition. However, near the free surface,

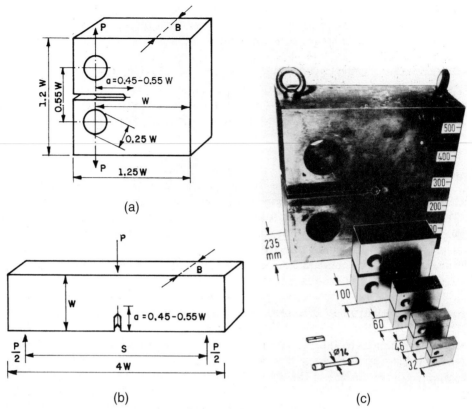

(a)

(b)   (c)

Figure 9.8 Typical ASTM standard plane-strain fracture toughness test specimens. (a) Compact tension. (b) Bending. (c) Photograph of specimens of various sizes. Charpy and tensile specimens are also shown, for comparison purposes. (Courtesy of MPA, Stuttgart)

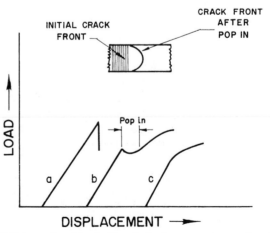

Figure 9.9 Schematic of typical load–displacement curves in a K_{Ic} test.

plastic deformation is much more pronounced than at the center, and it approaches the conditions of plane stress. Consequently, the plane-strain fracture advances much more in the central portion of the plate thickness, and in regions of material near the surfaces of the specimen, the failure eventually is by shear.

When the test piece becomes even thinner, the plane-stress condition prevails, and the load–displacement curve becomes as shown in Figure 9.9c. To make valid fracture toughness measurements in plane strain, the influence of the free surface, which relaxes the constraint, must be maintained small. This enables the plastic zone to be constrained completely by elastic material. The crack length must also be maintained greater than a certain lower limit.

Figure 9.10 shows the plastic zone at the crack front in a plate of finite thickness. At the edges of the plate ($x_3 \rightarrow \pm B/2$), the stress state approaches that of plane stress. At the center of a sufficiently thick plate, the stress state approaches that of plane strain. This is so because the ε_{33} component of strain is equal to zero at the center, as the material there in that direction is constrained, whereas near the edges the material can yield in the x_3 direction, so ε_{33} is different from zero.

Up to this point, the sample size and the crack length have been discussed in a qualitative way. The lower limits on width, thickness, and crack length all depend on the extent of plastic deformation through the $(K_{Ic}/\sigma_y)^2$ factor. In view of the lack of knowledge about the exact size of the plastic zone for the crack in mode I (the crack opening mode), it is very difficult to determine the lower limits of dimension of the test piece theoretically. These lower limits above which K_{Ic} remains constant are determined by means of trial tests. Samples of dimensions smaller than those limits tend to overestimate the K_{Ic} limit.

Preferably, in fracture toughness tests, the crack is introduced by fatigue from a starter notch in the sample. The fatigue crack length should be long enough to avoid interference in the crack-tip stress field by the shape of the notch. Under an applied load, the crack opening displacement can be measured between two points on the notch surfaces by various types of transducers. Figure 9.11 shows an assembly for measuring displacement in a notched specimen. Electrical resistance measurements have also been used

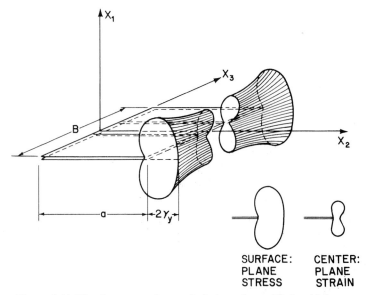

Figure 9.10 Plastic zone at the crack tip in a plate of finite thickness.

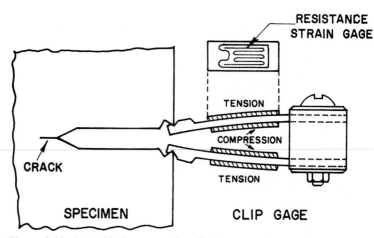

Figure 9.11 Assembly for measuring displacement in a notched specimen.

to detect crack propagation. Calibration curves are utilized for converting displacement measurements and resistance measurements into crack extension.

The load–displacement curves generally show a gradual deviation from linearity, and the "pop-in" step is very small (Figure 9.12). The procedure used in the analysis of load–displacement records of this type can be explained by means of the figure. Let us designate the linear-slope part as OA. A secant line, OP_5, is drawn at a slope 5% less than that of line OA. The point of intersection of the secant with the load–displacement record is called P_5. We define the load P_Q, for computing a conditional value of K_{Ic}, called K_Q, as follows: If the load on every point of curve before P_5 is less than P_5, then $P_5 = P_Q$ (case I in the figure). If there is a load more than P_5 and before P_5, this load is considered to be P_Q (cases II and III in the figure). In these cases, if $P_{\max}/P_Q > 1.1$, the test is not valid; K_Q does not represent the K_{Ic} value, and a new test needs to be done. After determining the point P_Q, we calculate the value of K_Q according to the known equation for the geometry of the test piece used. A checklist of points is given in Table 9.1 and Figure 9.13 shows schematically the variation of K_c, with the flaw size, specimen thickness, and specimen width. The stress intensity factor is calculated by using the equation

$$K_I = f\left(\frac{a}{W}\right)\frac{P}{B\sqrt{W}}. \tag{9.4}$$

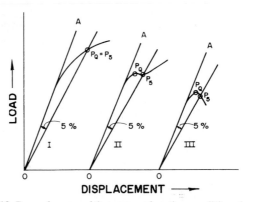

Figure 9.12 Procedure used for measuring the conditional value K_Q.

TABLE 9.1 Checklist for the K_{Ic} Test

1. Dimensions of test piece
   a. Thickness, $B \geq 2.5 (K_{Ic}/\sigma_Y)^2$
   b. Crack length, $a \geq 2.5 K_{Ic}/\sigma_Y)^2$
2. Fatigue precracking
   a. $K_{max}/K_{Ic} \leq 0.6$
   b. Crack front curvature $\leq$ 5% of crack length
   c. Inclination $\leq 10°$
   d. Length between 0.45 W and 0.55 W, where W is the width of the test sample
3. Characteristics of load–displacement curve. This is effectively to limit the plasticity during the test and determines whether the gradual curvature in the load–displacement curve is due to plastic deformation or crack growth.
   a. $P_{max}/P_Q \leq 1.1$

The function $f(a/W)$ has a different form for each specimen geometry. For the compact specimen (Figure 9.8a),

$$f\left(\frac{a}{W}\right) = \frac{2 + \dfrac{a}{W}}{\left(1 - \dfrac{a}{W}\right)^{3/2}} \left[0.886 + 4.64\left(\frac{a}{W}\right) - 13.32\left(\frac{a}{W}\right)^2 + 14.72\left(\frac{a}{W}\right)^3 - 5.60\left(\frac{a}{W}\right)^4\right]. \quad (9.5)$$

For the single-edge notched-bend specimen loaded in three-point bending (Figure 9.8b),

$$f\left(\frac{a}{W}\right) = \frac{3 \dfrac{S}{W} \sqrt{\dfrac{a}{W}}}{2\left(1 + 2\dfrac{a}{W}\right)\left(1 - \dfrac{a}{W}\right)^{3/2}} \left[1.99 - \frac{a}{W}\left(1 - \frac{a}{W}\right)\left\{2.15 - 3.93\left(\frac{a}{W}\right) + 2.7\left(\frac{a}{W}\right)^2\right\}\right]. \quad (9.6)$$

The preceding expressions are polynomial fits to functions.

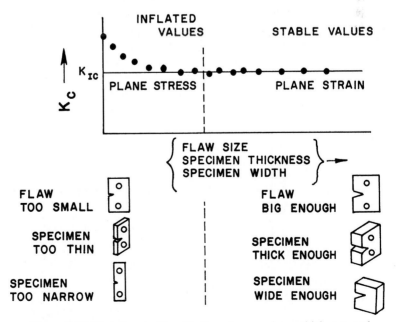

Figure 9.13 Variation in K_c with flaw size, specimen thickness, and specimen width.

EXAMPLE 9.1

Estimate the minimum specimen thickness for a valid plane-strain fracture toughness test for material having the following properties:

$$\text{Yield stress } \sigma_y = 400 \text{ MPa,}$$

$$\text{Fracture toughness } K_{Ic} = 100 \text{ MPa}\sqrt{\text{m}}.$$

Solution:

The minimum thickness of the specimen is $2.5 \ (K_{Ic}/\sigma_y)^2 = 2.5 \ (1/4)^2 = 0.156$ m = 156 mm.

9.4 CRACK OPENING DISPLACEMENT TESTING

For crack opening displacement (COD) testing, the test sample for determining δ_c is a slow-bend test specimen similar to the one used for K_{Ic} testing. A clip gage is used to obtain the crack opening displacement. During the test, one obtains a continuous record of the load P versus the opening displacement Δ (Figure 9.14). In the case of a smooth P–Δ curve, the critical value, Δ_c, is the total value (elastic + plastic) corresponding to the maximum load (Figure 9.14a). In case the P–Δ curve shows a region of increase in displacement at a constant or decreasing load, followed by an increase in load before fracture, one needs to make auxiliary measurements to determine that this behavior is associated with crack propagation. Should this be so, Δ_c will correspond to the first instability in the curve. If the P–Δ curve shows a maximum, and Δ increases with a reduction in P, then either a stable crack propagation is occurring or a "plastic hinge" is being formed. The "Δ_c" in this case (Figure 9.14b) is the value corresponding to the point at which a certain specified crack growth has started. If it is not possible to determine this point, one cannot measure the COD at the start of crack propagation. However, we can measure, for comparative purposes, an opening displacement δ_m, computed from the clip gage output Δ_m, corresponding to the first load maximum. The results in this case will depend on the geometry of the specimen.

Experimentally, we obtain Δ_c, the critical displacement of the clip gage. We need to obtain δ_c, the critical CTOD. Various methods are available, all based on the hypothesis that the deformation occurs by a "hinge" mechanism around a center of rotation at a depth of $r(W - a)$ below the crack tip, (Figure 9.15), where W is the width, and a is the length. Experimental calibrations of the crack using specimens of up to 50 mm in thickness, have

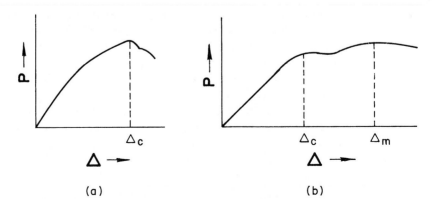

Figure 9.14 Schematic of Load P versus crack opening displacement Δ.

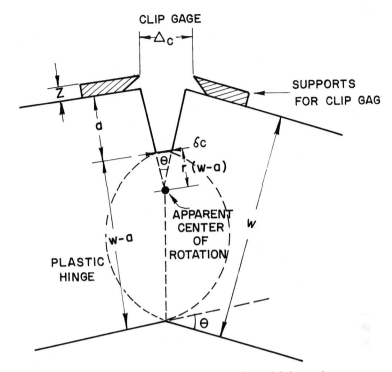

Figure 9.15 "Plastic hinge" mechanism of deformation.

shown that, for COD in the range 0.0625 to 0.625 mm, δ_c can be obtained to a very good approximation from the relation

$$\delta_c = \frac{(w - a)\Delta_c}{w + 2a + 3z}$$

This relation is derived on the basis of the assumption that the deformation occurs by a hinge mechanism about a center of rotation at a depth of $(w - a)/3$ below the crack tip (i.e., $r = \frac{1}{3}$). However, r can be smaller for smaller values of Δ_c. Note that $r \approx 0$ in the elastic case (very limited plastic deformation at the crack tip), and $r \approx \frac{1}{3}$ for a totally plastic ligament.

9.5 J-INTEGRAL TESTING

J_{Ic} defines the onset of crack propagation in a material in which large-scale plastic yielding makes direct measurement almost impossible. Thus, one can use J-integral testing to find the value of K_{Ic} for a very ductile material from a specimen of dimensions too small to satisfy the requirements of a proper K_{Ic} test.

ASTM standard E819-89 provides a procedure for determining J_c, the critical value of J. As pointed out in Chapter 7, the physical interpretation of the J-integral is related to the area under the curve of the load versus the load-point displacement for a cracked sample. Both compact tension and bend specimens can be used. The ASTM standard requires at least four specimens to be tested. Each specimen is loaded to different amounts of crack extensions (Figure 9.16). One calculates the value of J for each specimen from the expression

$$J = \frac{2A}{Bb},$$

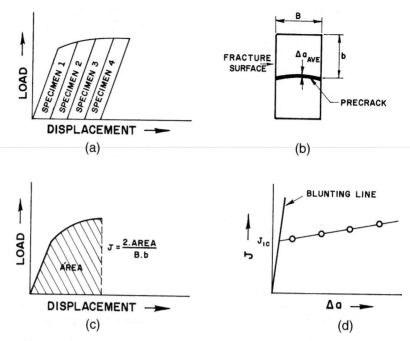

Figure 9.16 Method for determining J_{Ic}. (a) Load identical specimens to different displacements. (b) Measure the average crack extension by heat tinting. (c) Calculate J for each specimen. (d) Plot J versus Δ_a to find J_{Ic}.

where A is the area under load versus the load-point displacement curve, B is the specimen thickness, and b is the uncracked ligament. The value of J so derived is plotted against Δa, the crack extension of each specimen. One way of obtaining Δa is to heat-tint the specimen after testing and then break it open. When the specimen is heated, the crack surfaces oxidize. Next, a "best line" through the J points and a "blunting line" from the origin are drawn. This blunting line (indicating the onset of crack blunting due to plastic deformation) is obtained from the equation

$$J = 2\sigma_{\text{flow}} \Delta a, \tag{9.7}$$

where $\sigma_{\text{flow}} = (\sigma_y + \sigma_{UTS})/2$, in which σ_y is the yield stress and σ_{UTS} is the ultimate tensile stress.

The intercept of the J line and the blunting line gives J_{Ic}. J_{Ic} is related to K by

$$J_{Ic} = \frac{K_{Jc}^2}{E}. \tag{9.8}$$

9.6 FLEXURE TEST

The flexure or bend test is one of the easiest tests to do and is very commonly resorted to, especially with brittle materials that behave in a linear elastic manner. A very small amount of material is required, and preparation of the sample is relatively easy. The following assumptions are made in analyzing the flexure behavior of materials. We assume that the Euler–Bernoulli theory is applicable to a freely supported beam. (The beam is not clamped at any point.) That the Euler–Bernoulli theory is applicable means that plane sections remain plane, deformations are small, stress varies linearly with thickness, and there is no Poisson contraction or expansion. The condition of small deformation comes from the Euler–Bernoulli assumption that the specimen beam is bent into a circular arc. The condition of a small deformation can be easily violated if the material is deformed in a non-linear, viscoelastic, or plastic manner. Then stress gradients across the vertical section of the beam will not be linear.[3]

The two basic governing equations for a simple beam elastically stressed in bending are

$$\frac{M}{I} = \frac{E}{R} \tag{9.9}$$

and

$$\frac{M}{I} = \frac{\sigma}{y}, \tag{9.10}$$

where M is the applied bending moment, I is the second moment of area of the beam section about the neutral plane, E is Young's modulus of elasticity of the material, R is the radius of curvature of the bent beam, and σ is the tensile or compressive stress on a planar distance y from the neutral plane.

For a uniform circular section of beam,

$$I = \frac{\pi d^4}{64}, \tag{9.11}$$

where d is the diameter of the section.

For a uniform, rectangular section of beam,

$$I = \frac{bh^3}{12}, \tag{9.12}$$

where b is the width of the beam and h is the height of the beam.

Bending takes place in the direction of the depth; that is, h and y are measured in the same direction. Also, for a beam with a symmetrical section with respect to the neutral plane, replacing $h/2$ (or $d/2$) for y in Equation (9.10) gives the stress at the beam surface.

In the elastic regime, stress and strain are related by Hooke's law,

$$\sigma = E\varepsilon. \tag{9.13}$$

From Equations (9.9), (9.10), and (9.13), we obtain the following simple relation, valid in the elastic regime:

$$\varepsilon = \frac{y}{R}. \tag{9.14}$$

Figure 9.17 shows the elastic normal stress distribution through the thickness when a beam is bent. The stress and strain vary linearly with the thickness y across the section,

---

[3]The student should recall discussions on beam deflections in courses on mechanics of materials.

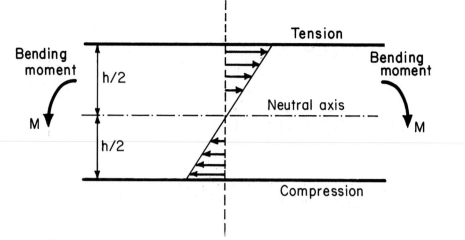

Figure 9.17 Normal stresses along a section of beam for linearly elastic material.

with the neutral plane ($y = 0$) representing the zero level. The material on the outside or above the neutral plane of the bent beam is stressed in tension, while that on the inside or below the neutral plane is stressed in compression. Thus, the elastic strain ε in a beam bent to a radius of curvature R varies linearly with distance y from the neutral plane across the beam thickness.

Two main types of flexure tests are three-point and four-point bend tests. Another variant of flexure tests is the so-called interlaminar shear stress (ILSS) that is used in fiber reinforced composites. We describe these briefly.

9.6.1 Three-Point Bend Test

In the three-point bend test, the load is applied at the center point of the beam, and the bending moment M increases from the two extremities to a maximum at the center point. (See Figure 9.18a.) In this case,

$$M = (P/2)(S/2) = PS/4,$$

while the moment of inertia, for a beam of a rectangular section, is

$$I = bh^3/12.$$

Using Equation 9.10, we can obtain the maximum stress in the outermost layer ($y = h/2$) as

$$\sigma_{max} = 3PS/2bh^2. \tag{9.15}$$

9.6.2 Four-Point Bending

Four-point bending is also called pure bending, since there are no transverse shear stresses on the cross sections of the beam in the inner span. For an elastic beam bent

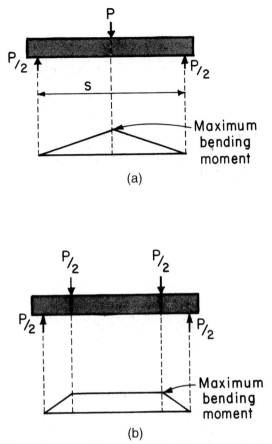

Figure 9.18 Application of loads and bending moment diagrams for (a) three-point bending and (b) four-point bending tests.

at four points, the bending moment is constant in the inner span (see Figure 9.18b) and is given by

$$M = \frac{\sigma_{max} I}{h/2} = \frac{P}{2} \cdot \frac{S}{4} \tag{9.16}$$

where I is the moment of inertia, $h/2$ is the distance from the neutral axis to the outer surface, and σ_{max} is the normal stress on a transverse section of the same outer fiber. The maximum stress in a rectangular beam undergoing a four-point bend is

$$\sigma_{max} = 3PS/4bh^2, \tag{9.17}$$

where S is the outer span.

9.6.3 Interlaminar Shear Strength Test

The interlaminar shear strength test is also known as the short-beam shear test. It is commonly used with fiber reinforced composites, with the fiber length parallel to the length of a three-point bend bar. In such a test, the maximum shear stress occurs at the midplane and is given by

$$\tau_{max} = \frac{3P}{4bh}. \tag{9.18}$$

The maximum tensile stress occurs at the outermost surface and is given by Equation 9.15. Dividing Equation 9.18 by Equation 9.15, we get

$$\frac{\tau_{max}}{\sigma_{max}} = \frac{h}{2S}. \tag{9.19}$$

Equation 9.19 says that if we make the load span S very small, we can maximize the shear stress τ such that the specimen fails under shear with a crack running along the midplane. Thus, if we deliberately make the span very small (hence the name, "short beam"), then it is likely that failure will occur under shear. A word of caution is in order about the interpretation of this test: The test becomes invalid if the fibers fail in tension before shear-induced failure occurs. The test will also be invalid if shear and tensile failure occur simultaneously. It is advisable to examine the fracture surface after the test, to make sure that the crack is along the interface and not through the matrix.

The three- and four-point bending tests are extremely useful for determining the strength of brittle materials, especially of ceramics. If brittle materials are tested in tension as described in Chapter 3, the alignment of the grips is very critical. Slight misalignments cause major stress inhomogeneities, which significantly affect the strength of the material. Flexure tests, on the other hand, are simple and reliable. The four-point bending test presents the following advantage over the three-point bending test: The material over the inner span is subjected to a constant stress; therefore, we have a larger volume being tested than in a three-point bend test. The resultant strength is called *bend strength, flexural strength* or, commonly, but erroneously, *modulus of rupture* (MOR).

A miniature bend test was developed for testing small specimens with 3-mm diameters. A disk instead of a bar was used. The deflection of the disk was measured as it was deformed by a punch (with a sphere at its tip), while the edges were held in place by a die. This miniature disk-bend test has been used to obtain yield and fracture stresses in small research specimens.

EXAMPLE 9.2

In a three-point bend test (see Figure E9.2), the specimen has a span of 50 mm and a load P of 50 N. The width and height are 5 mm each. Draw the moment and shear force diagrams. Find the maximum moment and maximum stress.

Solution: We have

$$\text{Load, } P = 50 \text{ N}, \qquad \text{Span, } S = 50 \text{ mm} = 5 \times 10^{-3} \text{ m}.$$

In three-point bending, the maximum bending moment occurs at the midpoint of the beam and is given by

$$\frac{P}{2} \times \frac{S}{2} = \frac{PS}{4} = \frac{50 \text{ N} \times 10^{-3} \text{ m}}{4} = 625 \times 10^{-3} \text{ N} \cdot \text{m}.$$

Note that the maximum stress also occurs along the centerline of the specimen; that is, the whole of the specimen is not subjected to a uniform stress, as would be the case in a tensile test. We have

$$\text{Maximum stress (Equation 9.15)} = \frac{3 \cdot PS^2}{bh^2} = \frac{3 \cdot 50 \times (50 \times 10^{-3})^2}{5 \times 10^{-3}(5 \times 10^{-3})^2}$$

$$= 3 \text{ MPa}.$$

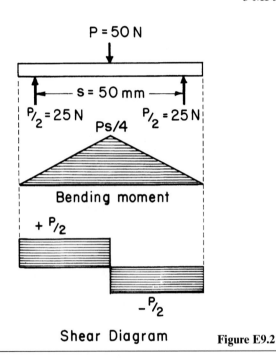

Figure E9.2

EXAMPLE 9.3

It is generally known that a given material will show a higher strength in a three-point bend test than in an axial tension test. Consider a rod of a square cross section and side a. If the span in the bend test is S, show that the ratio of the bend strength to the tensile strength is

$$\sigma_{bend}/\sigma_{ten} = 3S/a.$$

Solution: Uniaxial tension

$$\text{Force} = P,$$

$$\text{Cross-sectional area} = a^2,$$

$$\text{Tensile strength } \sigma_{ten} = P/a^2.$$

Three-point bending
The maximum stress in three-point bending is $\sigma_{bend} = \sigma_{max} = M_{max}y_{max}/I$, where

$$I = bh^3/12 = a^4/12,$$

$$M = PS/2,$$

$$y_{max} = a/2.$$

Hence,

$$\sigma_{\text{bend}} = P(S/2)(a/2)(12/a^4) = 3PS/a^3 = (3S/a)(P/a^2).$$

Thus, the maximum stress in bending is $3S/a$ times the tensile stress. Generally, $S \gg a$, so the difference can be very large, indeed!

9.7 FRACTURE TOUGHNESS TESTING OF BRITTLE MATERIALS

In brittle materials—especially ceramics—the strength is largely determined by the size and sharpness of flaws and by the resistance of cracks to propagation. Since plasticity is very limited in such materials, the size of the specimen can be reduced much more than in metals. Recall that the thickness B of the test specimen should exceed $2.5(K_{Ic}/\sigma_y)^2$. This ensures a plastic zone size that is small with respect to B, and therefore, the state of plane strain can be assumed. We will estimate the minimum acceptable specimen thickness for a typical ceramic, alumina, for which

$$K_{Ic} \approx 4 \text{ MPa m}^{1/2},$$

$$\sigma \approx 400 \text{ MPa}.$$

For this specimen,

$$B \geq 2.5 \times 10^{-4} \text{ m}.$$

Therefore, the minimum thickness is very small, and microstructural inhomogeneities limit the size of the specimen. We next discuss the most common methods of testing brittle materials (Figure 9.19).

A double-cantilever specimen (DCB) with a precrack of size a is illustrated in Figure 9.19(a). Three possible loading configurations are shown: wedge loading, applied load P, and applied moment M. A groove is machined into the specimen to guide the propagation of the crack. The three loading methods provide essentially three relationships, between K_I, the stress intensity factor, and the crack length.

A double-torsion specimen is very convenient for determining the fracture toughness of ceramics at high temperatures. It requires only the application of a compressive load P (Figure 9.19b). The stress intensity K_I does not depend on the length of the crack for $0.25L < a < 0.75L$. The fracture toughness is given by

$$K_I = PW_m \left[\frac{3}{Wt^3 t_1(1 - \nu)\xi} \right]^{1/2}, \tag{9.20}$$

where ξ is a geometrical factor that depends on the thickness of the specimen, t_1.

The notch bend test (Figure 9.19c) is analogous to the same test applied to metals. A notch is cut into the brittle material. A crack "pops in" during loading and then grows with P. This technique requires only small specimens.

9.7.1 Chevron Notch Test

The main advantage of the chevron notch test is that the critical stress intensity factor can be determined from the maximum load without resorting to precracking and crack length measurement. The test requires that the specimen undergo stable crack growth before

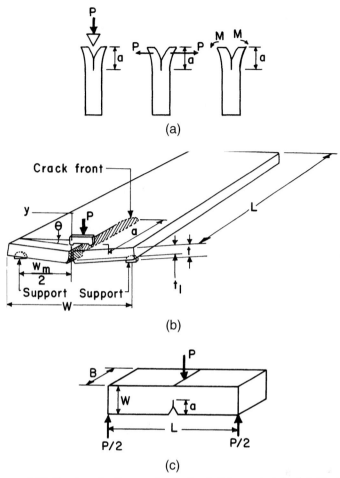

Figure 9.19 Fracture-testing methods for brittle materials. (a) Double-cantilever beam (DCB). (b) Double torsion. (c) Notch flexure.

reaching the maximum load, as indicated by the load–displacement curve deviating slightly from the initial linear part before final fracture.

The notches in the samples can be conveniently made with a low-speed diamond saw. The dimensions of the specimen should obey the following guidelines, as recommended in various references:[4]

$$S/W = 4, \quad W/B = 1.5, \quad \alpha_0 = a_0/W \geq 0.3, \quad \text{and} \quad \theta = 60°$$

Here, S is the span, W is the height of the specimen, B is the width of the specimen, and θ is the included angle. Figure 9.20a shows a schematic of the test arrangement and the details of the notch plane. The test can be performed in a universal testing machine at a constant crosshead speed. The chevron tip length, a_0, can be measured from optical micrographs of broken specimens, as shown in Figure 9.20b. The critical stress intensity factor can be obtained from the relationship

$$K_{Ic} = \frac{P_{max}}{B\sqrt{W}} Y_c(\alpha_0),$$

[4]See, for example, S.-X. Wu, *Eng. Fracture Mech.*, 19 (1984), 221.

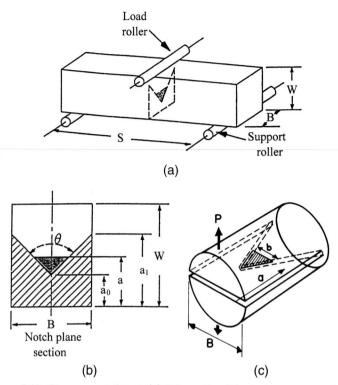

Figure 9.20 Chevron notch test. (a) Schematic of the test arrangement and the details of the notch plane. (b) The chevron tip length, a_0, can be measured from optical micrographs of broken specimens. (c) Chevron short-rod specimen.

where P_{max} is the maximum load, B is the width of the specimen, W is the height of the specimen, and Y_c is a dimensionless coefficient[5] given by

$$Y_c(\alpha_0) = 5.639 + 27.44\alpha_0 + 18.93\alpha_0^2 - 43.42\alpha_0^3 + 338.9\alpha_0^4$$

for the geometry of the specimen in this study (i.e., $\theta = 60°$ and $W/B = 1.5$).

In another variant of the chevron test, the chevron-notched short-rod specimen shown in Figure 9.20 which has been standardized by ASTM (E1304-89), has a wedge inserted into a slit that is cut in it, leaving a thin layer of ceramic with a V-shape. A crack is initiated at the tip of the wedge; the width of the crack increases as the crack moves forward. The wedge also guides the crack as it grows. The load that opens the crack can be supplied by applying tension to the two sides or by an ingenious bladder mechanism. In this mechanism, a bag containing a fluid is inserted into the slit. The fluid is then pressurized, creating a crack opening force P. The fracture toughness of the specimen is determined from

$$K_{Ic} \approx 22\, P_c B^{-3/2},$$

[5]See S.-X. Wu, *Chevron-Notched Specimens: Testing and Stress Analysis,* J. H. Underwood, S. W. Freiman, and F. I. Baratta (eds.) (Philadelphia: ASTM, 1984), p. 176.

where P_c is the maximum load for crack propagation and B is the diameter of the short rod. This technique has also been extended to metals (with a different equation). This geometry of the specimen does not require any fatigue precracking; this is a considerable advantage, because fatigue precracking can be complicated and "tricky," especially in ceramics.

9.7.2 Indentation Methods for Determining Toughness

Hardness indentations can generate cracks in brittle materials; two such examples are shown in Figure 9.21. Tensile stresses are generated under conical and pyramidal indentations. These tensile stresses can generate cracks, and the length of the cracks can be used to calculate a fracture toughness. A second use of such cracks is as initiation sites for fracture in the conventional bending test. The very attractive feature of these

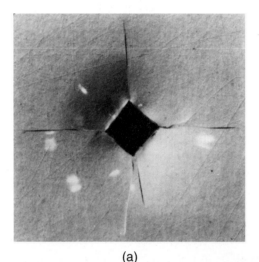

(a)

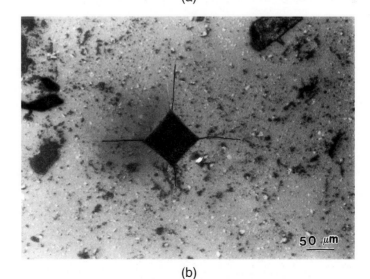

(b)

Figure 9.21 Fractures produced by hardness indentations in (a) AsS_3 glass (courtesy of B. R. Lawn and B. J. Hockey) and (b) Al_2O_3.

microhardness-induced cracks is that they are very small and on the same scale as cracks naturally occurring in ceramics (<1 mm).

Palmqvist was the first to recognize that indentation cracks could be used to obtain quantitative estimates of the fracture toughness of brittle materials.[6] Later, detailed studies by Lawn, Wilshaw, Evans, and coworkers laid the foundation for indentation fracture toughness tests.[7] A simple dimensional analysis shows that the hardness of a material (i.e., the material's resistance to plastic deformation) is given by

$$H = \frac{P}{\alpha c^2},$$

where c is the diagonal of the impression and P is the load. The area of the impression is $0.5c^2$, setting the value of the parameter α to $(1/1.854)$. For Vickers indentation (see Section 3.8.1). In a similar way, the toughness of the material is related to the load and crack size by

$$K_c = \frac{P}{\beta a^{3/2}}.$$

This gives the correct units for K_c: $Nm^{-3/2}$ or $Pa\ m^{1/2}$. The factor β incorporates a complex elasto-plastic interaction that will not be discussed here. It is important to emphasize that the crack is not always produced during the indentation period, but can be generated during unloading. There are elastic stresses caused by the indentation, producing compressive tangential components of stress; there is also plastic deformation, creating residual stresses on unloading. It is these residual stresses, with a tensile tangential component, that drive the crack. The problem can be analyzed as an internal cavity pressurized in an infinite body. This generates compressive radial stresses σ_{rr} and tensile tangential stresses $\sigma_{\theta\theta}$. The tangential stresses decay with $1/r_2$. On the other hand, a crack with length $2a$ forms, under ideal circumstances, a semicircle under the indentation, as shown in Figure 9.22. The stress intensity factor, in its turn, is given by

$$K_r = Y\sigma_{\theta\theta}\sqrt{\pi a}.$$

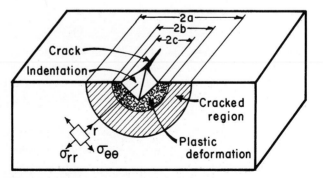

Figure 9.22 Schematic representation of indentation generating a plastic deformation region and a semicircular crack.

[6]S. Palmqvist, *Jernkontorets Ann.*, 141 (1957) 300; *Arch. Eisenhuttenwies.*, 33 (1962) 629.

[7]B. R. Lawn and T. R. Wilshaw, *J. Mater. Sci.*, 10 (1975) 1049; A. G. Evans and T. R. Wilshaw, *Acta Met.*, 24 (1976) 939; A. G. Evans and E. A. Charles, *J. Am. Cer. Soc.*, 59 (1976) 371.

Since

$$\sigma_{\theta\theta} = \frac{kP}{a^2},$$

it follows that

$$K_r = \frac{kY\pi^{1/2}P}{a^{3/2}} = \frac{k'P}{a^{3/2}},$$

where k' is a parameter. It has been shown that the size of the indentation depends on the hardness of the material and on the Young's modulus E. The following functional relationship has been found:

$$k' = \delta\left(\frac{E}{H}\right)^{1/2},$$

where δ is a geometrical factor that depends on the indentation. Thus,

$$K_r = \delta\left(\frac{E}{H}\right)^{1/2}\frac{P}{a^{3/2}}$$

Anstis et al.[8] take $\delta = 0.016 \pm 0.004$ (for a Vickers indentation). The fracture toughness of the material is the residual stress intensity factor at which the crack stops growing. Hence,

$$K_{Ic} = \delta\left(\frac{E}{H}\right)^{1/2}\frac{P}{a^{3/2}}.$$

Sometimes, equilibrium conditions are not established until after the load is removed. Slow growth of the cracks can then take place, and the measurement of a depends on the time interval involved. Sometimes, no well-defined radial cracks are formed. In that case, the load P should be adjusted so that well-developed cracks are generated—that is, cracks for which $a > 2c$. Figure 9.23 compares conventional and indentation fracture toughnesses for a number of ceramics. The error bars show the variability of the measurements. It can be seen that the results agree within 30%. The great advantage of indentation fracture toughness tests over conventional tests is that comparative tests with various materials can be carried out readily, providing relative values.

A second manner in which indentation is used is to generate a "starter" crack for the three-point bending test. A Knoop indenter is preferred, and a sharp crack is generated at the center of the specimen and in the side opposite the one where P (the center load) is applied.

[8]G. R. Anstis, P. Chantikul, B. R. Lawn, and D. B. Marshall, *J. Am. Cer. Soc.,* 64 (1981) 533.

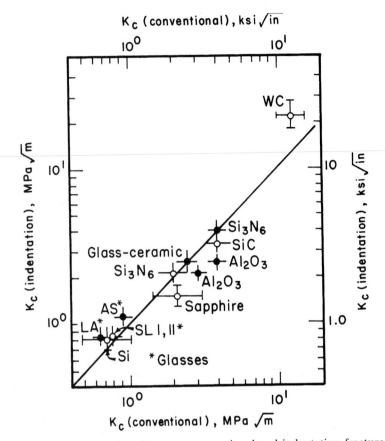

Figure 9.23 Comparison between conventional and indentation fracture toughness determinations for glasses and ceramics. (From G. R. Anstis, P. Chankitul, B. R. Lawn, and D. B. Marshall, *J. Am. Cer. Soc.,* 64 (1981) 533, Figure 5, p. 537)

EXAMPLE 9.4

Estimate the fracture toughness of the alumina specimen shown in Figure 9.21b. The indentation was caused by a load of 10 kgf using a Vickers diamond indenter attached to a uniaxial testing machine. The alumina specification is AD 95.

Solution: We measure

$$2c = 14 \text{ mm},$$

$$2a = 36.5 \text{ mm}.$$

From the magnification marker, we establish the magnification: $160\times$. Thus,

$$2c = 0.087 \text{ mm},$$

$$2a = 0.228 \text{ mm},$$

$$P = 10 \text{ kgf} = 102 \text{ N},$$

and we have

$$K_r = 0.016\left(\frac{E}{H}\right)^{1/2}\frac{P}{a^{3/2}}.$$

From Table 2.8,

$$E = 365 \text{ GPa}.$$

H is the hardness in N/m^2. We have

$$H = \frac{10 \times 1.85}{7.6 \times 10^{-3}} = 2{,}434 \text{ kg/mm}^2$$

$$= 23.85 \text{ GPa}.$$

Also,

$$K_r = 0.016\left(\frac{365}{23.85}\right)^{1/2}\frac{10^2}{(0.114 \times 10^{-3})^{3/2}}$$

$$= 5.13 \text{ MPa m}^{1/2}.$$

SUGGESTED READINGS

T. L. ANDERSEN. *Fracture Mechanics,* 2d ed. Boca Raton, FL: CRC, 1995.

R. W. HERTZBERG. *Deformation and Fracture Mechanics of Engineering Materials,* 4th ed. New York: John Wiley, 1996.

B. LAWN. *Fracture of Brittle Solids,* 2d ed. Cambridge, U.K.: Cambridge Univ. Press, 1993.

S. T. ROLFE AND J. M. BARSOM. *Fracture and Fatigue Control in Structures.* Englewood Cliffs, NJ: Prentice-Hall, 1977.

EXERCISES

9.1 A Charpy machine with a hammer weighing 200 N has a 1-m-long arm. The initial height h_0 is equal to 1.2 m. The Charpy specimen, (See Figure 9.2), absorbs 80 J of energy in the fracturing process. Determine:
   (a) The velocity of the hammer upon impact with the specimen.
   (b) The velocity of the hammer after breaking the specimen.
   (c) An average strain rate in the specimen.
   (d) The final height attained by the specimen.

9.2 Estimate the fraction of cleavage area in the four specimens shown in Figure 9.3.

9.3 Schematically show how the Charpy energy vs. temperature curve would be translated if the tests were carried at a low strain rate (approximately 10^{-2} s^{-1}).

9.4 If, instead of Charpy specimens with standard thickness equal to 10 mm, you would test specimens with reduced thickness (e.g., 5 mm) and increased thickness (e.g., 30 mm) what changes would you expect in the Charpy energy value, normalized to the thickness of the specimen.

9.5 The load–displacement curve, obtained from a fracture toughness test on metal sample is shown in Figure E9.5. The dimensions are as follows:

Crack length $a = 10$ mm,
Specimen thickness $B = 15$ mm,
Specimen width $W = 25$ mm,
Span $S = 50$ mm.

Use the recommended procedure to determine the K_{Ic} from this curve. Check whether this is a valid K_{Ic} test.

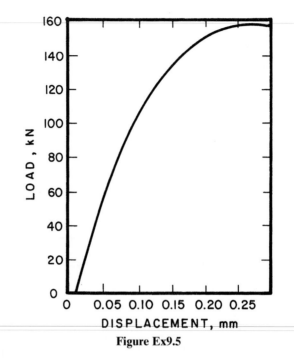

Figure Ex9.5

9.6 A thermoplastic polymer has a plane-strain fracture toughness $K_{Ic} = 15$ MPa$\sqrt{\text{m}}$ and a yield stress $\sigma_y = 80$ MPa. Estimate the requirements for dimensions of a fracture toughness specimen for this material.

9.7 Two samples of 0.45%C steel, one quenched and the other normalized, were tested for fracture toughness in a three-point bend test. The dimensions of the sample and the load–deflection curves for the two are shown in Figure E9.7. Determine K_{Ic} for the two samples, and establish whether the tests are valid. Verify whether the plane-strain conditions are met. Which steel would you expect to show a higher toughness? Does the result match your expectation? Given:

	QUENCHED	NORMALIZED
	$\sigma_y = 1,050$ MN/m²	$\sigma_y = 620$ MN/m²
	$a_1 = 13.7$ mm	$a = 9.3$ mm
Precrack	$a_2 = 11.6$ mm	$a_2 = 8.9$ mm
lengths:	$a_3 = 9.6$ mm	$a_3 = 9.4$ mm

9.8 A notched polymer specimen was tested for fracture toughness in a three-point bend test. The relevant dimensions of the specimen are:

Thickness $B = 5$ mm,

Width $W = 15$ mm,

Crack length $a = 1$ mm.

The load–deflection curve was linear until fracture occurred at 150 N. Compute K_{Ic} for this material.

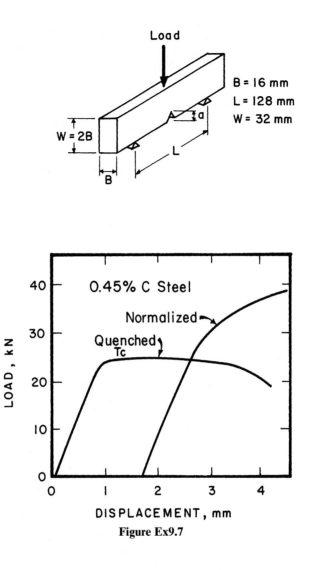

Figure Ex9.7

9.9 A compact tension specimen of a polymer with the following dimensions was used in a fracture toughness test:

$$\text{Thickness } B = 5 \text{ mm},$$

$$\text{Width } W = 50 \text{ mm},$$

$$\text{Crack length } a = 20 \text{ mm}.$$

Assuming a linear displacement curve to failure at a load of 200 N, compute K_{Ic} for this polymer.

9.10 A rectangular bar of ceramic 3 mm thick, 4 mm wide, and 60 mm long fractures in a four-point bend test at a load of 310 N. If the span of fixture is 50 mm, what is the modulus of rupture (or flexure strength) of the bar?

9.11 Norton NC-132 hot pressed Si_3N_4 has the following tensile strengths for the given tests:

Three-point bending:	930 MPa
Four-point bending:	720 MPa
Uniaxial tension:	550 MPa

Comment on the flaw sizes necessary to produce these failure stresses.

9.12 Calculate the tensile stresses generated by a load of 200 N acting on a specimen of SiC (a rectangular section of 5 mm height and 10 mm width) subjected to (a) three-point and (b) four-point bending. The span width is 50 mm and, for the four-point bending setup, the inner span is 25 mm. If the specimens are prenotched, with a notch depth of 1 mm, what are the stress intensity factors?

9.13 A cylindrical structural component with diameter 100 mm is subjected to a force of 100 kN at a distance of 500 mm from the clamp. The material used for the cylinder has a yield stress of 600 MPa. Will it yield plastically under the loading configuration shown in Figure E9.13. Use $I = \pi r^4/4$ for the moment of inertia of a cylindrical shaft.

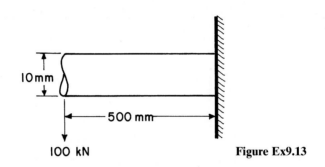

Figure Ex9.13

9.14 In sample of $MoSi_2$, an indentation made by a Vickers indenter gave the impression shown in Figure E9.14 under a load of 1 kN. Compute the hardness H of $MoSi_2$. Taking E for $MoSi_2$ to be 300 GPa, compute the fracture toughness of the sample.

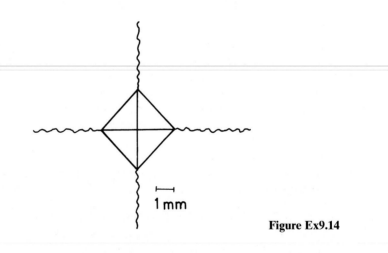

Figure Ex9.14

9.15 Estimate the fracture toughness for AsS_3 glass shown in Figure 9.21, knowing that the indentation was made with load $P = 10$ N. Young's modulus for this glass is $E = 75$ GPa. Assume same magnification for the two micrographs.

9.16 A chevron short-rod specimen with a diameter of 5 cm (Al_2O_3) was tested, and the critical load P_c was equal to 2,000 N. Determine the fracture toughness of the specimen.

10

Solid Solution, Precipitation, and Dispersion Strengthening

10.1 INTRODUCTION

A *solution* can be defined as a homogeneous mixture of two or more substances. Generally, one thinks of a solution as liquid, but gaseous or solid forms are possible as well. Indeed, we can have solutions of gases in a gas, gases in a liquid, liquids in a liquid, solids in a liquid, and solids in a solid. A solution can have one or more solutes dissolved in a solvent. The *solute* is the substance that is dissolved; the *solvent* is the substance in which the solute is dissolved. In a solution, there is always less solute than solvent. There are two kinds of solid solutions: substitutional and interstitial. Figure 10.1 shows examples of each in a schematic manner. Figure 10.1a is of brass, which is a substitutional solid solution of zinc (the solute) in copper (the solvent). We call such an alloy substitutional because the solute atoms merely substitute for the solvent atoms in their normal positions. In a substitutional solution, the atomic sizes of the solute and solvent atoms are fairly close. The maximum size difference is approximately 15%. When the atomic sizes of the solute and solvent are very different, as in the case of carbon or nitrogen in iron, we get an interstitial solid solution. Figure 10.1b shows such a solid solution of carbon in iron. We call these solutions interstitial solid solutions because the solute atoms occupy interstitial positions in the solvent lattice.

In this chapter, we first focus our attention on the phenomenon of *solid solution* and the *strengthening* that can be obtained by this process. Simply put, the phenomenon can be regarded as one form of restricting dislocation motion in crystalline materials, especially metals. We then extend this idea to precipitation and dispersion strengthening. Precipitates can be formed in certain alloys in the solid state. One starts with a solid solution at a high temperature, quenches it to a low temperature, and then ages it at an intermediate temperature to obtain a finely distributed precipitate. During aging, precipitates appear in a variety of sequences, depending on the alloy system under consideration. *Precipitation strengthening* has to do with the interaction of dislocations with precipitates, rather than

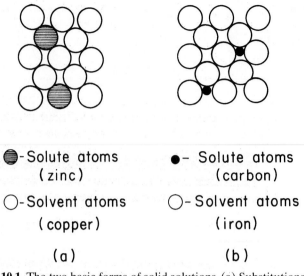

Figure 10.1 The two basic forms of solid solutions. (a) Substitutional solid solution of zinc in copper to form brass. (b) Interstitial solid solution of carbon in iron to form steel. The interstitial solid-solution carbon atoms are shown in the face-centered cubic form of iron.

having to do with single atoms of solutes. A logical extension of this idea is to artificially disperse hard ceramic phases in a soft metallic matrix, instead of obtaining them via a precipitation process. The mobility of dislocations is then restricted by these hard particles, and the alloy is strengthened. This process is called *dispersion strengthening*.

10.2 SOLID-SOLUTION STRENGTHENING

Dislocations are moderately mobile in pure metals, and plastic deformation occurs by means of dislocation motion (i.e., by shear). A very versatile method of obtaining high strength levels in metals would be to restrict this rather easy motion of dislocations. We saw earlier that grain boundaries (Chapter 5) and stress fields of other dislocations (Chapter 6) can play this restrictive role at low temperatures and increase the strength of the material. When the dislocation mobility in a solid is restricted by the introduction of solute atoms, the resultant strengthening is called *solid-solution-hardening,* and the alloy is called a *solid solution.* An example of the strengthening that can be achieved by solid solution is shown in Figure 10.2a, in which we plot the increase in yield stress of steel as a function of the content of the solute. Note that solutes such as carbon and nitrogen, which go into interstitial positions of the iron lattice, have much larger strengthening effects than substitutional atoms such as manganese. We shall explain this shortly. In order to analyze the phenomenon of hardening due to the presence of solute atoms, we must consider the increase in the stress necessary to move a dislocation in its slip plane in the presence of discrete barriers to the motion of dislocations. Conceptually, it is useful and easier to think in terms of an energy of interaction between the dislocation and the barrier (e.g., a solute atom or a precipitate). In the case of substitutional solutions, for a stationary dislocation, the interaction energy is the change in energy of the system consisting of a crystal and a dislocation when a solvent atom is removed and substituted with a solute atom. Knowing the interaction energy U, we can calculate the force dU/dx necessary to move a

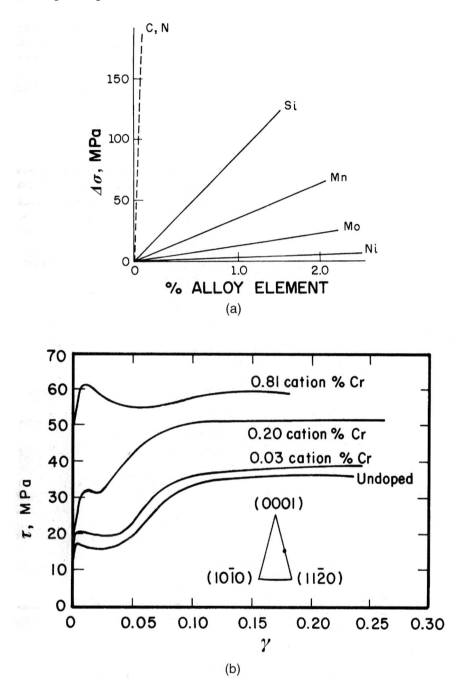

Figure 10.2 (a) Increase in strength, $\Delta\sigma$, of steel as a function of content of solute. The solid lines represent substitutional solute additions, while the dashed line represents interstitial solute additions. (After F. B. Pickering and T. Gladman, ISI Special Report 81, Iron and Steel Inst., (London: 1963), p. 10). (b) Increase in strength of sapphire (monocrystalline alumina) with small additions of chromium at 1400°C (Adapted from K. P. D. Lagerlof, B. J. Pletka, T. E. Mitchell, and A. H. Heuer, Radiation Effects, 74 (1983) 87).

dislocation a distance dx normal to its length. In ceramics, solutes can also exercise a strengthening effect, as demonstrated by Figure 10.2b for monocrystalline alumina with additions of chromium. This increase manifests itself at high temperatures, where the ceramics become relatively ductile

A dislocation has a stress field associated with it. (See Chapter 4.) Solute atoms, especially when their sizes are too large or too small in relation to the size of the host atom, are also centers of elastic strain. A solute atom is said to be a *point source* of dilation. A vacancy (i.e., a vacant lattice site) can also be considered a point source of (negative) dilation. Consequently, the stress fields from these sources (dislocations and point defects) can interact and mutually exert forces. Such an interaction due to size difference is called the *elastic misfit interaction* or *dilational misfit interaction*. Other types of interactions, such as electrical and chemical elastic modulus mismatch, are also possible. Each of these interactions represents an energy barrier to dislocation motion.

10.2.1 Elastic Interaction

In the case of a positive edge dislocation, there is an extra half plane above the slip plane. Hence, there will be a compressive stress above the slip plane and a tensile stress under it (Chapter 4). Because a solute atom placed randomly in a crystal has a stress field around it, this stress field would be minimized if the solute atom were to move to the dislocation. For the case of an interstitial atom of carbon in iron, the minimum-energy position at an edge dislocation is the dilated region near the core. A substitutional atom that is smaller than the solvent atom will tend to move to the compressive side. On the other hand, if a solute atom is larger than the solvent atom, it will be expected to move to the tensile side. Substitutional atoms such as Zn in Cu give rise to a completely symmetrical spherical distortion in the lattice, which corresponds to the elastic misfit problem associated with inserting a ball in a bigger or smaller hole; that is, the substitutional solute atom acts as a point source of dilation of spherical symmetry. It is important to note that such spherically symmetric stress fields due to substitutional impurity atoms can interact only with defects that have a hydrostatic component in their stress fields, as happens to be the case with an edge dislocation (see Eqns. 4.6). Screw dislocations, by contrast, have a stress field of a pure shear character; that is, the hydrostatic component of a screw dislocation is zero (see Eqns. 4.4 and 4.5). Therefore, to a first approximation, there is no interaction between screw dislocations and substitutional atoms, such as Zn in Cu or Mn in Fe. Interstitial atoms such as carbon or nitrogen in α-iron, however, not only produce a dilational misfit (in volume), but also induce a tetragonal distortion. Both carbon and nitrogen occupy interstitial positions at the face centers and/or the midpoints of the edges of the body-centered cubic structure (Figure 10.3). Carbon atoms occupy the midpoints of <001> edges. In Figure 10.3a, we indicate the positions of the carbon atoms by crosses. Figure 10.3b depicts the tetragonal distortion produced when a carbon atom moves to one of the cube edges of iron. The cubic shape changes to tetragonal, producing a tetragonal distortion along that particular <001> axis. Note that the figure shows the tetragonal distortion produced by a carbon atom in the iron cube; the carbon atom does not have an elongated form! The strain field due to this tetragonal distortion will interact with hydrostatic as well as shear stress fields. The important effect of the tetragonal distortion is that the interstitial atoms such as C and N in iron will interact and form atmospheres at both edge and screw dislocations and will lead to a more effective impediment to the movement of dislocations than in the case of substitutional atoms. (See Figure 10.2.)

We now derive an expression for the energy of interaction between an edge dislocation and a point source of expansion, such as an oversized (or undersized) solute atom.

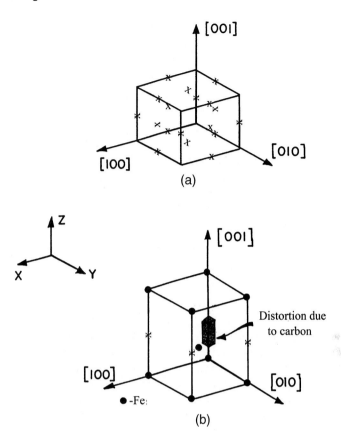

Figure 10.3 (a) Positions of interstitial atoms in the cube. (b) Carbon atom shown as a producer of a tetrogonal distortion.

This was first done by Cottrell[1] and Bilby.[2] Let σ_p be the hydrostatic component of the stress field of a dislocation, and let ΔV be the change in volume induced by the introduction of a solute atom of radius $r_0(1 + \varepsilon)$ in a cavity of radius r_0, where ε is positive. Then, for ε very small, we can write the change in volume as

$$\Delta V = \left(\frac{4}{3}\right)\pi r_0^3(1 + \varepsilon)^3 - \frac{4}{3}\,\pi r_0^3 = \frac{4}{3}\,\pi r_0^3[(1 + \varepsilon)^3 - 1],$$

so that

$$\Delta V \approx \frac{4}{3}\,\pi r_0^3 3\varepsilon,$$

or

$$\Delta V = 4\pi r_0^3\varepsilon. \tag{10.1}$$

[1]A. H. Cottrell, in Conference on Strength of Solids, Physical Society, 1968, p. 30, London.
[2]B. A. Bilby, Proc. Phys. Soc., A63 (1950) 191.

The stress field of an edge dislocation is given in rectangular coordinates in Chapter 4 (Eqns. 4.6). In cylindrical coordinates, we have (the student can do this as an exercise).

$$\sigma_{rr} = \sigma_{\theta\theta} = -\frac{Gb}{2\pi(1-\nu)}\frac{\sin\theta}{r},$$

$$\sigma_{zz} = -\frac{\nu Gb}{\pi(1-\nu)}\frac{\sin\theta}{r},$$

$$\sigma_{r\theta} = \frac{Gb}{2\pi(1-\nu)}\frac{\cos\theta}{r},$$

$$\sigma_{\theta z} = \sigma_{zr} = 0.$$

The hydrostatic pressure σ_p is, by definition, equal to $-\frac{1}{3}(\sigma_{rr} + \sigma_{\theta\theta} + \sigma_{zz})$. Thus, the hydrostatic stress associated with an edge dislocation, obtained from the preceding stress field, is

$$\sigma_p = \frac{1+\nu}{1-\nu}\frac{Gb}{3\pi}\frac{\sin\theta}{r}. \tag{10.2}$$

If we wish to convert this expression into rectangular coordinates, we need only use the relationship $r = (x^2 + y^2)^{1/2}$ and $\sin\theta = y/(x^2 + y^2)^{1/2}$.

The interaction energy (U_{int}) was defined by Eshelby[†] for a general ellipsoid of volume V in which both deviatoric and hydrostatic components of strain are generated and a general external stress field σ_{ij} as

$$U_{int} = V\sigma_{ij}(\varepsilon_{ij})_T,$$

where $(\varepsilon_{ij})_T$ is the strain tensor due to the transformation. For the simplified case of the solute atom, the stress is σ_p and the strain is ΔV per unit volume. We can calculate U_{int} in the following manner. Figure 10.4 shows the coordinates of a solute atom in the strain field of a dislocation. The elastic interaction energy due to misfit, U_{misfit}, for a solute atom at (r, θ) and at the dislocation origin $(0, 0)$ can be obtained from Equations (10.1) and (10.2) as

$$U_{misfit} = \sigma_p\Delta V = \frac{1+\nu}{1-\nu}\frac{Gb}{3\pi}\frac{\sin\theta}{r}4\pi\varepsilon r_0^3 = A\frac{\sin\theta}{r}, \tag{10.3a}$$

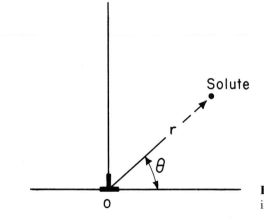

Figure 10.4 (r, θ)-coordinates of a solute atom in the strain field of an edge dislocation.

[†]J. D. Eshelby, *Proc. Roy. Soc. 241* (1957) 376; *A252* (1959) 561.

where

$$A = \frac{4}{3} \frac{1+\nu}{1-\nu} Gb\varepsilon r_0^3.$$

The force exerted by the solute on the dislocation is

$$F = -\partial U_{\text{misfit}}/\partial r = A \sin \theta/r^2. \qquad (10.3b)$$

Hence, solute atoms are attracted to dislocations and form what is called a "Cottrell atmosphere" around them, pinning them. This is especially true for interstitials, which tend to have a high mobility. Calculations similar to the foregoing can be carried out for interstitials that cause nonspherical distortions. Equation 10.3a is derived on the basis of linear elasticity theory; thus, it will not be valid at the dislocation core region, where linear elasticity does not apply. This is a great omission, as the binding energy will be a maximum precisely at the dislocation core. Therefore, the reader is forewarned that the interaction energy just determined is only an estimate. Consider again Equation 10.3a and Figure 10.4. The interaction energy U is positive in the region above the slip plane $(0 < \theta < \pi)$ and is negative below the slip plane $(\pi < \theta < 2\pi)$ for large solute atoms. This means that a solute atom larger in size than the matrix atom (i.e., ΔV positive) will be repelled by the compressive side of the edge dislocation and will be attracted by the tensile side, as the interaction energy will be negative there. For a solute atom of smaller size than the matrix atom (i.e., ΔV negative), the interaction energy will be negative in the upper part $(0 < \theta < \pi)$ of the dislocation, and the solute will be attracted there. In both cases, solute atoms will migrate to the dislocation, which will result in a reduction in the free energy of the system.

It is possible to estimate the increase in stress required to move a dislocation from the number of solute atoms surrounding it. Let the maximum force F_{max} between a dislocation and a solute atom be given by Equation 10.3b. When r reaches a sufficiently low value, we have (from Peach–Koehler's equation, Eqn. 4.17c, Chapter 4)

$$\Delta\tau = F_{\text{max}}/bL = A \sin \theta/r^2 bL,$$

where L is the spacing of solute atoms that "pin" a dislocation. A number of assumptions can be made to establish F_{max} and L. We shall make the very simple assumption that all solute atoms move to certain distance b from the dislocation. If C is the concentration of solute atoms per unit volume and ρ is the dislocation density (equal to the dislocation length per unit volume), then the spacing between solute atoms along a dislocation is

$$L = \rho/C.$$

Thus;

$$\Delta\tau = A \sin \theta \, C/r^2 b\rho.$$

If $r \approx b$ and $\sin \theta \approx 1$, we get

$$\Delta\tau = A \, C/b^3\rho.$$

Figure 10.2 shows such dependence of change in yield stress on solute content. On the other hand, if the solute atoms form a rigid network in the lattice, the average spacing between solute atoms is

$$L' \approx C^{-1/3}.$$

The spacing L of solute atoms along the plane of a moving dislocation is determined by several factors, including the angle of bowing out of the dislocation between obstacles. This misfit interaction energy is such a case can be calculated using statistics.

We have for screw dislocations

$$U_{\text{misfit}} = \varepsilon_{ij} \, \sigma_{ij}^{\text{screw}} \, \Omega,$$

where $\sigma_{ij}^{\text{screw}}$ represents the stress field associated with a screw dislocation and Ω is the specific volume given by $\Delta V = 3\Omega\varepsilon$, in which ε is the misfit parameter. The resultant force exerted by the solute atom of the dislocation has an equation similar in form to that in the case of a substitutional solute atom, but with the misfit parameter replaced by $(\varepsilon_{11} - \varepsilon_{22})/3$. A substitutional solute atom produces an isotropic strain field, i.e., $\varepsilon_{11} = \varepsilon_{22} = \varepsilon_{33}$, and it does not interact with a screw dislocation. However, for C in α-Fe, one has the extensional strain in the [100] direction, $\varepsilon_{11} = 0.38$, while the contractional strain along each of the two orthogonal directions, [010] and [001], is -0.026. Therefore, C in α-Fe hinders both edge and screw dislocations. It turns out that interstitial atoms with $(\varepsilon_{11} - \varepsilon_{22})$ as much as unity can show solubility in BCC metals. The reason for this is that metals can accommodate a greater uniaxial distortion than isotropic distortion by solute atoms, since the electron energy depends mainly on the specific volume.

10.2.2 Other Interactions

Besides the dilation and elastic misfit interactions there are other sources of dislocation–solute interactions: interactions due to a difference in modulus between the solute and the solvent, electrical interaction, chemical interaction, and local-order interaction due to the fact that a random atomic arrangement may not be the minimum-energy state in a solid solution. All these interactions will further hinder dislocation motion in a solid solution. Generally, however, their contributions are less important than the size effect described earlier.

10.3 MECHANICAL EFFECTS ASSOCIATED WITH SOLID SOLUTIONS

Many important mechanical effects are associated with the phenomenon of solid solution. In the case of steels, solute-dislocation interaction leads to a migration of interstitial solute atoms to a dislocation, where they form an atmosphere around it. This solute atmosphere, called the *Cottrell atmosphere,* has the effect of locking–in the dislocation, making it necessary to apply more force to free the dislocation from the atmosphere. This results in the well-known phenomenon of a pronounced yield drop in annealed low-carbon steels. A word of caution is in order here. Temperature is an important variable in the migration of solute atoms to a dislocation. If the temperature is too low, the solute may not be able to diffuse to allow a redistribution of solute atoms to dislocations. Such a redistribution may be thermodynamically expected, but if the temperature is too low, it will not occur in a reasonable length of time. At very high temperatures ($>0.5T_m$, where T_m is the melting point in kelvin), the mobility of foreign atoms will be much higher than that of dislocations, with the result that they will not restrict dislocation motion. In the range of temperatures where solute atoms and dislocations are about equally mobile, strong interactions with dislocations occur. The serrated stress–strain curve (or the Portevin–Le Chatelier effect) is another manifestation of this. We next describe some technologically important effects of solid-solution hardening.

EXAMPLE 10.1

Why are substitutional solid solutions more common than interstitial solid solutions?

Solution: Substitutional solid solutions are more common than interstitial ones mainly because of the atomic size limitations. Substitutional solid solubility can be quite appreciable—up to a difference of 14% in the atomic diameters of two metals. Copper (atomic radius = 0.128 nm), for example, can dissolve up to about 35% of zinc (atomic radius = 0.1331 nm) atoms in a substitutional manner. Cu and Ni (atomic radius = 0.1246 nm) have complete miscibility, from 0 to 100%. In the case of interstitial solid solutions, a small atom (C, N, or H, for instance) has to lodge itself in the interstices of the solvent metal atoms. C in γ-Fe (FCC) has available larger sized interstices than it does in α-Fe (BCC), although there are more interstices available in the latter. This is the reason that C has a comparatively greater solubility in γ-Fe than in α-Fe. In general, however, the range of interstitial hole sizes available is not very large—hence, the less common occurrence of interstitial solid solutions.

10.3.1 Well-Defined Yield Point in the Stress–Strain Curves

A schematic stress–strain curve exhibiting a well-defined yield point is shown in Figure 10.5a. Characteristically, annealed low-carbon steels show such stress–strain behavior. According to the theory of Cottrell and Bilby, the dislocations in annealed steels ($\rho \sim 10^7 \text{ cm}^{-2}$) are locked–in by the interstitial solute atoms (carbon). When stress is applied to such a steel in a tensile test, it must exceed a certain critical value to unlock the dislocations. The stress necessary to move the unlocked dislocations is less than the stress necessary to free them—hence the phenomenon of a sharp yield drop and the appearance of an upper and lower yield point in the tensile stress–strain curve. The solute atoms segregate to the dislocations because this results in a decrease in the free energy. Given proper conditions for atomic diffusion, one would expect complete segregation of solute atoms to dislocations. Figure 10.6 shows schematically a Cottrell atmosphere of C atoms at a dislocation core in iron. In iron, atoms of carbon and nitrogen diffuse easily at ambient temperatures, but in many substitutional alloys one has to resort to treatments at higher temperatures.

Figure 10.5b shows how a steel containing 0.008% C reacts to one-hour aging treatments. The progressive formation of a yield point with a subsequent plateau is clearly seen[3] After the dislocations have freed themselves from the Cottrell atmosphere, all the tensile curves in Figure 10.5b become identical.

10.3.2 Plateau in the Stress–Strain Curve and Lüders Band

After the load drop corresponding to the upper yield point, there follows a plateau region in which the stress fluctuates around a certain value. The elongation that occurs in this plateau is called the *yield-point elongation.* (See Figure 10.5.) It corresponds to a region of

---

[3]R. Foley, personal communication.

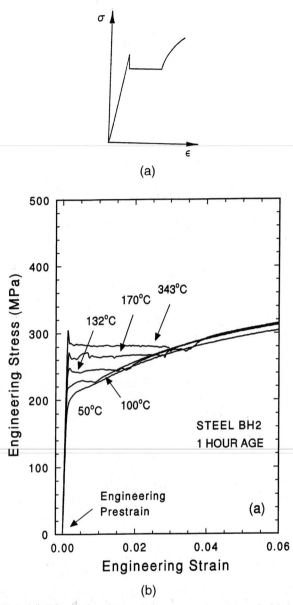

Figure 10.5 (a) Schematic stress–strain curve of an annealed low-carbon steel showing the yield-point phenomenon. (b) Low-carbon steel in a temper-rolled condition and annealed for one hour between 100°C and 34.3°C). (Courtesy of R. Foley)

nonhomogeneous deformation. In a portion of the tensile sample where there is a stress concentration, a deformation band appears such as that indicated in Figure 10.7a. As the material is deformed, this band propagates through the test sample. An intermediate position is indicated in Figure 10.7b. The deformation is restricted to the interface. This deformation band is known as the *Lüders band*. In the plateau region of the stress–strain curve, there could be two or more such bands. Sometimes Lüders bands are visible to the naked eye. After the formation of the last band, the stress–strain curve resumes its normal trajectory of strain-hardening. Knowing the cross-sectional areas A_1 and A_2 in Figure 10.7b, one can determine the number of Lüders bands from the yield-point elongation. One can also determine, from the strain rate of the sample, the speed of propagation of

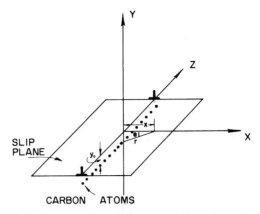

Figure 10.6 Cottrell atmosphere in iron consisting of an edge dislocation and a row of carbon atoms.

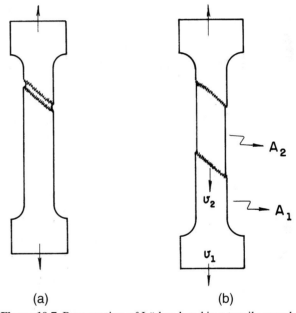

(a)   (b)

Figure 10.7 Propagation of Lüders band in a tensile sample.

these bands. An aspect of great technological importance is the formation of Lüders bands during the stamping of low-carbon steels, with the consequent irregularities in the final thickness of the sheet (see Section 3.9.2). This problem is tackled, in practice, in two ways:

1. By changing the composition of the alloy to eliminate the yield point. The addition of aluminum, vanadium, titanium, niobium, or boron to steel leads to the formation of carbides and nitrides as precipitates, which serve to remove the interstitial atoms from the solid solution.

2. By prestraining the sheet to a strain greater than the yield point strain such that the strains during the stamping operations occur in the strain-hardening region.

The explanation for the formation of Lüders bands is intimately related to the cause of the appearance of the well-defined yield point. The unlocking of dislocations that occurs at the upper yield point is, initially, a localized phenomenon. The unlocked dislocations move at a very high speed, because the stress required to unlock them is much higher than the stress required to move them, until they are stopped at grain boundaries. The stress concentration due to the dislocations that accumulate at grain boundaries unlocks the dislocations in the neighboring grains.

10.3.3 Strain Aging

As pointed out in the preceding sections, prestraining the steel to a strain greater than the yield strain will result in the removal of the yield point. However, if we let the sample rest before retesting, the yield point will return. This phenomenon is known as *strain aging*. Figure 10.8 shows the result of experiments done with an annealed austenitic alloy of composition $Fe - 31\% Ni - 0.1\% C$. The tensile test was stopped three times, each time for 3 hours, after three different strains: $\varepsilon = 0.08, 0.18,$ and 0.27. The test was stopped simply by turning off the machine. Initially, the sample did not show a well-defined yield point. However, on reloading after the 3-hour rest, the stress–strain curve showed clearly the appearance of a yield point followed by a plateau—i.e., a horizontal load-drop region—and, finally, a return to the original trajectory. The dashed lines indicate the values of stress at which the test was stopped. Note that, on reloading, the yield stress of the alloy increased for the three strains. The term "aging" is normally used when a precipitate forms. (See Section 10.4.) However, this is not the case in the example at hand. As the test was alternatively carried out and interrupted at ambient temperature, interstitial atoms would migrate to dislocations during the interruptions, locking the dislocations. On reloading, the dislocations were unlocked, and a well-defined yield point appeared. The experiments were carried out under identical conditions, but keeping the test sample unloaded for three hours. The well-defined yield point reappeared, but it was less marked. The above experiment indicates that the applied stress has an accelerating effect on the strain-aging process. Generally, low-carbon steels show strain aging.

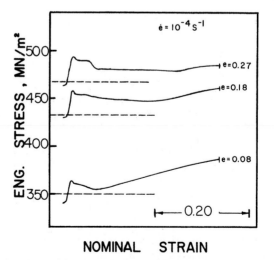

Figure 10.8 Reloading curves after stopping a test for three hours at nominal strains of 0.08, 0.18, and 0.27. The dashed lines indicate the stresses at which the test was stopped. Note the formation of a well-defined yield point in the three cases (Reprinted with permission from M. A. Meyers and J. R. C. Guimarães, *Metalurgia—ABM,* 34 (1978) 707).

Another commonly observed effect due to strain aging is an enhancement of the work-hardening rate, leading to an increase in the ultimate tensile strength of the material. This effect is sometimes referred to as *dynamic strain aging,* because it occurs concurrently with plastic deformation. In some cases, even the plot of flow stress vs. temperature shows a hump. The hump in the curve of ultimate tensile strength (UTS) vs. temperature for a nickel-based superalloy, Inconel 600, is shown in Figure 10.9. This hump is caused by solute atoms that have a mobility higher than the dislocations and that, therefore, can continue to "drag" them, leading to increased work-hardening. This enhanced work-hardening leads to a higher UTS. Note that the yield stress does not show such a hump.

10.3.4 Serrated Stress–Strain Curve

Under certain conditions, some metallic alloys show irregularities in their stress–strain curves that can be caused by the interaction of solute atoms with dislocations, by mechanical twinning, or by stress-assisted ("burst"-type) martensitic transformations. The first type (i.e., due to solute–dislocation interaction) has been called the *Portevin–Le Chatelier* effect. It generally occurs within a specific range of temperatures and strain rates. The solute atoms, being able to diffuse through the test sample at a speed greater than the displacement speed of the dislocations (imposed by the applied strain rate), "chase" the dislocations, eventually locking them. With increasing load, the unlocking of dislocations causes a load drop with the formation of small irregularities in the stress–strain curve.

Irregularities in the stress–strain curves of a nickel–iron-based superalloy, Inconel 718, tested in tension at 650°C after different processing schedules, are shown in Figure 10.10. Interactions between the solute atoms are more intense in the metal's undeformed condition. The stabilization treatment of 554°C for one hour produced more pronounced load drops. The Portevin–Le Chatelier effect is dependent on the density of dislocations, strain rate, concentration and mobility of solute atoms, and other factors. The effect occurs in a region where there is inverse strain-rate sensitivity (i.e., if the strain is increased, the flow stress decreases). This relationship is due to the interplay between solute atoms and dislocations. Under normal conditions—that is, in the absence of solute—the flow stress

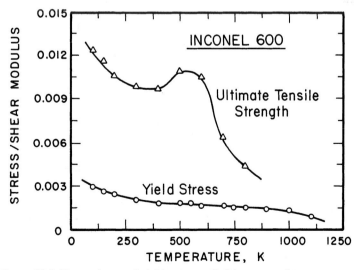

Figure 10.9 Dependence of yield stress and ultimate tensile stress on temperature for Inconel 600, a nickel-based superalloy. The hump in the curve due to dynamic strain aging is usually evident only at large strains (after R. A. Mulford and U. F. Kocks, *Acta Met.,* 27 (1979) 1125).

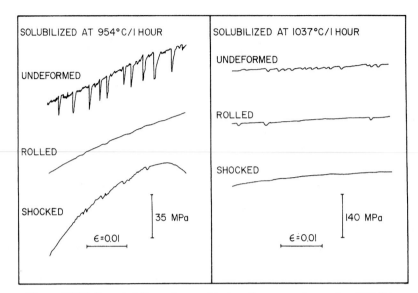

Figure 10.10 Serrated flow observed in tensile test performed at 650°C in Inconel 718 (a nickel–iron-based superalloy) solubilized at two temperatures. The undeformed, cold-rolled (19.1% reduction) and shock-loaded (51 GPa peak pressure) conditions are shown (From M. A. Meyers, Ph.D. dissertation, 1974).

increases with strain rate. In the Portevin–Le Chatelier regime, an increase in strain rate frees the dislocations from the solute atoms.

10.3.5 Snoek Effect

Interstitial solute atoms such as carbon and nitrogen can, under the action of an applied stress, migrate in the α-Fe lattice. Such short-range migrations of C or N can result in an anelastic or internal friction effect, called the Snoek effect after the person who discovered it. As mentioned earlier, carbon or nitrogen atoms occupy the octahedral interstices located at the centers of the cube edges and at the centers of the cube faces. If we apply a stress along the z, or [001], direction, the octahedral interstices along the x- and y-axes will contract, while the ones in the z-direction will expand. Given the right time and temperature, the interstitial atoms will move to sites along the z-axis. Such a change of site leads to a reduction in strain energy. On the other hand, a stress applied in the [111] direction will not result in a change of site, because all three of the cube's directions will be equally stressed and, on average, equally occupied by the carbon atoms. Such a movement of interstitials, when stress is applied along a cube direction and at levels less than the yield stress, can cause strain to lag behind stress; that is, the material will show the phenomenon of *internal friction*. The effect of this internal friction is commonly measured by a torsional pendulum. The angle of lag is called δ, and $\tan \delta$ is taken as a measure of the internal friction. Mathematically,

$$\tan \delta = \frac{\log \text{ decrement}}{\pi} = Q^{-1},$$

where the logarithmic decrement is the ratio of successive amplitudes of the swing of the pendulum. If the amplitude decays to $1/n$ of its original value in time t, then

$$\tan \delta = Q^{-1} = \frac{\ln(1/n)}{\pi \nu t},$$

where ν is the vibrational frequency of the pendulum.

Only the interstitials that occupy the normal sites in an undistorted lattice will contribute to internal friction. Interstitials in the strain fields of a dislocation or a substitutional solute atom, or those at a grain boundary, will have their behavior altered. Thus, the Snoek effect can be used to measure C or N concentration in *high-purity* ferrite, i.e., BCC α-Fe. Would you expect to observe the Snoek effect in γ-Fe?

10.3.6 Blue Brittleness

Carbon steels heated in the temperature range of 230 and 370°C show a notable reduction in elongation. This phenomenon is due to the interaction of dislocations in motion with the solute atoms (carbon or nitrogen) and is intimately connected with the Portevin–Le Chatelier effect. We classify it separately because of its distinct importance. When the temperature and the strain rate are such that the speed of the interstitial atoms is more than that of the dislocations, the latter are continually captured by the former. This results in a very high strain-hardening rate and strength with a reduction in elongation. With increasing strain rates, the effect occurs at higher temperatures, as diffusivity increases with temperature. Called *blue brittleness,* this effect refers to the coloration that the steel acquires due to the oxide layer formed in the given temperature range. In the range of temperature and strain rate in which the material is subjected to dynamic aging, the strain-rate sensitivity is also affected, tending to increase linearly with temperature. However, in the presence of dynamic aging, the strain-rate sensitivity becomes very small, and the yield stress becomes practically independent of the strain rate.

10.4 PRECIPITATION- AND DISPERSION-HARDENING

Precipitation-hardening, or age-hardening, is a very versatile method of strengthening certain metallic alloys. Two important alloy systems that exploit this strengthening technique are aluminum alloys and nickel-based superalloys. Figure 10.11 shows examples of precipitates in some systems. Figure 10.11 (a) shows a typical example of an Al–Cu alloy, with θ (CuAl$_2$) precipitates at the grain boundaries and θ' (Cu$_2$Al) precipitates in the grain interiors, Figure 10.11(b) shows Al$_3$Li precipitates in an Al–Li alloy, and Figure 10.11 (c) shows γ' (Ni$_3$Al) precipitates and aged carbides in a nickel-based superalloy. The aging treatment involves the precipitation of a series of metastable and stable precipitates out of a homogeneous, supersaturated solid solution. Various metastable structures offer different levels of resistance to dislocation motion. Figure 10.12 shows the variation in hardness with aging time in the aluminum–copper system. Also shown are the different types of precipitate that occur during the aging treatment. Peak hardness or strength corresponds to a critical distribution of coherent or semicoherent precipitates.

In dispersion-hardening, we incorporate hard, insoluble second phases in a soft metallic matrix. Here, it is important to distinguish dispersion-strengthened metals from particle-reinforced metallic composites. The volume fraction of dispersoids in dispersion-strengthened metals is generally low, 3–4% maximum. The idea is to use these small, but hard, particles as obstacles to dislocation motion in the metal and thus strengthen the metal or alloy without affecting its stiffness is any significant way. In the case of metallic particulate composites, the objective is to make use of the high stiffness of particles such as alumina to produce a composite that is stiffer than the metal alone. Improvements in strength, especially at high temperatures, also result, but at the

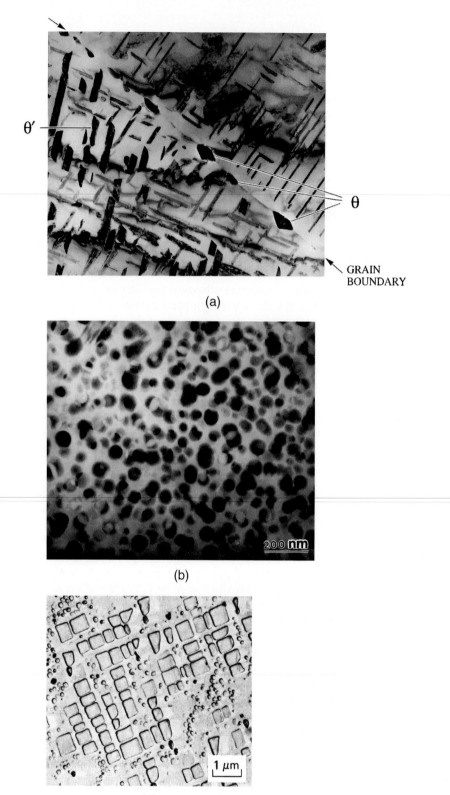

Figure 10.11 (a) θ precipitates (at grain boundaries) and θ' precipitates (in grain interior) in Al–Cu alloy. (Courtesy K. S. Vecchio). (b) Al$_3$Li precipitates in Al–Li alloy (TEM, dark field). (Courtesy K. S. Vecchio) (c) γ' precipitates and aged carbides in a superalloy. (courtesy of R. N. Orava)

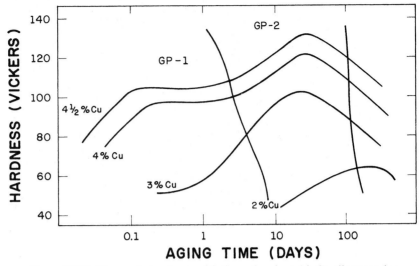

Figure 10.12 Change in hardness with time of various Al–Cu alloys aged at 130°C (Adapted with permission from H. K. Hardy and T. J. Heal, *Prog. Metal Phy.,* 5 (1954) 195).

expense of ductility and toughness. Examples of dispersion-strengthened systems include Al_2O_3 in Al or Cu, ThO_2 in Ni, and more. TD Nickel is the name of an oxide-dispersion-strengthened nickel. Very small spherical particles (20–30 nm in diameter) of thorium dioxide (ThO_2) are dispersed in nickel matrix by powder metallurgical processing. Dispersion-strengthened copper is made by an internal oxidation technique. An alloy of copper and a small amount of aluminum is melted and atomized into a fine powder. Heating the powder under oxidizing conditions leads to an in situ conversion of aluminum into alumina. Any excess oxygen in the copper is removed by heating the powder in a reducing atmosphere. The powder is then consolidated, followed by conventional metalworking. We give some examples of dispersion-hardened systems later in this chapter. Suffice it here merely to point out that dispersion-hardened systems have one great advantage over those hardened by precipitation, viz., the stability of the dispersoids. Thus, dispersion-hardened systems maintain high strength at high temperatures, at which precipitates tend to dissolve in the matrix. Figure 10.13 illustrates the differences between strengthening by precipitation and by dispersion-hardening. Nickel-based superalloys IN792 and MAR M-200 are precipitation-hardened by γ'' or γ' precipitates having compositions of Ni_3Nb and Ni_3Al, respectively. The TD nickel, on the other hand, contains a fine dispersion of ThO_2, a high-melting-point oxide that is insoluble in the matrix. At lower temperatures (up to 1,000°C), precipitation hardening is more effective; however, at approximately 1,100°C, the precipitates dissolve in the matrix and the strength is drastically reduced. The dispersoids continue to be effective strengtheners at still higher temperatures.

The strengthening in these systems, hardened by either precipitates or dispersoids, has its origin in the interaction of dislocations with the particles. In general, the interaction depends on the dimensions, strength, spacing, and amount of the precipitate. The detailed behavior, of course, differs from system to system. Let us first describe the phenomenon of precipitation-, or age-, hardening. The supersaturated solid solution is obtained by sudden cooling from a sufficiently high temperature at which the alloy has a single phase. The heat treatment that causes precipitation of the solute is called *aging*. The process can be applied to a number of alloy systems. Although the specific behavior varies with the alloy, the alloy must, at least:

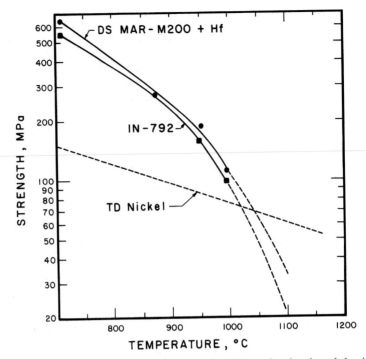

Figure 10.13 Comparison of yield strength of dispersion-hardened thoria-dispersed (TD) nickel with two nickel-based superalloys strengthened by precipitates (IN-792) and directionally solidified (DS) Mar M 200.

1. Form a monophase solid solution at high temperatures.
2. Reject a finely dispersed precipitate during aging, i.e., the phase diagram must show a declining solvus line.*

Figure 10.14(a) shows a part of the phase diagram of the Al–Cu system in which precipitation-hardening can occur, while Figure 10.14(b) shows the phase diagram of the Al–Li system. Lithium is interesting in that its addition to aluminum results in a lowering of the density, as well as a substantial increase in the modulus of the alloy. Both of the systems shown in Figure 10.14 fulfill the prerequisites for precipitation-hardening to occur. The precipitation treatment consists of the following steps:

1. *Solubilization.* This involves heating the alloy to the monophase region and maintaining it there for a sufficiently long time to dissolve any soluble precipitates.
2. *Quenching.* This involves cooling the single-phase alloy very rapidly to room temperature or lower so that the formation of stable precipitates is avoided. Thus, one obtains a supersaturated solid solution.
3. *Aging.* This treatment consists of leaving the supersaturated solid solution at room temperature or at a slightly higher temperature. It results in the appearance of fine-scale precipitates.

Table 10.1 presents some precipitation-hardening systems, with the precipitation sequence and the equilibrium precipitate shown. Although the behavior of different systems varies in detail, one may write the general aging sequence as follows:

supersaturated solid solution → transition structures → aged phase.

---

*The solvus line is the locus of points representing the limit of solid solubility as a function of temperature..

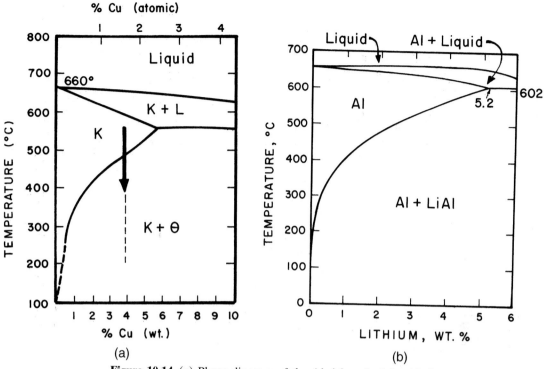

Figure 10.14 (a) Phase diagram of the Al-rich end of the Al–Cu system.
(b) Phase diagram of the Al-rich end of Al–Li system.

TABLE 10.1 Some Precipitation-Hardening Systems

Base Metal	Alloy	Sequence of Precipitates
Al	Al-Ag	Zones (spheres) ———— γ' (plates) ——— γ(Ag,Al)
	Al-Cu	Zones (disks) ——— θ'' (disks) ——— θ' ——— θ(CuAl$_2$)
	Al-Zn-Mg	Zones (spheres) ——— M' (plates) ——— (MgZn$_2$)
	Al-Mg-Si	Zones (rods) ——— β' ——— (Mg$_2$Si)
	Al-Mg-Cu	Zones (rods or spheres) ——— S' ——— S(Al$_2$CuMg)
	Al-Li-Cu	Zones——>θ''——>θ'——>θ(CuAl$_2$) Ti(CuAl$_2$Li) δ'——>δ(AlLi)
Cu	Cu-Be	Zones (disks) ——— γ' — γ(CuBe)
	Cu-Co	Zones (spheres) ——— β
Fe	Fe-C	ε-Carbide (disks) ——— Fe$_3$C("laths")
	Fe-N	α'' (disks) ——— Fe$_4$N
Ni	Ni-Cr-Ti-Al	γ' (cubes) ——— γ(Ni$_2$Ti, AR)

In the initial stages of the aging treatment, zones that are coherent with the matrix appear. These zones are nothing but a clustering of solute atoms on certain crystallographic planes of the matrix. In the case of aluminum–copper, the zones are a clustering of copper atoms on [100] planes of aluminum. The zones are transition structuring and are referred to as *Guinier–Preston zones,* or *GP zones,* in honor of the two researchers who first discovered them. We call them zones rather than precipitates in order to emphasize the fact the zones represent a small clustering of solute atoms that has not yet taken the form of precipitate particles. The GP zones are very small and have a very small lattice mismatch with the aluminum matrix. Thus, they are coherent with the matrix; that is, the lattice planes cross the interface in a continuous manner. Such coherent interfaces have very low energies, but there are small elastic coherency strains in the matrix. As these coherency strains grow, the elastic energy associated with them is reduced by the formation of semicoherent zones where dislocations form at the interface to take up the strain. Further growth of the semicoherent zones, or precipitates, results in a complete loss of coherency: An incoherent interface forms between the precipitate and the matrix.

The precipitate produced during the aging treatment can be coherent, semicoherent, or incoherent with the matrix (Figure 10.15). Coherency signifies that there exists a one-to-one correspondence between the precipitate lattice and that of the matrix. (See Figure 10.15a and b). A semicoherent precipitate signifies that there is only a partial correspondence between the two sets of lattice planes. The lattice mismatch is accommodated by the introduction of dislocations at the noncorrespondence sites, as shown in Figure 10.15c. Figure 10.16 shows such interfacial dislocations at semicoherent interfaces. An incoherent interface, shown in Figure 10.15d, implies that there is no correspondence between the two lattices. Such an interface is also present in dispersion-hardened systems.

The shape of the aging curve (see Figure 10.12) can be explained as follows. Immediately after quenching, only solid-solution-hardening is present. As GP zones form, hardness or strength increases because extra stress is needed to make dislocations shear the coherent zones. The hardness increases as the size of the GP zones increases, making it ever more difficult for the dislocations to shear the zones. As time goes on, incoherent equilibrium precipitates start appearing, and the mechanism of Orowan bowing (see Section 10.5) of dislocations around the particles becomes operative. The peak hardness or strength is associated with a critical dispersion of coherent or semicoherent precipitates. Further ag-

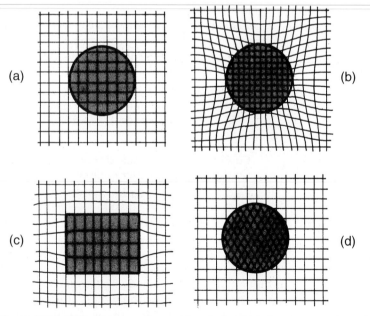

Figure 10.15 Different crystallographic relationships between matrix and second phase. (a) Complete coherency. (b) Coherency with strained, but continuous, lattice planes across the boundary. (c) Semicoherent, partial continuity of lattice planes across the interface. (d) Incoherent equilibrium precipitate, θ; no continuity of lattice planes across the interface.

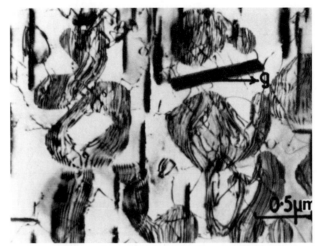

Figure 10.16 Interfacial dislocations formed in a semicoherent precipitate. (From G. C. Weatherly and R. B. Nicholson, *Phil. Mag.,* 17 (1968), 801) Figure 7, p. 813).

ing results in an increase in the interparticle distance, and a lower strength results as dislocation bowing becomes easier.

EXAMPLE 10.2

Dispersion-hardened materials have a stable microstructure at high temperatures, compared to precipitation-hardened materials. Is there any advantage, then, to strengthening by precipitation over that by dispersion of strong, inert particles?

Solution: In general, because it is very hard, a dispersion-hardened material can be very difficult to machine or work. A precipitation-hardened material, on the other hand, can be machined or worked before it is given the aging treatment, i.e., when it is soft. After machining, one can give the material the appropriate aging treatment to get the maximum strength and hardness.

EXAMPLE 10.3

Figure E10.3 shows the Al–Mg phase diagram. For an alloy with 5% Mg by weight, calculate the Al_2Mg (β) equilibrium volume fraction of precipitate if the densities of Al and Al_2Mg are 2.7 and 2.3 g/cm^3, respectively.

Solution: Basically, we have to calculate the volume fraction of Al_2Mg at room temperature (25°C). However, in this phase diagram, the data are given to 100°C. We assume that there is not much change between 100°C and 25°C. Applying the lever rule,* we find that the fraction of β (Al_2Mg) by weight is

$$\frac{5 - 1}{35 - 1} = 0.12.$$

---

*The lever rule allows us to compute the relative phase amounts in a two-phase alloy. The student should consult any book on materials engineering.

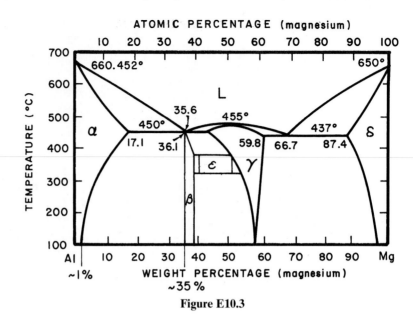

Figure E10.3

Changing this into the volume fraction (divide the respective mass fraction by the density of each component) we have

$$f = \frac{0.12/2.3}{0.88/2.7 + 0.12/2.3} \left(= \frac{V_{Al_2Mg}}{V_{Al} + V_{Al_2Mg}} \right)$$

$$= 0.14 \quad \text{or} \quad 14\%.$$

10.5 DISLOCATION–PRECIPITATE INTERACTION

Finely distributed precipitates present an effective barrier to the motion of dislocations. Two of the important models that explain the strengthening due to precipitates respectively assert that (1) dislocations cut through the particles in the slip plane and (2) dislocations circumnavigate around the particles in the slip plane. Depending on both the nature of the precipitate and the crystallographic relationship between the precipitate and the matrix, we can have two limiting cases:

1. *The precipitate particles are impenetrable to the dislocations.* Orowan pointed out that if a ductile matrix has second-phase particles interpenetrating the slip plane of dislocations, an additional stress will be necessary to make a dislocation expand between the particles. The applied stress should be sufficiently high to bend the dislocations in a roughly semicircular form between the particles. If so, the dislocations will extrude between the particles, leaving dislocation loops around them, as per the mechanism shown schematically in Figure 10.17a. Under an applied shear stress τ, the dislocation bows in between the precipitate particles until segments of dislocation with opposite Burgers vector cancel each other out, leaving behind dislocation loops around the particle. An example of Orowan bowing is shown in a transmission electron micrograph in Figure 10.17(b). The material is an Al–0.2% Au alloy, solution annealed, followed by 60 hours at 200°C and 5% plastic deformation. At points marked A in this figure, one can see dislocations pinned by the pre-

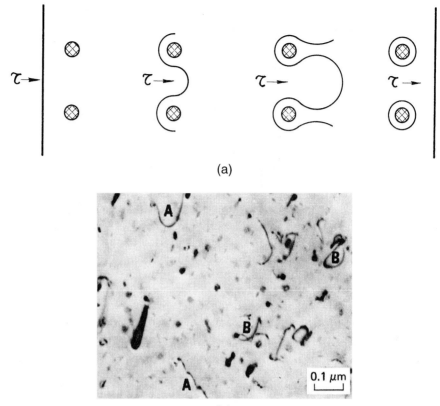

(a)

(b)

Figure 10.17 (a) The Orowan model (after E. Orowan, in *Internal Stresses in Metals and Alloys* London: Institute of Metals, 1948, p. 451) (b) Obstruction of dislocation motion by uniformly distributed nonshearing particles in an aluminum alloy (transmission electron microscope) (Courtesy of M. V. Heimendahl).

cipitates and Orowan bowing of dislocation segments. At point B, the dislocations have left the slip plane and formed prismatic dislocation loops. The dislocations in the micrograph are characteristically very short and have been severely impeded in their movement.

Now, the stress necessary to bend a dislocation to a radius r is given roughly by (see Section 4.2.4, Eqn. 4.17d)

$$\tau \approx Gb/2r. \tag{10.4}$$

Let x be the average separation between two particles in the slip lane. Then a dislocation, under a shear stress τ, must be bent to a radius on the order of $x/2$ for it to be extruded between the particles instead of cutting them. The shear stress to do this is given by making $r = x/2$ in Equation 10.4; that is,

$$\tau \approx \frac{Gb}{2r} = Gb/x. \tag{10.5}$$

Should the stress necessary to cause the particle shear be greater than Gb/x (rigorously speaking, $2T/bx$, where T is the dislocation line tension), the dislocation will bow between the particles rather than shear them. This, in essence, is the Orowan model of strengthening due to dispersion or incoherent precipitates. The increase in the yield stress due to the presence of particles is given by Equation 10.5, so that, as long as there is no particle shear, the total yield stress for an alloy strengthened by a dispersed phase or an incoherent precipitate is given by

$$\tau_y = \tau_m + Gb/x, \tag{10.6}$$

where τ_m is the critical shear stress for matrix yielding in the absence of a precipitate. Note that more precise formulations of the Orowan stress have been made, involving more accurate expressions for the dislocation line tension T and taking into account the effect of the finite particle size on the average interparticle spacing.

2. *The precipitate particles are penetrable to dislocations;* that is, the particles are sheared by dislocations in their slip planes. If the extra stress (in addition to τ_m) necessary for particle shear is less than that for bending the dislocation between the particles ($=Gb/x$), the particles will be sheared by dislocations during yielding, and we can write

$$\tau_y < \tau_m + Gb/x.$$

Thus, we see that the strength of the particle and the crystallographic nature of the particle/matrix interface will determine whether dislocations will cut the particles. In internally oxidized alloys (e.g., Cu + SiO$_2$), where the obstacles to dislocation motion are small, very hard ceramic particles with a very high shear modulus and an incoherent interface, the initial flow stress is controlled by the stress necessary to extrude the dislocations between the hard and impenetrable particles, as per the Orowan mechanism. Since the shear strength of the obstacles is generally very much higher than that of the matrix, very large stresses will be required for particle shear to occur. The initial yield stress is then controlled by the interparticle spacing. Such behavior is also shown by precipitation-hardened alloys when the equilibrium precipitate is an intermetallic compound (e.g, CuAl$_2$ in the system Al–Cu). However, in the initial stages of aging, the small precipitates or zones are coherent with the matrix and thus can be sheared by dislocations. A vivid example of particle (Ni$_3$Al) shear by dislocation is illustrated in Figure 10.18 by a Ni-19% Cr-6% Al alloy aged at 750° for 540 hours and strained 2%.

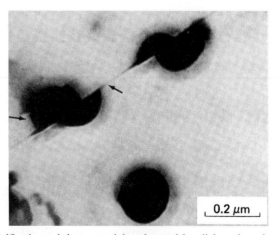

Figure 10.18 γ'-precipitate particles sheared by dislocations in a Ni-19% Cr-69% Al alloy aged at 750°C for 540 hours and strained 2%. The arrows indicate the two slip-plane traces (transmission electron microscopy) (Courtesy of H. Gleiter).

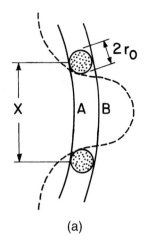

(a)

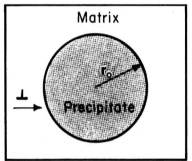

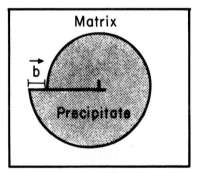

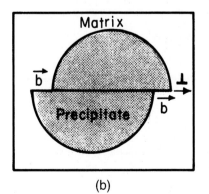

(b)

Figure 10.19 (a) Dislocation at two successive positions *A* and *B*. (b) Dislocation shearing precipitate. (c) An array of precipitates in a cubic arrangement.

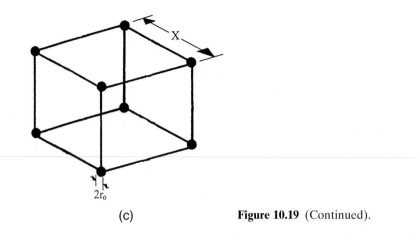

(c)   **Figure 10.19** (Continued).

Consider the encounter of a dislocation such as that at A in Figure 10.19c with spherical particles (radius r_0 in the slip plane) that are interpenetrating the slip planes. Let x be the separation between particles in the slip plane. If the particles are not strong enough to support the Orowan stress, the dislocation will cut the particles (of radius r_0) by moving from position A to position B, as shown in Figure 10.19a. The passage of a matrix dislocation, which generally will not be a slip dislocation for the precipitate, will result in a faulted plane or an interface, say, of specific energy γ. Figure 10.19b shows schematically the formation of such a plane. The increase in the energy of the particle is $\pi r_0^2 \gamma$. Then, in the absence of any thermally activated process, the stress τ_{shear} necessary to move a dislocation of length x from A to B by particle shear can be obtained as follows. A dislocation under an applied stress τ_{shear} has a force per unit length of $\tau_{shear} b$ on it; therefore, the dislocation of length x will have a force of $\tau_{shear} bx$ on it. When the dislocation moves through a distance equal to the particle diameter $(2r_0)$, the work done by this force is $\tau_{shear} bx\, 2r_0$. This work must equal the surface energy of the interface created by cutting of the particle, viz., $\pi r_0^2 \gamma$. Thus, we can write

$$\tau_{shear} bx\, 2r_0 = \pi r_0^2 \gamma,$$

or

$$\tau_{shear} = \pi r_0 \gamma / 2bx. \qquad (10.7a)$$

The particle spacing x is shown in Figure 10.19c: a simplified array of precipitates in a cubic arrangement. It is possible to express the volume fraction of precipitate as

$$f = \left(\frac{4}{3}\pi r_0^3\right)/x^3. \qquad (10.7b)$$

It is assumed that each corner precipitate contributes one-eighth to the total volume; the eight corner precipitates together count as one. Expressing Equation 10.7b in terms of r_0/x, we have

$$r_0/x = [(3/4\pi)f]^{1/3}. \qquad (10.7c)$$

Inserting equation 10.7c into 10.7a yields

$$\tau_{shear} = \text{constant } f^{1/3}.$$

Hence, the stress required to shear precipitates is a function only of the volume fraction of the particle that has been transformed. However, in the first stages of precipitation, the volume fraction transformed increases with time until it reaches the equilibrium value, which is determined from the phase diagram.

We discuss next the transition between the particle shear and Orowan bowing mechanisms. In the initial stages, as precipitation or aging continues, the precipitate particles increase in size and volume. As the size and amount of particles increase, more work needs to be done by the dislocation in shearing the particles. It turns out that [see V. Gerold and H. Haberkorn Phys. Stat. Solidi, 16 (1966) 675] the shear strength τ of the alloy depends on the particle radius r and the particle volume fraction f according to the proportionality

$$\tau \propto \sqrt{rf}.$$

With aging of the precipitates, both r and f increase. Soon, however, a stage is reached in which the precipitate volume fraction does not increase any more. Actually, the maximum precipitate volume fraction is dictated by the alloy phase diagram. The precipitate size, however, continues to increase on further aging, because larger particles tend to grow at the expense of smaller particles. This growth is called *precipitate coarsening.*[†] The thermodynamic driving force for precipitate coarsening is the decrease in surface area, and thus, surface energy of the precipitate with increasing size. In the initial stages of aging, both r and f increase, and the strength of the alloy increases. This, however, does not go on indefinitely, because, as precipitate coarsening occurs, the interparticle distance x increases. In fact, x becomes so large, that an alternative deformation process begins, viz., dislocation bowing or looping around the particles via the Orowan mechanism. This happens because the shear stress required to bow the dislocation between the particles is less than that required to shear them.

The stress necessary to bend a dislocation in between the particles, τ_{Orowan}, is given by Equation 10.5. If $\tau_{\text{shear}} > \tau_{\text{Orowan}}$, the dislocation will expand between the precipitate particles, and if $\tau_{\text{shear}} < \tau_{\text{Orowan}}$, the particles will be cut. Whether or not the particles will be sheared depends on r_0, the particle size, and on γ, the specific interface energy. For coherent precipitates—for example, the GP zones in Al alloys—the values of γ are expected to be on the order of the magnitude of antiphase domain boundaries, with a maximum value of about 100 mJ/m². Thus, from such values of γ, we can estimate that only very small particles ($2r_0 < 50$ nm) will be cut. With aging, the second-phase particles grow in size, so that the average spacing between them also increases (for a given precipitate volume fraction), and τ_{Orowan}, representing the stress necessary to bend a dislocation between particles, decreases monotonically (Figure 10.20) By contrast, the stress necessary to cut the particles increases with the particle size or aging time, because the fraction transformed increases until it reaches saturation. The figure shows the curves for dislocation bowing in between the particles and particle shear for two different equilibrium volume fractions, $f_2 > f_1$. These different volume fractions are obtained by changing the composition of the alloy. For instance, by increasing the Cu content of an Al–Cu alloy, we increase f. The corresponding transitions between the cutting and bowing mechanisms occur at r_{c1} and r_{c2}, respectively. The yield stress of the alloy as a function of aging time will then follow the dashed curve, a resultant of the two mechanisms. The transition between the shearing of the particle and bowing between the particles will occur at r_c, given by Equation 10.7a.

10.6   PRECIPITATION IN MICROALLOYED STEELS

Steels form one of the most important groups of engineering materials. Over the years, the physical and process metallurgy of steels has continually evolved to meet newer

[†]It is also called *Ostwald ripening.*

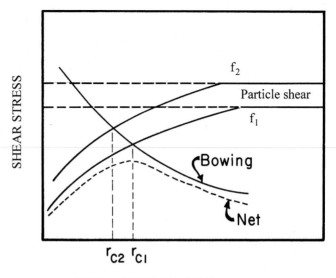

Figure 10.20 Competition between particle shear and dislocation bowing mechanisms. f_1 and f_2 represent two volume fractions of precipitates $(f_2 > f_1)$; the different volume fractions can be obtained by using different alloy compositions.

demands and challenges. The development of microalloyed steels in the second half of the 20th century can be regarded as one of the greatest metallurgical achievements ever. This success can be attributed, in a large measure, to a clearer understanding of structure–property relations in low-carbon steels. Of course, the final product resulted from a fruitful combination of physical, mechanical, and process metallurgy. Microalloyed steels[4] have largely substituted the mild steel* as the basic structural material. A microalloyed steel is a low-carbon steel (0.05 to 0.2% C, 0.6 to 1.6% Mn) that contains about 0.1% of elements such as Nb, V, or Ti. Some other elements (e.g., Cu, Ni, Cr, and Mo) may also be present in small proportions (up to about 0.1%). Elements such as Al, B, O, and N have significant effects as well. Table 10.2 lists some important second phases generally encountered in HSLA steels. Microalloyed steels are usually subjected to what is called a *controlled-rolling* treatment. Controlled rolling is nothing but a sequence of deformations by hot rolling at certain specific temperatures, followed by controlled cooling. The main objective of this treatment is to obtain a fine ferritic grain size. The ferritic grain size obtained after austenitization and cooling depends on the initial austenitic grain size, because ferrite nucleates preferentially at the austenite grain boundaries. The ferrite grain size also depends on the transformation temperature of the reaction austenite $(\gamma) \rightarrow$ ferrite (α). Lower transformation temperatures favor the nucleation rate that results in a large number of ferritic grains and, consequently, in a very small ferritic grain size (5 to 10 μm). Thus, to obtain a maximum of grain refinement, the controlled-rolling procedure modifies the hot-rolling process with a view toward exploiting the capacity of the microalloying elements so as to retard the recrystallization of the deformed austenite grains. The microalloyed additions result in the precipitation of second-phase particles during the austenitization treatment, and these particles impede the growth of the austenite grains. Hence, precipitates of, say, carbides or carbonitrides of Nb, V, or Ti can inhibit or retard

---

[4] There is some confusion about the terminology in the literature. Earlier, the commonly accepted term was "high-strength low-alloy (HSLA) steels;" later the term "microalloyed steels" gained wider acceptance.

*Mild steel is the term used to denote a carbon steel with a carbon content between 0.1 to 0.3%.

TABLE 10.2 Important Precipitates in High-Strength Low-Alloy Steels

Element(s)	Main Precipitates
Niobium	$Nb(C,N)$, Nb_4C_3
Vanadium	$V(C,N)$, V_4C_3
Niobium + molybdenum	$(Nb,Mo)C$
Vanadium + nitrogen	VN
Copper + niobium	$Cu,Nb(C,N)$
Titanium	$Ti(C,N)$, TiC
Aluminum + nitrogen	AlN

the growth of these grains, resulting in a posterior ferrite grain refinement. The Hall–Petch relationship between the yield stress and the ferrite grain size (see Section 5.3) indicates the strengthening that is possible through grain size refinement. Besides grain size strengthening, some strengthening occurs due to carbide precipitation. In summary, then, during hot rolling, the fine carbide particles form in austenite and control its recrystallization. The result is a fine ferritic grain size. Secondly, carbides of Nb, V, or Ti precipitate during and soon after the $\gamma \rightarrow \alpha$ transformation and lead to a precipitation strengthening of the ferrite. Together, these two strengthening methods lead to steels with yield strengths in the range 400 to 600 MPa and with good toughness.

We saw earlier in this chapter that, in general, when the precipitates are dispersed through the matrix with very small interparticle spacing (≤ 10 nm), the stress required to extrude the dislocations in between the particles will be very high, and dislocations will shear the particles. However, there is little evidence of such shear of precipitates in steels. The carbides, nitrides, and carbonitrides are very hard (Diamond Pyramid Hardness 2,500 to 3,000), and the presence of such hard particles in a matrix means that dislocations will be able to cut them only when they are extremely small. The critical particle size, which corresponds to a transition between the Orowan mechanism and the particle shear mechanism, decreases with an increase in the particle hardness, as shown schematically in Figure 10.21.

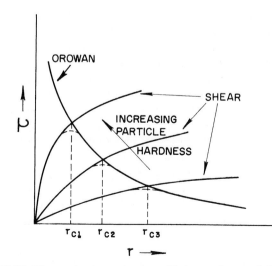

Figure 10.21 Change in shear stress τ with increasing particle radius and hardness.

EXAMPLE 10.4

Make a schematic of the precipitate volume fraction f as a function of time. Explain your diagram in terms of nucleation and growth of precipitate from a supersaturated solid solution.

Solution: The precipitates nucleate in the matrix after an initial incubation period, t_0, the time required to form stable nuclei. Following nucleation, the precipitate particles grow in size over time. Such nucleation and growth processes generally show very fast kinetics at first and then finish slowly because of the depletion of solute in the matrix. Figure E10.4 shows the desired schematic. Note the logarithmic time scale.

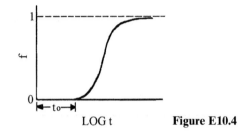

LOG t   **Figure E10.4**

EXAMPLE 10.5

Consider a precipitation-strengthened aluminum alloy. After an appropriate heat treatment, the microstructure of the alloy consists of precipitates with a mean spacing of 0.2 μm. Compute the shear stress required for Orowan bowing of dislocations in this material.

For aluminum, we have the following data:
Lattice parameter $a_0 = 0.4$ nm,
Shear modulus = 30 GPa.

Solution: The shear stress for Orowan bowing is given by

$$\tau \approx \frac{Gb}{l}.$$

The Burgers vector for aluminum (FCC) is

$$b = \frac{\sqrt{2}}{2}a = \frac{\sqrt{2}}{2}0.4 \text{ nm} = 0.29 \text{ nm}$$

We can assume that a dislocation bowing around the precipitates becomes unstable when it becomes a semicircle—that is, when its radius is equal to half the interparticle spacing. Thus,

$$r = \frac{x}{2} = 0.1 \ \mu\text{m},$$

$$\tau = \frac{30 \ 10^9 \cdot 0.28 \ 10^9}{0.1 \ 10^{-6}} = 84.85 \text{ MPa}.$$

EXAMPLE 10.6

Consider a dispersion-strengthened alloy with average interparticle spacing of λ. If N_v is the number of particles per unit volume, d is the mean particle diameter, and f is the volume fraction of particles, then show that

$$\lambda \approx d[(1/2\,f)^{1/3} - 1].$$

Solution: Let r be the mean radius of the particle, i.e., $2r = d$. Then the number of particles per unit volume, $N_v = f/(4\pi r^3/3)$, or $1/N_v = 4\pi r^3/3f$.
Taking the cube root of both sides, we obtain

$$1/(N_v)^{1/3} = (4\pi r^3/3f)^{1/3} = r(4\pi/3f)^{1/3} = 2r(\pi/6f)^{1/3}.$$

Now, the interparticle spacing λ is the average center-to-center spacing between two particles, less a particle diameter, i.e.,

$$\lambda = 1/(N_v)^{1/3} - d.$$

Hence, $\lambda = d(\pi/6f)^{1/3} - d \approx d[(1/2\,f)^{1/3} - 1].$

EXAMPLE 10.7

For the alloy Al_2Mg, calculate the critical spacing of precipitates at which the mechanism of hardening changes from particle shear to particle bypass. Take

$$r_{Al2Mg} = 1{,}400 \text{ mJm}^{-2},$$
$$\text{Atomic radius (Al)} = 0.143 \text{ nm},$$
$$G_{Al} = 26.1 \text{ GPa}.$$

Solution: Assume that the precipitates are arrayed as SC (simple cubic). Therefore, the spacing x of precipitates is equal to the edge length, and for Al_2Mg in a cubic arrangement,

$$V_{Al2Mg} = 4/3\ \pi r_{Al2Mg}.$$

From Equation 10.7b, we calculate the volume fraction for precipitates:

$$f = 4/3(\pi r_{Al2Mg}^3)/x^3.$$

Thus,

$$r_{Al2Mg} = (3f/4\pi)^{1/3}x = 0.323x.$$

For particle shear,

$$\tau_{shear} = (\pi r_{Al2Mg}\gamma_{Al2Mg})/2bx.$$

For bypass,

$$\tau_{\text{Orowan}} = Gb/x.$$

The critical condition is obtained by setting $\tau_{\text{shear}} = \tau_{\text{Orowan}}$. Hence,

$$(\pi r_{\text{Al}_2\text{Mg}} \gamma_{\text{Al}_2\text{Mg}})/2bx = Gb/x.$$

Now, $r_{\text{Al}_2\text{Mg}} = 0.323x$ and $b = 2r_{\text{Al}}$ because Al is FCC. Substituting these values into the preceding equation, we get

$$(0.323\pi\, \gamma_{\text{Al}_2\text{Mg}}\, x)/4r_{\text{Al}}x = 2Gr_{\text{Al}}/x,$$

or

$$
\begin{aligned}
x &= (8Gr_{\text{Al}}^2)/0.323\,\pi\gamma_{\text{Al}_2\text{Mg}} \\
&= (8 \times 26.1 \times 10^9 \times 0.143 \times 10^{-9})/(0.323\pi\,1{,}400 \times 10^{-7}/10^{-4}) \\
&= 3.0 \times 10^{-9}\ \text{m} \\
&= 3.0\ \text{nm}.
\end{aligned}
$$

EXAMPLE 10.8

Steel is one of the most important engineering materials. Consider the different strengthening mechanisms discussed in this and earlier chapters, and make a list of different contributions to the strength of low-carbon steel.

Solution: Here is a list of the various possible contributions to the strength of steel:

1. Lattice friction stress, or the Peierls–Nabarro stress, σ_i.
2. Solid-solution strengthening, $k\sqrt{C}$, where k is a constant and C is the solute concentration.
3. Dislocation or strain hardening, $\alpha Gb\sqrt{\rho}$, where α is a constant approximately equal to 0.5, G is the shear modulus, b is the Burgers vector, and ρ is the dislocation density. (This contribution is discussed in Chapter 6.)
4. Grain-size strengthening, $k_y d^{-0.5}$, where k_y is the Hall–Petch constant and d is the grain size. (This contribution is discussed in Chapter 5.)
5. A precipitation-hardening contribution if there are any precipitates present, such as carbides of iron, niobium, titanium, or vanadium.

SUGGESTED READINGS

V. Gerold, in *Dislocations in Solids,* Vol. 4, F. R. N. Nabarro (ed.). New York: Elsevier/North Holland, 1979), p. 219.

A. Kelly and R. B. Nicholson (eds.). *Strengthening Methods in Crystals.* Amsterdam: Elsevier, 1971.

D. T. Llewellyn. *Steels: Metallurgy and Applications,* 2d ed. Oxford: Butterworth-Heinemann, 1992.

J. W. Martin *Precipitation Hardening.* Oxford: Pergamon Press, 1968.

Microalloying '75, Union Carbide, New York, 1977.

E. Nembach and D. G. Neite. *Precipitation Hardening of Superalloys by Ordered γ'-Particles,* Progress in Materials Science Series, Vol. 29. Oxford: Pergamon Press, 1985, p. 177.

A. K. Vasudevan and R. D. Doherty, eds., *Aluminum Alloys: Contemporary Research and Applications.* Boston: Academic Press, 1989.

EXERCISES

10.1 Compute the hydrostatic stress, in terms of an (r, θ) coordinate, associated with an edge and a screw dislocation in an aluminum lattice. Take $G = 26$ GPa and $b = 0.3$ mm. For a given r, what is the maximum value of this stress?

10.2 Consider the copper–zinc system that is used to make a series of brasses. The radius of solute zinc atom is 0.133 nm, while that of solvent copper atom is 0.128 nm. Calculate the dilational misfit ΔV for this alloy. Compute the hydrostatic stress σ_p for an edge and a screw dislocation in the system. Use the two quantities σ_p and ΔV to obtain the dilational misfit energy for the alloy. Also, compute the force exerted by a solute atom of zinc on a dislocation in copper.

10.3 The interaction energy between an edge dislocation (at the origin) and a solute atom (at r, θ) is given by

$$U = \frac{A}{r} \sin \theta,$$

where A is a constant. Transforming into Cartesian coordinates, plot lines of constant energy of interaction for different values (positive and negative) of $A/2U$. On the same graph, plot the curves for the interaction force. Indicate by arrows the direction in which the solute atoms, with ΔV positive, will migrate.

10.4 Consider a metal with shear modulus $G = 40$ GPa and atomic radius $r_0 = 0.15$ nm. Suppose the metal has a solute that results in a misfit of $\varepsilon = (R - r_0)/r_0 = 0.14$. Compute the elastic misfit energy per mole of solute.

10.5 Estimate the amount of solute (atomic percent) necessary to put one solute atom at each site along all the dislocations in a metal. Assume that 1 mm³ of a metal contains about 10^6 mm of dislocation lines.

10.6 Compute the condensation temperature T_c for the following cases:
(a) Carbon in iron with C_0 (average concentration) $= 0.01\%$ and U_i (interaction energy) $= 0.08$ aJ (0.5 eV).
(b) Zinc in copper with $C_0 = 0.01\%$ and $U_i = 0.019$ aJ (0.12 eV).

10.7 One of the Hume-Rothery rules for solid solutions[5] is that the solubility of solute B in solvent A becomes negligible when the atomic radii of A and B differ by more than 15%. Plot the maximum solubility (atomic percent) of Ni, Pt, Au, Al, Ag, and Pb as a function of the ratio of solute and solvent (Cu) radii, and verify that the solid solubility in Cu drops precipitously at a size ratio of about 1.15. Use the following data:

$$r_{Ni} = 0.1246 \text{ nm} \qquad r_{Al} = 0.143 \text{ nm}$$
$$r_{Ag} = 0.1444 \text{ nm}$$
$$r_{Pt} = 0.139 \text{ nm} \qquad r_{Pb} = 0.1750 \text{ nm}$$
$$r_{Cu} = 0.1278 \text{ nm} \qquad r_{Au} = 0.1441 \text{ nm}.$$

10.8 A steel specimen is being tested at a strain rate of 3×10^{-3} s⁻¹. The cross-sectional length is 0.1 m. A Lüders band forms at the section, with an instantaneous strain of 0.2. What is the velocity of propagation of the two Lüders fronts?

[5] See W. Hume-Rothery and G. V. Raynor, *The Structure of Metals and Alloys* (London: Institute of Metals, 1956), p. 97.

10.9 An overaged, precipitation-hardenable alloy has a yield strength of 500 MPa. Estimate the interparticle spacing in the alloy, given that $G = 30$ GPa and $b = 0.25$ nm.

10.10 Consider a unit cube of a matrix containing uniform spherical particles (with radius r) of a dispersed second phase.

    (a) Show that the average distance between the particles is given by

$$\Lambda = \frac{1}{(N)^{1/3}} - 2r,$$

where N is the number of particles per unit volume.

    (b) Compute Λ for a volume fraction f of particles equal to 0.001 and $r = 10^{-6}$ cm.

10.11 For a precipitation-hardenable alloy, estimate the maximum precipitate size that can undergo shear by dislocations under plastically strain. Take matrix shear modulus $= 35$ GPa, Burgers vector $= 0.3$ nm, and specific energy of precipitate–interface created by shear $= 100$ mJ/m^2.

10.12 An aluminum alloy contains 2% volume fraction of a precipitate that results in $\varepsilon = 5 \times 10^{-3}$. Determine the average spacing l between precipitates above which there will be a significant contribution to strength due to the difference in atomic volume of the matrix and the precipitate. Below this critical value of l, what will be the mechanism controlling yielding?

10.13 Calculate the critical radius of precipitates for which an Al–Mg alloy containing 10% Mg will be strengthened by Orowan looping instead of particle shear. Use the following data:

$$\gamma_{Al_2Mg} = 1.4 \text{ J/m}^2,$$

$$G_{Al} = 26.1 \text{ GPa},$$

$$r_{Al} = 0.143 \text{ nm}.$$

10.14 An Al–Cu alloy with 4% weight Cu is aged to form θ precipitates (Cu Al$_2$).

    (a) Using Figure 10.14a, determine the volume fraction of CuAl$_2$. Take $\rho_{Al} = 2.7 \times 10^3$ kg/m^3 and $\rho_{Cu} = 8.9 \times 10^3$ kg/m^3.

    (b) Establish the stresses required for precipitate shearing and bypass by dislocations as a function of precipitate radius, given that $\gamma_{CuAl_2} = 2.7$ J/m^2, $G_{Al} = 26.1$ GPa, and $r_{Al} = 0.143$ nm.

11

Martensitic Transformation

11.1 INTRODUCTION

In this chapter, we discuss one important means of altering the mechanical response of metals and ceramics: martensitic transformation. Martensitic transformation is a highly effective means of increasing the strength of steel. An annealed medium-carbon steel (such as AISI 1040) has a strength of approximately 100 MPa. By quenching (and producing martensite), the strength may be made to reach about 1 GPa, a tenfold increase. The ductility of the steel is, alas, decreased.

A quite different effect is observed in ceramics. If a ceramic undergoes a martensitic transformation during the application of a mechanical load, the propagation of cracks is inhibited. For example, partially stabilized zirconia has a fracture toughness of approximately 7 MPa $m^{1/2}$. An equivalent ceramic not undergoing martensitic transformation would have a toughness less than or equal to 3 MPa $m^{1/2}$.

An additional, and very important, effect associated with martensitic transformations is the "shape memory effect." Alloys undergoing this effect "remember" their shape prior to deformation. The three effects just described have important technological applications.

11.2 STRUCTURES AND MORPHOLOGIES OF MARTENSITE

Quenching has been known for over 3,000 years and is, up to this day, the single most effective mechanism known for strengthening steel. However, it is only fairly recently that the underlying mechanism has been studied in a scientific manner and really become understood. Initially attributed to a beta phase supposedly existing in the Fe–C system, the strengthening effect is now known to be due to a metastable phase: *martensite*. The investigations leading to the understanding of the mechanisms governing, and factors affecting, martensitic transformations have posed a great challenge to researchers over the past 50 years. Out of a confusing maze of apparently contradictory phenomena, order has appeared. Martensiticlike transformations have been identified in a great number of systems, including pure metals, solid solutions, intermetallic compounds, and ceramics. In order to assess the mechanical behavior of martensite and take advantage of its unique responses in technological applications, one has to understand the fundamental aspects of the transformation. Toward that end, Table 11.1 presents a number of systems in which martensiticlike transformations have been observed.

The original use of the martensitic transformation was exclusively to harden steel. Other developments have led to its use in different contexts. In transformation-induced plasticity (TRIP) steels, the martensitic transformation occurs during deformation and strengthens the regions ahead of a crack or near a neck, the ductility of the material is enhanced, while the strength level remains high. This results in great toughness. Ceramics (zirconia) are toughened through the same principle; the fracture toughness of partially stabilized zirconia can be as high as three times that of conventional ceramics.

Another use of the martensitic transformation is the shape-memory effect. Upon being plastically deformed, the material undergoes internal changes in the configuration of the martensite plates. Heating will recompose the initial shape. This effect is discussed in detail in Section 11.5. More complex processing procedures involving the martensitic transformation, such as ausforming and maraging, have been developed for steels.

A martensitic transformation is a lattice-distortional, virtually diffusionless structural change having a dominant deviatory component and an associated change in shape such that strain energy dominates the kinetics and morphology of the transformation. The requirement that there be no diffusion stems from thermodynamics: The driving energy required for martensitic transformation is much higher (in the case of irreversible martensites, especially) than that needed for diffusional decomposition (such as precipitation or spinodal decomposition). Hence, as the alloy is cooled, the latter transformations would take place at a higher temperature, where the free-energy difference between the two phases is not very large. Figure 11.1 shows the free energies of the parent and martensitic phase as a function of temperature. At T_0, the equilibrium temperature, the two phases have the same free energy. M_s is the highest temperature at which martensite forms spontaneously. The critical free energy required for the martensitic transformation is ΔF_{cr} and is around 1,200 kJ/mol for Fe–Ni and Fe–C alloys. Hence, if a diffusion-induced transformation competes with the martensitic transformation, the cooling in the region where $T_0 \to M_s$ has to be fast enough to avoid the former transformation. On the other hand, if T_0 is low enough, there is essentially no diffusion, and slow cooling will produce martensite. Upon heating above A_s, the martensite reverts to austenite. For irreversible martensites, the gap in which $M_s \to A_s$ is a few hundred kelvins; for reversible martensite, the gap is of a few tens of kelvin.

Martensite can exhibit a variety of morphologies, depending on the composition of the alloy, the conditions in which it is formed, and its crystalline structure. The three most common morphologies are the lenticular (lens shaped), the lath (a large number of blocks juxtaposed in a shinglelike arrangement), and the acicular (needle shaped). These are shown in Figures 11.2 through 11.5. Lenticular martensite occurs in Fe–Ni and Fe–Ni–C alloys with approximately 30% Ni and in Fe–C alloys with over 0.6% C. The center region is called the *midrib* and etches preferentially. The substructure is characterized by twins, dislocations, or both. In the particular case of Figure 11.2a, the region adjacent to the midrib is twinned, and the external parts are dislocated. Figure 11.2b shows lenticular martensite in a Cu–Al–Ni alloy. This material exhibits the shape-memory effect. Lath martensite, on the other hand, is quite different, consisting of small, juxtaposed blocks that are arranged in packets separated by low-angle grain boundaries. Each packet is composed of blocks with a thickness varying between a few micrometers and a few tens of micrometers; the

TABLE 11.1 Systems in which Martensitic or Quasi-martensitic Transformation Occurs[1]

Alloy	Structure Change
Co, Fe–Mn, Fe–Cr–Ni	FCC→HCP
Fe–Ni	FCC→BCC
Fe–C, Fe–Ni–C, Fe–Cr–C, Fe–Mn–C	FCC→BCT
In–Tl, Mn–Cu	FCC→FCT
Li, Zr, Ti, Ti–Mo, Ti–Mn	BCC→HCP
Cu–Zn, Cu–Sn	BCC→FCT
Cu–Al	BCC→distorted HCP
Au–Cd	BCC→orthorhomic
ZrO_2	tetragonal→monoclinic

([1]Adapted with permission from V. F. Zackay, M. W. Justusson, and D. J. Schmatz, *Strengthening Mechanisms in Solids,* (Metals Park, Ohio: ASM, 1962), p. 179).

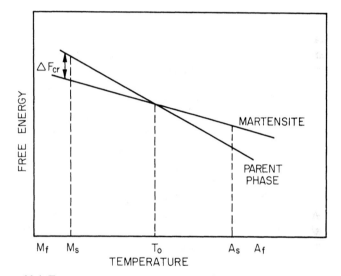

Figure 11.1 Free energy versus temperature for austenitic and martensitic phases. M_s, M_f, A_s, and A_f marked on abscissa.

blocks make specific angles with their neighbors. There is a repetitive pattern in each packet, leading to a 360° rotation and a resultant periodicity. Low-carbon steels and Fe–Ni alloys with less than about 30% nickel exhibit this morphology, shown in Figure 11.3.

In steels, there is a significant difference in the mechanical properties of twinned and dislocated martensites. Figure 11.4 shows a medium-carbon steel (0.3% C) that can exhibit both lath (dislocated) and lenticular (twinned) martensites. It can be seen that twinned martensite gives poor toughness, which is consistent with what we learned about twinning in Chapter 5. Mechanical twinning can give rise to microcracks, which are initiation sites for failure of the material (see Fig. 8.6). The example of Figure 11.4, from Thomas' research, is a wonderful illustration of how the microstructure (in this case, inside the martensite lenses and laths) can have a dramatic effect on mechanical properties. This fact is often overlooked by engineers.

Acicular martensite is shown in Figure 11.5. This form occurs in austenitic stainless steel (Fe–Cr–Ni alloys) after deformation. Needles form at the intersection of the slip bands (either dislocations, stacking faults, twins, ε-martensite, or a combination thereof).

Figure 11.2(a) Lenticular martensite in an Fe–30% Ni alloy. (Courtesy of J. R. C. Guimarães)

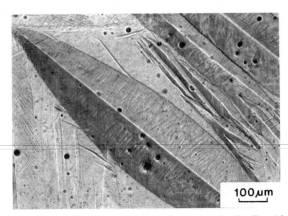

Figure 11.2(b) Lenticular (thermoelastic) martensite in Cu–Al–Ni alloy. (Courtesy of R. J. Salzbrenner)

Since the intersection of these bands is a thin "tube," the martensite forming in it has this specific shape (marked by arrows in the figure). Acicular martensite has the BCC or BCT structure and has a marked effect on the strength and work-hardening ability of the alloy. Other martensite morphologies have been observed also. ε-martensite is HCP and forms in plates. It can be produced in steel by subjecting the metal to a high pressure (>13 GPa) or in austenitic stainless steels by deformation. After substantial plastic deformation, sheaves of fine parallel laths were observed to form along the austenite slip bands in austenitic Fe–Ni–C alloys. Yet another morphology is the butterfly martensite, so called because two lenses form in a coupled manner; the resultant microstructure resembles a number of butterflies. The plastic deformation accompanying the martensite, constrained by the surrounding matrix, can occur by either slip or twinning. Examples of twinned martensite are shown in Figure 11.6 (for an Fe alloy with 22.5 wt. % Ni and 4 wt. % Mn) and Figure 11.7 (for a U–Re alloy). The transmission electron micrograph (Figure 11.6a) and dark-field picture (Figure 11.6b) show the group of twins inside a martensite lens. Crystallographic analysis through the electron diffraction pattern of Figure 11.6c reveals

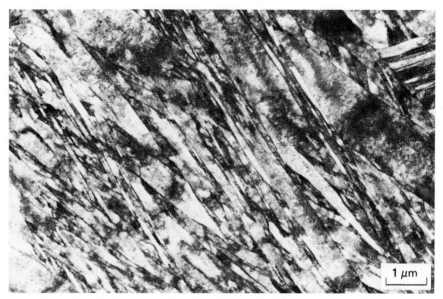

Figure 11.3 Lath martensite. (Reprinted with permission from C. A. Apple, R. N. Caron, and G. Krauss, *Mel-Trans.*, 5 (1974) 593)

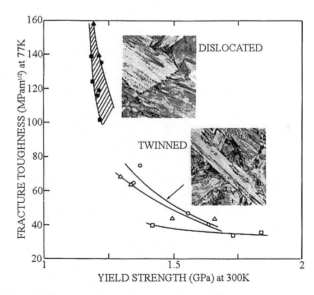

Figure 11.4 Comparison of mechanical properties between twinned and dislocated martensite in medium-carbon (0.3% C) steel. (Courtesy G. Thomas)

the habit and twinning planes. In the case of the U–Re system (Figure 11.7), the twins propagate from the lenses into the matrix; the two martensite lenses are indicated by *M*.

In spite of these differences in morphology, some unique features are common to all martensites. The most important is the existence of an *undistorted and unrotated plane*. The crystallographic orientation relationship between parent and martensite phases is such that there *always* is a plane that has the same indices in the two structures. This undistorted and unrotated plane is called the *habit plane;* it is usually a plane with irrational indices.

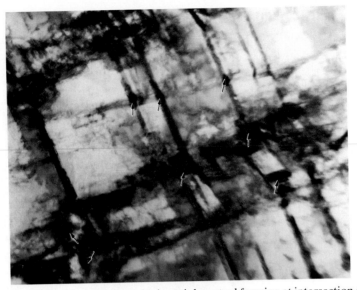

Figure 11.5 Acicular martensite in stainless steel forming at intersection of slip bands (TEM, courtesy G. A. Stone)

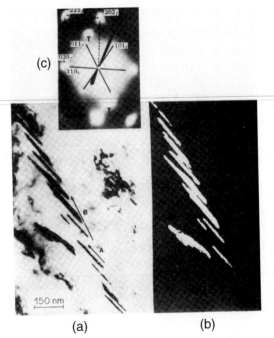

Figure 11.6 (a) Transmission electron micrograph showing a group of twins inside martensite transformed at $-140°C$ and 2 GPa. (b) Dark-field image of twins on $(112)_b$ plane; (c) Stereographic analysis for habit (in FCC) and twin (in BCC) planes. (From S. N. Chang and M. A. Meyers, *Acta Met.*, 36 (1988) 1085)

Figure 11.7 Martensite lenses being transversed by twins, which produce self-accommodation. (Courtesy of A. R. Romig)

For a steel with 1.4% carbon, Kurdjumov and Sachs found the following relationships for habit plane (225):[1]

$$(111)_A \| (011)_M$$
$$[10\bar{1}]_A \| [01\bar{1}]_M.$$

Steels with less than 1.4% carbon exhibit the same relationship. This specific martensite is known as (225). Nishiyama investigated the Fe–Ni–C alloys and steels with carbon content greater than 1.4% and obtained the following relationship for habit plane (259):[2]

$$(111)_A \| (011)_M$$
$$[11\bar{2}]_A \| [01\bar{1}]_M.$$

11.3 STRENGTH OF MARTENSITE

The martensitic transformation has the ability to confer a great degree of strength on steels; other alloys do not seem to have such strong martensites. The strength of martensite in steel is dependent on a number of factors, the most important being the carbon content of the steel. While the Rockwell C hardness of iron increases from 5 to 10 when it is transformed to martensite, it increases from 15 to 65 when the carbon content is 0.80% (eutectoid steel). The origin of the high hardness of martensite has been the object of great controversy in the past. It is now fairly well established that there is no single, unique mechanism responsible for it. Rather, a number of strengthening mechanisms operate, most of which we have described in Chapters 5, 6, and 10. Nevertheless, the relative importance of these strengthening mechanisms and their interactions are still the object of controversy. It seems that interstitial solution-hardening and substructure strengthening (work-hardening) are the most important ones.

Most metals exhibit a dependence of yield stress on grain size; the martensite lenses divide and subdivide the grain when they form. Hence, a small-grained alloy produces small martensitic plates, whereas a large-grained alloy produces a distribution of sizes

[1]G. Kurdjumov and G. Sachs, *Z. Phys.*, 64 (1930) 325.

[2]Z. Nishiyama, *Sci Rep. Tohoku Univ.*, 2B (1934) 627.

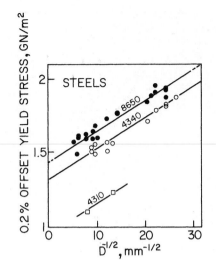

Figure 11.8 Effect of prior austenite grain size on the yield stress of three commercial martensitic steels. (Adapted with permission from R. A. Grange, *Trans. ASM,* 59 (1966) 26)

whose mean is much larger. This is shown in Figure 11.8. Three commercial steels (AISI 4310, 4340, and 8650) exhibit a dependence of yield stress on prior austenitic grain size. The slope of the Hall–Petch plot seems to be the same for the three. However, for the range of grain sizes usually encountered, the contribution of grain size is not very important: The grain sizes are equal to 0.1 mm or more. Only in steels that have undergone thermomechanical processing to reduce the austenitic grain size is this strengthening mechanism of significance.

The contribution of substitutional solid-solution elements to the strength of ferrous martensites is relatively unimportant; additionally, it is difficult to separate it from other indirect effects, such as the change in M_s, and stacking-fault energy due to the addition of these elements.

On the other hand, interstitial solutes (carbon and nitrogen, for instance) can play an important effect. If we regard martensite as a supersaturated solution of carbon in ferrite, a great portion of its strength could be ascribed to solution-hardening. Foreman and Makin developed an equation of the following form to express the effect of the solute concentration C on the yield stress of the alloy if only the interaction between dislocations and single-atom obstacles is considered[3]

$$\tau_0 = \left(1 - \frac{\phi'}{5\pi}\right) G \left(\frac{F_{\max}}{2T}\right)^{3/2} (3C)^{1/2}. \tag{11.1}$$

Here, $F_{\max}$ is the maximum force exerted by the obstacle on the dislocation, T is the tension of the dislocation line, G is the shear modulus, and ϕ' is the angle turned through by the dislocation immediately before it frees itself from the obstacle. The interesting aspect of this equation is that the yield stress should increase with the square root of the solute concentration. And indeed, results obtained by Roberts and Owen confirm Equation 11.1, as can be seen in Figure 11.9. These researchers used alloys with very low M_s (below 77 K), to avoid any secondary effect of the carbon atoms, such as precipitation-hardening or the formation of a Cottrell atmosphere. The fact that the room-temperature tests exhibit the same slope as the ones conducted at 77 K shows that even at room temperature, solid-solution-hardening is operating and effectively strengthening martensite.

[3]A. J. E. Foreman and M. J. Makin, *Phil Mag.,* 14 (1966) 191.

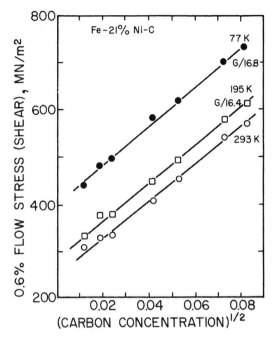

Figure 11.9 Plot of 0.6% proof stress (one-half of tensile stress) versus $C^{1/2}$ for Fe–Ni–C lath martensite at various temperatures. The slopes are shown as fractions of the shear modulus, which is denoted by G. (Adapted with permission from M. J. Roberts and W. J. Owen, *J. Iron Steel Inst.*, 206 (1968) 37)

Snoek ordering consists of the reorientation of a system of point defects of tetragonal or lower symmetry that are randomly distributed in the stress field of a dislocation (see Section 10.3.5). Single jumps of carbon atoms can organize the atoms in such a way as to minimize their energy. Snoek ordering can take place in a much shorter time interval than does the formation of a Cottrell atmosphere, because no long-range diffusion is required.

The formation of a Cottrell atmosphere, on the other hand, requires that the atoms diffuse toward regions in the dislocation in which their strain energy will be minimized. Carbon atoms produce tetragonal distortions and shear stresses; hence, they seek regions around both edge and screw dislocations in which the shear strains cancel each other. Cottrell atmospheres produce both static and dynamic aging. A manifestation of the latter is the serrated flow (the Portevin–Le Chatelier effect: see also Chapter 10, Section 10.3.4).

Carbon atoms have also been shown to exhibit a clustering behavior. Carbon-rich regions have been identified by transmission electron microscopy in steels that had been exposed to temperatures no higher than ambient temperature. These clusters do not change the crystalline structure of the martensite, but produce periodic strain fields, resulting in a "modulated" structure. In this sense, the clustering is closer to a spinodal decomposition than to a precipitation reaction. If the martensite is aged at higher temperatures, cementite and other metal carbides are precipitated. The latter process is called *tempering*.

Frequently, precipitation is observed in martensite. Quenched carbon steels with M_s above room temperature may contain precipitates that form during cooling. In certain low-carbon steels these precipitates have been identified as cementite. It seems that carbon is a more efficient strengthener as a precipitate than in solid solution. The contribution of precipitates in ferrous martensites exceeds that of a solid solution. A very important contribution is that of strain hardening. In twinned martensite, a very fine array of twins 5 to 9 nm thick presents a very effective barrier for additional deformation. These fine twins are the most important factor in the strength of martensite. When martensite is dislocated, the density of dislocations is typically 10^{10} to 10^{11} cm^{-2}; the substructure resembles that of BCC steel that has been heavily deformed by conventional means.

The contributions to the strength of the martensite in a 0.4% carbon steel can be distributed as follows:[4]

Boundary strengthening	620 MPa
Dislocation density	270 MPa
Solid solution of carbon	400 MPa
Rearrangement of carbon in quench (Cottrell atmosphere Snoek effect, clustering, precipitation)	750 MPa
Other effects	200 MPa
Total	2,240 MPa

Williams and Thompson point out that these effects are not necessarily additive; however, this simplified scheme shows the various contributions.

Yet another source of strengthening is the intrinsic resistance of the lattice to dislocation motion (Peierls–Nabarro stress). This type of resistance accounts for the temperature dependence of yield stress in martensite. Iron exhibits a strong temperature dependence of yield stress at low temperatures, as do other BCC metals. This same behavior is observed in martensite, independent of the existence of precipitates and solutes.

11.4 MECHANICAL EFFECTS

A martensite lens introduces macroscopic strains in the lattice surrounding it. This is best seen by making fiducial marks on the surface and transforming the material. The fiducial marks will be distorted by the strains. The strains introduced by a martensite lens can be decomposed into a dilatational and a shear strain. The dilatational strain is perpendicular to the midrib plane, and the shear strain is parallel to the midrib plane. In ferrous alloys, the dilatation is approximately 0.05 and the shear strain γ is about .020. Figure 11.10 shows a fiducial mark made on the surface of a hypothetical alloy. The shear direction is such that the plane is not distorted. Hence, $\tan \theta = \gamma$, and θ is equal to 11°. The strain matrix can be expressed as

$$\begin{pmatrix} \varepsilon_{11} & \varepsilon_{12} & \varepsilon_{13} \\ \varepsilon_{12} & \varepsilon_{22} & \varepsilon_{23} \\ \varepsilon_{13} & \varepsilon_{23} & \varepsilon_{33} \end{pmatrix} = \begin{pmatrix} 0 & 0 & 0 \\ 0 & 0 & 0.10 \\ 0 & 0.10 & 0.05 \end{pmatrix}. \tag{11.2}$$

Recall that $\varepsilon_{23} = \gamma_{23}/2$. These strains are well beyond the elastic limit of the matrix, and there is plastic deformation in the region surrounding the martensitic lens. This is reflected in Figure 11.10 by the distortion of the fiducial line.[†]

The dilatational and shear stresses and strains imposed by the martensitic transformation interact with externally applied stresses, and very special responses ensue. The effects of externally applied tensile, compressive, and hydrostatic stresses are shown in Figure 11.11. The uniaxial tension and compression increase M_s, whereas hydrostatic compression lowers it. The explanation is that, under the effect of the applied stress, the me-

---

[4]See J. C. Williams and A. W. Thompson, in *Metallurgical Treatises*, J. K. Tien and J. F. Elliott (eds.) (Warrendale, PA: TMS-AIME, 1981), p. 487.

[†]Fiducial line is an imaginary straight line initially drawn, before transformation.

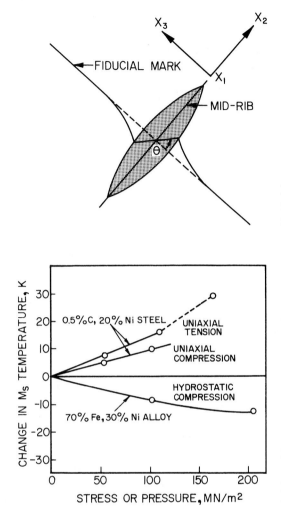

Figure 11.10 Distortion produced by martensite lens on fiducial mark on surface of specimen.

Figure 11.11 Change in M_s temperature as a function of loading condition. (Adapted with permission from J. R. Patel and M. Cohen, *Acta Met.,* 1 (1953) 531)

chanical work done by the transformation, which can be decomposed into the dilatational and shear components, $\sigma\varepsilon$ and $\tau\gamma$, is either increased or decreased:

$$W = \sigma\varepsilon + \tau\gamma \tag{11.3}$$

The hydrostatic stress counters the lattice expansion produced by martensite, but does not affect the shear stress. Hence, a greater amount of free energy is required to trigger the transformation. Referring to Figure 11.1 we can see that a greater ΔF will require a lowering of M_s. For the tensile test, the applied stress can be decomposed into a normal (positive) stress and a shear stress, both of which aid the transformation. The shear stress aids the martensite variants aligned with the direction of maximum shear (45° to the tensile axis). These variants will form preferentially; hence, the free-energy requirement is decreased and M_s is increased. In the compressive test, the normal portion of the stress is negative and counters the dilatational stress of the transformation, whereas the shear stress favors it. (There are always favorably oriented variants.) Since the shear stress term dominates the expression (because of the greater shear strain γ), the tensile stress should be more effective in increasing M_s than the compressive stress is. This is exactly what is shown in Figure 11.11.

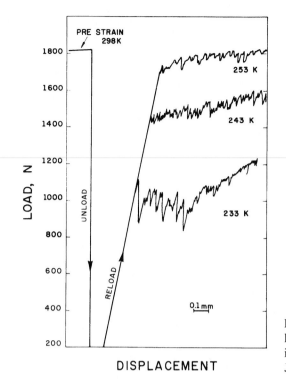

Figure 11.12 Tensile curves for Fe–Ni–C alloy above M_s, showing martensite forming in elastic range (stress assisted). (Courtesy of J. R. C. Guimarães).

Another experimental procedure consists of conducting tensile tests at temperatures above M_s. When the stress level reaches the value at which martensite forms at the test temperature, a significant load drop is observed. Figure 11.12 shows this effect. The load drop is due to the shear strain of the martensite, which produces an instantaneous increase in strain of the martensite, which produces an instantaneous increase in length of the specimen. As the difference between the test temperature and M_s increases, the stress at which martensite starts forming increases; this can be directly inferred from Figure 11.12. In Figure 11.13 the yield stress is plotted as a function of temperature; when martensite forms in the elastic line, the stress at which it forms is equal to the yield stress (as in Figure 11.12, for instance). The temperature dependence of the stress for martensite transformation is clearly shown by the three straight lines in Figure 11.13. At M_s, as expected, martensite

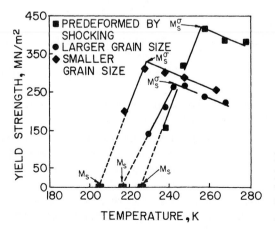

Figure 11.13 Temperature dependence of the yield strength of Fe–31% Ni–0.1% C (Adapted with permission from J. R. C. Guimarães, J. C. Gomes, and M. A. Meyers, *Supp. Trans. Japan Inst. of Metals*, 17 (1976) 41).

forms spontaneously, without any stress. The plots of yield stress versus temperature for three different conditions have inverted-V shapes. At the point marked M_s^σ, the slope of the curve changes, and above this temperature, the yield stress is produced by conventional dislocation motion; hence, it shows the regular increase with decreasing temperature. Between M_s and M_s^σ, on the other hand, we have *stress-assisted martensite* establishing yield, and the temperature dependence is inverted, leading to a yield stress of zero at M_s. It is worth noting that the three alloys in Figure 11.13 have the same composition, but different processing histories. M_s temperature is affected by grain size. M_s increases with increasing grain size and predeformation; in the case shown in the figure, predeformation was accomplished by shock loading.

The formation of strain-induced martensite occurs in the temperature range above M_s^σ in Figure 11.13. Substantial plastic deformation, in which the substructure has to be sensitized, is required before the first martensite forms. This kind of martensite is called *strain induced*, to differentiate it from *stress-assisted* martensite. Figure 11.14 illustrates the effect of strain-induced martensite on the stress–strain curve of an austenitic stainless steel at −50°C. The austenite has a low yield stress and work-hardening rate; the transformation to martensite is also shown (right-hand column). The experimentally obtained curve reflects the fact that an increase in martensite volume fraction is accompanied by plastic strain; the simple rule-of-mixture curve is higher than the experimental curve because there are complex synergistic processes between the two phases (α and γ).

Strain-induced martensite is responsible for a very beneficial property: the transformation-induced plasticity (TRIP) effect. Remarkable combinations of high strength and toughness have been obtained in TRIP steels. The high strength is due to

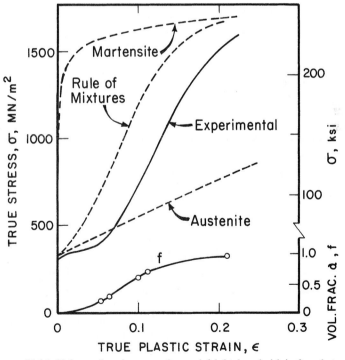

Figure 11.14 Volume fraction transformed (right-hand side), f, and stress (left-hand side) as a function of plastic strain for an austenitic (metastable) steel deformed at −50°C; experimental and idealized stress–strain curves for austenite, martensite, and mixture. (After R. G. Stringfellow, D. M. Parks, and G. B. Olson, *Acta Met.*, 40 (1992) 1703)

work-hardening, carbide precipitation, and dislocation pinning by solutes during thermo-mechanical treatment. The high toughness comes from a combination of high strength and high ductility. The latter is a direct consequence of the strain-induced martensite transformation. If a certain region in the metal is severely deformed plastically, strain-induced transformation takes place, increasing the local work-hardening rate and inhibiting an incipient neck from further growth. On the other hand, if a crack has already formed, martensitic transformation at the crack tip will render its propagation more and more difficult. In Section 11.6, the toughening of a ceramic (ZrO_2) by stress-assisted martensite will be described.

Another mechanical aspect of importance is the fracture of martensite. Fracture is usually initiated in a martensitic alloy along the martensite–austenite or martensite–martensite boundaries. Indeed, upon investigating the fracture surfaces of Fe–31% Ni–0.1% C alloy, Chawla et al.[5] found that the density of dimples increased as a function of the amount of martensite in the cross section; the same result was obtained by decreasing the grain size. Hence, the dimple size was tied to the density of interfaces. In carbon-free or low-carbon steels, martensite is fairly soft, and the fracture is, consequently, ductile. In high-carbon steels, on the other hand, martensite is hard and brittle, and the fracture surface takes a cleavage appearance, with the fracture path traversing the plates (or laths). Of great importance in the initiation of fracture is the existence of microcracks in the structure. Marder et al.[6] found a great number of microcracks in Fe–C martensites; when the grain size was decreased, the incidence of microcracks decreased. The microcracks were formed when one lens impinged on another. Figure 11.15 shows how these cracks occur. The microcracks act as stress-concentration sites when the specimen is loaded; they are initiation sites for macrocracks.

Tempering of martensite[†] in steels is performed to improve toughness. However, the tempering process might induce embrittlement. *Temper martensite embrittlement* (TME) results from the segregation of impurities to the previous austenitic grain boundaries, providing a brittle path for propagation of the fracture. The fracture takes on the intergranular morphology. Temper embrittlement (TE) is caused by the impurities antimony, phosphorus, tin, and arsenic (less than 100 ppm required) or larger amounts of silicon and manganese. TME and TE occur in different ranges of temperatures; TME is a much more rapid process.

11.5 SHAPE-MEMORY EFFECT

The *shape-memory effect* (SME) is the unique property that some alloys possess according to which, after being deformed at one temperature, they recover their original shape upon being heated to a second temperature. The built-in memory is produced by the martensitic transformation. The effect was first discussed by the Russian metallurgist Kurdjumov. In 1951, Chang and Read[7] reported its occurrence in an In–Ti alloy. However, wide exposure of this property came only after the development of the nickel–titanium alloy by the Naval Ordnance Laboratory (NiTiNOL) in 1968.[8] Since then, research activity in this field has been intense, and a number of β-phase SME alloys have been investigated, including AgCd, AgZn, AuCd, CuAl, CuZn, FeBe, FePt, NbTi, NiAl, and ternary alloys. The Nitinol family of alloys has found wide technological application, and adjustments in composition can be made to produce M_s temperatures between −273 and 100°C. This is an ex-

[5]K. K. Chawla, J. R. C. Guimarães, and M. A. Meyers, *Metallography,* 10 (1977) 201.

[6]A. R. Marder, A. D. Benscoter, and G. Krauss, *Met. Trans.,* 1 (1970) 1545.

[7]L. C. Chang and T. A. Read, *Trans. Met. Soc. AIME,* 191 (1951) 49.

[8]W. J. Buehler and F. E. Wang, *Ocean Eng.,* 1 (1968)

[†]Tempering consists of heating the martensitic structure to an intermediate temperature.

tremely helpful feature, and alloys are tailored for specific applications. In the majority of SME alloys the high-temperature phase is a disordered β phase (body-centered cubic), while the martensitic phase is an ordered BCC structure with a superlattice or orthorhombic structure. Two separate mechanical effects characterize the response of SME alloys: *pseudoelasticity and strain-memory effect.* We describe these next, in connection with tensile and compressive tests.

Pseudoelasticity, or superelasticity, is the result of stress-induced martensitic transformation in a tensile test that reverts to the parent phase upon unloading. The individual martensite plates do not grow explosively, as in the ferrous martensites, and little irreversible damage is done to the lattice. The shear strain of one plate is accommodated by neighboring plates. The complex motion of the interfaces between the martensite plates along the various variants and within the same martensite plate takes place by the displacement of the interfaces between the different twins. Figure 11.16a shows the pseudoelastic effect for a Cu–Al–Ni alloy with $M_s = -48°C$. The test was conducted at 24°C (72°C above M_s). At A, stress-induced martensite starts to form. At B, the martensitic transformation has been completed, and any straining beyond that point will produce irreversible plastic deformation or fracture. Upon unloading, the martensite reverts to the parent phase between C and D. Further unloading results in the return to the original length of the specimen. The pseudoelastic strain exceeds 6%. The magnitude of the pseudoelastic strain can be calculated from a knowledge of the habit plane of the martensite (and its orientation with respect to the tensile axis) and the magnitude of the shear strain for the transformation. Since the habit plane of martensite is irrational, it

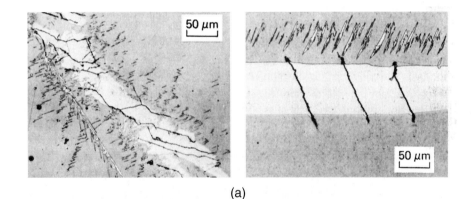

(a)

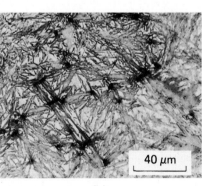

(b)

Figure 11.15 Microcracks generated by martensite. (a) Fe–8%, Cr–1% C (225 martensite sectioned parallel to habit plane). (Courtesy of J. S. Bowles, University of South Wales) (b) Carburized steel. (Reprinted with permission from C. A. Apple and G. Krauss, *Met. Trans.,* 4 (1973) 1195)

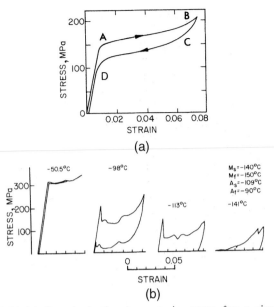

Figure 11.16 (a) Pseudoelastic stress–strain curve for a single-crystal Cu–Al–Ni, alloy at 24°C (72°C above M_s). (b) Dependence on temperature of stress–strain characteristics along the characteristic transformation temperatures. Strain rate: 2.5×10^{-3} min^{-1}. (Reprinted with permission from C. Rodriguez and L. C. Brown, in *Shape Memory Effects*, (New York: Plenum Press, 1975), p. 29)

has a multiplicity of 24, and there is always a habit plane oriented very closely to the plane of maximum shear.

The pseudoelastic (or superelastic) effect is illustrated in a very simplified fashion in Figure 11.17. A specimen with initial length L_0 is compressed. Stress-induced martensitic transformation takes place, and the austenite-martensite interfaces are glissile; that is, they can move under the applied stress. In Figure 11.17c, two martensite lenses exist. They continue to grow in Figure 11.17d. When the stress is decreased, they shrink in the same order as the initial growth. When the stress is reduced to zero, all martensite has disappeared, and the specimen has returned to the original length L_0. Figure 11.17g shows the corresponding stages on a stress–strain curve, the loop of Figure 11.16a. The stress–strain curve returns to the origin after the load is removed.

It is not sufficient for the temperature at which testing is conducted to be above M_s to obtain the pseudoelastic effect, as shown in Figure 11.16b. These tests were conducted on a Cu–Al–Ni alloy. The temperatures A_s and A_f (austenite start and finish, respectively) are also important. If the testing temperature is below A_s, the martensite will not revert to austenite upon unloading; the tests conducted at −141°C and −113°C show this irreversibility. For the test conducted at −50.5°C and −98°C, total reversibility is obtained, since this temperature is above A_f (−90°C). Another observation that can be made in Figure 11.16b is that the stress at which martensite forms increases with increasing temperature.

When the deformation is irreversible (at −113°C and −141°C in Figure 11.16), the effect receives the name *strain-memory effect*. Additional heating is required to return the martensite to its original shape, since the deformation temperature is below A_s. Upon heating, the original dimensions will be regained, as the martensite interfaces move back to retransform the lattice. The sequences in which the plates form and in which they disappear are inverted: The first plate to form is the last to disappear.

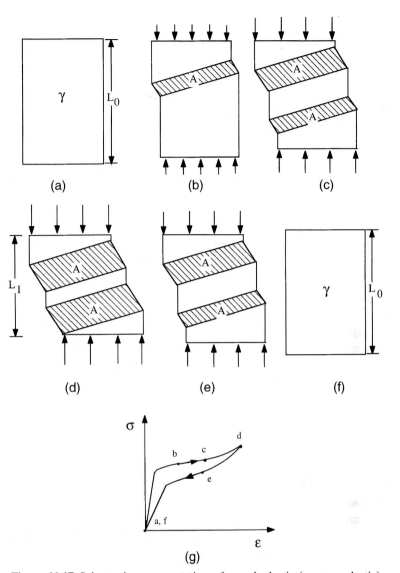

Figure 11.17 Schematic representation of pseudoelastic (or superelastic) effect. (a) Initial specimen with length L_0. (b, c, d) Formation of martensite and growth by glissile motion of interfaces under increasing compressive loading. (e) Unloading of specimen with decrease in martensite. (f) Final unloaded configuration with length L_0. (g) Corresponding stress–strain curve with different stages indicated.

The strain-memory effect is also obtained when deformation is imparted at temperatures below M_s. This is actually the procedure used in most technological applications. In that case, the structure consists of thermally induced martensite; it is present in such a way that all variants occur. When the external stress is applied, the variants that have shear strains aligned with the applied shear strain tend to grow, and the unfavorably oriented variants shrink. Figure 11.18 shows schematically how this takes place. Only two variants are shown, for simplicity. The variant that favors the applied tensile strain grows at the expense of the unfavorably oriented one. Hence, all unfavorable variants disappear, and the favorable variant takes over the structure. On heating, the structure reverts to the original one, composed of equal distribution of the two variants, giving the strain recovery.

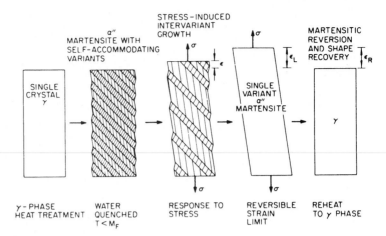

Figure 11.18 Sequence showing how growth of one martensite variant and shrinkage of others results in strain ε_L (Courtesy of R. Vandermeer).

Figure 11.19 shows the strain-memory effect for compressive, tensile, and flexure loading. Only two martensite variants are shown: *A* and *B*. In the schematic drawing, variant *B* favors tensile strains, whereas variant *A* produces compression in the direction of loading. Under compressive stresses, variant *A* grows at the expense of *B*. Under tensile loading, the opposite is true. And under bending, variant *B* grows in the outside, while variant *A* grows in the inside. The situation in a real material is much more complex, and polycrystalline effects come into play. Nevertheless, the simple scheme of Figure 11.19 shows the essential features of the strain-memory effect. Upon heating, the three specimens return to the original shape by the reverse motion of the martensite interfaces. Further heating would make the martensite revert to austenite.

When the strain-memory effect is obtained above M_s, a fully austenitic structure gradually becomes martensitic under stress. This is shown in the schematic representation of Figure 11.20. Only one variant of martensite is depicted. The loading stage is similar to that for the superelastic effect. However, upon unloading, the martensite remains in the material, and heating is required to return the martensite to its original dimensions. The reverse transformation occurs in the same order as the martensite transformation, and the specimen "remembers" its original slope.

Other potential benefits of the shape-memory effect involve the increased damping capacity of the material, which can become very large because of the work required to form the martensite. Circuit breakers, pseudoelastic wires for support in brassieres, overheat protection systems, sensors in heating and ventilation, components in the Hubble telescope, pseudoelastic dental arch wires, a pseudoelastic scoliosis correction system (a biomedical application), and porous pseudoelastic tissue are additional examples of applications of the pseudoelastic and strain-memory effects.

Structures containing their own sensors, actuators, and computational or control capabilities are called "smart," "adaptive" or "intelligent" structures. Alloys with good strain-memory effects play an important role in the design of these structures, and novel uses are being continually introduced. Indeed, the shape-memory effect has found some unique uses. One is as a tight coupling for pneumatic and hydraulic lines. The F-14 jet fighter tube couplings are made of Nitinol that is fabricated at room temperature with a diameter 4% less than that of the tubes which will be joined. Then, the couplings are cooled below M_s ($-120°C$) and expanded mechanically until their diameter is 4% larger than those of the tubes. They are held at this temperature until they are placed over the tube ends. Allowed to warm, they will shrink to their initial diameter; impeded by the tube, they will provide

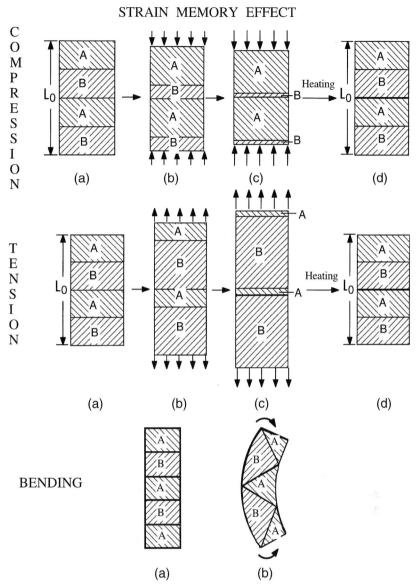

Figure 11.19 Schematic representation of strain-memory effect in compression, tension, and bending. Variant *A* favors a decrease in dimension in the direction of its length, whereas variant *B* favors an increase in dimension.

a tight fit. Electrical connectors that are opened and closed by changes in temperature are another application. Orthopedic and orthodontic aids have also been made of SME alloys, and Nitinol seems to react well in the body fluid environment. The pen-drive mechanism in recorders is a very successful application of the SME; over 600,000 of these new drives are now in service.

11.6 MARTENSITIC TRANSFORMATION IN CERAMICS

By far the most important, but not the only, martensitic transformation in a ceramic is the tetragonal-to-monoclinic transformation exhibited by zirconia (ZrO_2). This transformation

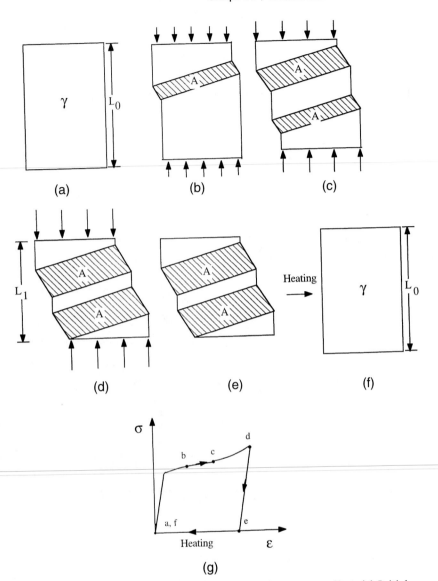

Figure 11.20 Schematic representation of strain-memory effect. (a) Initial specimen with length L_0. (b, c, d) Formation of martensite and growth by glissile motion of interfaces under increasing compressive stresses. (e) Unloading of specimen. (f) Heating of specimen with reverse transformation. (g) Corresponding stress–strain curve with different stages indicated.

leads to a significant enhancement in the toughness of ceramics if ZrO_2 is used either alone or as a distributed phase in other ceramics, such as alumina. Garvie et al. reported a very significant increase in tensile rupture strength (from 250 to 650 MPa) and work of fracture (equivalent to toughness) for tetragonal zirconia, in comparison with monoclinic zirconia.[9] They attributed this increase in strength to a martensitic transformation occurring during deformation, in a manner analogous to the TRIP effect. The three most common ways in which this transformation is used are as follows:

[9]R. C. Garvie, R. H. Hannink, and R. T. Pascal, *Nature*, 258 (1975) 703.

1. Tetragonal zirconia polycrystals (TZPs), which are nearly single-phase polycrystalline ceramics. TZPs are fabricated from fine-grained zirconia powders by sintering.

2. Partially stabilized zirconia (PSZ), in which tetragonal-ZrO_2 is a precipitate phase and the matrix is cubic zirconia. The highest toughnesses reported in PSZ are around 18 MPa m$^{1/2}$.

3. Zirconia-toughened alumina (ZTA), in which zirconia is a dispersed phase in the alumina matrix. ZTA materials are fabricated by cosintering Al_2O_3 and ZrO_2 powders. ZTA materials are relatively tough (K_{Ic} up to approximately 14 MPa m$^{1/2}$) and have high strength (1–2 GPa). This represents a significant enhancement in comparison with pure Al_2O_3 (K_{Ic} about 3 MPa m$^{1/2}$).

Figure 11.21a shows lenticular PSZ precipitates in cubic zirconia. The lens plane corresponds to the {100} planes of the cubic phase; thus, there are three possible variants for the precipitates. The tetragonal lens is coherent with the cubic matrix, and the

(a)

(b)

Figure 11.21 (a) Lenticular tetragonal zirconia precipitates in cubic zirconia (PSZ). (b) Equiaxial ZrO_2 particles (bright) dispersed in alumina (ZTA) (Courtesy of A. H. Heuer).

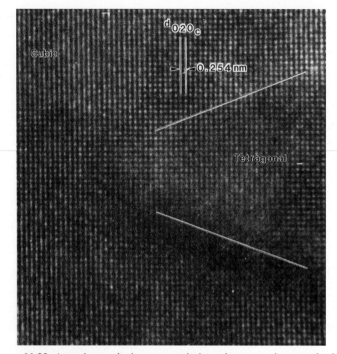

Figure 11.22 Atomic-resolution transmission electron micrograph showing extremity of tetragonal lens in cubic zirconia; notice the coherency of boundary (Courtesy of A. H. Heuer).

atomic-resolution TEM of Figure 11.22 shows the correspondence between the planes; the (100) of the tetragonal and cubic phases are parallel. The lenses are shaped approximately as oblate spheroids with an aspect ratio of 5. Figure 11.21b shows ZrO_2 particles (bright) in an alumina ceramic.

Zirconia has three allotropic forms: cubic, tetragonal, and monoclinic. Figure 11.23 shows the ZrO_2–MgO phase diagram. In pure zirconia, only very small particles (approximately 60 nm) can be retained at room temperature in the tetragonal structure. By using a stabilizing compound such as magnesia (MgO), calcia (CaO), yttria (Y_2O_3), or ceria (CeO_2), it is possible to retain the tetragonal phase, generally stable only between 1,240 and 1,400°C, at room temperature. The range of MgO additions for which this occurs is shown in the phase diagram by hatching. The tetragonal-to-monoclinic transformation, which takes place martensitically under applied stress, has a dilational (about 4–6%) and a shear (approximately 14%) component. The martensitic nature of the transformation is evident in the transmission electron micrograph of Figure 11.24. The martensite lenses form a zigzag pattern between two larger lenses; this is a typical autocatalytic nucleation sequence, in which one lens, impinging on a boundary, generates the defects that nucleate the subsequent lens. The process continues, leading to the characteristic pattern. The martensite shown in the figure was generated through rapid solidification.

The increase in toughness due to the martensitic tetragonal-to-monoclinic transformation can be qualitatively explained as follows. In the regions surrounding a propagating crack, the stresses induce the transformation, which has dilatational and shear strain components. These strains work against the stress field generated by the crack, decreasing the overall stress intensity factor and, thereby, increasing the toughness. Figure 11.25a illustrates this behavior in ZTA; Figure 11.25b shows the effect in PSZ. The gray grains correspond to tetragonal ZrO_2, whereas the black grains are transformed to the monoclinic phase. The crack, advancing from left to right, triggers the transformation; more black dots

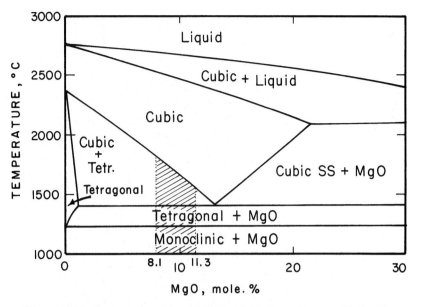

Figure 11.23 ZrO_2-rich portion of ZrO_2–MgO phase diagram. Notice the three crystal structures of ZrO_2: cubic, monoclinic, and tetragonal.

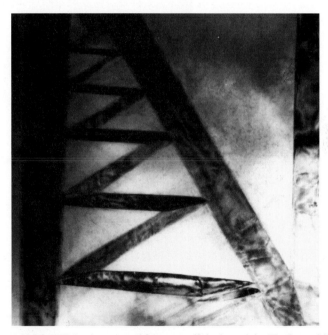

Figure 11.24 TEM of martensitic monoclinic lenses in Zr_2O_2 stabilized with 4 wt% Y_2O_3 and rapidly solidified; the zigzag pattern of lenses is due to autocatalysis (Courtesy of B. A. Bender and R. P. Ingel).

surround the crack, leading to its arrest. In Figure 11.25b, the lenticular tetragonal Zr_2O_3 precipitates in the cubic matrix are transformed to monoclinic in the region surrounding the crack. They appear as brighter lenses in the TEM because of favorable transmission conditions.

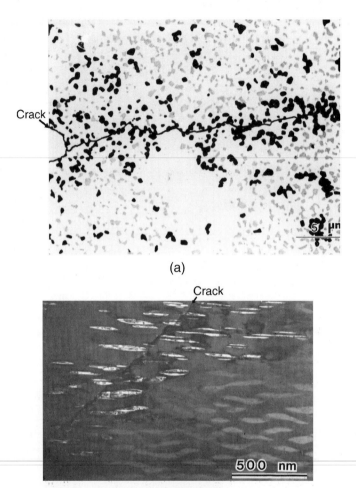

(a)

(b)

Figure 11.25 (a) Zirconia-toughened alumina (ZTA) traversed by a crack. The black regions represent monoclinic (transformed) zirconia, the gray regions tetragonal (untransformed) zirconia. (Courtesy of A. H. Heuer) (b) Partially stabilized zirconia (PSZ) lenticular precipitates transformed from tetragonal to monoclinic in the vicinity of a crack. Notice the brighter transformed precipitates (Courtesy of A. H. Heuer).

SUGGESTED READINGS

J. W. CHRISTIAN. "The Strength of Martensite," in *Strengthening Methods in Crystals*, A. KELLY and R. B. NICHOLSON (eds.). Amsterdam: Elsevier, 1971, p. 261.

J. W. CHRISTIAN. *The Theory of Transformations in Metals and Alloys*, 2d ed. Elmsford, NY: Pergamon Press, 1981.

D. J. GREEN, R. H. J. HANNINK, and M. V. SWAIN. *Transformation Toughening of Ceramics*. Boca Raton, FL, CRC, 1989.

A. H. HEUER. "Fracture-Tough Ceramics," in *Frontiers in Materials Technologies*, M. A. MEYERS and O. T. INAL (eds.). Amsterdam: Elsevier, 1985, p. 265.

A. H. HEUER, F. F. LANGE, M. V. SWAIN, and A. G. EVANS. "Transformation Toughening: An Overview." *J. Am. Cer. Soc.,* 69 (1986) i–iv.

G. KRAUSS. *Principles of Heat Treatment of Steel.* Metals Park, OH: ASM, 1980.

12

Intermetallics

An intermetallic is a compound phase of two or more normal metals (ordered or disordered). Interest in intermetallics waned in the 1960s and 1970s. However, the demand for materials that are strong, stiff, and ductile at high temperatures has led to a resurgence of interest in intermetallics, especially silicides and ordered intermetallics such as aluminides. A testimony to this resurgence was the appearance in 1994 on the subject of a two-volume set by J. H. Westbrook and R. L. Fleischer, *Intermetallic Compounds: Principles and Practice* (New York: John Wiley). Intermetallic aluminides and silicides can be very oxidation and corrosion resistant, because they form strongly adherent surface oxide films. Also, intermetallics span a wide range of unusual properties. An important example outside the field of high-temperature materials involves the exploitation of martensitic transformations, exotic colors, and the phenomenon of shape memory in gold-based intermetallics in jewelry making. In what follows, we describe the silicides first and then the ordered intermetallics.

12.1 SILICIDES

About 300 intermetallic compounds melt at temperatures above 1,500°C. A survey of some silicide intermetallics for high-temperature applications showed that, based on criteria such as availability, phase changes in the temperature range of interest, and oxidation resistance, Ti_5S_3 and $MoSi_2$ seem to be the most promising materials: Ti_5Si_3 has the lowest density of all intermetallics, and $MoSi_2$ has a superior oxidation resistance. For service at temperature up to 1,600°C, one needs characteristics such as high strength, creep resistance, fracture toughness, oxidation resistance, and microstructural stability. Figure 12.1 shows a plot of melting point vs. density for intermetallics having $0.8T_m = 1,600$°C. Here

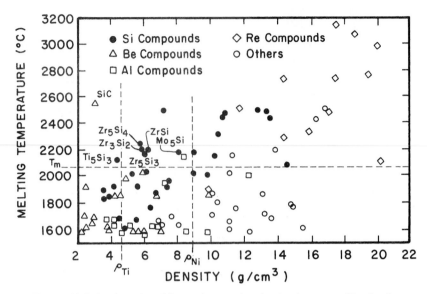

Figure 12.1 A plot of melting point vs. density for intermetallics having $0.8 T_m = 1,600°C$. (After P. J. Meschter and D. S. Schwartz, *J. of the Minerals, Metals and Materials Soc.* (Nov. 1989), p. 52)

we are assuming that intermetallics retain their strength up to temperatures of 80% T_m, the melting point of the material in K. This puts a lower limit on T_m equal to 2,067°C for a service temperature of 1,600°C. Also selected in the plot is an upper limit on density, viz., the density of nickel.

Molybdenum disilicide is a very promising intermetallic because of several of its characteristics. MoSi$_2$ has a tetragonal crystal structures, a high melting point, a relatively moderate density (6.31 g cm^{-3}), an excellent oxidation resistance, and a brittle-to-ductile transition at around 1,000°C, which can result in high toughness at the service temperature. Among the problems associated with MoSi$_2$ are its rather low low-temperature toughness and low high-temperature strength. MoSi$_2$ shows a catastrophic oxidation behavior around 500°C. In the literature, this problem has been termed pesting and is described as the retention of MoO$_3$ as an oxidation product at the grain boundaries. The expansion in volume accompanying the formation of MoO$_3$ results in severe microcracking. Among the efforts to ameliorate these problems, perhaps the most promising one is to use the approach of making a composite with MoSi$_2$ as a matrix. Table 12.1 summarizes the advantages and limitations of MoSi$_2$.

TABLE 12.1 Advantages and Limitations of Monolithic MoSi$_2$

Advantages

Moderate density: 6.24 g/cm^3
High melting point: 2,020°C
Outstanding oxidation resistance at $<$ 1,700°C
Potential upper temperature limit: 1,600°C
Deforms plastically above 1,000–1,200°C
Amenable to electrodischarge machining (EDM)

Limitations

Low room-temperature fracture toughness (3.0 MPa m$^{1/2}$)
Low strength and creep resistance at elevated temperatures
  (e.g., 140 MPa at 1,200°C)

12.2 ORDERED INTERMETALLICS

In the simple description of crystal structure of metals given in Chapter 1, we tacitly assumed a *random* atomic arrangement of A and B atoms in a unit cell of a metallic alloy consisting of atoms of species A and B. When A *and* B are arranged in a random manner, we have a *disordered alloy*. In such an alloy, equivalent crystallographic planes are statistically identical. Truly random—that is, completely disordered—alloys are not common, but there are many alloy systems that come close to having a random or disordered distribution of species A and B. It turns out that in a vast number of alloy systems, it is energetically favorable for atoms A and B to segregate to preferred lattice sites. Generally, such an ordered arrangement of atoms is obtained below a critical temperature T_c and in certain well-defined atomic proportions, i.e., stoichiometric compounds such as AB_3, AB, etc. Among examples of these systems, one may cite CuAu, Cu_3Au, Mg_3Cd, FeCo, FeAl, and aluminides of Ni and Ti. When the bonding is not totally metallic, but is partly ionic in nature, such an alloy is called an *intermetallic compound*. Table 12.2 gives a summary of important characteristics of some intermetallics. In what follows, we examine (1) the differences in the dislocation behavior in ordered alloys vis-à-vis disordered, or ordinary, alloys, (2) the effect of ordering on mechanical behavior, and (3) efforts to enhance the low-temperature ductility of ordered alloys, with a special emphasis on nickel aluminide (Ni_3Al), which has some very unusual properties.

TABLE 12.2 Properties of Some Intermetallic Compounds

Intermetallic Compound	Crystal Structure	Melting Point (°C)	Density (g/cm³)	Young's Modulus (GPa)
FeAl	B2(Ordered BCC)	1250-1400	5.6	263
Fe_3Al	DO_3	1540	6.7	
NiAl	B2 (Ordered FCC)	1640	5.9	206
Ni_3Al	$L1_2$ (Ordered FCC)	1390	7.5	337
TiAl	$L1_0$ (Ordered tetragonal)	1460	3.9	94
Ti_3Al	DO_{19} (Ordered CPH)	1600	4.2	210
$MoSi_2$	Tetragonal	2020	6.31	430

EXAMPLE 12.1

Molybdenum disilicide shows a phenomenon called *pesting*. Describe this phenomenon and indicate some means of overcoming it.

Solution: It has been observed that at about 500°C, $MoSi_2$ shows an accelerated oxidation. A product of this oxidation is MoO_3, which is accompanied by a rather large change in volume. This catastrophic oxidation can result in severe microcracking. Among some of the proposed remedies to overcome pesting are the following:

- Preoxidize the $MoSi_2$, to form a continuous SiO_2 surface film.
- Minimize porosity, to minimize the formation of MoO_3 at the pore surfaces.
- Use alloying to alter the oxidation characteristics of $MoSi_2$.
- Use metal coatings.

EXAMPLE 12.2

There is some interest in the use of gold-based intermetallic alloys in the jewelry industry. Can you describe some other possibilities in this area?

Solution: Platinum-based intermetallic alloys represent another possibility. Platinum is a soft metal like gold, but has a silverlike color. Thus, platinum alloys based on a $PtAl_2$ intermetallic can be of interest because they show a higher hardness than Pt and they can range in color from orange through pink to the yellow of pure gold.[1]

12.2.1 Dislocation Structures in Ordered Intermetallics

There are some very important differences between the dislocation structures observed in common metals and those in ordered intermetallics. In FCC metals, dislocations split into partials, and the partials are separated by a stacking-fault ribbon. (See Chapter 4). The partials, however, are confined to a single slip plane and do not have parallel Burgers vectors. As illustrated in Figure 12.2, the characteristic dislocation structure in an ordered alloy consists of two superpartial dislocations separated by a faulted region or what is also called an *antiphase boundary* (APB). Figure 12.2b shows an example of partial dislocations separated by a faulted region of 5 nm width in Ni_3Al deformed at 800°C.

An ordered structure results in some interesting characteristics. The ordered state in A_3B-type alloys is a low-energy state, so the movement of dislocations and vacancies results in a destruction of the local order; that is, a higher energy state is produced. Thus, activities such as dislocation motion and vacancy migration are subject to some restrictions. For example, in ordered structures, the dislocations must travel in pairs—a leading dislocation and a trailing dislocation. The passage of a leading dislocation destroys the order, while the passage of a trailing dislocation restores it. Also, thermally activated phenomena, such as diffusion via vacancies, suffer retardation.

Let us consider the ordered Ni_3Al intermetallic. The $L1_2$ crystal structure of Ni_3Al is shown in Figure 12.3a. The aluminum atoms are located at corners of cubes, while the Ni atoms are at the centers of the faces of the cube. Figure 12.3b shows a (111) slip plane and the slip direction <010>, consisting of two $\frac{1}{2}$<010> vectors, in Ni_3Al. Note that the APB in between the two superpartials lies partly on the (111) and partly on the (110) face. Interestingly, the superpartials in this case have the same Burgers vectors (along the screw direction). These superpartials can extend to any slip plane that contains the dislocation line or Burgers vector. The APB can be transferred from one plane to the other by cross-slip of the superpartial screws. This situation allows the superpartials to reduce the energy of the intermetallic by extending to the plane with a minimum APB energy, because the configurational energy decreases with decreasing APB energy. When a pair of screw partials is fully transferred from the (111) plane to the cross-slip plane (010), we get what is called *Kear-Wilsdorf lock*.[2] We can estimate the energy change associated with this lock in the following way. The APB, a kind of stacking fault, results in an energy increase that is proportional to the quantity $(E_{AA} + E_{BB} - 2E_{AB})$, where E_{AA}, E_{BB}, and E_{AB} are the bonding energies of AA, BB, and AB pairs, respectively. The superpartials of a pair repel each other elastically, but are held together by the APB. If r is the separation between two superpartials, the interaction energy is given by $-K \ln r$, where K is a constant involving elastic constants of materials and the character of the dislocation. If γ is the surface energy of the

---

[1] J. Hurly and P. T. Wedepohl, *J. Mater. Sci.,* 28 (1993) 5648.

[2] B. H. Kear and H. G. F. Wilsdorf, *Trans. AIME,* 224 (1962) 382).

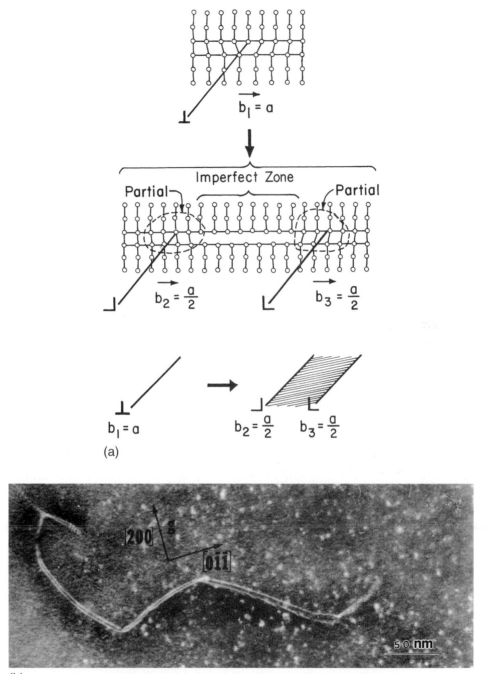

(a)

(b)

Figure 12.2 (a) The characteristic dislocation structure in an ordered alloy consists of two superpartial dislocations, separated by a faulted region or an antiphase boundary (APB). (b) Superpartial dislocations separated by approximately 5 nm in Ni_3Al deformed at 800°C; $b = [110]$ and superpartials $b_1 = b_2 = \frac{1}{2}[110]$. (Courtesy of R. P. Veyssiere)

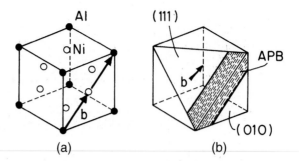

Figure 12.3 (a) The L1$_2$ crystal structure of Ni$_3$Al. The aluminum atoms are located at the corners of a cube, while the Ni atoms are at the centers of the faces. (b) A (111) slip plane and the slip direction <010>, consisting of two 1/2<110> vectors, in Ni$_3$Al. Note that the APB in between the two super-partials lies partly on the (111) and partly on the (010) face.

APB, then the energy of an APB of width r is γr. Thus, the energy of a pair of superpartials with an APB of width r can be written as

$$E(r) = \gamma r - K \ln r. \tag{12.1}$$

At the equilibrium separation r_0, the two components of the energy balance, and we can write

$$dE/dr = 0 = \gamma - K/r_0, \tag{12.2}$$

or

$$\gamma = K/r_0.$$

From Equations 12.1 and 12.2, we get, at $r = r_0$, the equilibrium energy

$$E(r_0) = (K/r_0)r_0 - K \ln (K/\gamma).$$

or

$$E(r_0) = K[1 + \ln (\gamma/K)]. \tag{12.3}$$

Applying Eqn (12.3) to the primary and cross slip planes, we can write:

For primary slip

$$E_p(r_0) = K[1 + \ln (\gamma_p/K)]$$

For cross slip

$$E_c(r_0) = K[1 + \ln (\gamma_c/K)]$$

Hence

$$\Delta E(r_0) = E_c(r_0) - E_p(r_0) = K[\ln (\gamma_c/K) - \ln (\gamma_p/K)]$$

Thus, the energy associated with the Kear–Wilsdorf lock can be written as

$$\Delta E = K \ln \lambda,$$

where $\lambda = \gamma_c/\gamma_p$, in which the subscripts c and p represent the cross-slip and primary planes, respectively. Kear–Wilsdorf locks harden the intermetallic because they inhibit slip; as λ decreases, the tendency to form these locks increases.

There are some other differences between intermetallics and common alloys. Generally, common disordered alloys show an isotropic behavior, whereas most intermetallic compounds have anistropic elastic properties. This can result in excessive elastic strain on certain planes; in particular, it can introduce shear stresses perpendicular to screw dislocation lines. These dislocations not only will repel each other along the radial directions, but also will exert a torque on each other.[3]

12.2.2 Effect of Ordering on Mechanical Properties

Mechanical properties of an alloy are altered when it has an ordered structure. We define the degree of long-range order (LRO) by means of a parameter

$$S = \frac{r - f_A}{1 - f_A}, \tag{12.4}$$

where r is the fraction of A sites occupied by A atoms and f_A is the fraction of A atoms in the alloy. Thus, S goes from 0 (completely disordered) to 1 (perfectly ordered). The various dislocation morphologies observed in ordered alloys have been summarized by Marcinkowski and are presented in Table 12.3.

A superdislocation (i.e., closely spaced pairs of unit dislocations bound together by an antiphase boundary) in a perfectly ordered crystal and a single dislocation in a completely disordered crystal will both experience less friction stress than either of them will experience at an intermediate degree of order S. Thus, qualitatively, one would expect a yield stress maximum at an intermediate degree of order (i.e., the change in yield stress is not directly related to the degree of ordering). For example, Cu_3Au crystals show a lower yield stress when fully ordered than when only partially ordered. Experiments showed that this results from the fact that the maximum in strength is associated with a critical domain size. Short-range order (SRO) results in a distribution of neighboring atoms that is not random. Thus, the passage of a dislocation will destroy the SRO between the atoms across the slip plane. The stress required to do this is large. A crystal of Cu_3Au in the quenched state (SRO) has nearly double the yield stress of that in the annealed (LRO). The maximum in strength is exhibited by a partially ordered alloy with a critical domain size of about 6 nm. The transition from deformation by unit dislocations in the disordered state to deformation by superdislocations in the ordered state gives rise to a peak in the curve of flow stress versus degree of order.

[3]M. H. Yoo, *Acta Met.*, 35 (1987) 1559.

TABLE 12.3 Dislocation Morphologies in Some Ordered Alloys[a]

Superlattice Type (*Strukturbericht* Designation)	Chemical Designation	Unit Cell Dimensions	Alloy Types	Superlattice Dislocation Type	Burgers Vector of Each Dislocation	Antiphase Boundary Type[b]
B2	CsCl	a_0	NiAl, AgMg AuZn		$a_0\langle100\rangle$	None
			CuZn, FeCo FeAl, FeRh NiAl, AgMg AuZn		$\frac{1}{2}a_0\langle111\rangle$	NN
DO$_3$	Fe$_3$Al	a_0	Fe$_3$Al, Fe$_3$Si Fe$_3$B		$\frac{1}{4}a_0\langle111\rangle$	NN NNN
L1	Cu$_3$Au	a_0	Cu$_3$Au, Ni$_3$Mn Ni$_3$Al, Ni$_3$Fe Cu$_3$Pd, Ni$_3$Ti Ag$_3$Mg, Ni$_3$Ta Ni$_3$Si, Cu$_3$Pt Ni$_3$Ga, Ti$_3$Al		$\frac{1}{6}a_0\langle112\rangle$	NN NN + SF
DO$_{19}$	Mg$_3$Cd	a_0 c_0	Mg$_3$Cd, TiAl		$\frac{1}{6}a_0\langle10\bar{1}0\rangle$ $\frac{1}{2}a_0\langle2\bar{1}\bar{1}0\rangle$	NN SF NNN
L1$_0$	CuAu	a_0 c_0	CuAu, CoPt FePt, TiAl		$\frac{1}{6}a_0\langle112\rangle$	NN NN + SF SF

[a]Adapted with permission from H. J. Marcinkowski, in Treatise on Material Science and Technology, Vol. 5 (New York: Academic Press, 1976, p. 181).

[b]NN, nearest neighbor, NNN, next-nearest neighbor, SF, stacking fault.

The presence of atomic order leads to a marked change in the flow curve of the alloy. Figure 12.4 shows the flow curves of a fully ordered FeCo alloy at low temperatures, where the order is not affected. Stage I is associated with a well-defined yield point. This is followed by a high linear work-hardening stage, II. Finally, there occurs stage III, with nearly zero work-hardening. The stress–strain curves of the same alloy in the disordered state are shown in Figure 12.5. The curves in Figure 12.4 (ordered) are markedly different from the ones in Figure 12.5 (disordered). The sharp yield point and stage II are absent in

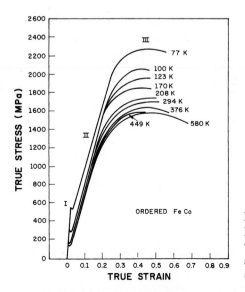

Figure 12.4 Stress–strain curves of ordered FeCo alloys at different temperatures. (Adapted with permission from S. T. Fong, K. Sadananda, and M. J. Marcinknowski, *TransAIME*, 233 (1965) 29)

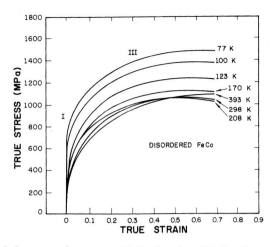

Figure 12.5 Stress-strain curves of fully disordered FeCo alloys at different temperatures. (Adapted with permission from S. T. Fong, K. Sadananada, and M. J. Marcinkowski, *TransAIME*, 233 (1965) 29)

the disordered alloy, which goes straight into stage III after gradual yielding. Fully ordered alloys deform by means of the movement of superlattice dislocations at rather low stresses. However, the superdislocations must move as a group in order to maintain the ordered arrangement of atoms. This makes cross-slip difficult. Long-range order thus leads to high strain-hardening rates and frequently, to brittle fracture.

Figure 12.6 shows this effect of ordering on uniform elongation of FeCo–2% V at room temperature. The ductility of the alloy decreases with increasing LRO. Mg_3Cd is the only known exception to this tendency toward brittleness, because of a restricted number of slip systems or less easy cross-slip.

Ordered alloys such as FeCo and Ni_3M obey the Hall–Petch relationship between flow stress and grain size; viz., (see Chapter 5)

$$\sigma = \sigma_0 + kD^{-1/2},$$

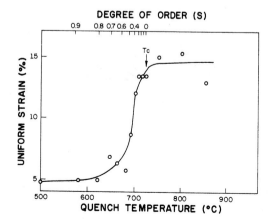

Figure 12.6 Effect of atomic order on uniform strain (ductility) of Fe-Co–2V at 25°C. (Adapted with permission from N. S. Stoloff and R. G. Davies, *Acta Met.,* 12 (1964) 473)

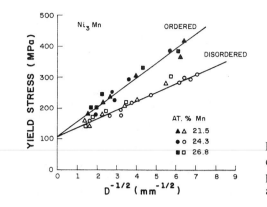

Figure 12.7 Hall-Petch relationship for ordered and disordered alloys. (Adapted with permission from T. L. Johnston, R. G. Davies, and N. S. Stoloff, Phil Mag., 12 (1965) 305)

where σ is the flow stress at a given strain, σ_0 and k are constants for that strain, and D is the grain diameter. In these alloys, long-range order increases k, as shown in Figure 12.7 for Ni_3Mn. This increase in k with long-range order can be explained by the change in the number of slip systems with order, since the ease of spreading of slip across boundaries is controlled by the degree of order.

The effect of atomic ordering on fatigue behavior is shown in Figure 12.8 in the form of S–N curves for ordered and disordered Ni_3Mn. The improved fatigue performance in the ordered state is explained by less ease of cross-slip and a decrease in slip-band formation in that state. Slip bands lead to the formation of extrusions and intrusions on the sample surface, which in turn lead to fatigue crack nucleation. (See Chapter 14.)

Gamma-prime-strengthened superalloys are an example of the effect of ordering on strength. The Ni_3Al precipitate produces very low coherency stresses and is coherent with the austenitic matrix. The strengthening effect is clearly evident in Figure 12.9b, which shows the strength of the austenitic matrix and Ni_3Al separately, and the strength of MarM-200, composed of 65 to 85% gamma prime (the ordered Ni_3Al). In ordered structures, it is energetically favorable for dislocations to move in groups, forming antiphase boundaries between them as seen in Section 12.2. The equilibrium distance between the pairs, as well as their form, was found to depend on the particle size, particle distribution, energy of the antiphase boundary, elastic constants, and external shear stress. The preceding parameters are part of the equations derived by Gleiter and Hornbogen[4] for the in-

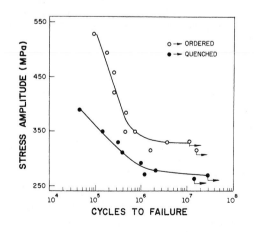

Figure 12.8 Effect of atomic order on fatigue behavior of Ni_3Mn. (Adapted with permission from R. C. Boettner, N. S. Stoloff, and R. G. Davies, *Trans. AIME,* 236 (1968) 131)

[4]H. Gleiter and E. Hornbogen, *Phys. Status Solids,* 12 (1965), 235, 251.

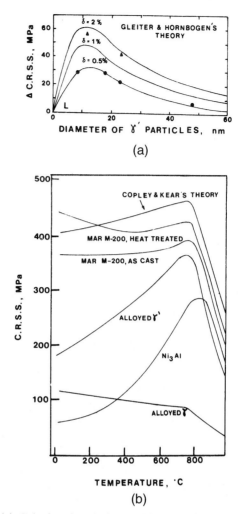

(a)

(b)

Figure 12.9 (a) Calculated and observed increase in the critical resolved shear stress (CRSS) in an Ni-Cr-Al alloy as a function of the diameter of the precipitate; full lines represent calculations ($\bullet$ $\delta = 0.5\%$ Al; $\blacktriangle$ $\delta = 1.8\%$ AJ) δ is atomic percent aluminum. (Adapted with permission from H. Gleiter and H. Hornbogen, Phys. Status Solids, 12 (1965) 235. (b) Effect of temperature on CRSS for Ni_3Al, γ, and Mar M-200 superalloy ($\gamma + \gamma'$). (Adapted with permission from S. M. Copley and B. H. Kear, Trans. TMS-AIME, 239 (1967) 987)

crease in the critical resolved shear stress, $\Delta\tau$ or $\Delta CRSS$. The results of calculations are compared with observed results for a Ni-Cr-Al alloy in Fig. 12.9(a), where δ is the atomic percent aluminum. The experimental results are marked by dots and triangles; they refer to 0.5 and 1.8% aluminum, respectively. The correlation is good, and maximum strengthening is obtained for particles having a diameter of 100 Å.

Another outstanding property of Ni_3Al and some other intermetallics is the increase in yield stress with temperature. As is seen from Figure 12.9b, the yield stress increases by a factor of 5 when the temperature is raised from ambient temperature to 800°C. This temperature dependence is unique and contrary to what would be expected on the basis of thermally activated motion of dislocations. (We discuss this and other aspects of ordered intermetallics in Section 12.2.3.) Thus, in spite of the normal temperature dependence of the austenite (also shown in Figure 12.9b), the alloy MarM-200 exhibits a constant yield stress up to 800°C; the decrease in the flow stress of γ is compensated for by the increase of γ' (Ni_3Al). It is interesting to note that other ordered alloys, such as Cu_3Au and Ir_3Cr, do not exhibit this unique behavior, while Ni_3Ge, Ni_3Si, Co_3Ti, and Ni_3Ga do. High voltage TEM work on Ni_3Ge has shown dramatic changes in dislocation configuration. For Ni_3Ge, it was found that the substructure at $-196°C$ consisted roughly of an equal number of edge and screw dislocations, while at 27°C it consisted mostly of screw dislocations aligned along 101]; this is shown in Figure 12.10. Thus, the decreased mobility of screw dislocations with increasing temperature was responsible for the strengthening effect. At temperatures above the one providing maximum strength, the change in slip plane from {111} to {100} would be responsible for the decrease in strength. This explanation is different from the one previously provided.

It is this very unusual behavior—the increase in flow stress with temperature—that makes nickel aluminides very attractive for high-temperature applications. Ni_3Al remains ordered up to its melting point (1,400°C) and also shows an increasing yield strength with temperature. A decrease in yield strength occurs at very high temperatures due to the start of thermally activated slip on {100} planes, and not due to disorder.

Long-range ordered alloys of the Ni_3Al type show some important and unique features alluded to earlier, such as an increasing yield stress with increasing temperature. The problem with these alloys, however, is their lack of ambient temperature ductility. Figure

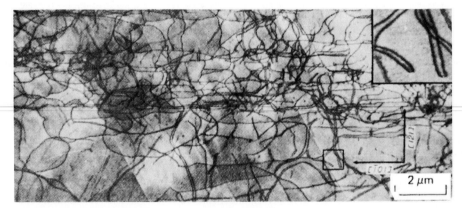

(a)

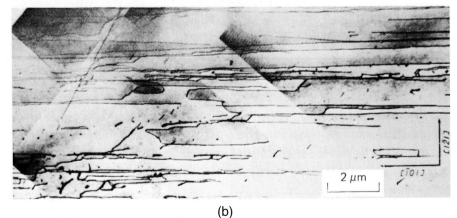

(b)

Figure 12.10 Effect of deformation temperature on the dislocation arrangement in the {111} primary slip plane of ordered Ni_3Ge. (a) $T = -196°C$, $\varepsilon_p = 2.4\%$. (b) $T = 27°C$, $\varepsilon_p = 1.8\%$. (Courtesy of H.-r. Pak.)

12.11 shows yield stress as a function of temperature for Ni₃Al-based alloys, Hastelloy-X, and type 316 stainless steel. It is not surprising that L1₂-type intermetallics are major candidates for use at elevated temperatures, about 900–1000°C.

The anomalous yield behavior of Ni₃Al has been the subject of a number of investigations. At temperatures $T < T_p$, slip occurs mainly on the octahedral {111} planes, while at temperatures $T > T_p$, slip becomes dominant on cubic {100} planes. T_p is the temperature corresponding to the maximum in strength ≈800°C. Sun and Hazzledine used weak-beam TEM to identify dislocation structures with low mobilities in Ni₃Al-type ordered intermetallics.[5] They observed that in the region of yield stress anomaly, a king mechanism unlocks the Kear–Wilsdorf locks described earlier. This mechanism leads to the formation of special kink configurations with switched superpartials, as well as the formation of what are called APB tubes. In summary, the increase in yield strength below T_p is related to the formation of K–W locks (lowering of λ) while the decrease in yield strength above T_p is attributed to the change of slip from {111} <110> to {100} <110>. (See Fig. 12.3.)

12.2.3 Ductility of Intermetallics

As we have seen, many alloy systems of the general composition A_3B have an ordered structure formed by regular stacking of closed-packed layers. The stacking sequence, however, can range from the more common cubic or hexagonal to less common and more complex transition structures with unit cells extending over 15 layers. Such intermetallics are generally quite brittle at low temperatures, which makes their processing very difficult. There are two common causes of brittleness in intermetallics:

(1)   The crystal structure is of low symmetry; that is, not enough slip systems are available for general plastic deformation to occur. As is well known, one needs at least five independent systems for an arbitrary change in shape to occur.

(2)   Enough slip systems are available, but there are crack propagation paths along the grain boundaries that are easy to take and that will cause embrittlement.

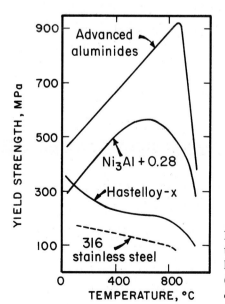

Figure 12.11 Yield stress as a function of test temperatures for Ni₃Al based aluminide alloys. Hastelloy-X, and type 316 stainless steel. (Adapted from C. T. Liu and J. O. Stiegler, *Science,* 226 (1984) 636).

[5]Y. Q. Sun and P. M. Hazzledine, in *High Temperature Ordered Intermetallic Alloys* (Dordrecht, the Netherlands, Kluwer, 1992), p. 177.

Generally, ordered hexagonal alloys have very limited ductility and ability to process while ordered cubic alloys have good ductility.

Alloying   Various researchers have tried to make ordered intermetallics more ductile by different approaches. Baker and Munroe classify these attempts into four categories: microalloying, macroalloying, processing-induced microstructure control, and fiber reinforcement.[6] We summarize these efforts next.

Microalloying   An examination of the Ni–Al phase diagram shows four intermetallics: $NiAl_3$, Ni_2Al_3, NiAl, and Ni_3Al. Ni_3Al is nothing but γ', the strengthening phase in many Ni-based superalloys meant for high-temperature use as described earlier. Ni_3Al has an $L1_2$ crystal structure with Al atoms at the cube corners and Ni atoms at the face-centered positions. (See Figure 12.3a.) Single-crystal Ni_3Al is very ductile at and below room temperature. Its ductility decreases with temperature until the peak in yield strength occurs. In polycrystalline form, nickel aluminide has practically no ductility at room temperature. Ni_3Al does possess five independent slip systems <111> {110}, which is the condition for generalized plastic flow, as per the von Mises criterion. Instead of high ductility, polycrystalline Ni_3Al shows intrinsic grain-boundary weakness, as evidenced by its tendency toward brittle, intergranular fracture at room temperature. It turns out that boron is a very effective dopant for restoration of ductility in Ni_3Al. Boron-free polycrystals fracture without any plastic yielding, and very small additions of boron can lead to dramatic results. As little as 0.05% wt. % B can improve the strain to failure from nearly 0 to 50% and can alter the fracture mode from intergranular to transgranular. Figure 12.12 shows this restoration of room-temperature ductility in Ni_3Al as a function of boron content. Note the very small amount of boron required to do the job. As the figure reveals, boron-doped Ni_3Al shows a broad maximum in strength. The poor ductility of intermetallics and the effect of boron are generally explained in terms of environmental effects, especially moisture. (See shortly).

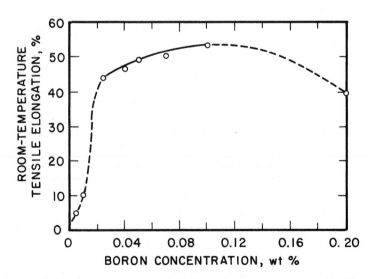

Figure 12.12 Plot showing the restoration of room-temperature ductility in Ni_3Al as a function of boron content. (After K. Aoki and O. Izumi, *Nippon Kinzoku Takkasishi*, 43 (1979) 1190).

[6]I. Baker and P. R. Munroe, *Journal of Metals*, Feb. 1988, p. 28.

Intergranular failure at room temperature also has been attributed to the segregation of impurities such as sulfur to grain boundaries. In one experiment, a decrease in ductility was measured as the sulfur content of the Ni_3Al increased from 32 to 176 ppm by weight. Auger electron spectroscopy showed that sulfur did indeed segregate to grain boundaries. Attempts at purifying Ni_3Al to restore its ductility have not worked in practice.

Macroalloying   Macroalloying additions (less than 1 at.%) have been used to introduce modifications in intermetallics that lead to enhanced ductility. Such modifications include changing the crystal structure to one of higher symmetry, promoting the operation of additional or different slip systems, and other changes.

Alloys of the $(Ni, Co, Fe)_3V$ system can have ordered structure (cubic, hexagonal, or more complex transitional). The ordered hexagonal structure is too brittle for processing. Macroalloying can be used to create a window in the composition space in the $(Ni, Co, Fe)_3V$ system that has an intrinsically ductile, ordered cubic structure. An important parameter in the characterization of intermetallics is the electron concentration (e/a), which is the number of valence electrons per atom. Ordered structures of the type A_3B are built by stacking close-packed layers. It turns out that the stacking sequence is influence by the atomic radius ratio (R_a/R_0) and the electron concentration. In the $(Ni, Co, Fe)_3V$ system, nickel, cobalt, and iron have about the same atomic size. Thus, these elements influence the crystal structure through their electronic effects. If a portion of Co in $(Fe, Co)_3V$ is replaced by an equal number of Ni and Fe atoms, then we shall have altered the composition of the compound, but not the e/a ratio. In Co_3V, a six-layer stacking sequence occurs, with one-third of the layers having a hexagonal character ($ABABAB \ldots$) and two-thirds of the layers having a cubic character ($ABCABC \ldots$). Replacing Co by Ni gives a higher e/a ratio and a predominantly hexagonal stacking first, followed by fully hexagonal stacking. A reduction in the e/a ratio to 7.89 or less gives an cubic ordered structure ($L1_2$). Thus, one can choose a suitable combination of Ni, Co, and Fe to obtain the desirable cubic ordered structure (the same as that of Cu_3Au).

Titanium-based alloys are lighter than Co-based and Ni-based superalloys. However, the service temperature of Ti-based alloys is less than 500°C. TiAl and Ti_3Al, the aluminides of titanium, have lower densities, higher stiffness, and higher use temperatures than nickel aluminides do. Titanium aluminides show good oxidation resistance up to 900°C, but have poor ductility and strength at low temperatures. Additions of β-stabilizing elements such as Nb, Mo, and W can result in some improvement in ductility in Ti_3Al. However, such macroalloying additions of heavier elements are accompanied by a penalty on density.

Processing–induced microstructural control   Polycrystalline nickel aluminides are brittle at room temperature. Single-crystal nickel aluminides, however, are ductile. One very straightforward approach would be to use single crystals of these materials. Another approach is to combine grain refinement with another ductility-enhancing feature, such as a martensitic transformation. A fine grain size would result in slip homogenization, eliminate grain boundary segregation, and allow enough deformation for the martensitic transformation to be induced. In principle, such a technique should work for any intermetallic. In practice, though, the grain size required for ductility may be too small (perhaps less than 1 μm) and thus difficult to achieve. An example of the beneficial effect of fine grain size is the use of rapid solidification technology to produce very fine grains in Ti_3Al. The reader should be cautioned, however, that although a grain refinement can lead to improvement in low-temperature properties, it can result in rather poor creep properties at high temperature because of the grain-boundary-related creep processes. (See Chapter 13.)

Ordered iron aluminides based on Fe_3Al also offer oxidation resistance and low material cost, but have limited ductility at ambient temperatures. In addition, the strength drops drastically above 600°C. Sikka used thermomechanical processing to improve room-temperature ductility in iron aluminides.[7] A suitable combination of melting practice, processing, heat treatments, and test conditions resulted in 15–20% room-temperature elongation values. The recipe involves an alloy lean in alloying elements, vacuum melting, an unrecrystallized or only slightly recrystallized microstructure, oil quenching after heat treatment, higher-than-normal strain rates, and a moisture-free environment.

Fiber Reinforcement   This approach involves the use of fibers to toughen the intermetallics. The idea is the same as that in ceramic matrix composites—viz., provide a weak interface ahead of a propagating crack, and thus bring into play a variety of energy-dissipating processes such as crack deflection, fiber pullout, etc. (See Chapter 15.)

Environmental effects in intermetallics   There is evidence that the poor ambient ductility encountered in ordered intermetallics is due mainly to environmental effects.[8] Both moisture and hydrogen, at levels found in ambient air, are thought to be responsible for inducing embrittlement in ordered intermetallics. In the case of water vapor, the phenomenon involves the reaction of reactive elements in the intermetallics with the ambient water vapor, to form an oxide (or hydroxide) and generate atomic hydrogen, which leads to a loss of ductility accompanied by a change in fracture mode from transgranular to intergranular. In the case of H_2, atomic hydrogen is produced by dissociation of physisorbed hydrogen molecules on intermetallic surfaces. It would thus appear that the main reason for the efficacy of boron in rendering Ni_3Al more ductile is that boron suppresses the environmental embrittlement, possibly by slowing diffusion of hydrogen.

Iron aluminides based on Fe_3Al are also sensitive to environmental effects. A major problem again is the ever-present moisture in the air. The water vapor reacts with aluminum to produce hydrogen at the surface of the metal. This hydrogen is adsorbed in the aluminide during plastic deformation, leading to low ambient ductility.

EXAMPLE 12.3

What is the source of the excellent high-temperature oxidation resistance shown by aluminides of nickel, cobalt, and iron?

Solution: Although the aluminides are quite brittle, they readily form a layer of alumina at high temperatures. The alumina layer provides the excellent oxidation resistance up to 1,000°C. Such aluminides are used as coatings on gas turbine components. Kanthal alloys used for heating elements are also based on iron aluminides.

SUGGESTED READINGS

E. P. George, M. Yamaguchi, K. S. Kumar, and C. T. Liu. *Ann. Rev. Mater, Sci.,* 24 (1994) 409.

C. T. Liu, R. W. Cahn, and G. Sauthoff (eds). *High-temperature ordered intermetallic alloys— physical metallurgy/mechanical behavior.* Boston and Dordrecht: Kluwer, 1992.

N. S. Stoloff and R. H. Jones (eds). *Processing and Design Issues in High Temperature Materials.* Warrendale: TMS, 1997.

A. K. Vasudevan and J. J. Petrovic (eds). *High Temperature Structural Silicides.* Amsterdam: Elsevier, 1992.

J. H. Westbrook and R. L. Fleischer (eds.) *Intermetallic Compounds: Principles and Practice.* New York: John Wiley, 1994.

[7]V. K. Sikka, *Sampe Quarterly,* 22 (July 1991) 2.

[8]E. P. George, C. T. Liu, H. Lin, and D. P. Pope, *Mater. Sci. & Eng.,* A192 (1995) 277; E. P. George, C. T. Liu, and D. P. Pope, *Acta Met.,* 44 (1996) 1757.

EXERCISES

12.1 High-temperature applications of intermetallics in air would require oxidation resistance. Comment on some possible sources of such oxidation resistance in intermetallics.

12.2 The formation of a silica film at grain boundaries in $MoSi_2$ can lead to embrittlement. Suggest some means of avoiding this phenomenon.

12.3 Order and disorder transitions are commonly associated with metals, not with polymers. Why?

12.4 An intermetallic compound of Al and Mg has a stable range of 52Mg–48Al to 56Mg–44Al (on a weight basis). What atomic ratios do these compositions correspond to? The atomic weight of Al is 27, and that of Mg is 24.31.

13

Creep and Superplasticity

13.1 INTRODUCTION

The technological developments wrought since the early 20th century have required materials that resist higher and higher temperatures. Applications of these developments lie mainly in the following areas:

1. Gas turbines (stationary and on aircraft), whose blades operate at temperatures of 800–950 K. The burner and afterburner sections operate at even higher temperatures, viz. 1,300–1,400 K.
2. Nuclear reactors, where pressure vessels and piping operate at 650–750 K. Reactor skirts operate at 850–950 K.
3. Chemical and petrochemical industries.

All of these temperatures are in the range $(0.4–0.65)T_m$, where T_m is the melting point of the material.

The degradation undergone by materials in these extreme conditions can be classified into two groups:

1. *Mechanical degradation.* In spite of initially resisting the applied loads, the material undergoes anelastic deformation; its dimensions change with time.

2. *Chemical degradation.* This is due to the reaction of the material with the chemical environment and to the diffusion of external elements into the materials. Chlorination (which affects the properties of superalloys used in jet turbines) and internal oxidation are examples of chemical degradation.

540

This chapter deals exclusively with mechanical degradation. The anelastic or time-dependent deformation of a material is known as *creep*. A great number of high-temperature failures can be attributed either to creep or to a combination of creep and fatigue. Creep is characterized by a slow flow of the material, which behaves as if it were viscous. If a mechanical component of a structure is subjected to a constant tensile load, the decrease in cross-sectional area (due to the increase in length resulting from creep) generates an increase in stress; when the stress reaches the value at which failure occurs statically (ultimate tensile stress), failure occurs. The temperature regime, in kelvins, for which creep is important in metals and ceramics is $0.5T_m < T < T_m$, the melting of the material. This is the temperature range in which diffusion is a significant factor. A thermally activated process, diffusion shows an exponential dependence on temperature. Below $0.5T_m$, the diffusion coefficient is so low, that any deformation mode exclusively dependent on it can effectively be neglected.

In glasses and polymers, creep becomes important at temperatures above T_g, the glass transition temperature. At $T > T_g$, these materials turn rubbery or leathery, and viscoelastic and viscoplastic effects become important. Section 13.3 presents the various mechanisms responsible for creep. The critical temperature for creep varies from material to material; lead creeps at ambient temperature, whereas creep becomes important in iron above 600°C. In general, the phenomenon of creep is important at high temperatures. Some nickel-based superalloys can withstand temperatures as high as 1,500 K, and ceramics have temperature capabilities that are considerably higher (up to 2,000 K). Ice, on the other hand, also undergoes creep, which is responsible for the slow flow of glaciers.[1] Even the earth's mantle is subjected to creep, giving it an effective viscosity.

Creep in rocks has been at the center of controversy concerning the nature of geological processes on the planet Venus. The maximum height of mountains on Venus has been calculated on the basis of rock creep, assuming a certain temperature and period of time. (The mountains are subjected to compressive stresses due to their own weight.) This maximum calculated height has been compared with actual topographic observations from the space probe *Galileo*. It happens that dry rock has a creep rate orders of magnitude lower than that of hydrated rock. Weertman has performed calculations for both dry and wet rock, each resulting in a value of 10^9 years for the period of active creep in the mountains. Based on this figure, the maximum height of mountains made of quartzite would be 0.12 km (wet) and 7.6 km (dry). The calculations were done for $T = 750$ K, the surface temperature on Venus. They help to elucidate the mechanisms involved in the formation of the planetary surface.

In spite of the fact that creep has been known since 1834, when Vicat conducted the first experiments assessing the phenomenon, it is only in the 20th century that systematic investigations have been conducted. The creep test is rather simple and consists of subjecting a specimen to a constant load (or stress) and measuring its length as a function of time, at a constant temperature. Figure 13.1 shows the characteristic curve; the ordinate shows the strain and the abscissa shows time. Three tests are represented in the figure; three constant loads corresponding to three engineering stresses, σ_a, σ_b, and σ_c, were used. The creep curves are usually divided into three stages: I, primary or transient; II, secondary, constant rate, or quasi viscous; and III, tertiary. This division into stages was made by Andrade, one of the pioneers in the study of creep. Stage II, in which the creep rate $\dot{\varepsilon}$ is constant, is the most important. It can be seen that $\dot{\varepsilon}_a > \dot{\varepsilon}_b > \dot{\varepsilon}_c$ as a consequence of the relationship $\sigma_a > \sigma_b > \sigma_c$. This creep rate is also known as the *minimum creep rate,* because

[1] See J. Weertman, *Ann. Rev. Earth Plan. Sci.,* 11 (1983) 215.

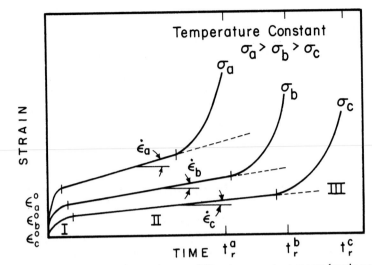

Figure 13.1 Creep strain vs. time at difference constant stress levels and temperature.

it corresponds to the inflection point of the curve. (See Figure 13.1.) In stage III there is an acceleration in the creep rate, leading to eventual rupture of the specimen.

In Figure 13.1, the rupture times t_r^a, t_r^b, and t_r^c increase with decreasing stress. The strains ε^0 are called *instantaneous* strains and correspond to the strains at the instant of loading. In Figure 13.2, the engineering stress was kept constant and the temperature was varied. Since the tests are conducted in tension, the stress rises as the length of the specimen increases, because of the reduction in area. The dashed lines in Figures 13.1 and 13.2 represent the constant stress curves. Initially they are identical, because $\varepsilon_e = 0$. As the specimen increases in length, the stress increases and so does the creep rate, at a constant load. The failure times under constant stress and constant load can be drastically different. The curves shown in Figure 13.2 have been expressed mathematically as

$$\varepsilon_t = \varepsilon^0 + \varepsilon[1 - \exp(-mt)] + \dot{\varepsilon}_s t \tag{13.1}$$

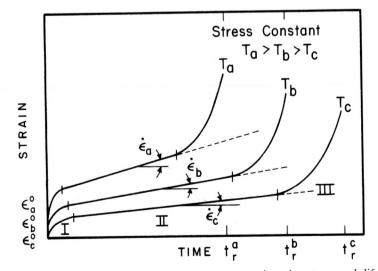

Figure 13.2 Creep strain vs. time at a constant engineering stress and different temperatures.

where ε^0 is the instantaneous strain (the strain at the instant of application of load), $\dot{\varepsilon}_s t$ is a linear function of time, depicting stage II, and the term $\varepsilon[1 - \exp(-mt)]$ represents stage I in which m is the exponential time parameter and ε is the strain that stage.

From a fundamental point of view, there are significant differences between the constant-load and constant-stress creep tests. Andrade realized this important difference and built a constant true-stress creep machine that used a weight which dropped gradually into a fluid as the specimen extended.[2] Thus, by Archimedes' principle, the force exerted by the weight decreased with displacement. The shape of the weight was such that a constant stress on the specimen was ensured. For this type of machine, the load should decrease with an increase in length in such a way that the true stress remains constant.

Another important difference between the two tests is that the onset of stage III is greatly retarded at constant stress. The dashed lines in Figure 13.2 show the trajectory that a constant true-stress test would follow.

From an engineering point of view, the creep test at constant load is more important than the one at constant stress because it is the load, not the stress, that is maintained constant in engineering applications. On the other hand, fundamental studies should be conducted at constant stress, with the objective of elucidating the underlying mechanisms. The reason for this is that the study of the evolution of the substructure of an alloy under increasing stress would be excessively complex.

The essential components and principles of operation of a constant-stress creep-testing machine are shown in Figure 13.3. This system contains a variable lever arm, which is a curved line that acts as a cam in such a manner that the force acting on the specimen is a function of its length. Two positions are shown in the figure. If the initial and current cross sections of the specimen are A_0 and A_1, respectively, then

$$\sigma_0 A_0 l_1 = P l_2 \tag{13.2}$$

and

$$\sigma_1 A_1 l_1 = P l_2'$$

where P is the load and $l_1, l_2,$ and l_2' are lever arms defined in the figure. At constant stress, $\sigma_0 = \sigma_1$; since the volume of the specimen is constant (for stages I and II of creep),

$$A_0 L_0 = A_1 L_1,$$

where L_0 and L_1 are the initial and current lengths of the specimen, respectively. Thus,

$$\frac{L_1}{L_0} = \frac{l_2}{l_2'}. \tag{13.3}$$

The exact shape of the lever arm can be established in such a manner that Equation 13.3 is obeyed. The astute student will certainly be able to obtain the mathematical description for this curved surface.

It is important to recognize that, even at a constant stress, the creep curve will deviate from linearity at a certain point. This can be due to several causes, the most important being the formation of internal flaws such as cavities (known as creep cavitation) and necking of the specimen. The minimum creep rate, or slope of stage II of creep, is a very important parameter. This stage, also known as steady-state creep, is usually represented by the equation

---

[2] E. N. da L. Andrade, *Proc. Roy Soc. (London),* A84 (1911) 1.

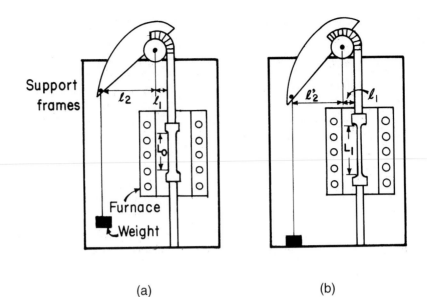

(a)   (b)

Figure 13.3 Creep machine with variable lever arms to ensure constant stress on specimen; note that l_2 decreases as the length of the specimen increases. (a) Initial position. (b) Length of specimen has increased from L_0 to L_1.

$$\dot{\varepsilon}_s = \frac{AGb}{kT} D_0 \exp(-Q_C/RT)\left(\frac{b}{d}\right)^p \left(\frac{\sigma}{G}\right)^n, \tag{13.4}$$

where A is a dimensionless constant, D_0 is a frequency factor, G is the shear modulus, b is the Burgers vector, k is Boltzmann's constant, T is the absolute temperature, σ is the applied stress, d is the grain size, p is the inverse grain-size exponent, n is the stress exponent, Q_c is the appropriate activation energy, and R is the gas constant. This equation is known as the Mukherjee–Bird–Dorn equation.[3] It will be shown in Section 13.3 that the activation energy for diffusion is often equal to the activation energy for creep ($Q_c = Q_D$). The diffusion coefficient is

$$D = D_0 \exp\left(-\frac{Q_D}{RT}\right)$$

and

$$\dot{\varepsilon}_s = \frac{AGbD}{kT}\left(\frac{b}{d}\right)^p \left(\frac{\sigma}{G}\right)^n. \tag{13.5}$$

Essentially, Equations 13.4 and 13.5 express the steady-state creep rate as a function of the applied stress, temperature, and grain size. In this chapter we will use d to designate the grain size, to differentiate it from D, the diffusion coefficient. Equation 13.4 is also a fundamental equation in superplasticity. Figure 13.4 illustrates the application of the Dorn–Mukherjee–Bird equation to metals (aluminum and aluminum alloys) and ceramics. This is usually done by plotting a normalized strain rate ($\dot{\varepsilon}kT/DGb$) vs. a normalized stress (σ/G). The agreement with the equation is excellent, and the slope of these plots enables the exponent n to be determined. For both cases, it is approximately equal to 5. The

[3] A. K. Mukherjee, J. E. Bird, and J. E. Dorn, *Trans. ASM*, 62 (1964) 155.

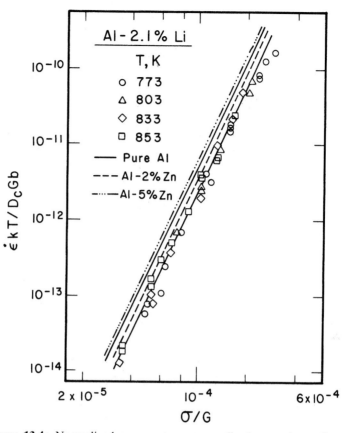

Figure 13.4a Normalized creep rate vs. normalized stress (according to Mukherjee–Bird–Dorn equation) for (a) aluminum, Al–Zn, and Al–Li solid solutions (from K.-T. Park, E. J. Lavernia, and F. A. Mohamed, *Acta Met. Mat.*, 38 (1990) 1837, Figure 11, p. 1844) and (b) various ceramics (from A. H. Chokshi and T. G. Langdon, *Matls. Sci. and Techn.*, 7 (1991) 577, Figure 1, p. 579).

exponent, in its turn, can provide information on the fundamental mechanism of creep. This will be discussed at length in Sections 13.3–13.7. In ceramics, *n* is observed to be in two ranges: 1–3 or 5–7. The significance of these results will be discussed later.

Equation 13.4 is important because it enables strain to be predicted in a specimen under creep conditions, once the various parameters that describe its creep response are established. The creep rate is dependent on stress, temperature, grain size, and other material parameters.

Another test, commonly used in place of the creep test, is the *stress-rupture* (or *creep-rupture*) test. This consists of an accelerated creep test that leads to rupture. It is usually carried out at a constant load, for the sake of simplicity. The important parameter obtained from the test is the time to rupture, whereas in the regular creep test, the minimum creep rate is the experimental parameter sought.

The sections that follow deal with several important aspects of creep. Section 13.2 describes the extrapolation methods used to obtain the response to creep at very large times after conducting more accelerated tests. Theories of creep are described in Sections 13.3–13.7. The very helpful deformation-mechanism maps called Weertman–Ashby maps are presented in Section 13.8. And some important heat-resisting alloys are described in Section 13.9. Section 13.10 treats polymers and Section 13.11 discusses superplasticity.

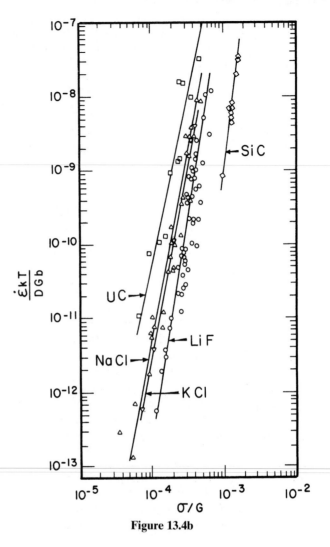

Figure 13.4b

13.2   *CORRELATION AND EXTRAPOLATION METHODS*

The central theme of materials science and engineering is the structure–property–performance triangle. In creep, the correlation between properties and performance is very critical, because in certain applications we want to know the performance during an extended period (20 or more years), while the properties (secondary creep rate or stress-rupture life) are known for a shorter period. In general, industrial equipment operating at a high temperature is designed to have a certain lifetime. For jet turbines, 10,000 hours (about 1 year) is a reasonable value. For stationary turbines, the weight of the components is not so critical, and a life of 100,000 hours (about 11 years) is the goal. For nuclear reactions, for obvious reasons, we use the criterion of 350,000 hours (40 years). A great number of advanced alloys are used in these projects, and the engineer does not have on hand the results of lengthy tests. Hence, several extrapolation methods have been developed that seek to predict the performance of alloys based on tests conducted over a shorter period. The number of parametric methods developed exceeds 30; the three most common are the Larson–Miller, Manson–Haferd, and Sherby–Dorn methods.

In 1952, Larson and Miller proposed a method that correlates the temperature T (in kelvins) with the time to failure t_r, at a *constant* engineering stress σ.[4] The Larson–Miller equation has the form

$$T(\log t_r + C) = m, \tag{13.6}$$

where C is a constant that depends on the alloy, m is a parameter that depends on stress, and rupture time. Hence, if C is known for a particular alloy, one can find m in a single test. From this result, one can then find the rupture times at any temperature, *as long as the same engineering stress is applied.* Thus, the following procedure is adopted. If we want to know the rupture time at a certain stress level σ_a and temperature T_a, we conduct the test at $T_b > T_a$ and stress level σ_a. Substituting these values into Equation 13.6, we find m. The latter test has a short duration, because the time to rupture decreases with temperature at a constant stress. Figure 13.5 shows schematically the family of lines for different levels of stress. This figure is the graphic representation of Equation 13.6. It can be seen that C does not depend on the stress; it is the intersection between the various lines. On the other hand, each line has a different slope m, which is dependent on the stress.

The value of C is unaltered by the units, as long as the unit of time is hours. However, m is dependent on units. In the older literature, use is made of English units (Rankine), and a conversion has to be made. At a certain stress level, we need only two data points to establish C and m. Since the value of C is constant for an alloy, we can build a "master plot" that represents the creep rupture response of an alloy over a range of temperatures and stresses. As an example, Figure 13.6 shows the master plot for the ferrous alloy S-590. The data were obtained between 811 and 1,089 K and fall on one single line, due to the correct choice of C: 17 log(hours). From this plot, we can obtain the time to rupture at any temperature and stress.

Soon after Larson and Miller proposed their parameter, Manson and Haferd presented the results of their experiments, which disagreed with Equation 13.6 on the following points:[5]

1. The family of lines intersects not on the ordinate axis ($1/T = 0$), but at a specific point (t_a, T_a).
2. A better linearization is obtained if the results are plotted as $\log t_r$ versus T instead of $\log t_r$ versus $1/T$. This led Manson and Haferd to propose the following equation:

$$\frac{\log t_r - \log t_a}{T - T_a} = m. \tag{13.7}$$

Equation 13.7 is represented graphically in Figure 13.7. We use the same extrapolation procedure as that of Larson and Miller to obtain rupture times at different times and temperatures. T_a, t_r, and m are parameters to be established for a given material. T_a and t_r are constant, and m depends on the stress. In Figure 13.7, three stresses are shown, leading to three lines with different slopes $m_c > m_b > m_a$. The times t_r and t_a are usually expressed in hours. As with the Larson–Miller parameter, the early literature (up to 1980) usually reports value for the Manson-Haferd parameter in the English system, whereas the more recent literature uses SI units.

Another method that has found considerable success is the Orr–Sherby–Dorn method,[6] based on fundamental studies conducted by Sherby, Dorn, and coworkers with the objective of understanding creep better. The method is based on the fundamental result found by them these researchers, viz.,

[4] F. R. Larson and J. Miller, *Trans. ASME,* 74 (1952) 765.

[5] S. S. Manson and A. M. Haferd, *NACA TN,* 2890, March 1958.

[6] R. L. Orr, O. D. Sherby, and J. E. Dorn, *Trans. ASM,* 46 (1954) 113.

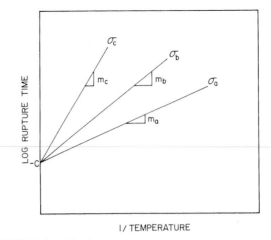

Figure 13.5 Relationship between time to rupture and temperature at three levels of engineering stress, σ_a, σ_b, and σ_c, using Larson–Miller equation ($\sigma_a > \sigma_b > \sigma_c$).

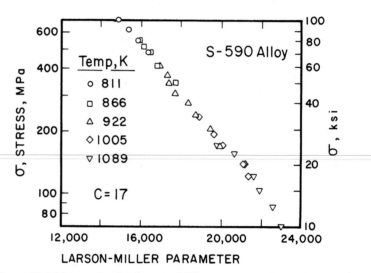

Figure 13.6 Master plot for Larson–Miller parameter for S-590 alloy (an Fe-based alloy) ($C = 17$). (From R. M. Goldhoff, *Materials in Design Eng.*, 49 (1959) 93)

$$\ln t_r - \frac{Q}{kT} = m, \tag{13.8}$$

where Q is the activation energy of diffusion (or creep), m is the Sherby–Dorn parameter, and t_r is the time to rupture. Figure 13.8 shows the graphical representation of this parameter. It differs from the Larson–Miller parameter in that the isostress lines are parallel. Equation 13.8 has a certain fundamental justification. Monkman and Grant[†] and others observed that, for a great number of alloys, the minimum creep rate $\dot{\varepsilon}_s$ was inversely proportional to the rupture time t_r, or

$$\dot{\varepsilon}_s t_r = k'. \tag{13.9}$$

[†] F. C. Monkman and N. J. Grant, *Proc. ASTM*, 56 (1956) 593.

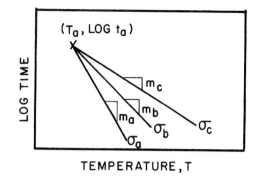

Figure 13.7 Relationship between time rupture and temperature at three levels of stress, σ_a, σ_b, and σ_c, using Manson–Haferd parameter ($\sigma_a > \sigma_b > \sigma_c$).

Applying Equation 13.4, which states that creep is a thermally activated mechanism and that the minimum creep rate increases exponentially with temperature at the same value of stress, and combining the preexponential terms, we have

$$\dot{\varepsilon}_s = A' \exp(-Q_c/kT). \tag{13.10}$$

Substituting Equation 13.9 into Equation 13.10 yields

$$t_r = \frac{k'}{A'} \exp(Q_c/kT),$$

or, taking the logarithm of both sides,

$$\ln t_r - \ln \frac{k'}{A'} = \frac{Q_c}{kT}.$$

Converting to logarithms to the base 10 and setting $\log k'/A' = m$, we get

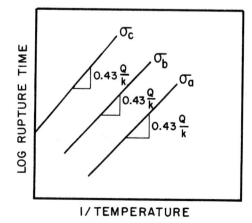

Figure 13.8 Relationship between time to rupture and temperature at three levels of stress, $\sigma_a > \sigma_b > \sigma_c$, using Sherby–Dorn parameter.

$$2.3\left(\log t_r - \log \frac{k'}{A'}\right) = \frac{Q_c}{kT}$$

$$\log t_r - m = 0.43\frac{Q_c}{kT}.$$

The slope of the lines in Figure 13.8 is $0.43Q_c/k$, which is equal to $0.43Q_D/k$. If we know the activation energy for diffusion and one point on the line, we have all the other points. The activation energy for self-diffusion can be obtained from the diffusion coefficients at two different temperatures. A thermally activated process, the diffusion obeys the equation

$$D = D_0 \exp(-Q_D/kT), \tag{13.11}$$

where D is the diffusion coefficient at T.

In this and the previous section, Q, the activation energy, is expressed as energy (joules per atom). If Q is expressed per mole, or atom gram, then R (the gas constant) should be used instead of k (Boltzmann's constant). The value of R is 8.314 J/(mole-K). In Figure 13.8, the slope would be $(0.43Q)/8.314$, or $Q/19.3$ (when Q is expressed in J/mole).

Table 13.1 presents estimated values for the parameters of the three equations for a number of engineering alloys.

TABLE 13.1 Some Values of Constants for Time–Temperature Parameters[1]

Material	Sherby–Dorn Q kJ/mole	Larson–Miller C	Manson–Haferd	
			T_a, K	log t_a
Various steels and stainless steels	≈400	≈20	—	—
Pure aluminum and dilute alloys	≈150	—	—	—
S-590 alloy (Fe base)	350	17	172	20
A-286 stainless steel	380	20	367	16
Nimonic 81A (Ni base)	380	18	311	16
1Cr-1Mo-0.25V steel	460	22	311	18

[1] Adapted from N. E. Dowling, *Mechanical Behavior of Materials* (Englewood Cliffs, NJ: Prentice Hall, 1993), p. 699, Table 15.1.

EXAMPLE 13.1

The alloy Inconel 718 was diligently tested by graduate student M. A. Meyers at 820 MPa and a temperature of 650°C. Three conditions of the alloy were tested: undeformed, cold rolled, and shock hardened (by explosives). After days of patient data collecting (this was in the 70' prior to automated data recording), he obtained the curves shown in Fig. E13.1. Using the Larsen-Miller parameter, determine the times to rupture if this alloy will be used at (a) 550°C and the same stress and (b) 650°C and 600 MPa. Take $C = 18$.

Solution: (a) We use the equation

$$T(\log t_r + C) = m$$

with $t_r \approx 110$ hours for the undeformed condition, $t_r \approx 130$ hours for the rolled condition, and $t_r \approx 200$ hours for the shocked condition. We have

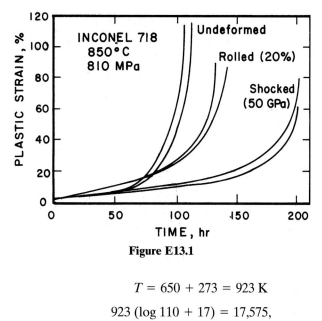

Figure E13.1

$$T = 650 + 273 = 923 \text{ K}$$

$$923 (\log 110 + 17) = 17{,}575,$$

$$923 (\log 130 + 17) = 17{,}642,$$

$$923 (\log 200 + 17) = 17{,}814.$$

At 550°C, T = 823 K, and $t_r = 22.6 \times 10^4$ hours for the undeformed condition, $t_r = 27.3 \times 10^3$ hours for the rolled condition, and $t_r = 44.2 \times 10^3$ hours for the shocked condition.

(b) No result can be obtained in this case because the stress has to be constant, for the application of the Larsen-Miller equation to two conditions.

EXAMPLE 13.2

Calculate the predicted time to rupture for the undeformed Inconel 718 superalloy, using the Sherby–Dorn correlation method.

Solution: From Table 13.1, (for the Ni-based alloy Nimonic, fairly similar to Inconel 718):

$$Q \approx 380 \text{ kJ/mole.}$$

We have to obtain the Sherby–Dorn parameter m:

$$\log t_r - m = 0.43 \frac{Q_c}{kT},$$

$$m = \log t_r - 0.43 \frac{Q_c}{kT}.$$

Since the activation energy is expressed in J per mole, we have to use R (= 8.314 J/mol. K) instead of k (Boltzmann's constant). Thus,

$$m = \log 110 - 0.43 \times \frac{380 \times 10^3}{8.314 \times 923}$$

$$= 2.04 - 21.29 = -19.25.$$

Applying Sherby-Dorn's equation to 550°C (823 K) yields

$$\log t_r = m + 0.43 \frac{Q_c}{kT} = 19.25 + \frac{0.43 \times 380 \times 10^3}{8.314 \times 823},$$

or

$$t_r = 42.7 \times 10^3 \text{ hours.}$$

EXAMPLE 13.3

Calculate the time to rupture at 650°C and 100 MPa stress for a 1Cr–1Mo–0.25V steel, according to the Larson–Miller, Sherby–Dorn, and Manson–Haferd methods, if this alloy underwent rupture in 20 hours when tested in tension at the same stress level at a temperature of 750°C.

Solution: The Larson–Miller equation is $T(\log t_r + C) = m$. From Table 13.1, $C = 22$. Thus, at 750°C, $T = 750 + 273 = 1{,}023$ K and $t_r = 20$ hours. Therefore,

$$m = 1023 \times (\log 20 + 22) \approx 2.4 \times 10^4.$$

At 650°C, $T = 650 + 273 = 923$ K, and we have

$$923 (\log t_r + 22) = 2.4 \times 10^4,$$

so that

$$\log t_r = \frac{2.4 \times 10^4}{923} - 22$$

and

$$t_r = 6.7 \times 10^3 \text{ hours.}$$

The Sherby–Dorn equation is $\ln t_r - Q/(kT) = m$. From Table 13.1, $Q = 460$ kJ/mole. Because Q uses moles, we must use R instead of k. At 750°C, $T = 1{,}023$ K and $t_r = 20$ hours. Thus,

$$m = \ln 20 - \frac{460 \times 10^3}{8.314 \times 1023} = -51.1.$$

At 650°C, $T = 923$ K, and we obtain

$$\ln t_r = m + \frac{Q}{kT}$$

$$= -51.1 + \frac{460 \times 10^3}{8.314 \times 923},$$

so that

$$t_r = 6.9 \times 10^3 \text{ hours.}$$

The Manson–Haferd equation is $(\log t_r - \log t_a)/(T - T_a) = m$. From Table 13.1, $T_a = 311$ K, so that $\log t_a = 18$. At 750°C, $T = 1{,}023$ K, and it follows that $t_r = 20$ hours. Therefore,

$$m = \frac{\log 20 - 18}{1{,}023 - 311} = -0.023.$$

At 650°C, $T = 923$ K, and we have

$$\frac{\log t_r - \log t_a}{T - T_a} = m,$$

$$\frac{\log t_r - 18}{923 - 311} = -0.023,$$

$$\log t_r = 3.924,$$

$$\underline{t_r = 8.4 \times 10^3 \text{ hours.}}$$

13.3 FUNDAMENTAL MECHANISMS RESPONSIBLE FOR CREEP

The history of progress in our understanding of creep can be divided into two periods: before and after 1954. In that year, Orr et al. introduced the concept that the activation energy for creep and diffusion are the same for an appreciable number of metals (more than 25).[7] Figure 13.9 shows their results. The activation energy for diffusion is connected to the diffusion coefficient by Equation 13.11. Note that several mechanisms can be responsible for creep; the rate-controlling mechanism depends both on the stress level and on the temperature, as will be seen in Sections 13.4–13.7. For temperatures below $0.5T_m$, half the melting point of the material, in kelvins, the activation energy for creep tends to be lower than that for self-diffusion, because diffusion takes place preferentially along dislocations (pipe diffusion), instead of in bulk. Figure 13.10 shows the variation in Q_C/Q_D for some metals and ceramics. The activation energy for diffusion through dislocations is considerably lower than that for bulk diffusion.

For the temperature range $T > 0.5T_m$, the mechanisms responsible for creep can be conveniently described as a function of the applied stress. The creep mechanisms can be divided into two major groups: boundary mechanisms, in which grain boundaries and, therefore, grain size, play a major role; and lattice mechanisms, which occur independently of grain boundaries. In Equation 13.4, the exponent $p = 0$ for lattice mechanisms, and $p \geq 1$ for boundary mechanisms.

13.4 DIFFUSION CREEP

Diffusion creep tends to occur for $\sigma/G \leq 10^{-4}$. (This value depends, to a certain extent, on the metal.) Two mechanisms are considered important in the region of diffusion creep. Nabarro and Herring proposed the mechanism shown schematically in Figure 13.11a.[8] It involves the flux of vacancies inside the grain. The vacancies move in such a way as to produce an increase in length of the grain along the direction of applied (tensile) stress. Hence, the vacancies move from the top and bottom region in the figure to the lateral regions of the grain. The boundaries perpendicular (or close to perpendicular) to the loading

[7] R. L. Orr, O. D. Sherby, and J. E. Dorn, *op. cit.*, 113.

[8] F. R. Nabarro, "Deformation of Crystals by the Motion of Single Ions," *Report of a Conference on Strength of Solids, Physical Society,* London, 1948, p. 75; and C. Herring, *J. Appl. Phys.,* 21 (1950), 437.

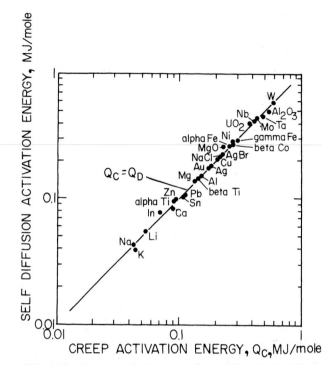

Figure 13.9 Activation energies for creep (stage II) and self-diffusion for a number of metals. (Adapted with permission from O. D. Sherby and A. K. Miller, *J. Eng. Mater. Technol.*, 101 (1979) 387)

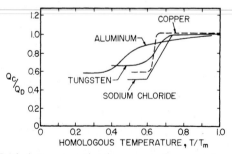

Figure 13.10 Ratio between activation energy for secondary creep and activation energy for bulk diffusion as a function of temperature. (Adapted with permission from O. D. Sherby and A. K. Miller, *J. Eng. Mater. Technol.*, 101 (1979) 387).

direction are distended and are sources of vacancies. The boundaries close to parallel to the loading direction act as sinks.

Nabarro and Herring developed a mathematical expression connecting the vacancy flux to the strain rate. They started by supposing that the "source" boundaries had a concentration of vacancies equal to $C_0 + \Delta C$ and the sink boundaries a concentration C_0. They assumed that

$$\Delta C = \frac{C_0 \sigma}{kT},$$

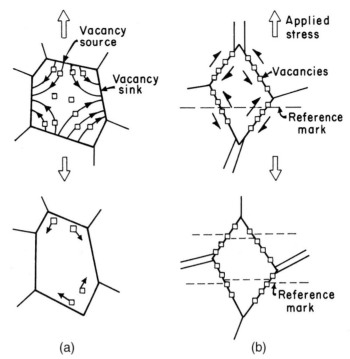

Figure 13.11 Flow of vacancies according to (a) Nabarro–Herring and (b) Coble mechanisms, resulting in an increase in the length of the specimen.

where σ was the applied stress and C_0 the equilibrium vacancy concentration. The flux of vacancies is therefore given by

$$J = k'D_\ell \left(\frac{\Delta C}{x}\right) = k''D_\ell \left(\frac{\Delta C}{d}\right),$$

where x is the diffusion distance, which is a direct function of the grain size (approximately equal to $d/2$), D_ℓ is the lattice diffusion coefficient, d is the grain diameter, and k' and k'' are proportionality constants ($k'' = 2k'$). The strain rate is related to the increase in grain size d in the direction of the applied stress:

$$\dot{\varepsilon} = \frac{1}{d}\frac{dd}{dt}.$$

The change in grain length, dd/dt, can be obtained from the shift of vacancies, each having a volume Ω:

$$\frac{dd}{dt} = J\Omega.$$

Thus, the following equation can be obtained for the creep rate:

$$\dot{\varepsilon}_{\text{NH}} = k''\frac{\Omega D_\ell C_0 \sigma}{d^2 kT}.$$

("NH," of course, denotes Nabarro–Herring.) Expressing this equation in the format of Equation 13.5 (making $\Omega = 0.7b^3$), we have

$$\dot{\varepsilon}_{\text{NH}} = A_{\text{NH}} \frac{D_\ell Gb}{kT} \left(\frac{b}{d}\right)^2 \left(\frac{\sigma}{G}\right). \tag{13.12}$$

A_{NH} is typically equal to 10–15.

Coble proposed the second mechanism explaining diffusion creep.[9] It is based on diffusion in the grain boundaries instead of in the bulk. This diffusion results in sliding of the grain boundaries. Hence, if a fiducial scratch is made on the surface of the specimen prior to creep testing, the scratch will show a series of discontinuities (at the grain boundaries) after testing if Coble creep is operative.

Figure 13.11b shows, in a schematic manner, how the flow of vacancies along a boundary generates shear. Notice that there is also additional accommodational diffusion necessary. Coble creep leads to the relationship

$$\dot{\varepsilon}_C = A_C D_{\text{gb}} \frac{Gb}{kT} \left(\frac{\delta}{b}\right)\left(\frac{b}{d}\right)^3\left(\frac{\sigma}{G}\right), \tag{13.13}$$

where A_c is typically equal to 30–50, δ is the effective width of the grain boundary for diffusion, and D_{gb} is the grain-boundary diffusion coefficient.

Note that in Equations 13.12 and 13.13, the strain rate is proportional to the stress—that is, $n = 1$. Also, the strain rate goes as d^{-2} for Nabarro–Herring creep and as d^{-3} for Coble creep. This enables researchers to differentiate between the two mechanisms: They establish the creep rates for specimens with different grain sizes and find the exponent on the grain size. A practical way of having an alloy with high resistance to Nabarro–Herring or Coble creep is to increase the size of the grains. This method is used in superalloys; a fabricating technique called *directional solidification* has been developed to eliminate virtually all grain boundaries perpendicular and inclined to the tensile axis.

Harper and Dorn observed another type of diffusional creep in aluminum.[10] This occurred at high temperatures and low stresses, and the creep rates were over 1,000 times greater than those predicted by Nabarro–Herring. (Also, little Coble creep was observed.) The two researchers concluded that creep occurred exclusively by dislocation climb (see Fig. 13.14). Harper–Dorn creep is governed by an equation of the form

$$\dot{\varepsilon}_{\text{HD}} = A_{\text{HD}} \frac{D_\ell Gb}{kT} \left(\frac{\sigma}{G}\right). \tag{13.14}$$

The parameter A_{HD} is typically equal to 10^{-11}. Since no grain boundaries are involved in this creep, the grain size does not appear in the equation. For Harper–Dorn creep to make a significant contribution, the grain size of the material has to be large ($>400\ \mu$m); otherwise, Nabarro–Herring and Coble creep dominate.

In metals, Harper–Dorn creep has been observed in a number of systems. In ceramics, there is little evidence for this type of diffusion creep mechanism. Ceramics in general have small grain sizes, which favor other creep mechanisms. The stable, small grain size and the limited number of slip systems, as well as high Peierls–Nabarro stress, lead to the prominence of Nabarro–Herring and Coble creep. Diffusion in ceramics is more complex than in metals, because either one or two ionic species might participate, and in the case of

[9] R. L. Coble, *J. Appl. Phys.,* 34 (1963) 1679.

[10] J. Harper and J. E. Dorn, *Acta Met.,* 5 (1957) 654.

multicomponent ceramics, more than one cation or ion might be involved. Figure 13.12 shows the different domains of creep in alumina as a function of grain size and temperature; the main ion is shown for each domain.

13.5 DISLOCATION CREEP

In the stress range $10^{-4} < \sigma/G < 10^{-2}$, creep tends to occur by dislocation glide, aided by vacancy diffusion (when an obstacle is to be overcome). This mechanism should not be confused with Harper–Dorn creep, which relies exclusively on dislocation climb. Orowan proposed that creep is a balance between work-hardening (due to plastic strain) and recovery (due to exposure to high temperatures). Hence, at a constant temperature, the increase in stress is

$$d\sigma = \left(\frac{\partial \sigma}{\partial \varepsilon}\right)_{t,\sigma} d\varepsilon + \left(\frac{\partial \sigma}{\partial t}\right)_{\sigma,\varepsilon} dt, \tag{13.15}$$

where $(\partial\sigma/\partial\varepsilon)_{t,\sigma}$ is the rate of hardening, and $(\partial\sigma/\partial t)_{\sigma,\varepsilon}$ is the rate of recovery, of the material. The strain rate $\dot{\varepsilon}$ can be expressed as a ratio between the rate of recovery and the rate of hardening.

In the mid-1950s, Weertman developed a pair of theories of the minimum creep rate based on dislocation climb as the rate-controlling step.[11] In his first theory, Weertman presented Cottrell–Lomer locks as barriers to plastic deformation; his second theory applies to HCP metals, in which these barriers do not exist. Hence, he assumed different barriers, depending on the material. Figure 13.13 shows schematically how the mechanism based on Cottrell–Lomer locks operates. Dislocations are pinned by obstacles, but overcome them by climb, aided by either interstitial or vacancy generation or destruction. The obstacles are assumed to be Cottrell–Lomer locks, which are formed by dislocations that intersect and react (see Chapter 4 Sec. 4.2.5). Figure 13.13a shows dislocations pinned between the locks and climbing over them. Note that dislocations are continuously generated by the Frank–Read source in the horizontal plane, and the ones overcoming the obstacle are replaced by others. To calculate the creep rate, we have to find the rate of escape of the dislocations from the locks. The height h a dislocation has to climb in order to pass through a lock is the position at which the applied stress on the dislocation, due to the other dislocations in the pileup, is equal to the repulsive force due to the stress field of the lock. Other obstacles (shown in Figure 13.13b) can have the same effect.

The stress exerted by a dislocation due to the pileup effect is given in Section 4.2.8 and is (Eqn. 4.20)[†]

$$\sigma^* = \tilde{n}\sigma, \tag{13.16}$$

where $\tilde{n}$ is the number of dislocations in the pileup (we use $\tilde{n}$ here to avoid confusion with n, the exponent in power-law creep) and σ is the stress applied on one dislocation. Now, taking the stress field around a dislocation as a function of distance (Section 4.2.4) and equating it to Equation 13.16, Weertman arrives at (this is slightly different from Eqn. 4.6)

$$h = \frac{Gb}{\tilde{n}\sigma 6\pi(1 - \nu)}. \tag{13.17}$$

[11] J. Weertman, *J. Appl. Phys.*, 26 (1955) 1213; 28 (1957) 362.

[†] Shear stresses were converted into normal stresses.

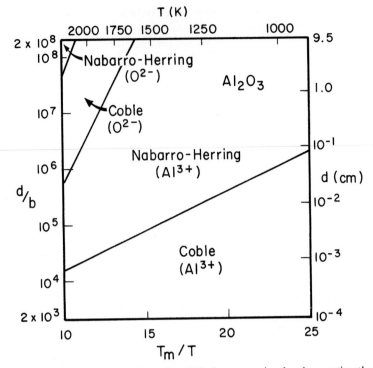

Figure 13.12 Different regimes for diffusion creep in alumina; notice that cations (Al^{3+}) and anions (O^{2-}) have different diffusion coefficients, leading to different regimes of dominance. (From A. H. Chokshi and T. G. Langdon, *Defect and Diffusion Forum*, 66–69 (1989) 1205, Figure 8, p. 1217)

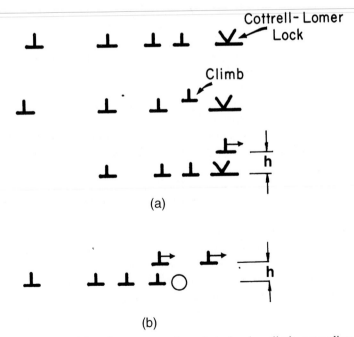

Figure 13.13 Dislocation overcoming obstacles by climb, according to Weertman theory. (a) Overcoming Cottrell–Lomer locks. (b) Overcoming an obstacle.

The rate of climb is determined by the rate at which the vacancies arrive at or leave the dislocation. (Weertman did the derivation for vacancies and not interstitials.) For the concentration gradient of vacancies, Weertman obtained a rate of climb

$$r = \frac{N_0 D_\ell \tilde{n} \sigma b^5}{kT},\tag{13.18}$$

when $(\tilde{n}\sigma b^3)/kT < 1$. N_0 is the equilibrium concentration of vacancies and D_ℓ is the diffusion coefficient at the test temperature T. With a known climb height and rate of climb, it is possible to calculate the rate of creep. If M is the number of active Frank–Read sources per unit volume, L is the distance the edge portion of a dislocation loop moves after breaking away from a barrier, and L' is the portion the screw moves, then the creep rate is given by

$$\dot{\varepsilon} = \frac{r}{h} LL'M = \frac{6\pi(1 - \nu)\tilde{n}^2 b^4 N_0 LL'M D_\ell \sigma^2}{kGT}.\tag{13.19}$$

If we assume, to a first approximation, that $\tilde{n}$, the number of dislocations in a pileup, and M, the number of Frank–Read sources per unit volume, are proportional to σ, we can recast Equation 13.19 in the Mukherjee–Bird–Dorn format as

$$\dot{\varepsilon} = A\left(\frac{D_\ell Gb}{kT}\right)\left(\frac{\sigma}{G}\right)^5.$$

The stress exponent 5 is characteristic of this regime. The term A incorporates the various parameters and proportionality coefficients. Power law creep with $n \approx 5$ has been observed at high stress levels in a number of ceramics, including KBr, KCL, LiF, NaCl, NiO, SiC, ThO_2, UC, and UO_2. (See Figure 13.4b.) As in the case of metals, the substructure is characterized by subgrains with misorientations of approximately $2°$.

Dislocation climb is shown schematically in Figure 13.14. Under compressive loads, vacancies are attracted to the dislocation line (Figure 13.14a). Once a row of vacancies has joined the dislocation line, the line is effectively translated upwards. Thus, the dislocation moves perpendicularly to the Burgers vector during climb. In tension (Figure 13.14b), the opposite occurs: Vacancies move away from the dislocation line, and the dislocation effectively moves down.

Creep behavior with a stress exponent $n \approx 3$ is observed in a number of ceramics, such as Al_2O_3, BeO, Fe_2O_3, MgO, and ZrO_2 ($+ 10\%$ Y_2O_3). In this case, few or no solutes are present, and those that are do not play a role.

Dispersion-strengthened alloys are characterized by an exponent higher than 7 and by a high activation energy for creep. Dispersoids (see Chapter 10) are stable up to very high temperatures. Small particles, such as Y_2O_3 and ErO_2, are added to the alloy as dispersoids; this increases the high-temperature capability of these materials substantially, and the dispersoids act as effective barriers to dislocation motion.

Particle-reinforced composites (such as SiC and aluminum reinforced with aluminum oxide) exhibit the same effects: The stress exponent n and activation energy for creep are very high. This is illustrated in Figure 13.15, for an Al–$30°$ SiC composite. The slope in Figure 13.15 is given by

$$n = \frac{\partial \ln \dot{\gamma}}{\partial \ln \tau} = \frac{\partial \ln 2\dot{\varepsilon}}{\partial \ln \sigma/2} = \frac{\partial \ln \dot{\varepsilon}}{\partial \ln \sigma}.$$

Taking logarithms and derivatives of both sides, we have (at constant T)

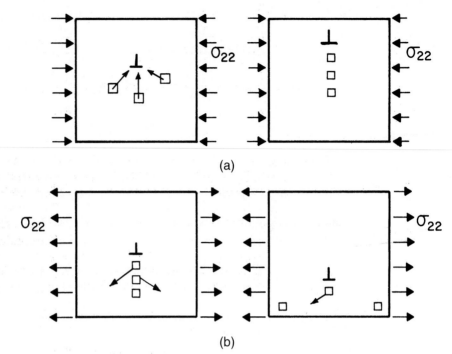

(a)

(b)

Figure 13.14 Dislocation climb (a) upwards, under compressive σ_{22} stresses, and (b) downwards, under tensile σ_{22} stresses.

$$\dot{\varepsilon} = A'\left(\frac{\sigma}{G}\right)^n = A''\sigma^n,$$

$$\ln \dot{\varepsilon}_s = \ln A'' + n \ln \sigma,$$

$$\partial \ln \dot{\varepsilon}_s = n\partial \ln \sigma,$$

$$n = \frac{\partial \ln \dot{\varepsilon}_s}{\partial \ln \sigma}.$$

Thus, the slope is equal to the stress exponent. In Figure 13.15, the slope n varies from 14.7 to 7.4 as the stress is increased. The activation energy for creep for this aluminum composite has a value of 270–500 kJ/mole; this is significantly higher than the activation energy for aluminum self-diffusion.

13.6 DISLOCATION GLIDE

Dislocation glide occurs for $\sigma/G > 10^{-2}$. At a certain stress level, the power law breaks down. Figure 13.16 presents the region in which the law ($n = 4$) breaks down, and n increases to 10; this occurs for $\dot{\varepsilon}_s/D > 10^9$. An analysis of the deformation substructure by transmission electron microscopy and showed that, at high stresses, dislocation climb was replaced by dislocation glide, which does not depend on diffusion.[12] Hence, when $\dot{\varepsilon}_s/D > 10^9$, thermally activated dislocation glide is the rate-controlling step; this is the same deformation mode as the one in conventional deformation at ambient temperature. Kestenbach et al. observed that the substructure changed from equiaxial subgrains to dislocation tangles and

[12] H.-J. Kestenbach, W. Krause, and T. L. da Silveira, Acta Met. *26* (1978) 661.

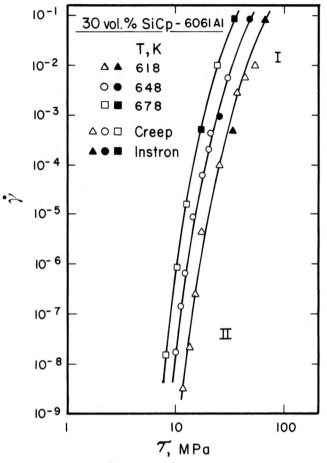

Figure 13.15 Shear stress vs. shear strain rate in an aluminum (6061) with 30 vol % SiC particulate composite in creep. (From K.-T. Park, E. J. Lavernia, and F. A. Mohamed, *Acta Met. Mat.,* 38 (1990) 2149, Figure 7, p. 2153)

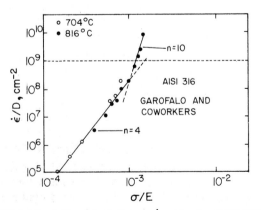

Figure 13.16 Power relationship between $\dot{\varepsilon}$ and σ for AISI 316 stainless steel. (Adapted with permission from S. N. Monteiro and T. L. da Silveira, *Metalurgia-ABM,* 35 (1979) 327)

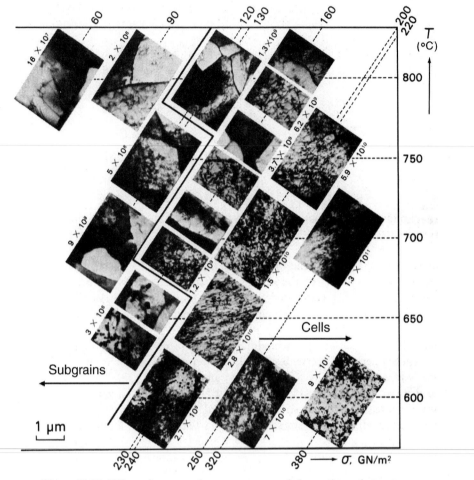

Figure 13.17 Effect of stress and temperature on deformation substructure developed in AISI 316 stainless steel in middle of stage II. (Reprinted with permission from H.-J. Kestenbach, W. Krause, and T. L. da Silveira, *Acta Met.*, 26 (1978) 661)

elongated subgrains when the stress reached a critical level. A similar effect is observed when the temperature is decreased and the stress is maintained constant. Figure 13.17 shows the substructures at various values of stress and temperatures for secondary creep.

13.7 GRAIN-BOUNDARY SLIDING

Grain-boundary sliding usually does not play an important role during primary or secondary creep. However, in tertiary creep it does contribute to the initiation and propagation of intercrystalline cracks. Another deformation process to which it contributes significantly is superplasticity; it is thought that most of the deformation in superplastic forming takes place by grain-boundary sliding.

The grain-boundary sliding rate is controlled by the accommodating processes where the sliding surface deviates from a perfect plane. One can readily see that we cannot have a perfect plane defined by the boundaries between different grains; we cannot look separately at the sliding between two grains having a common interface. The requirements of strain compatibility are such that we have to model the interface as sinusoidal, as is de-

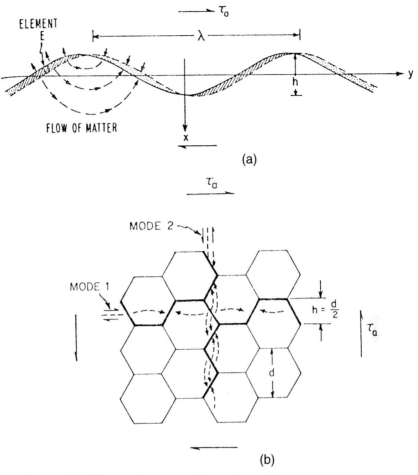

Figure 13.18 (a) Steady-state grain-boundary sliding with diffusional accommodations. (b) Same process as in (a), in an idealized polycrystal; the dashed lines show the flow of vacancies. (Reprinted with permission from R. Raj and M. F. *Ashby, Met. Trans.,* 2A (1971) 1113)

picted in Figure 13.18. The applied stress τ_a can produce sliding only if it is coupled with diffusional flow that transports material (or vacancies) over a maximum distance of λ, the wavelength of the irregularities. Figure 13.18b shows the same effect in a polycrystalline aggregate. The individual grain boundaries are translated by a combination of sliding and diffusional flow under the influence of the applied stress.

The manner in which the individual grains move and change their relative positions by sliding and diffusional accommodation is shown in Figure 13.19. The sliding of grains under the influence of σ, coupled with minor changes in shape, makes possible the sequence *a-b-c*, which results in a strain of 0.55; the unique features of this mechanism is that the sequence is accomplished with relatively little strain *within* the grains.

13.8 DEFORMATION-MECHANISM (WEERTMAN–ASHBY) MAPS

Deformation mechanism maps also named after two people who first introduced them, are a graphical description of creep, representing the ranges in which the various deformation modes are rate-controlling steps in the stress-versus-temperature space. Weertman–Ashby

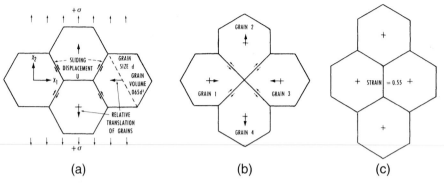

Figure 13.19 Grain-boundary sliding assisted by diffusion in Ashby–Verrall's model. (Reprinted with permission from M. F. Ashby and R. A. Verrall, *Acta Met.*, 21 (1973) 149)

plots assume, for simplicity, that there are some independent and distinguishable ways by which a polycrystal can be deformed, but still retain its crystallinity:

1. Above the theoretical shear strength, plastic flow of the material can take place without dislocations, by simple glide of one atomic plane over another.
2. Movement of dislocations by glide.
3. Dislocation creep, including glide and climb, both being controlled by diffusion.
4. Nabarro–Herring creep.
5. Coble creep.

The theories developed for these different modes of deformation without loss of crystallinity propose constitutive equations that are used in the establishment of the ranges involved. Figure 13.20 shows a typical map for silver. The theoretical shear stress is ap-

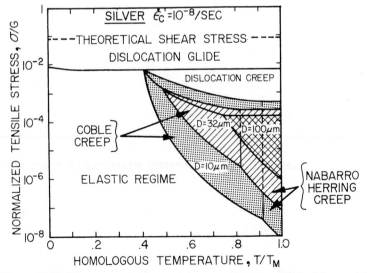

Figure 13.20 Weertman–Ashby map for pure silver, established for a critical strain rate of 10^{-8} s^{-1}; it can be seen how the deformation-mechanism fields are affected by the grain size. (Adapted with permission from M. F. Ashby, *Acta Met.*, 20 (1972) 887)

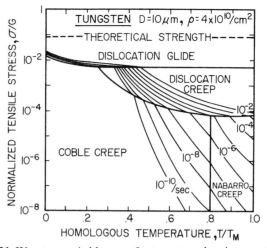

Figure 13.21 Weertman–Ashby map for tungsten, showing constant strain-rate contours. (Reprinted with permission from M. F. Ashby, *Acta Met.*, 20 (1972) 887)

proximately equal to $G/20$ and is practically independent of temperature. A small dependence on temperature is exhibited by G and is built into the ordinate of the figure. For values of σ/G between 10^{-1} and 10^{-2}, slip by dislocation movement is the controlling mode at all temperatures. It can be seen that the grain size affects the extent of the fields. Three grain sizes are represented: 10, 32, and 100 μm. The fields also depend on the strain rate. The map shown in Figure 13.21 was made for a strain rate of 10^{-8} s^{-1}. The Coble and Nabarro–Herring mechanisms, especially, are affected by the grain size, because of their nature.

Deformation-mechanism maps have technological applications. Consider, for example, a turbine blade operated in a temperature and stress range that are known. The specific stress–temperature profile can be plotted on a deformation-mechanism map in the form of a line. Different parts of the blade undergo different deformation modes. These modes, the rate of creep of each portion, and the respective constitutive equation can be read from the map. Multiaxial stress states can be resolved by calculating the maximum shear stress or the effective stress. A strengthening mechanism is helpful only if it retards the creep rate in the correct portion of the map. For instance, dispersion-hardening is effective in controlling dislocation glide and climb, but cannot effectively stop Nabarro–Herring or Coble Creep.

From the deformation-mechanism map, we can, in addition to determining the dominant mechanism for a certain combination of stress and temperature, find the strain rate (creep rate) that will result. For this, we have to apply the appropriate constitutive equations and plot the constant strain-rate contours. This is shown in Figure 13.21 for tungsten. The lines allow ready identification of the creep rate. The region in Figure 13.20 consisting of the elastic regime is occupied by Coble creep in Figure 13.21. The reason for this is that Figure 13.20 applies to one constant strain rate (10^{-8} s^{-1}), whereas Figure 13.21 is built for a whole range of strain rates. Hence, at a strain rate of 10^{-8} s^{-1}, the metal might respond elastically, whereas at a strain rate orders of magnitude lower, Coble creep becomes significant.

Similar maps can be built for ceramics, and a representative map is shown in Figure 13.22. The different domains, as well as the curves for constant strain rates (from 10^{-10}/s to 1/s), are illustrated in the plot. Note that the different diffusing ions (Al^{+3} and O^{2-}) have to be considered.

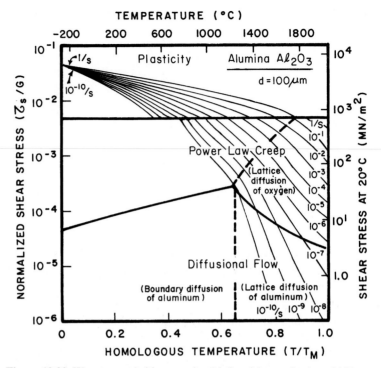

Figure 13.22 Weertman–Ashby map for Al_2O_3 with a grain size of 100 μm. (Adapted from H. J. Frost and M. F. Ashby, *Deformation-Mechanism Maps,* (Tarry Town, NY Pergamon Press, 1982), p. 100, Figure 14.3)

13.9 HEAT-RESISTANT MATERIALS

High-temperature materials can be classified into two groups: metals and ceramics. High-temperature alloys are, in their turn, classified into superalloys and refractory alloys. The latter are alloys of elements with high melting points, such as tantalum, molybdenum, and tungsten. The superalloys are usually alloys developed for elevated temperature service, usually based on group VIIIa elements, where relatively severe mechanical stressing is encountered and where high surface stability is frequently required.

The development of superalloys was initiated in the 1930s, and their first use was in turbo superchargers of reciprocating airplane engines. The introduction of the turbine in the 1940s was a strong motivator for subsequent developments. Superalloys encompass the nickel, iron, cobalt, and iron–nickel systems. The majority of authors does not include chromium-based alloys in this group. The maximum service temperature (temperature capability) has increased continuously in the past; it can be around 1,200°C. The life of turbines has increased from 5,000 to over 20,000 hours. The combined effects of high stresses, high temperatures, and long times have required improvements in the following properties:

1. *Short-term mechanical properties:* yield stress, ductility.
2. *Long-term mechanical properties:* low- and high-cycle fatigue, creep, creep-fatigue.
3. *Hot corrosion resistance:* The principal deterioration processes are oxidation, chlorination, sulfidation, and carburization.

Nickel-based superalloys are the most important group; most commercial nickel-based alloys have more than 10 constituent elements and over 10 trace elements. These can be divided into the following categories, depending on the function and position of the element in the periodic chart:

1. Elements that form substitutional solid solutions in the austenitic matrix: cobalt, iron, chromium, vanadium, molybdenum, tungsten.
2. Elements that form precipitates: aluminum, titanium, niobium, tantalum. Figure 13.23 shows the cuboidally shaped γ' precipitates Ni_3Al, Ni_3Ti, and $Ni_3(Al, Ti)]$ that are aligned along specific planes of the austenitic matrix.
3. Carbide-forming elements: chromium, molybdenum, tungsten, vanadium, niobium, tantalum, titanium.
4. Elements that segregate along the grain boundaries: magnesium, boron, carbon, zirconium.

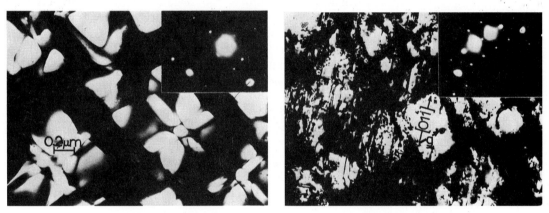

Figure 13.23 Transmission electron micrograph of Mar M-200; notice the cuboidal γ' precipitates (Courtesy of L. E. Murr)

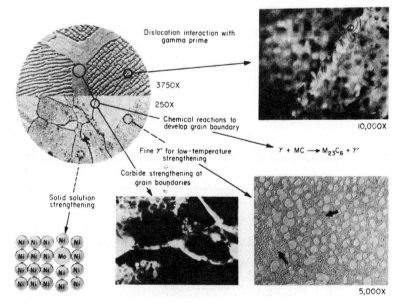

Figure 13.24 Major microstructural strengthening mechanisms in nickel-based superalloys. (Reprinted with permission from C. T. Sims and W. C. Hagel (eds.), *The Superalloys* New York: Wiley, 1972) p. 33)

5.   Elements forming protective and adherent oxides: chromium, aluminum.

6.   Rare-earth elements.

The microstructure of superalloys reflects the concern of using all possible strengthening mechanisms to retard creep. Figure 13.24 on page 567 is a composite of these features. One has to retard the movement of dislocations. This is achieved by substitutional solid solution atoms and by a great volume percentage of the Ni$_3$(Ti,Al) phase γ'. The grain boundaries are strengthened by precipitation of M$_{23}$C$_6$ carbides on them. Secondary γ', very fine, is precipitated in the space between neighboring primary γ', which is larger. One also wants to carefully avoid the topologically close-packed (TCP) phases R and σ, that occur accidentally and after long exposure to high temperatures, embrittling the alloy.

Figure 13.25 shows the stress-rupture properties of a number of nickel-based superalloys. The stress required for rupture in 1,000 hours is plotted against the temperature. The load-bearing ability in the upper range of the use of these superalloys is only a fraction of the one at lower temperatures. The range 800 to 1,000°C is a very critical one.

The use of single-crystal turbine blades represents a significant technological development: Grain-boundary sliding is eliminated by this technique, and an increase in temperature capability of approximately 50°C over that of polycrystalline superalloys is achieved. The method of producing single-crystal turbine blades involves investment cast-

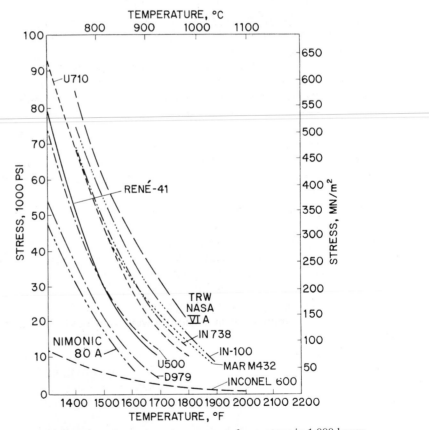

Figure 13.25 Stress versus temperatures curves for rupture in 1,000 hours for selected nickel-based superalloys. (Reprinted with permission from C. T. Sims and W. C. Hagel (eds.), *The Superalloys* (New York: Wiley, 1972), p. vii)

ing in a controlled thermal environment. An added advantage is that the composition of the alloy is simpler than in polycrystalline superalloys.

High-performance ceramics are prime candidates for structural components in advanced automotive gas turbine engines. Operating conditions for such components involve high mechanical and thermal stresses at elevated temperatures; hence, the creep resistance of these materials is of great importance. Excessive creep deformation can affect the dimensional stability of the component, ultimately leading to a loss of function. Generally, there is a high resistance to slip and diffusional mechanisms in high-performance ceramics such as SiC and Si_3N_4. Damage mechanisms such as cavitation, solution of silicon nitride into the glassy phase at the grain boundary, and grain-boundary sliding are associated with creep in these ceramics. As mentioned earlier in this chapter, different creep mechanisms give different creep activation energies. Cavitation can result in a reduction in the strength of the material and lead to time-dependent failure (i.e., creep rupture). Viscous flow in any glassy grain-boundary phase can lead to excessive creep deformation. Silicon nitride is a good example of a high-performance ceramic material to use for illustrating some of the unusual problems that must be faced before their full potential can be realized. Silicon nitride has excellent short-term strength and fracture toughness. Generally, hot, isostatically pressed silicon nitride shows superior properties to silicon nitride processed by pressureless sintering or uniaxial hot pressing. The amount and nature of any densification aids can significantly affect creep behavior. Yttria is a common densification aid used in silicon nitride. Another important variable is the testing method (compression, bending, or tensile testing) used for creep. Although tensile testing of ceramics is not very common, it has been used to study the high-performance ceramics.

Studies aimed at evaluating the long-term mechanical performance of silicon nitride (trade designation NT154) have shown that cavitation along two-grain junctions controls both creep deformation and creep rupture strength.[13] Silicon nitride is available with different purity levels. Ferber et al. observed that the creep and creep rupture behavior of silicon nitride (NT 164) were significantly improved, compared to that of the commercial material (NT 154), if one could ensure the absence of cavitation along two-grain junctions.[14] These authors attributed the growth of cavities to the following processes, which occur in a sequence: (1) solution of silicon nitride into the intergranular phase at the cavity boundary, (2) transport of the dissolved species along the grain boundary, and (3) precipitation of the species at low-stress sites remote from the cavity. Yet another factor in the high-temperature behavior of nonoxide ceramics such as SiC and Si_3N_4 is their oxidation resistance in air. Oxidation of silicon nitride, rather than creep, was observed to initiate a stress- oxidation damage zone in the material. Finally, we reiterate the importance of the testing method. Wiederhorn et al. found asymmetric behavior of Si_3N_4 in creep:[15] A linear response was obtained in compression, but a power-law response held in tension, with the creep exponent n in the range $2 < n < 5$. These researchers observed minimal cavitation in compression; in tension, however, cavities formed at multigrain junctions, and the creep strain was proportional to the volume fraction of cavities. Thus, cavitation is responsible for creep strain in tension, but not in compression. This discussion should bring home to the reader some important differences between creep mechanisms in metals and in nitrogen ceramics. In metals, a lot of creep strain can occur, but not much of it is due to cavitation. Also, the tertiary creep of metals is absent in silicon nitride.

[13] D. C. Cranmer, B. J. Hockey, S. M. Wiederhorn, and R. Yeckley, *Ceram. Eng. Sci. Proc.*, 12 (1991) 1862.

[14] M. K. Ferber and M. G. Jenkins, *J. Am. Ceram. Soc.*, 75 (1992) 2453; and M. K. Ferber, M. G. Jenkins, and T. A. Nolan, *J. Am. Ceram. Soc.*, 77 (1994) 657.

[15] S. M. Wiederhorn, B. J. Hockey, W. E. Luecke, R. Krause, and J. French, unpublished results.

Ceramics and ceramic composites possess a higher temperature capability than metals. Whereas ceramics tend to be brittle, the addition of reinforcing fibers adds toughness to them. Temperatures approaching 2,000°C can be reached with acceptable creep rates. As mentioned earlier, chemical degradation becomes very important at these temperatures, especially for nonoxide ceramics. (Oxides are, obviously, immune to oxidation.) A high-temperature material of great promise is $MoSi_2$, especially as a matrix for high-temperature structural composites (see Chapter 12).

13.10 CREEP IN POLYMERS

As mentioned in the preceding sections, creep is a thermally activated process and thus becomes important at high temperatures. The term *high temperature* is a relative one; is more convenient to use the term *homologous temperature*, $T_H = T/T_m$, where T is the temperature of interest in kelvins and T_m is the melting of the material in kelvins. Typically, creep becomes a significant deformation mode for metals at a homologous temperature greater than 0.4 and for ceramics at a homologous temperature greater than 0.5. (In the case of amorphous polymers, one uses the glass transition temperature T_g rather than the melting point T_m.) At low temperatures, most metals and ceramics show time-independent deformation. In general, polymers show a much larger dependence on time and temperature than metals and ceramics do; that is, polymers show creep effects at much lower stresses and temperatures. This stems from their weak van der Waals interchain forces. In polymers, time-dependent deformation becomes important even at room temperature. Two terms are used to describe the time-dependent behavior of polymers: *creep* and *stress relaxation.* In creep, one applies a constant stress, and the strain response is measured as a function of time. In stress relaxation, one applies a constant strain, and the response is measured in terms of a decrease in stress as a function of time. We have discussed some aspects of these two phenomena in Chapter 2 in connection with viscoelasticity.

For a glassy, viscoelastic polymer subjected to a constant stress σ_0, there is an initial elastic strain recovery, followed by a slow, time-dependent recovery. This viscoelastic response can be modeled as a spring and a dashpot in series (also called the Maxwell model), as shown in Figure 13.26. An application of stress to this system results in a strain e in the system. This strain is the sum of two contributions, and we can write

$$\varepsilon = \varepsilon_1 + \varepsilon_2, \tag{13.20}$$

where e_1 is the strain in the spring and e_2 is the strain in the dashpot. The stresses in the spring and the dashpot are identical, because the two are in series, i.e.,

$$\sigma = \sigma_1 = \sigma_2. \tag{13.21}$$

Then, we can write the following relationships for the elastic and the viscous case,

$$\frac{d\sigma}{dt} = E\frac{d\varepsilon_1}{dt}, \qquad \sigma = \eta\frac{d\varepsilon_2}{dt}. \tag{13.22}$$

From Equations 13.20 and 13.21, we get

$$\frac{d\varepsilon}{dt} = \frac{d\varepsilon_1}{dt} + \frac{d\varepsilon_2}{dt} = \frac{1}{E}\frac{d\sigma}{dt} + \frac{\sigma}{\eta}. \tag{13.23}$$

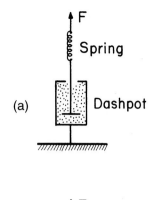

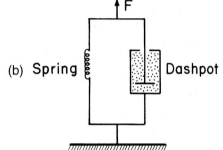

Figure 13.26 Spring–dashpot analogs (a) in series and (b) in parallel.

Note that the series, or Maxwell, model does not correctly predict the behavior of a viscoelastic material under constant stress or creep conditions (i.e., $\sigma = \sigma_0$), because, in this case,

$$\frac{d\varepsilon}{dt} = \frac{\sigma_0}{\eta}. \tag{13.24}$$

That is, the Maxwell model for creep or constant-stress conditions predicts that the strain increases linearly with time. (See Figure 13.27a.) Most polymers, however, show de/dt increasing with time. The Maxwell model is more realistic in the case of a stress relaxation test, during which we impose a constant strain $e = e_0$ and $de/dt = 0$. Under these conditions, Equation 13.23 can be written as

$$0 = \frac{1}{E}\frac{d\sigma}{dt} + \frac{\sigma}{\eta},$$

or

$$\frac{d\sigma}{\sigma} = -\left(\frac{E}{\eta}\right)dt.$$

We can integrate this expression to get

$$\sigma = \sigma_0 \exp\left(-\frac{Et}{\eta}\right). \tag{13.25}$$

The quantity η/E is referred to as the *relaxation time* τ, and we can rewrite Equation 13.25 as

$$\sigma = \sigma_0 \exp(-t/\tau). \tag{13.26}$$

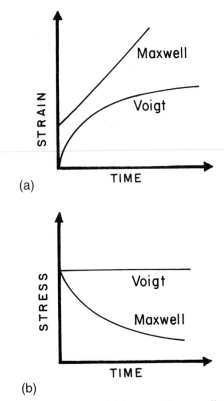

Figure 13.27 (a) Strain–time and (b) stress–time predictions for Maxwell and Voigt models.

Equation 13.26 says that the stress decays exponentially with time, as shown in Figure 13.27b (Maxwell). This is quite reasonable for many polymers; however, the process of stress relaxation does not go on indefinitely in real materials.

In another model, called the Voigt model, the spring and the dashpot are arranged in parallel (Figure 13.26b). This means that the strains in the two components are identical, i.e.,

$$\varepsilon = \varepsilon_1 = \varepsilon_2, \tag{13.27}$$

and the stresses in the two components add, to give the stress on the system, i.e.,

$$\sigma = \sigma_1 + \sigma_2. \tag{13.28}$$

It can be shown from Equations 13.27 and 13.28 that

$$\frac{d\varepsilon}{dt} = \frac{\sigma_0}{\eta} - E\frac{\varepsilon}{\eta}. \tag{13.29}$$

Let us now examine the predictions of the Voigt model for creep, or constant stress loading, and for stress relaxation. For the constant-stress situation, $\sigma = \sigma_0$, Equation 13.29 becomes

$$\frac{d\varepsilon}{dt} + E\frac{\varepsilon}{\eta} = \frac{\sigma_0}{\eta}.$$

This differential equation has the solution

$$\varepsilon = \left(\frac{\sigma_0}{E}\right)\left[1 - \exp\left(\frac{-Et}{\eta}\right)\right]. \qquad (13.30)$$

Remembering that the quantity η/E is the relaxation time τ, we find that the variation in strain with time at a constant stress (creep) is given by

$$\varepsilon = \left(\frac{\sigma_0}{E}\right)\left[1 - \exp\left(\frac{-t}{\tau}\right)\right]. \qquad (13.31)$$

This relationship is shown in Figure 13.27a; the prediction of the Voigt model is quite realistic, because $\varepsilon \rightarrow \sigma_0/E$ as $t \rightarrow \infty$.

For the stress relaxation case, we have an imposed constant strain $\varepsilon = \varepsilon_0$, and therefore, $de/dt = 0$. The Voigt model predicts that

$$\frac{\sigma}{\eta} = \frac{E\varepsilon_0}{\eta},$$

or

$$\sigma = E\varepsilon_0.$$

This linear elastic response, however, shown in Figure 13.27b, does not conform to reality.

The molecular weight of a polymer can affect its creep behavior. The strain response of a polymer as a function of time, $\varepsilon(t)$, is shown in Figure 13.28. Also shown is the effect of molecular weight. The effect of increasing the degree of cross-linking is in the same direction as that of increasing the molecular weight. Both tend to promote secondary bonding between chains and thus make the polymer more creep resistant. Compared to glassy polymers, semicrystalline polymers tend to be more creep resistant. Polymers containing aromatic rings in the chain are even more creep resistant. Both increased crystallinity and the incorporation of rigid rings add to the thermal stability, and thus to the creep resistance, of a polymer.

In a constant-stress test of the kind just described, a parameter of interest is the *creep compliance J*. This is the ratio of strain to stress. Since the strain will be a function of time, the compliance will also be a function of time. Thus,

$$J(t) = \varepsilon(t)/\sigma_0. \qquad (13.32)$$

From Equation 13.32 and 13.31, we can write the creep compliance as

$$J(t) = \varepsilon(t)/\sigma_o = (1/E)[1 - \exp(-t/\tau)]. \qquad (13.33)$$

If one plots a series of creep compliances as a function of time, both on logarithmic scales, over a range of temperature, one gets the curve shown in Figure 13.29a. It turns out that such individual plots can be superposed by horizontal shifting (along the log-time axis) by an amount log a_t, to obtain a master curve shown in Figure 13.29b. In Figure 13.29a, we use arrows to indicate the horizontal shift of data to obtain a master curve corresponding to a reference temperature of the polymer. This figure shows that, when creep compliance is a measured at a series of temperatures, with the glass transition temperature T_g as the reference temperature, then curves above T_g are shifted to the right, while curves below T_g are shifted to the left.

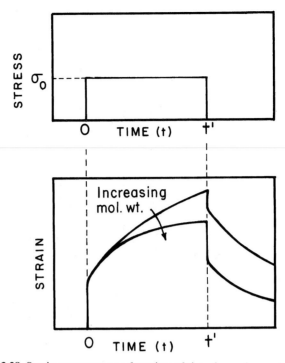

Figure 13.28 Strain response as a function of time for a glassy, viscoelastic polymer subjected to a constant stress σ_0. Increasing the molecular weight or degree of cross-linking tends to promote secondary bonding between chains and thus make the polymer more creep resistant.

As discussed earlier, a thermally activated process shows a dependence of its on temperature that can be described by an Arrhenius-type expression. When viscous flow occurs in a polymer, the network structure breaks and re-forms locally. The thermal energy for such viscous flow is available above the glass transition temperature T_g. Below T_g, the thermal energy is not high enough for the breaking and remaking of the bonds, and the material does not flow easily. In the viscoelastic range, time and temperature have similar effects on polymers. There are two easy ways of studying such behavior. In the first of these, we can impose a constant deformation on the polymer and follow the resultant stress. This will give us a stress relaxation modulus as a function of time. The other technique involves the application of a constant stress and measuring the deformation as a function of time. This will give us a curve of compliance vs. time. A very useful principle called time–temperature superposition allows us to take the data at one temperature and superimpose them on data taken at another temperature by a shift along the log-time axis. This principle is of great practical use, inasmuch as obtaining data over a full range of creep compliance or stress relaxation behavior can involve years. The principle allows one to shift data taken over shorter time spans, but at different temperatures, to obtain a master curve that covers longer time spans. Williams, Landel, and Ferry found that the natural logarithm of a_T (the time-shift factor) follows a simple expression, viz.,

$$\ln a_T = -C_1(T - T_s)/(C_2 + T - T_s),$$

where C_1 and C_2 are constants and T_s is a reference temperature for a given polymer.[16] If we take the reference temperature to be the glass transition temperature T_g, then

---

[16] M. L. Williams, R. F. Landel, and J. D. Ferry, *J. Amer. Chem. Soc.*, 77 (1955) 3701.

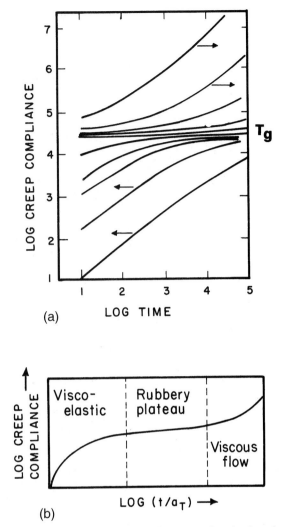

LOG CREEP COMPLIANCE

LOG TIME

(a)

LOG CREEP COMPLIANCE

| Visco- | Rubbery |
| elastic | plateau |

Viscous
flow

LOG (t/a_T) ⟶

(b)

Figure 13.29 (a) A series of creep compliances vs. time, both on logarithmic scales, over a range of temperature. (b) The individual plots in (a) can be superposed by horizontal shifting (along the log-time axis) by an amount log a_T, to obtain a master curve corresponding to a reference temperature T_g of the polymer. (c) Shift along the log-time scale to produce a master curve. (Courtesy of W. Knauss) (d) "Experimentally" determined shift factor.

$C_1 = 17.5$ K and $C_2 = 52$ K. If the reference temperature T_s is taken to be about 50°C above T_g, then $C_1 = 20.4$ K and $C_2 = 101.6$ K

Figure 13.29a shows a series of creep compliances as a function of time over a range of temperatures. The arrows indicate the direction of shift with respect to a reference temperature—for example, the glass transition temperature. The amount of shift can be calculated by the Williams–Landel–Ferry expression. The master curve for creep, obtained by superposing horizontally shifted curves, is shown in Figure 13.29b. Another way of treating this problem is shown in Figure 13.29c, where we plot the stress relaxation modulus as a function of time, both on logarithmic scales. The "experimentally" determined time-shift factor, as a function of temperature, is shown in Figure 13.29d. (Compare this with the master curve shown in Figure 13.29c.)

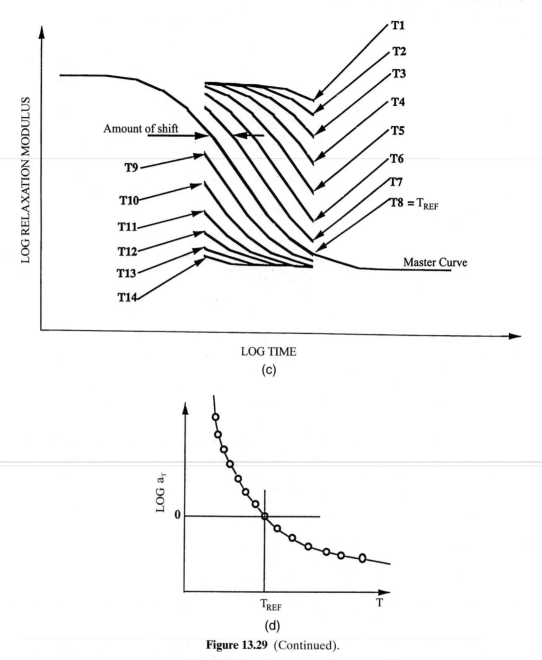

Figure 13.29 (Continued).

Now recall the model in which spring and dashpot are arranged in parallel, i.e., the Voigt model. The model is used to explain the stress relaxation behavior of a polymer. We impose a constant strain ε_0 and follow the drop in stress $\sigma(t)$ as a function of time. (See Figure 13.30.) Instead of a compliance term, we now have a stress relaxation modulus, given by

$$E(t) = \sigma(t)/\varepsilon_0.$$

In the case of stress relaxation also, one can obtain a master curve, as shown schematically in Figure 13.31. Also shown in the figure is the effect of cross-linking and molecular weight. Stress relaxation in polymers is of great practical significance when the polymers are used

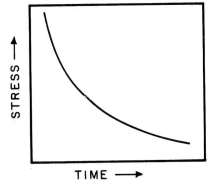

Figure 13.30 A constant imposed strain ε_0 results in a drop in stress $\sigma(t)$ as a function of time.

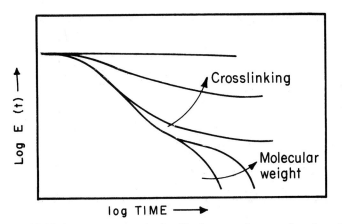

Figure 13.31 A master curve obtained in the case of stress relaxation, showing the variation in the reduced modulus as a function of time. Also shown is the effect of cross-linking and molecular weight.

in applications involving gaskets and seals. At times, this effect can be exploited beneficially: for example, in a situation where residual stresses are not desirable, we can incorporate a polymer to undergo easy stress relaxation in response to residual stresses.

EXAMPLE 13.4

Data on stress relaxation modulus vs. time for polyisobutylene (also known as chewing gum) are shown in Figure E13.4a. The data span a range of 10^{-2} to 10^2 hours in time. Obtain the curve of the time-shift factor for a reference temperature of 298 K. Obtain a master curve for polyisobutylene based on time–temperature superposition of data.

Solution: By using the Williams–Landel–Ferry expression

$$\log a_T = -17.5(T - T_{\text{ref}})/[52 + (T - T_{\text{ref}})]$$

for a temperature range of $-80°C$ to $80°C$, we obtain the curve of the time-shift factor vs. temperature shown in Figure E13.4b. Using this time-shift curve, we superimpose the individual stress relaxation modulus curves given in

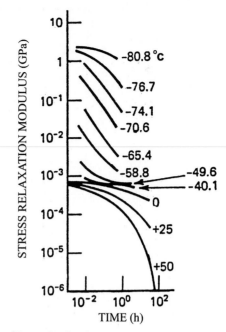

Figure E13.4a (From E. Catsiff and A. V. Tobolsky, *J. Polymer Sci.,* 19 (1956) 111)

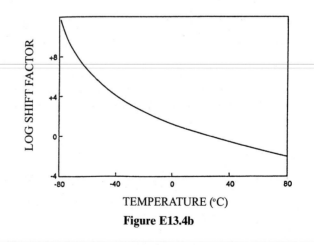

Figure E13.4b

the statement of the problem (Fig. E13.4a), to obtain the master curve shown in Figure E13.4c. Note that the time scale ranges from 10^{-14} to 10^2 hours.

EXAMPLE 13.5

The creep strain rate of a polymer is given by the expression

$$\dot{\varepsilon} = 4.5 \times 10^{11} \exp(-100 \text{ kJ}/RT),$$

where T is the temperature in kelvins and R is the universal gas constant. How much time will it take for a rod of this polymer to extend from 10 mm to 15 mm at 100°C?

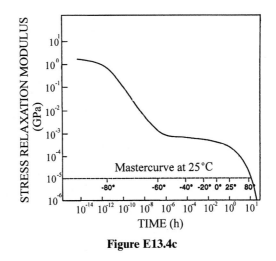

Figure E13.4c

Solution:

$$\dot{\varepsilon} = \Delta\varepsilon/\Delta t = 4.5 \times 10^{11} \exp[-100{,}000/(8.3 \times 373)]$$
$$= 4.2 \times 10^{-3}\, s^{-1},$$

Hence,

$$\Delta t = 0.5/(4.2 \times 10^{-3}) = 119\ s.$$

EXAMPLE 13.6

The activation energy for stress relaxation in a polymer is 50 kJ/mol. The relaxation time at 25°C is 90 days. What is the relaxation time at 100°C?

Solution:

$$1/\tau = A \exp(-Q/RT),$$
$$(1/\tau_{25})/(1/\tau_{100}) = \exp(-Q/R298)/\exp(-Q/R398),$$
$$\tau_{100} = \tau_{25} \exp[(Q/R)(1/398 - 1/298)]$$
$$\tau_{100} = 90 \exp[((50 \times 10^{3})/8.314)(1/398 - 1/298)]$$
$$= 90 \times 6.4 \times 10^{-3}$$
$$= 0.57\ \text{day}.$$

EXAMPLE 13.7

A nylon cord, used to tie a sack, has an initial stress of 5 MPa. If the relaxation time for this cord is 180 days, in how many days will the stress reduce to 1 MPa?

Solution:

$$\sigma = \sigma_0 \exp(-t/\tau),$$

$$1 \text{ MPa} = 5(-t/180),$$

$$t = -180 \ln 1/5,$$

$$t = 290 \text{ days}.$$

13.11 SUPERPLASTICITY

Some metallic alloys and ceramics show a peculiar behavior called *superplasticity*. This is the ability to flow, in tension, to very large elongations. Figure 13.32 shows a dramatic illustration of superplastic behavior. The specimen was extended at a temperature of 413 K, and a total strain of 48.5 was reached without failure. The phenomenon of superplasticity was observed for the first time in 1934. A great deal of activity has taken place in the area since then, and superplastic forming has become a successful industrial process.

Superplasticity has been obtained in a number of alloy systems, including titanium alloys (Ti–6% Al–4% V), iron-based alloys, and aluminum alloys. High-strength nickel-based superalloys have been found to exhibit superplastic behavior, and the process of "gatorizing" (supposedly named after an alligator living in the lake in front of the research institute) has been developed by Pratt and Whitney. The potential of superplastic forming is especially bright for titanium alloys, which are known to be very difficult to form, because of their HCP structure. Superplasticity has also been discovered in ceramics.

The basic reason that some materials can deform superplastically when others cannot is related to how they respond to changes in strain rate. The example of hot glass (above the transition temperature) comes to mind. We are all familiar with Coke bottles stretched to very high strains; these interesting items are sold in curio shops. Glass shows a Newtonian viscous behavior above a certain temperature. Fiberglass is formed in such a manner and can be pulled to extremely fine fibers. In a lamellar flow, Newtonian viscosity is defined by (see Eqn. 3.29)

$$\tau = \eta \frac{dv}{dy},$$

where dv/dy is the variation in velocity of the fluid with distance y, τ is the shear stress necessary to create the velocity gradient dv/dy, and η is the viscosity. The derivative dv/dy is equivalent to the shear strain rate $\dot{\gamma}$. (See Section 3.6.2.) Thus, we can write

$$\tau = \eta \dot{\gamma}. \tag{13.34}$$

The stress–versus–strain-rate relationship from many materials is not linear, but of the form (see Eqn. 3.23)

$$\sigma = K\dot{\varepsilon}^m, \tag{13.35}$$

where K and m are constants and m is called the strain-rate sensitivity. In general, m varies between 0.02 and 0.2, for homologous temperatures between 0 and 0.9 (90% of the melt-

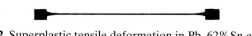

Figure 13.32 Superplastic tensile deformation in Pb–62%Sn eutectic alloy tested at 415 K and a strain rate of 1.33×10^{-4} s^{-1}; total strain of 48.5. (From M. M. I. Ahmed and T. G. Langdon, *Met. Trans. A*, 8 (1977) 1832)

ing point in K). Hence, one would have, at the most, an increase of 15% in the yield stress by doubling the strain rate. Comparing Equations 13.34 and 13.35, we see that a value of $m = 1$ will give a Newtonian viscous solid. Such a material would not undergo tensile instability and could be stretched indefinitely. Figure 13.33a shows schematically how a high value of m will inhibit tensile instability (necking) and, consequently, enhance plasticity in tension. The specimen is being deformed, in tension, at a velocity v. The length increases from L_0 to L_1 and then to L_2. At L_2, necking starts. If the material has a high value of m, this instability will be inhibited because of the localized strengthening effect. When the length is L_2, the strain rate over the specimen is

$$\dot{\varepsilon}_2 = \frac{v}{L_2}.$$

In the incipient neck region, which acts as a "minispecimen" embedded in the large specimen, one has

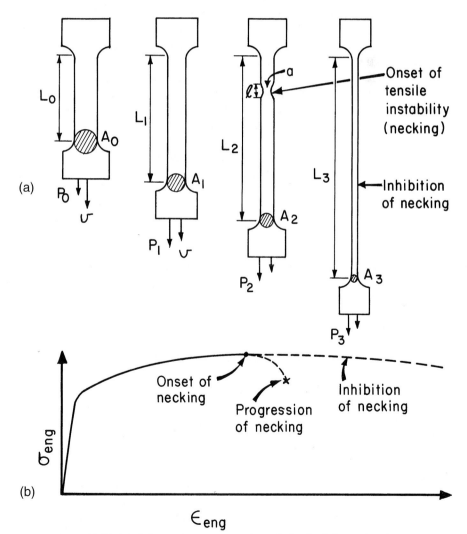

Figure 13.33 (a) Schematic representation of plastic deformation in tension with formation and inhibition of necking. (b) Engineering-stress–engineering-strain curves.

$$\dot{\varepsilon}_2' = \frac{v}{\ell}.$$

Since $\ell \ll L_2$, one has $\dot{\varepsilon}_2' \gg \dot{\varepsilon}_2$.

The strain rate sensitivity can be obtained from Equation 13.35 by applying that equation to two strain rates and eliminating K. When we do this, we obtain

$$m = \frac{\ln(\sigma_2'/\sigma_2)}{\ln(\dot{\varepsilon}_2'/\dot{\varepsilon}_2)}.$$

For a high value of m, the strength σ_2' in the neck region is much higher than σ_2, and further progression of plastic deformation at that region is halted. Mathematically,

$$\sigma_2' > \frac{P_2}{a}.$$

When m is low, σ_2' is not sufficiently higher, and

$$\sigma_2' < \frac{P_2}{a}.$$

a is the cross-sectional area in the neck region. The deformation continues to concentrate itself at the neck, with the attendant reduction in area caused by the constancy of volume. This leads to failure. Figure 13.33b shows the two alternative paths beyond the maximum in the engineering-stress–engineering-strain curve. Thus, one concludes that superplasticity is the result of the inhibition of necking as a result of a high value of m.

Under certain conditions of temperature and strain rate, some metals and ceramics exhibit an enhancement of m. Curve (a) in Figure 13.34 shows the stress as a function of strain rate for an Mg–Al eutectic alloy tested at 350°C. One can see that in region II, the stress rises more rapidly with strain rate. Curve (b) in the same figure shows the strain rate sensitivity m as a function of strain rate. The maximum, $m = 0.6$, occurs for a strain rate of

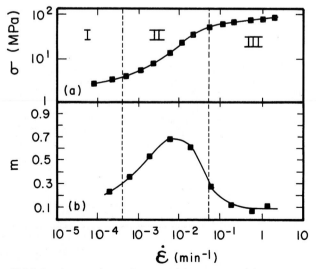

Figure 13.34 Strain-rate dependence of (a) stress and (b) strain-rate sensitivity for Mg–Al eutectic alloy tested at 350°C (grain size 10 μm). (After D. Lee, *Acta. Met.,* 17 (1969) 1057)

10^{-2} s^{-1}. $m = \partial\ln \sigma/\partial \ln \dot{\varepsilon}$ is the slope of curve (a) in the figure. Figure 13.35 shows the variation in $\partial L/L_0$ (the tensile fracture strain) with $\dot{\varepsilon}$ for a Zr–22% Al alloy. The maxima (at the three temperatures) in $\Delta L/L_0$ correspond roughly to the center of region II. Further proof of the effect of m on the extent of superplastic flow is provided by Figure 13.36, which contains data from several studies. Data for alloys of Fe, Mg, Pu, Pb, Sn, Ti, Zn, Zr are plotted, and the correlation is excellent. As m approaches unity, $\Delta L/L_0$ reaches extraordinarily high values.

The microstructural requirement for a high value of m is a small grain size. The testing temperatures are usually above $0.4T_m$, where T_m is the absolute melting point, and the strain rates in which superplasticity is observed are usually intermediate (10^{-4} s$^{-1} < \dot{\varepsilon} < 10$ s^{-1}). Superplasticity is usually enhanced by thermal cycling, i.e., straining the material sequentially at two different temperatures. All alloys that show structural superplastic behavior have a very fine grain size (<10 μm). For these small grain sizes, and at the deformation temperatures, most of the plastic deformation takes place by grain-boundary sliding, and not by the conventional dislocation mechanisms in the interior of the grains. Specimens deformed to very large strains routinely exhibit an equiaxial grain structure, in contrast with conventional deformation, in which the strain undergone by the individual grains is equal to the overall strain, and the grains assume an elongated shape. Grain-boundary sliding accounts for 50–70% of the overall strain. Superplastic materials may be

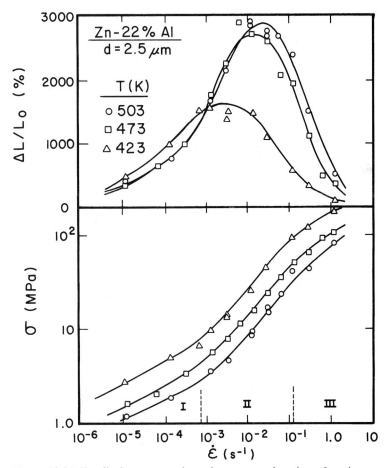

Figure 13.35 Tensile fracture strain and stress as a function of strain rate for Zr–22% Al alloy with 2.5-μm grain size. (After F. A. Mohamed, M. M. I. Ahmed, and T. G. Langdon, *Met. Trans. A*, 8 (1977) 933)

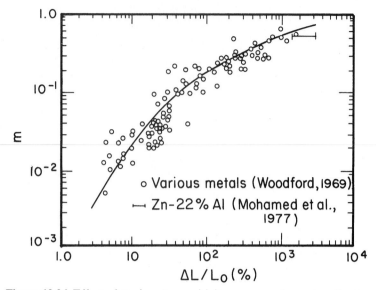

Figure 13.36 Effect of strain-rate sensitivity m on maximum tensile elongation for different alloys (Fe, Mg, Pu, Pb–Sr, Ti, Zn, Zr based). (From D. M. R. Taplin, G. L. Dunlop, and T. G. Langdon, *Ann. Rev. Mater. Sci.*, 9 (1979) 151, 180, Figure 16)

likened to sand: The granules retain their shape with plastic deformation. In contrast to sand, however, superplastic materials do not have interstices between the grains. Thus, some plastic accommodation must occur as the grains slide past each other. The contribution of grain-boundary sliding to plastic deformation is more substantial the greater the grain-boundary surface per unit volume. Since the grain-boundary surface is inversely proportional to the grain size, this explains why the contribution of grain-boundary sliding is less important in materials with larger grain sizes.

Table 13.2 shows the tensile elongation of a number of superplastic materials. It is interesting to note that superplasticity has been found in composites and ceramics. Sherby and co-workers obtained tensile elongations of 1,300% in an aluminum alloy: SiC whisker reinforced composite. The Sherby team used temperature cycling. These researchers also detected superplasticity in ultrahigh-carbon steel and were able to attribute the splendid properties of the Damascus sword to superplastic forming. Thus, the use of superplasticity is centuries old. For ceramics, superplasticity (in tension) is a technology with great potential. Wakai et al. obtained tensile elongations of 120% in an yttria-stabilized polycrystal containing 90% tetragonal zirconia and 10% cubic zirconia.[17] A grain size of 0.3 μm produced a strain rate sensitivity of $m = 0.5$ at 1450°C. This elongation was exceeded in work done by Nieh et al., who obtained a value of 350% at 1,550°C.[18] Nanocrystalline ceramics (Section 5.5) are especially attractive in this regard. Superplasticity was also obtained in an aluminosilicate and other ceramic systems.

One of the major problems in superplastic forming is the formation of voids at grain boundaries. Cavitation during superplastic forming results in a deterioration of the mechanical properties of parts formed by superplasticity. Cavities form because of incompatible deformation of adjacent grains and weaken the material. These voids can be reduced or eliminated by superimposing a hydrostatic stress upon the applied tensile stress. This is illustrated in Figure 13.37. The aluminum alloy shown exhibited considerable cavitation at a

[17] F. Wakai, S. Sakaguchi, and Y. Matsuno, *Advanced Ceramic Materials, Vol. 1, No.3* (1986) 259.

[18] T. G. Nieh, C. M. McNally, and J. Wadsworth, *Scripta Met.*, 22 (1988) 1297.

TABLE 13.2 Materials Exhibiting Very High Tensile Strains (Adapted from D. M. R. Taplin, G. L. Dunlap, and T. G. Langdon, *Ann. Rev. Mater. Sci.,* 9 (1979) 15)

Material	Maximum strain (%)
Al–33% Cu eutectic	1,500
Al–6% Cu–0.5% Zr	1,200
Al–10.7% Zn–0.9% Mg–0.4% Zr	1,500
Bi–44% Sn eutectic	1,950
Cu–9.5% Al–4% Fe	800
Mg–33% Al eutectic	2,100
Mg–6% Zn–0.6% Zr	1,700
Pb–18% Cd eutectic	1,500
Pb–62% Sn eutectic	4,850
Ti–6% Al–4% V	1,000
Zn–22% Al eutectoid	2,900
Al(6061)–20% SiC (whiskers)	1,400
Partially stabilized zirconia	120
Lithium aluminosilicate	400
Cu–10% Al	5,500
Zirconia	350
Zirconia + SiO_2	1,000

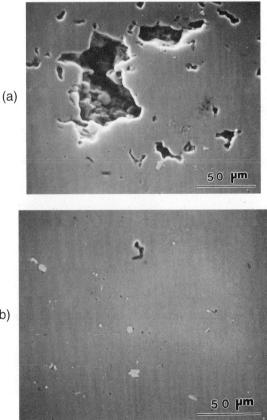

Figure 13.37 Cavitation in superplasticity formed 7475-T6 aluminum alloy ($\varepsilon = 3.5$) at 475°C and 5×10^{-4} s^{-1}. (a) Atmospheric pressure. (b) Hydrostatic pressure $P = 4$ MPa. (Courtesy of A. K. Mukherjee)

plastic strain of 350%, or 3.5. The application of a superimposed hydrostatic pressure of 4 MPa, through a gaseous medium, decreases the cavitation substantially. Otherwise, the cavitation would lead to premature failure. Figure 13.38a shows the effect of grain size on the elongation of an 7475 Al alloy. Figure 13.38b shows a number of specimens superplastically deformed up to failure. The initial specimen is at the bottom, and the effect of increasing the superimposed pressure is shown from bottom to top. The pressures (and elongations to failure) are, respectively, 330% (atmosphere), 720% (1.4 MPa); 830% (2.8 MPa), and 1330% (and no failure at 5.6 MPa).

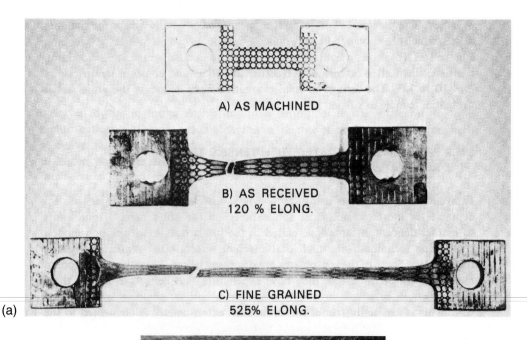

(a)

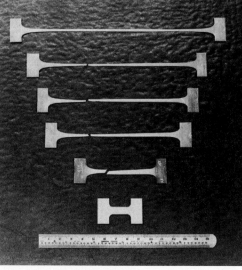

(b)

Figure 13.38 (a) Effect of grain size on elongation: (A) Initial configuration. (B) Large grains. (C) Fine grains (10 μm) (Reprinted with permission from N. E. Paton, C. H. Hamilton, J. Wert, and M. Mahoney, *J. Metal, 34* (1981) No. 8, 21, p. 25). (b) Failure strains increase with superimposed hydrostatic pressure (from 0 to 5.6 MPa). (Courtesy of A. K. Mukherjee)

SUGGESTED READINGS

W. R. Cannon and T. G. Langdon. "Creep of Ceramics." *J. Matls. Sci.,* (1983) 1 (Part 1); 23 (1988) 1 (Part 2).

A. H. Chokshi, A. K. Mukherjee, and T. G. Langdon, "Superplasticity in Advanced Materials," Matls. Sci and Eng. R: Reports, (1993) 237–274.

A. H. Chokshi and T. G. Langdon. "Characteristics of Creep Deformation in Ceramics." *Matls. Sci. and Techn.,* 7 (1991) 577.

A. G. Evans and T. G. Langdon. "Structural Ceramics." *Progr. Matls. Sci.,* 21 (1976) 171.

H. J. Frost and M. F. Ashby. *Deformation-Mechanism Maps.* Oxford: Pergamon Press, 1982.

F. Garofalo. *Fundamentals of Creep and Creep Rupture in Metals.* New York: Macmillan, 1965.

J. Gittus. *Creep, Viscoelasticity and Creep Fracture in Solids.* New York: Halsted Press (Wiley), 1975. *J. Eng. Mater. Technol.,* 101 (1979) 317 ff.

B. P. Kashyap, A. Arieli, and A. K. Mukherjee. "Microstructural Aspects of Superplasticity." *J. Matls. Sci.,* 20 (1985) 2661.

H. H. Kausch. *Polymer Fracture,* 2d. ed. Berlin and New York: Springer-Verlag, 1987.

F. R. N. Nabarro, and H. L. de Villiers. *The Physics of Creep.* UK: Taylor and Francis, 1995.

W. D. Nix and J. C. Gibeling. "Mechanisms of Time-Dependent Flow and Fracture of Metals," in *Flow and Fracture at Elevated Temperatures* (ed. R. Raj). Metals Park, Ohio: ASM, 1985.

J. P. Poirier. *Creep of Crystals: High Temperature Deformation Processes in Metals, Ceramics, and Minerals.* Cambridge, U.K.: Cambridge University Press, 1985.

O. D. Sherby and P. M. Burke. "Mechanical Behavior of Crystalline Solids at Elevated Temperature." *Progr. in Matls. Sci.,* 13 (1967) 325.

J. Weertman, and J. R. Weertman. "Mechanical Properties, Strongly Temperature Dependent," in *Physical Metallury,* 4th ed. (eds. R. W. Cahn and P. Haasen). 4th ed, New York, NY: Elsevier, 1995.

EXERCISES

13.1 A cylindrical specimen creeps at a constant rate during 10,000 hours when it is subjected to a constant load of 1,000 N. The initial diameter and length of the specimen are 10 and 200 mm, respectively, and the creep rate is 10^{-8} h^{-1}. Find:

(a) The length of the specimen after 100, 1,000, and 10,000 hours.

(b) The true and engineering strains after these periods.

(c) The true and engineering stresses after these periods.

13.2 Give three reasons why the extrapolation of creep data obtained over a short period can be dangerous over long periods.

13.3 By means of plots, show how isochronal stress-versus-strain curves can be constructed from creep curves for various stresses at a certain temperature.

13.4 T. E. Howson, J. E. Stulga, and J. K. Tien[19] obtained the following stress-rupture results for the superalloy Inconel MA 754 (a dispersion-strengthened alloy):

Temperature (°C)	Applied Stress (MPa)	Rupture Life (hours)
760	189.7	—
760	206.9	83.9
760	206.9	111.2
760	224.2	38.6
760	224.2	29.0

[19] T. E. Howson, J. E. Stulga, and J. K. Tien, *Met. Trans.,* 11A (1980) 1599.

Temperature (°C)	Applied Stress (MPa)	Rupture Life (hours)
760	241.4	6.9
760	258.7	1.8
746	206.9	320.8
774	206.9	65.0
788	206.9	33.2
982	110.4	195.1
982	113.8	136.6
982	113.8	106.9
982	116.5	27.6
982	117.3	106.3
982	120.7	13.0
982	120.7	39.0
996	110.4	52.6
996	110.4	41.3
1010	110.4	20.3
1010	110.4	41.7
1024	110.4	9.4

(a) Verify whether this alloy obeys a Larson–Miller relationship, and find C. Then prepare a master plot.

(b) Determine the predicted stress-rupture life if the alloy is stressed at 1,000°C and 50 MPa.

13.5 What is the predicted stress-rupture life of AISI 316 steel (18% ~ 8% Ni) at 800°C and 160 MPa? (See Figure Ex 13.5.)

13.6 Verify whether the data of Exercise 13.5 obey the Manson–Haferd correlation.

13.7 Assuming that pure silver creeps according to the Dorn equation, estimate the rupture time at 400°C when the silver subjected to a stress of 50 MPa, knowing that at 300°C and at the same stress level the rupture time is 2,000 hours.

13.8 In Exercise 13.4, verify how closely the Monkmon–Grant relationship is obeyed.

13.9 T. E. Howson, D. A. Mervyn, and J. K. Tien[20] studied the creep and stress-rupture response of oxide-dispersion-strengthened (ODS) superalloys produced by mechanical alloying. They determined that the activation energy for creep Q_c was 619 kJ/mol by conducting tests at a constant applied stress of 558.7 MPa at the three temperatures of 746, 760, and 774°C.

(a) The results shown in Figure Ex 13.9 were found for experimental alloy MA 6000 E at 760°C. Estimate the value of *n,* and discuss this value in terms of the micro-structure exhibited by the alloy (made by means of dispersion-strengthening by inert yttrium oxide dispersoids plus precipitation-strengthening by gamma prime).

(b) By applying Eq. 13.4, show how the activation energy can be found. Make the appropriate plot, and find the minimum creep rate at the aforementioned three temperatures. Note that the activation energy is given per mole.

13.10 Lead (melting point, 327.5°C) was tested at ambient temperature (23°C) and three different engineering stress levels: 8.5, 9, and 10 MPa. The curve that was obtained was shown in Figure Ex 13.10.

(a) From the temperature aspect, establish whether the room temperature is in the creep domain for Pb.

(b) Obtain the minimum creep rates for the three stress levels.

(c) Obtain parameters for the curve at 8.5 MPa, as expressed by Equation 13.1.

(d) Obtain the stress exponent in the Mukherjee–Bird–Dorn equation. Based on this value, what mechanism of creep do you expect?

[20] T. E. Howson, D. A. Mervyn, and J. K. Tien, *Met. Trans.,* 11A (1980) 1609.

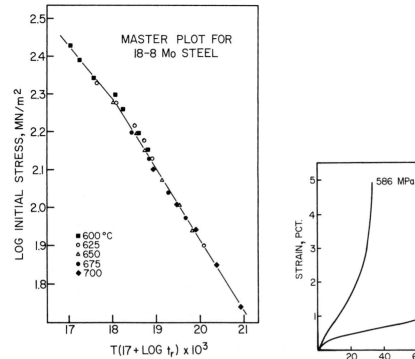

Figure Ex13.5 Master plot for Larson–Miller parameter for AISI 316 stainless steel. (Courtesy of H.-J. Kestenbach)

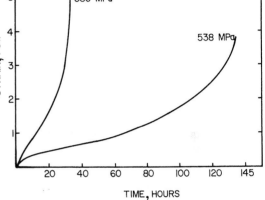

Figure Ex13.9

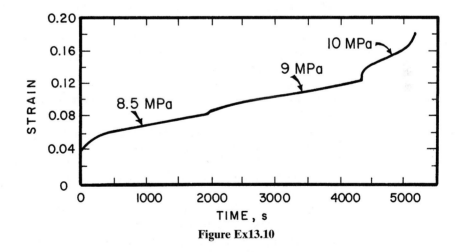

Figure Ex13.10

13.11 Using the results from the previous problem, predict the minimum creep rate for the same material if the test would be carried out at 10°C and stress levels of (a) 10 MPa and (b) 20 MPa. The self-diffusion coefficients for Pb are

$$D = 5.42 \times 10^{-8} \text{ cm}^2 \cdot \text{hr}^{-1} \text{ at } 250°C,$$

$$D = 2.92 \times 10^{-7} \text{ cm}^2 \cdot \text{hr}^{-1} \text{ at } 285°C.$$

13.12 Determine the slopes for the creep behavior of UO_2, shown in Figure Ex 13.12. Discuss the deformation mechanisms in the two regions.

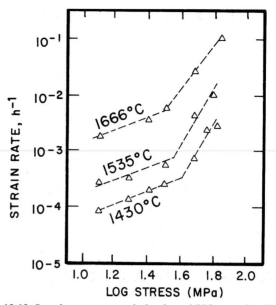

Figure Ex13.12 Steady-state creep behavior of UO_2 can be divided into two regimes with different stress exponents. The transition stress between the regimes decreases with increasing temperature. (From L. E. Poteat and C. S. Yust, in *Ceramic Microstructures,* R. M. Fulrath and J. A. Pask (eds.) (New York: John Wiley & Sons, (1968) 646, Fig. 31.1, p. 647)

13.13 Tungsten is being used at half its melting point ($T_m \approx 3{,}400°C$) and a stress level of 160 MPa. An engineer suggests increasing the grain size by a factor of 4 as an effective means of reducing the creep rate.

(a) Do you agree with the engineer? Why? What if the stress level were equal to 1.6 MPa?

(b) What is the predicted increase in length of the specimen after 10,000 hours if the initial length is 10 cm?

(*Hint:* Use a Weertman–Ashby map.)

13.14 Stress relaxation in a polymer results from molecular displacements. Thus, one would expect that the effect of temperature on stress relaxation would be similar to that of any other thermally activated process. As described in the text, an Arrhenius-type expression describes the temperature dependence of such phenomena. Thus, the relaxation time is the inverse rate, or

$$\frac{1}{\tau} \propto \exp(-Q/kT),$$

where the symbols have their usual significance. Describe how you would determine, from this expression, the activation energy Q of the molecular process causing the relaxation.

13.15 Curves of creep modulus (the inverse of creep compliance) vs. time for four different temperatures are shown in Figure Ex 13.15. Obtain a master curve for the polymer at a reference temperature of 101.6°C.

13.16 An amorphous polymer has a glass transition temperature of 100°C. A creep modulus of 1 GPa was measured after 1 hour at 75°C. Using the Williams–Landel–Ferry expression, determine the time required to reach this modulus at 50°C.

13.17 The viscosity of an amorphous polymer is 10^5 Pa.s at 190°C and 2×10^2 Pa.s at 270°C. At what temperature will the viscosity be 10^9 Pa.s?

13.18 What is the strain undergone by a polymer in tension at 67°C for one minute if the polymer's strain-rate response is given by $\dot{\varepsilon} = 4.5 \times 10^{28} \exp(-200 \text{ kJ}/RT)$?

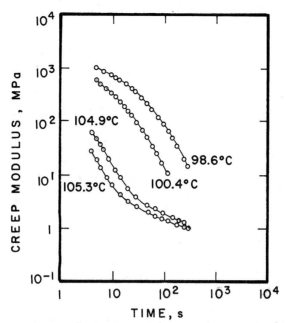

Figure Ex13.15 (After W. Kurz, J. P. Mercier, and G. Zambelli, *Introduction aux Sciences des Matériaux* (Lausanne, Switzerland: Presses, 1987), p. 287)

13.19 A nylon cord has an initial stress of 2 MPa and is used to tie a sack. If the relaxation time for this cord is 250 days, how many days will it take for the stress to drop to 0.1 MPa?

13.20 How much time will it take for a rod of polymer to extend from 20 mm to 30 mm at 120°C if it is deformed at a strain rate $\dot{\varepsilon} = 4.5 \times 10^{11} \exp(-100 \text{ kJ}/RT)$?

13.21 Find the initial stress for a nylon cord if the relaxation time for the cord is 100 days and in 50 days the stress is reduced to 1 MPa.

13.22 (a) Determine the strain-rate sensitivity in the superplastic range for the alloys shown in Figures 13.34a and 13.35, and explain the values encountered. (b) Why does the maximum in ductility vary with temperature in Figure 13.35?

13.23 Explain why the presence of voids decreases the maximum strain in superplastic deformation.

14

Fatigue

There is some confusion in the literature about the terminology pertaining to fatigue. We define fatigue as a degradation of mechanical properties leading to failure of a material or a component under *cyclic* loading. This definition excludes the so-called phenomenon of static fatigue, which is sometimes used to describe stress corrosion cracking in glasses and ceramics in the presence of moisture. Brittle solids (glasses and crystalline ceramics) undergo subcritical crack growth in an aggressive environment under static loads. Silica-based glasses are especially susceptible to this kind of crack growth in the presence of moisture. If a glassy phase exists at grain boundaries and interfaces, it will be susceptible to such an attack. Thus, static fatigue is more appropriately a stress corrosion phenomenon, rather than a cyclic stress-related phenomenon.

In general, fatigue is a problem that affects any structural component or part that moves. Automobiles on roads, airplanes (principally the wings) in the air, ships on the high sea constantly battered by waves, nuclear reactors and turbines under cyclic temperature conditions (i.e., cyclic thermal stresses), and many other components in motion are examples in which the fatigue behavior of a material assumes a singular importance. It is estimated that 90% of service failures of metallic components that undergo movement of one form or another can be attributed to fatigue. Often, a fatigue fracture surface will show some easily identifiable macroscopic features, such as beach markings. Figure 14.1 shows a schematic of the fracture surface of, say, a steel shaft that failed in fatigue. The main features of this kind of failure are a fatigue crack initiation site, generally at the surface; a fatigue crack propagation region showing beach markings; and a fast-fracture region where the crack length exceeds a critical length. Typically, the failure under cyclic loading occurs at much lower stress levels than the strength under monotonic loading.

In this chapter, we present a basic description of the various aspects of fatigue in different materials, followed by a brief examination of the various fatigue-testing techniques.

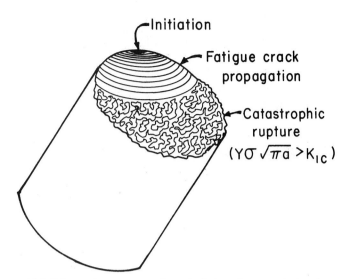

Figure 14.1 Schematic representation of a fatigue fracture surface in a steel shaft, showing the initiation region (usually at the surface), the propagation of fatigue crack (evidenced by beach markings), and catastrophic rupture when the crack length exceeds a critical value at the applied stress.

14.1 FATIGUE PARAMETERS AND S–N CURVES

We first define some important parameters that will be useful in the subsequent discussion of fatigue. These parameters, shown in Figure 14.2, are as follows:

cyclic stress range, $\quad\quad \Delta\sigma = \sigma_{max} - \sigma_{min}$,

cyclic stress amplitude, $\quad \sigma_a = (\sigma_{max} - \sigma_{min})/2$,

mean stress, $\quad\quad\quad\quad \sigma_m = (\sigma_{max} + \sigma_{min})/2$,

stress ratio, $\quad\quad\quad\quad R = \sigma_{min}/\sigma_{max}$,

where σ_{max} and σ_{min} are the maximum and minimum stress levels, respectively.

Traditionally, the behavior of a material under fatigue is described by the *S–N* (or *σ–N*) curves (Figure 14.3), where *S* (or *σ*) is the stress and *N* is the number of cycles to failure. Such an *S–N* curve is frequently called a *Wöhler curve,* after the German engineer who

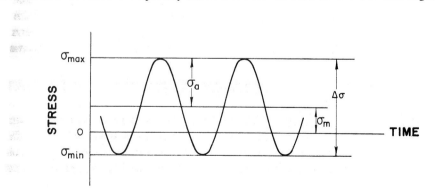

Figure 14.2 Fatigue parameters.

first observed that kind of fatigue behavior. For steels, in general, one observes a fatigue limit or endurance limit (curve *A* in Figure 14.3a), which represents a stress level below which the material does not fail and can be cycled indefinitely. Such an endurance limit does not exist for nonferrous metals (curve *B* in the figure). Polymeric materials show essentially similar *S–N* curves. Figure 14.3b shows a schematic of an *S–N* curve for a variety of polymers. Polymers that form crazes, such as polymethylmethacrylate (PMMA) and polystyrene (PS), may show a flattened portion in the very beginning, indicated as stage I in the figure. In region II, the stress is not high enough for crazes to form. Crazed regions are, of course, the sites of microcrack nucleation. They will form in the initial quarter of a tensile cycle in such materials. Recall that the crazes do not form in compression. Such a flat region does not exist for polymers that do not show craze formation, and the *S–N* curve for such polymers will be very similar to that of metals; that is, stage I is simply an extension of stage II. An example of an actual *S–N* curve showing the three stages in the case of polystyrene

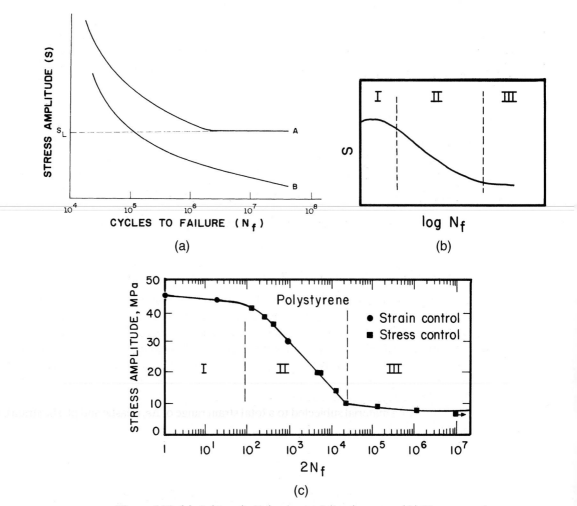

(a)

(b)

(c)

Figure 14.3 (a) *S* (stress)–*N* (cycles to failure) curves. (A) Ferrous and (B) nonferrous metals; S_L is the endurance limit. (b) *S–N* curves for polymeric materials. Polymers that form crazes, such as polymethylmethacrylate (PMMA) and polystyrene (PS), may show a flattened portion in the very beginning, indicated as stage I. (c) An example of an actual *S–N* curve showing the three stages in the case of polystyrene.

is presented in Figure 14.3c. Polystyrene shows extensive crazing at room temperature. Polycarbonate, on the other hand, does not show crazing at room temperature, and its S–N curve does not show stage I. No such S–N curves are available for ceramics, although, as we shall see later in the chapter, stable subcritical crack propagation under cyclic fatigue can occur in ceramics. Note that the relationship between S and N is not a single-valued function, but serves to indicate a statistical tendency. Also, the fatigue life determined in terms of S–N curves cannot be separated into the initiation and propagation parts of fatigue.

14.2 FATIGUE STRENGTH OR FATIGUE LIFE

Traditionally, fatigue life has been presented in the form of an S–N curve (Figure 14.3). With regard to this measure, *fatigue strength* refers to the capacity of a material to resist conditions of cyclic loading. However, in the presence of a measurable plastic deformation, materials respond differently to strain cycling than to stress cycling. Thus, one would expect that the fracture response of a material under cyclic conditions would show a similar difference. In this section, we treat fatigue life in terms of strain versus number of cycles to failure N_f or number of reversals to failure, $2N_f$. It is convenient to consider separately the elastic and plastic components of strain. The elastic component can be readily described by means of a relation between the true stress amplitude and the number of reversals (i.e., twice the number of cycles), viz.,

$$\Delta\varepsilon_e/2 = \sigma_a/E = (\sigma_f'/E)(2N_f)^b,$$

where $\Delta\varepsilon_e/2$ is the elastic strain amplitude, σ_a is the true stress amplitude, σ_f' is the fatigue strength coefficient (equal to the stress intercept at $2N_f = 1$), N_f is the number of cycles to failure, and b is the fatigue strength exponent. This relationship is an empirical representation of the S–N curve above the fatigue limit in Figure 14.3. On a log–log plot, it gives a straight line of slope b.

The plastic strain component is better described by the Manson–Coffin relationship,[1]

$$\Delta\varepsilon_p/2 = \sigma_a/E = \varepsilon_f'(2N_f)^c,$$

where $\Delta\varepsilon_p/2$ is the amplitude of the plastic strain, ε_f' is the ductility coefficient in fatigue (equal to the strain intercept at $2N_f = 1$), $2N_f$ is the number of reversals to failure, and c is the ductility exponent in fatigue. On a log–log plot, the Manson–Coffin relation gives a straight line of slope c. It has been observed that a smaller value of c results in a longer fatigue life. In the regime of high-strain, low-cycle fatigue, the Manson–Coffin relation assumes great importance. Experimentally, it is frequently more convenient to control the total strain. In many structural components, the material in a critical place (say, at a notch root) may be subjected, essentially, to strain control conditions due to the elastic constraint of the surrounding material. For a material subjected to a total strain range of $\Delta\varepsilon_t$ (elastic and plastic strain), we can determine the fatigue strength by a superposition of the elastic and plastic strain components, i.e.,

$$\Delta\varepsilon_t/2 = \Delta\varepsilon_e/2 + \Delta\varepsilon_p/2 = (\sigma_f'/E)(2N_f)^b + \varepsilon_f'(2N_f)^c.$$

Thus, we expect that the curve of the fatigue life, in terms of total strain, will tend to the plastic curve at large total-strain amplitudes, whereas it will tend to the elastic curve at low total-strain amplitudes, as shown schematically in Figure 14.4. An example of such a behavior from a real material (an 18%-Ni maraging steel) is shown in Figure 14.5.

---

[1] L. F. Coffin, *Trans. ASME,* 76 (1954) 931; S. M. Manson and M. H. Hirschberg, in *Fatigue: An Interdisciplinary Approach* (Syracuse, NY: Syracuse University Press, 1964), p. 133.

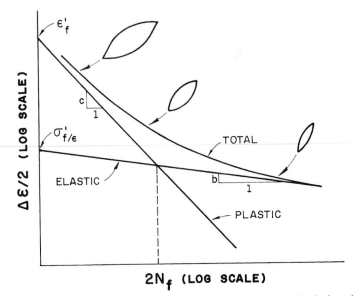

Figure 14.4 Superposition of elastic and plastic curves gives the fatigue life in terms of total strain. (Adapted with permission from R. W. Landgraf, in *ASTM STP* 467 (Philadelphia: ASTM, 1970), 3, p. 10)

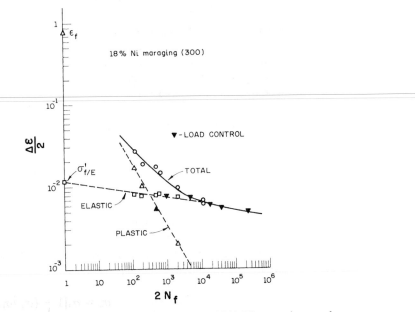

Figure 14.5 Fatigue life in terms of strain for an 18%-Ni maraging steel. (Adapted with permission from R. W. Landgraf, in *ASTM STP* 467, (Philadelphia: ASTM, 1970), 3, p. 25)

14.3 EFFECT OF MEAN STRESS ON FATIGUE LIFE

The mean stress σ_m can have an important effect on the fatigue strength of a material. A simple and crude way to demonstrate the effect of σ_m would be to present S–N curves of a given material for different values of σ_m on the same graph. Figure 14.6 shows such curves schematically. Note that, for a given stress amplitude σ_a, as the mean stress increases, the fatigue life decreases.

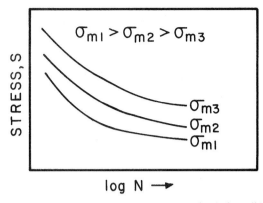

Figure 14.6 Effect of mean stress on S–N curves. The fatigue life decreases as the mean stress increases.

We can describe the effect of σ_m in a very simple manner. Suppose that the limiting value of any combination of stresses is σ_f, the monotonic true fracture stress. We can think of other arbitrary limits, such as the ultimate tensile stress σ_{UTS} or the yield stress σ_y, but σ_f is the maximum allowable true stress. Figure 14.7 shows a schematic plot of alternating stress σ_a, (or S) versus σ_m. Note that, for $\sigma_m = 0$, the alternating stress σ_a is a maximum and equal to σ_f. For $\sigma_a = \sigma_f$, the fatigue life is simply one-fourth of a cycle. For an ideal material, one would expect the relationship $\sigma_a + \sigma_m \leq \sigma_f$, the limiting value of any combination of stresses, to be valid. Thus, one can expect a straight line to join points A and B in Figure 14.7. Cyclic loading is not possible to the right of line AB. Then, in the presence of a mean stress of, say, $\sigma_f/3$, we will have a maximum allowable stress equal to $2\sigma_f/3$. Note that this description is an oversimplification of the real behavior of the material, inasmuch as it assumes that the damage produced in each cycle by cyclic plastic strain is independent and noncumulative.

Various empirical expressions have been proposed that take into account the effect of mean stress on fatigue life. Some of these are the following:

Goodman's relationship, which assumes a linear effect of mean stress between $\sigma_m = 0$ and σ_{UTS}:

$$\sigma_a = \sigma_0[1 - \sigma_m/\sigma_{UTS}].$$

Gerber's relationship, which assumes a parabolic effect of mean stress between $\sigma_m = 0$ and σ_{UTS}:

$$\sigma_a = \sigma_0[1 - (\sigma_m/\sigma_{UTS})^2].$$

Soderberg's relationship, which assumes a linear effect of mean stress between $\sigma_m = 0$ and σ_y:

$$\sigma_a = \sigma_0(1 - \sigma_m/\sigma_y).$$

In all these relationships, σ_m is the mean stress, σ_a is the fatigue strength in terms of stress amplitude when $\sigma_m \neq 0$, σ_0 is the fatigue strength in terms of stress amplitude when $\sigma_m = 0$, σ_{UTS} is the monotonic ultimate tensile strength, and σ_y is the monotonic yield stress. Figure 14.8 shows the three relations schematically. Experimentally, it has been observed that the great majority of data falls between the Gerber and Goodman lines. Thus,

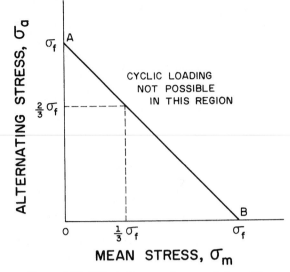

Figure 14.7 Effect of mean stress on fatigue life.

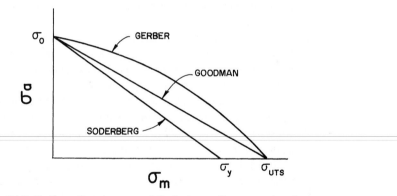

Figure 14.8 Gerber, Goodman, and Soderberg diagrams, showing mean stress effect on fatigue life.

the Goodman diagram represents a conservative estimate of the mean stress effect. Note that the three expressions involve uniaxial stresses. In most real-life situations, one encounters biaxial or triaxial situations. Hence, one needs to define stresses corresponding to σ_y or σ_{UTS} under multiaxial stress situations. A practical way around this is to use the concept of equivalent distortion energy—that is, to compute the distortion energy for uniaxial and multiaxial states and (a) assume that the cycling behavior of a material is equivalent when the material is cycled between two energy distortion values and (b) compute the maximum value of the mean stress by using von Mises yield criterion. (See Chapter 3.) Then we can find the failure conditions for a given multiaxial stress state.

14.4 CUMULATIVE DAMAGE AND LIFE EXHAUSTION

The discussion in the preceding sections was restricted to fatigue under simple conditions of constant amplitude, constant frequency, and so on. In real life, the service conditions are rarely so simple. Many components and structures are subject to a range of fluctuating loads, mean stress levels, and variable frequencies. Thus, it is of great importance to be able to predict the life of a component subjected to variable-amplitude conditions, starting from

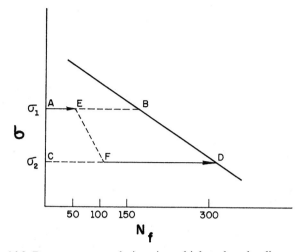

Figure 14.9 Damage accumulation, in a high-to-low loading sequence. (Adapted with permission from B. I. Sandor, *Fundamentals of Cyclic Stress and Strain* (Madison, WI: University of Wisconsin Press, 1972)

data obtained in simple constant-amplitude tests. The cumulative-damage theories attempt to do just that.

Basically, these theories consider fatigue to be a process of the material's accumulating damage until a certain maximum tolerable damage is reached. In other words, the phenomenon of fatigue is considered to be an exhaustion process of a material's inherent life (or ductility). A schematic fatigue life diagram, shown in Figure 14.9, elucidates the concept. At a constant stress of, say σ_1, the life of the material is 150 cycles, while at σ_2 it is 300 cycles. According to the cumulative-damage theory, in going from A to B or C to D, we gradually exhaust the material's fatigue life. That is, at points A and C, 100% of life at that level is available, while at points B and D, the respective lives are completely exhausted. If fatigue damage does, indeed, accumulate in a linear manner, each cycle contributes the same amount of damage at a given stress level. For example, on cycling the material from A to E, we exhaust one-third of the fatigue life available at σ_1. If we now change the stress level to σ_2, then the percentage of life already exhausted at σ_1 is equivalent to the percentage of life exhausted at σ_2. That is, one-third of fatigue life at σ_2 is equivalent to one-third of fatigue life at σ_1. Thus, in descending from E to F, we get from 50 to 100 cycles, and, as only one-third of fatigue life was exhausted at σ_1, two-thirds of fatigue life—that is, 200 cycles—is still available at σ_2. The same kind of change can be described for a low-to high-stress traversal.

This linear damage model does not concern itself with the physical picture of the fatigue damage. It does, however, give an empirical way of predicting the fatigue life after a complex loading sequence. The method is generally known as the *Palmgren–Miner rule* or, simply, *linear cumulative-damage theory*.[2] The Palmgren–Miner rule says that the sum of all life fractions is unity; that is,

$$\sum_{i=1}^{k} n_i/N_i = 1, \quad \text{or} \quad n_1/N_1 + n_2 N_2 + n_3/N_3 + n_4 N_4 + \ldots n_k/N_k = 1, \quad (14.1)$$

where k is the number of stress levels in the block spectrum loading; $N_1, N_2, \ldots, N_i$ are the fatigue lives corresponding to stress levels $\sigma_1, \sigma_2, \ldots \sigma_i$, respectively; and, $n_1, n_2, \ldots, n_i$ are

[2]A. Palmgren, *Z. Ver. Dtsch. Ing.,* 53 (1924) 339; M. A. Miner, *J. Appl. Mech.,* 12 (1945) 159.

the number of cycles carried out at the respective stress levels. This rule is obeyed by a series of materials if the underlying assumptions are satisfied. The principal assumption is that the damage accumulation rate at any level does not depend on the prior loading history of the material; in other words, the damage per cycle is the same at the beginning or at the end of fatigue life, at a given stress level. This implies that the magnitude and direction of the change in amplitude (from low to high or high to low) do not have an effect on fatigue life. We also assume that in each block the loading is totally reversible (i.e., $\sigma_m = 0$). The validity of these assumptions is problematic. For example, it is quite likely that, for blocks identical in size and amplitude, a change in load from high to low would be much more dangerous than one from low to high: A crack initiated at high loads can continue to grow at low loads, whereas in the reverse case, at low loads, perhaps the crack would not ever have formed.

EXAMPLE 14.1

The S–N curve of a material is described by the relationship

$$\log N = 10(1 - S/\sigma_{max}),$$

where N is the number of cycles to failure, S is the amplitude of the applied cyclic stress, and σ_{max} is the monotonic fracture strength—i.e., $S = \sigma_{max}$ at $N = 1$. A rotating component made of this material is subjected to 10^4 cycles at $S = 0.5\,\sigma_{max}$. If the cyclic load is now increased to $S = 0.75\,\sigma_{max}$, how many more cycles will the material withstand?

Solution:

For $S = 0.5\,\sigma_{max}$,

$$\log N_1 = 10\,(1 - 0.5) = 5.$$

Thus,

$$N_1 = 10^5 \text{ cycles.}$$

For $S = 0.75\,\sigma_{max}$,

$$\log N_2 = 2.5.$$

So

$$N_2 = 316 \text{ cycles.}$$

Using Palmgren–Miner's rule, we have

$$n_1/N_1 + n_2/N_2 = 10^4/10^5 + n_2/316 = 1$$

or

$$n_2 = 284 \text{ cycles.}$$

EXAMPLE 14.2

A microalloyed steel was subjected to two fatigue tests at ±400 MPa and ±250 MPa. Failure occurred after 2×10^4 and 1.2×10^6 cycles, respectively, at these two stress levels. Making appropriate assumptions, estimate the fatigue life at ±300 MPa of a part made from this steel that has already undergone 2.5×10^4 cycles at ±350 MPa.

Solution: We have

$$\Delta\sigma(N_f)^a = c,$$

$$800(2 \times 10^4)^a = 500(1.2 \times 10^6)^a,$$

$$\frac{800}{500} = 1.6 = \left(\frac{1.2 \times 10^6}{2 \times 10^4}\right)^a = (60)^a,$$

$$a = 0.115,$$

$$c = 800(2 \times 10^4)^{0.115} \text{ MPa}$$

$$= 2{,}498 \text{ MPa}.$$

At ±350 MPa,

$$N_{f1} = \left(\frac{c}{\Delta\sigma}\right)^{1/a} = \left(\frac{2{,}498}{700}\right)^{1/0.115}$$

$$= (3.57)^{8.7} = 6.4 \times 10^4 \text{ cycles.}$$

For 2.5×10^4 cycles,

$$\frac{N_1}{N_{f1}} = \frac{2.5 \times 10^4}{6.4 \times 10^4} = 0.39.$$

At ±300 MPa,

$$N_{f2} = \left(\frac{c}{\Delta\sigma}\right)^{1/a} = \left(\frac{2{,}498}{600}\right)^{1/0.115}$$

$$= (4.16)^{8.7} = 2.45 \times 10^5 \text{ cycles.}$$

From Palmgren–Miner's rule,

$$\frac{N_1}{N_{f1}} + \frac{N_2}{N_{f2}} = 1,$$

$$\frac{N_2}{N_{f2}} = 1 - \frac{N_1}{N_{f1}} = 1 - 0.39 = 0.61.$$

Therefore,

$$N_2 = 0.61 \times N_{f2}$$

$$= 0.61 \times 2.45 \times 10^5$$

$$= \underline{1.49 \times 10^5} \text{ cycles.}$$

14.5 *MECHANISMS OF FATIGUE*

In this section, we describe the physical mechanisms responsible for fatigue mechanisms—mainly, fatigue crack nucleation and propagation. We assume that our starting material does *not* have any preexisting crack or cracklike defects. The fracture mechanics approach, focusing only on the propagation of preexisting cracks, will be examined in Section 14.6.

14.5.1 Fatigue Crack Nucleation

Fatigue cracks nucleate at singularities or discontinuities in most materials. Discontinuities may be on the surface or in the interior of the material. The singularities can be structural (such as inclusions or second-phase particles) or geometrical (such as scratches or steps). The explanation of preferential nucleation of fatigue cracks at surfaces perhaps resides in the fact that plastic deformation is easier there and that slip steps form on the surface. Slip steps alone can be responsible for initiating cracks, or they can interact with existing structural or geometric defects to produce cracks. Surface singularities may be present from the beginning or may develop during cyclic deformation, as, for example, the formation of intrusions and extrusions at what are called the persistent slip bands (PSBs) in metals. These bands were first observed in copper and nickel by Thompson et al.[3] They appeared after cyclic deformation and *persisted* even after electropolishing. On retesting, slip bands appeared again in the same places. Later, the dislocation structure in the PSBs was investigated extensively. Figure 14.10a shows a TEM micrograph of a polycrystalline copper sample that was cycled to a total strain amplitude of 6.4×10^{-4} for 3×10^5 cycles. Fatigue cycling was carried out in reverse bending at room temperature and at a frequency of 17 Hz. The thin foil was taken 73 μm below the surface. Two parallel PSBs (diagonally across the micrograph) embedded in a veined structure in polycrystalline copper can be seen. The PSBs are clearly distinguished and consist of a series of parallel "hedges" (a ladder). These ladders are channels through which the dislocations move and produce intrusions and extrusions at the surface. Stacking-fault energy and the concomitant ease or difficulty of cross-slip play an important role in the development of the dislocation structure in the PSBs. Kuhlmann–Wilsdorf and Laird have discussed models for the formation of PSBs in metals.[4] They compared the deformation substructures produced by unidirectional and cyclic (fatigue) deformation and interpreted them in terms of the differences between the two modes of deformation. The principal differences are as follows:

1. Due to the much larger time spans of deformation in fatigue, the dislocation structures formed are much closer to the configurations having minimum energy than the ones generated by monotonic straining. That is, more stable dislocation arrays are observed after fatigue.
2. The oft-repeated to-and-fro motion in fatigue minimizes the buildup of surpluses of local Burgers vectors, which are fairly prevalent after unidirectional (monotonic) strain.
3. Much higher local dislocation densities are found in fatigued specimens.

The characteristic dislocation arrangements observed in FCC metals form in the following manner. In monocrystals, we first have uniform fine slip, followed by the formation of veins consisting of dense bundles of dislocation dipoles and other debris. After this,

[3]N. Thompson, N. J. Wadsworth, and N. Louat, *Phil. Mag.,* 1 (1956) 113.
[4]D. Kuhlmann-Wilsdorf and C. Laird, *Mater. Sci. Eng.,* 27 (1977) 137; and *Mater. Sci. Eng.,* 37 (1979) 111.

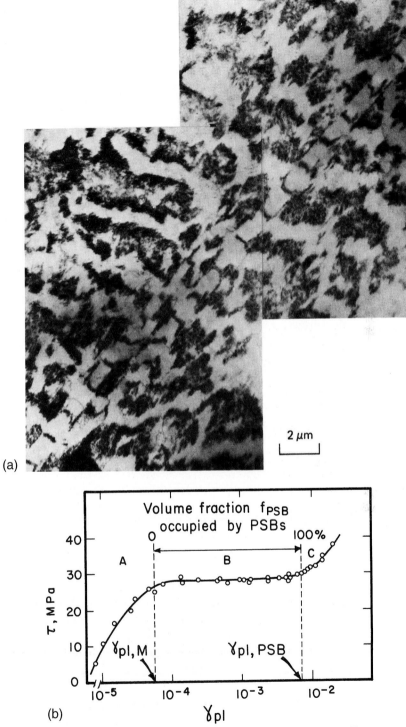

(a)

(b)

Figure 14.10 (a) Persistent slip bands in vein structure. Polycrystalline copper fatigued at a total strain amplitude of 6.4×10^{-4} for 3×10^5 cycles. Fatiguing carried out in reverse bending at room temperature and at a frequency of 17 Hz. The thin foil was taken 73 μm below the surface. (Courtesy of J. R. Weertman and H. Shirai) (b) Cyclic stress–strain curve for a single crystal of copper oriented for single slip (After H. Mughrabi, *Mater. Sci. & Eng.*, 33 (1978) 207)

PSBs are formed. They occur with the onset of saturation and are often associated with slight work-softening. There also seems to be a threshold strain for PSB formation, equal to 8×10^{-5} in the case of copper monocrystals. When subjected to strain-controlled cycling, an initially annealed metal hardens at first and then attains a saturation stress. If we plot this saturation stress against the applied plastic strain, we get another type of cyclic stress-strain curve, an example of which is shown for a single crystal of copper oriented for single slip in Figure 14.10b. The curve has three stages, one of which is a plateau region, and each stage is characterized by a distinct dislocation structure. At low strains in the plateau region, the structure consisted of a hard matrix containing a loop–patch dislocation structure and a soft PSB with dislocations in a ladderlike arrangement. At large strains in the plateau region (plastic shear strain greater than 2×10^{-3}), most of the matrix phase and a part of the PSB had a well-developed mazelike structure. In the case of a polycrystal, the grains in the softest orientation and with not much constraint from their neighbors deform and harden by the accumulation of dislocations. An example of such a structure in a Cu–Ni polycrystal is shown in Figure 14.11. Dislocation walls form on {100} planes.

The interface between the PSB and the matrix represents a discontinuity in the density and distribution of dislocations. Hence, one would expect PSBs to be the preferential sites for fatigue crack nucleation. The surface of a metal subjected to cyclic stressing will have PSB extrusions and intrusions. Recall that monotonic loading of a metal results in the formation of slip steps at the surface. On being subjected to cyclic loading, however, the surface of the metal will have intrusions and extrusions where PSBs emerge. A model for this form of nucleation is shown in Figure 14.12. During the loading part of the cycle, slip occurs on a favorably oriented plane, and during the unloading part of the cycle, reverse slip occurs on a parallel plane, because the slip on the original plane is inhibited due to hardening or, perhaps, the oxidation of the newly formed free surface. The first cyclic slip may create an extrusion or an intrusion at the surface. An intrusion may grow and form a crack by continued plastic deformation during subsequent cycles. An actual example of the formation of intrusions and extrusions in a sample of a copper sheet subjected to 15,000 cycles under a 60-MPa amplitude is shown in an SEM micrograph in Figure 14.13. Even during cyclic stressing in the tension–tension mode, this mechanism can function, as the plastic strain occurring at the peak load may lead to residual compressive stresses during the decreasing-load part of the cycle.

Twin boundaries can be important crack nucleation sites in hexagonal close-packed materials such as magnesium, titanium, etc., and their alloys. Inclusions and second-phase

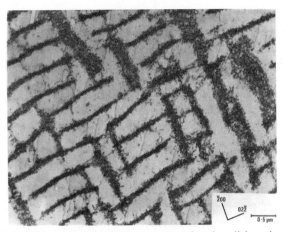

Figure 14.11 Well-developed maze structure, showing dislocation walls on {100} in Cu–Ni alloy fatigued to saturation. (From P. Charsley, *Mater. Sci. & Eng.*, 47 (1981) 181)

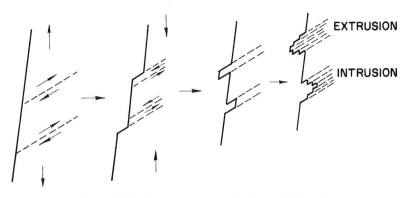

Figure 14.12 Fatigue crack nucleation at slip bands.

particles are commonly the dominant nucleation sites in materials of commercial purity—for example, aluminum, high-strength steels, and many polymers. Grain boundaries can become important nucleation sites at large strain amplitudes and at temperatures greater than about $0.5T_m$, where T_m is the melting point in kelvin, or in the presence of impurities that produce grain-boundary embrittlement (e.g., O_2 in iron). Some of these mechanisms are illustrated schematically in Figure 14.14.

Since most fatigue failures form at the surface of a material, the condition of the surface is very important. Indeed, polishing the surface can significantly increase the fatigue life of the material. A very important technological process to enhance fatigue life is *shot peening,* in which small metallic spheres are accelerated and hit the surface of the part. This bombardment by small particles puts a surface layer of the component in residual compression. This technique is used routinely in industry. Examples include compression coil springs in the automotive industry, wing skins for aircraft and other applications. Figure 14.15(a) shows, in a schematic fashion, a surface layer under residual compressive stress due to the cold-working from shot peening. The interior is under a small tensile stress as a result. The effect of shot peening on the endurance limits of steels with different ultimate tensile strengths is shown in Figure 14.15(b). For the as-forged components, the endurance limit is approximately 15% of the ultimate tensile strengths. Shot peening doubles the endurance limit, although it becomes less effective once the part is polished or ground, because the endurance limit is considerably increased by that process.

1 μm

Figure 14.13 SEM of extrusions and intrusions in a copper sheet. (Courtesy of M. Judelwicz and B. Ilschner)

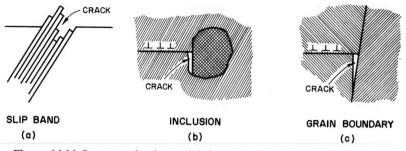

SLIP BAND   INCLUSION   GRAIN BOUNDARY
   (a)   (b)   (c)

Figure 14.14 Some mechanisms of fatigue crack nucleation. (After J. C. Grosskreutz, *Tech. Rep. AFML-TR-70-55,* (Wright–Patterson AFB, OH: Air Force Materials Laboratory), 1970)

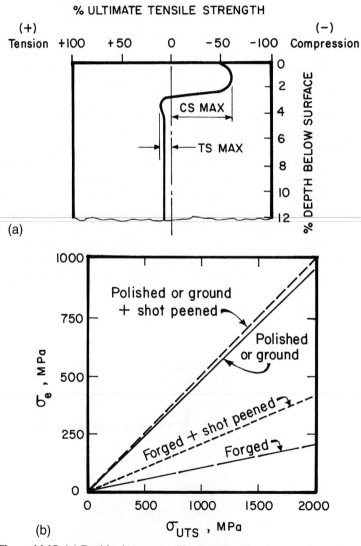

(a)

(b)

Figure 14.15 (a) Residual stress profile generated by shot peening of a surface. (b) Effect of shot peening on fatigue life of steels with different strength levels. (After Mann, 1967)

14.5.2 Fatigue Crack Propagation

At large stress amplitudes, a very large fraction (around 90%) of fatigue life is spent in the growth or propagation of a crack. For a component that contains a notch, this fraction becomes even larger. Inasmuch as in most real structures cracklike imperfections are present, the crack propagation part can be a very important aspect of fatigue.

A brief description of crack propagation, in terms of microstructural processes, follows. A few cracks nucleate at the surface and start propagating in a crystallographic shear mode (stage I) on planes oriented at approximately 45° to the stress axis. (See Figure 14.16.) During this stage, cracking occurs along the crystallographic slip planes, and the crack growth is on the order of a few micrometers or less per cycle. Little is known about crack propagation in this stage. Many consider the stage to be an extension of the nucleation process. Once a crack is initiated, say, at a slip band on the surface, it continues along the slip band until it encounters a grain boundary. These crystallographic cracks penetrate a few tenths of a millimeter in this mode. From there on, a dominant crack starts propagating in a direction perpendicular to the stress axis in the tensile mode. This is called stage II, and typically, the fracture surface shows striation markings. The ratio of the extent of stage I to stage II decreases with an increase in stress amplitude. The stress concentration at the tip of the crack causes local plastic deformation in a zone in front of the crack. With crack growth, the plastic zone increases in size until it becomes comparable to the thickness of the specimen. When this occurs, the plane-strain condition at the crack front in stage II does not exist any more, the crack plane undergoes a rotation, and the final part of rupture occurs in plane-stress or shear mode. Microscopic observations of fatigue fracture surfaces frequently show striations in stage II. Propagation occurs in a direction perpendicular to the tensile stress, and in a large number of metals and alloys (principally of Al and Cu), at high amplitudes, the fracture surface shows the characteristic striations. Such striations have been observed in polymers as well. Frequently, each striation is thought to represent one load cycle, and indeed, it has been observed by means of programmed amplitude fatigue tests that in many materials these striations do represent the crack front position in each cycle. An example of the variation in fatigue striations in a 2014-T6 aluminum alloy is presented in Figure 14.17. Figure 14.17a shows the striations in the early stage of fatigue life, while Figure 14.17b shows the striations in the late stage of cycling. Note the smaller striation spacing in the early stage, indicating a lower crack propagation rate than in the late stage. These micrographs were taken in a

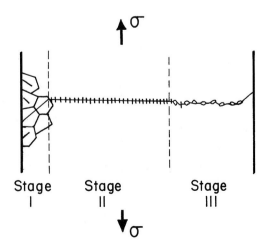

Figure 14.16 Stages I, II, and III of fatigue crack propagation.

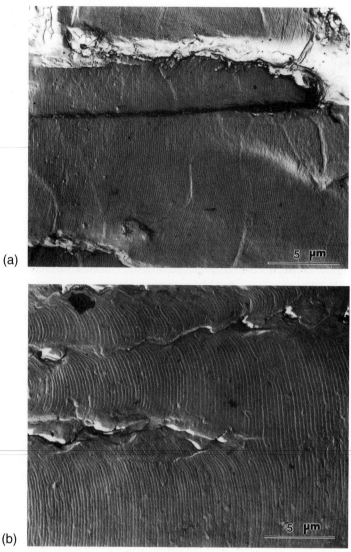

(a)

(b)

Figure 14.17 Fatigue striations in 2014-T6 aluminum alloy; two-stage carbon replica viewed in TEM. (a) Early stage. (b) Late stage. (Courtesy of J. Lankford)

TEM from a two-stage carbon replica of the surface of the aluminum alloy. The reason for this is that only TEM could provide the high-enough magnification for viewing the closely spaced striations in the early stage of fatigue. However, the reader is warned that such a correlation is not always available. If it were, one should be able to relate striation spacing to ΔK (see Section 14.6) and obtain a one-to-one correspondence between the macroscopic growth rate and ΔK. One cannot always do this, however, indicating that the crack front may have advanced by a combination of the formation of striations and other fracture mechanisms.

Care should be exercised in the interpretation of fatigue striations. It has been observed in an Fe–Si alloy that, whereas the fatigue striations were 2 μm apart, the actual advance of the crack front per cycle was only 10^{-9} m, or 2,000 times smaller![5] These results

[5]W. Yu, K. Esablul, and W. W. Gerberich, *Met. Trans.,* 15A (1984) 889.

show that, under certain conditions, the crack front remains "dormant" for many cycles, while damage accumulates in the material. At a certain point, the crack advances discontinuously. This phenomenon is very common in polymers at low values of ΔK. A craze forms gradually at the tip of the crack during fatigue. When the craze reaches a critical length, the crack advances through it. The process repeats itself periodically.

At higher values of ΔK, striations become less important in the overall crack propagation rate. One model of fatigue crack growth by a striation mechanism is shown in Figure 14.18. This model involves repetitive blunting and sharpening of the crack front. Figure 14.18a shows the situation at zero load. During the tensile part of the load cycle, plastic strains at the crack tip cause localized slip on planes of maximum shear. (See Figure 14.18b.) The situation at maximum tensile load is shown in Figure 14.18c. The start of the compressive cycle is shown in Figure 14.18d. The reversal of the loading direction, during compression, causes the crack faces to join (Figure 14.18e). However, the new surface created during the tensile part of the cycle is not completely "rehealed," due to slip in the reverse direction. Depending on the material and the environment, a large part of slip during compression occurs on new slip planes, and the crack tip assumes a bent form with "ears," as shown in Figure 14.18e. At the end of the compression half of the cycle, the crack tip is resharpened, and the propagation sequence of the next cycle is restarted. (See Figure 14.18f.) This model of plastic blunting and resharpening seems to be valid for any ductile material, including polymers. There is evidence that the crack propagates in a similar manner in stage I, but with only a group of slip planes at 45° operating. However, one must bear in mind that, although the presence of striations confirms a fatigue failure mechanism, an absence of striations does not necessarily preclude fatigue. In fact, a variety of other fracture modes are possible in fatigue. In single-phase materials, transgranular or intergranular fracture modes are possible, while any second phases that are present may lead to dual fracture modes. Figure 14.19 shows the various possible microscopic fracture modes in fatigue.

Fracture surfaces in polymers produced under fatigue conditions also show some characteristic features. Generally, two distinct regions are present: a region of smooth, slow crack growth surrounding the fracture initiation site and a rough region corresponding to

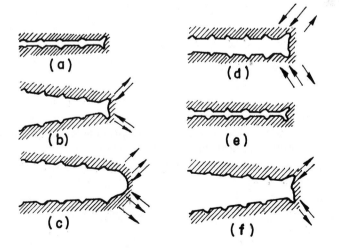

Figure 14.18 Fatigue crack growth by a plastic blunting mechanism. (a) Zero load. (b) Small tensile load. (c) Maximum tensile load. (d) Small compressive load. (e) Maximum compressive load. (f) Small tensile load. The loading axis is vertical (After C. Laird, in *Fatigue Crack Propagation*, ASTM STP 415, (Philadelphia: ASTM, 1967), 131.)

rapid crack growth. Sometimes, semicircular concentric bands are seen in the smooth region near the starting flaw. These bands are indicative of discontinuous crack growth, and the band width represents the extent of plastic zone or craze that developed ahead of the crack tip. Fracture surfaces produced by cyclic loading in polymers frequently show striations. There is some confusion on the use of the term *striation* in the literature on fatigue in polymers and metals. In metals, the term is used to denote markings on the fracture surface, without regard to any correlation between the striation spacing and the crack growth per cycle. In polymers, the term *striation* is used only when there is a one-to-one correlation between the striation spacing and the crack growth in each cycle. This stems from the fact that there are other types of discontinuous growth which result in fracture surface markings, but without any advance in the fatigue crack. In particular, in polymers there occur discontinuous growth bands (DGBs), which correspond to a burst of fatigue crack growth after some hundreds of fatigue cycles; that is, the crack tip remains stationary for some cycles and then undergoes an advance. DGBs resemble striations, but their spacing is much larger than the crack growth in a cycle. The formation of a DGB is thought to be due to the accumulation of damage ahead of the fatigue crack over many cycles, followed by a sudden jump by the crack. One model explains that a craze forms at the fatigue crack and that the crack initially grows along the craze–matrix interface, as shown in Figure 14.20, and then along the craze filled midrib until the crack is arrested. A repetition of this process results

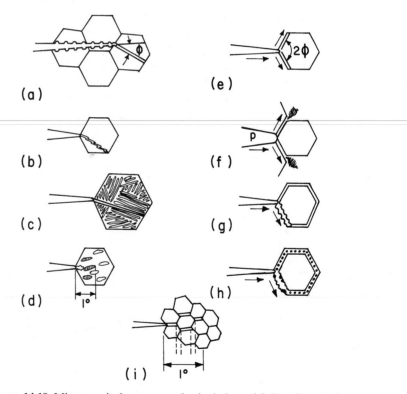

Figure 14.19 Microscopic fracture modes in fatigue. (a) Ductile striations triggering cleavage. (b) Cyclic cleavage. (c) α–β interface fracture. (d) Cleavage in an α–β phase field. (e) Forked intergranular cracks in a hard matrix. (f) Forked intergranular cracks in a soft matrix. (g) Ductile intergranular striations. (h) Particle-nucleated ductile intergranular voids. (i) Discontinuous intergranular facets. (Adapted from W. W. Gerberich and N. R. Moody, in *Fatigue Mechanisms*, ASTM STP 675, (Philadelphia: ASTM, 1979) 292.)

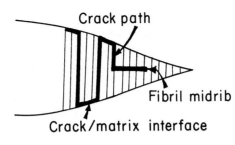

Figure 14.20 Discontinuous crack growth through a craze at the tip of a fatigue crack. (After L. Konczol, M. G. Schincker and W. Döll, *J. Mater. Sci.*, 19 (1984) 1604.)

in the appearance of dark bands on the fracture surface. Thus, the DGBs represent the successive positions of crack tips that have been blunted, advanced, and arrested repeatedly.

14.6 LINEAR ELASTIC FRACTURE MECHANICS APPLIED TO FATIGUE

The use of large monolithic structures has resulted in widespread application of fracture mechanics. In particular, the phenomenon of fatigue crack propagation can be analyzed in terms of linear elastic fracture mechanics. The basic assumption here is that cracks already exist in a structural component and that they will grow as the component gets used in service. In terms of fatigue crack growth studies, it is also implied that the fatigue life of a component is determined mostly by the crack growth under cyclic loading.

We can determine K_{Ic} or K_c for a given material in the laboratory and can use the data obtained to determine a failure locus in terms of a critical applied stress and a corresponding critical crack length, or vice versa. (See Figure 14.21.) For example, in Figure 14.21a, we can observe that, for a given crack length a_1, there is a critical failure stress σ_1

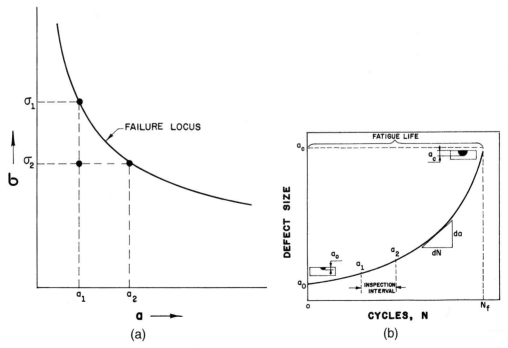

Figure 14.21 (a) Failure locus. (b) Schematic of crack length a as a function of number of cycles, N.

of the material. Conversely, for a given design stress σ_2, there is a critical crack length a_2. In principle, then, the region under the failure locus represents the safe region with respect to a catastrophic failure. Consider, for instance, a component containing a crack length of a_1, at a stress of σ_2, where $\sigma_2 < \sigma_1$. Under these conditions, the component will be safe because a_1 is smaller than the critical defect size a_2, which corresponds to the applied stress σ_2. This security is based, of course, on the assumption that loading is static and that the crack does not grow in service. But we know very well that cracks in structures do grow during service. An increase in crack length at σ_2, in service, from a_1 to a_2 will eventually lead to structural failure. Thus, although the fracture toughness of a material establishes the failure condition and the residual strength of a structural component, the component's service life or durability is mainly a function of its resistance to subcritical crack growth (i.e., its resistance to crack growth by fatigue, creep, stress corrosion, etc.).

As we pointed out in Chapter 7, linear elastic fracture mechanics accepts the pre-existence of cracks in a structural member. The model for the crack tip is the same as that described for nonfatigue regimes. The material containing a crack, under tension, has a small plastic zone at the crack tip, and this plastic zone is surrounded by a rather large elastic region. This being so, we focus our attention on the propagation of cracks under conditions of fatigue. Once again—and it is worth repeating—we do not concern ourselves here with the crack nucleation problem under fatigue. Under cyclic loading, a dominant crack grows, as a function of the number of cycles, from an initial size a_0 to a critical size a_c, corresponding to failure, as shown in Figure 14.21b. The basic problem is thus reduced to one of characterizing the growth kinetics of the dominant crack in terms of an appropriate driving force. From there, one can estimate the service life and/or schedule inspection intervals under designed loading conditions and service environments. Since crack growth starts from the most highly stressed region at the crack tip, we characterize the driving force in terms of the stress intensity factors at the tip—that is, the range of the stress intensity factor $\Delta K = K_{max} - K_{min}$, where K_{max} and K_{min} are the maximum and minimum stress intensity factors corresponding to the maximum and minimum loads, respectively. The crack growth rate per cycle, da/dN, can then be expressed as a function of the cyclic stress intensity factor at the crack tip, ΔK. Hence, if a mathematical equation describing the crack growth process and the appropriate boundary conditions is available, we can, in principle, compute the fatigue life (i.e., number of cycles to failure). Paris et al. proposed the following *empirical* relationship (known as the Paris–Erdogan relationship) for crack growth under cyclic conditions:[6]

$$da/dN = C(\Delta K)^m. \tag{14.2}$$

Here, a is the crack length, N is the number of cycles, ΔK is the cyclic stress intensity factor as defined earlier, and C and m are empirical constants that depend on the material, environment, and test conditions, such as the load ratio R, the test temperature, the waveform, etc. Another empirical relation connecting the parameters C and m is

$$C = A/(\Delta K_0)^m$$

where A and K_0 are some other material constants.

Since many variables affect the crack growth rate in fatigue, we can write, in a very general way,

$$da/dN \approx \Delta K/\Delta N = F(\Delta K, K_{max}, R, \text{frequency, temperature}, \ldots). \tag{14.3}$$

---

[6]P. C. Paris, M. P. Gomez, and W. P. Anderson, *The Trend in Engineering,* 13 (1961) 9; P. C. Paris and F. Erdogan, *J. Basic Eng., Trans. ASME,* 85 (1963) 528.

Clearly, one cannot obtain such an ideal and detailed characterization. In practice, one collects data under restricted conditions, but consistent with the applications in service. In principle, the rate equation 14.2 can be integrated to determine the service life N_f, or an appropriate inspection interval ΔN, for a structural component. We have

$$N_f = \int_{a_0}^{a_f} \frac{da}{F(\Delta K, \ldots)},$$

or

$$\Delta N = N_2 - N_1 = \int_{a_1}^{a_2} \frac{da}{F(\Delta K, \ldots)}. \tag{14.4}$$

Rewriting, we get

$$N_f = \int_{K_{f\max}}^{K_{f\max}} \frac{dK}{(dK/da)F(\Delta K, \ldots)}, \tag{14.5}$$

or

$$\Delta N = N_2 - N_1 = \int_{K_1}^{K_2} \frac{dK}{(dK/da)F(\Delta K, \ldots)}. \tag{14.6}$$

If we plot the logarithm of the crack growth rate da/dN against the logarithm of the alternating stress intensity factor $\Delta K = K_{\max} - K_{\min}$ at the crack tip, we get the kind of curve shown in Figure 14.22. The curve has a sigmoidal form with three regions. Region II is the one that shows the Paris–Erdogan type of power-law relation between da/dN and ΔK. The power-law region connects the upper and lower limiting regions. The lower limit on the cyclic stress intensity factor in region I denotes a threshold value below which the crack does not propagate. This limit is called the *threshold cyclic stress intensity factor* ΔK_{th}. The upper limit in region III indicates the conditions of accelerated crack growth rate associated with the start of final rupture.

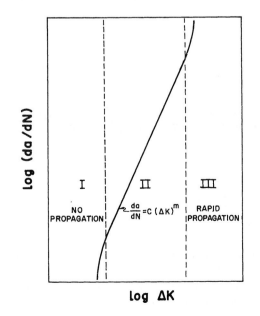

Figure 14.22 Schematic of crack propagation rate da/dN versus alternating stress intensity factor ΔK.

Many researchers (see, for example, Suresh (1991) in the Suggested Readings) have discussed the primary operating mechanisms and the important variables in the three stages of fatigue crack propagation:

Stage I: In this stage, the average crack growth per cycle is less than a lattice spacing. Crack propagation mechanisms are characteristic of a discontinuous medium. The microstructure of the material, the stress ratio R, and the environment have a large influence on the crack growth.

Stage II: This is the power-law regime, where the Paris–Erdogan relationship applies. Crack propagation mechanisms in stage II are characteristic of a continuous medium. The influence of the microstructure, R, the environment, the thickness of the material, etc. on crack growth is small.

Stage III: Crack propagation mechanisms in this stage are similar to those in the static mode (cleavage, intergranular, microvoid coalescence, etc.) In stage III, the microstructure, R, and the thickness of the material have a large influence on crack growth, but the influence of the environment is small.

The Paris–Erdogan power relationship (Equation 14.2) describes the crack propagation rate in stage II for a variety of materials–polymers, metals, and ceramics. It is very useful because of its extreme simplicity. For example, it has been observed experimentally that data points in the form of log (da/dN) versus log ΔK for a given material (with a constant microstructure) from three different samples—an edge crack in a compact-tension sample, a through-the-thickness central crack in a plate, and a plate containing a partial through-the-thickness crack—all fall on the same line. Also, there is experimental evidence that the stress level by itself does not influence the fatigue crack growth rate for levels below the general yielding stress. Thus, we can assume that the parameter ΔK describes uniquely the crack growth rates for many engineering applications. However, rather gross microstructural features of a material, such as the directionality imparted by aligned inclusions, can influence fatigue crack growth rates drastically, changing the value of m significantly. Figure 14.23 illustrates the directionality in the fatigue crack propagation rate in an

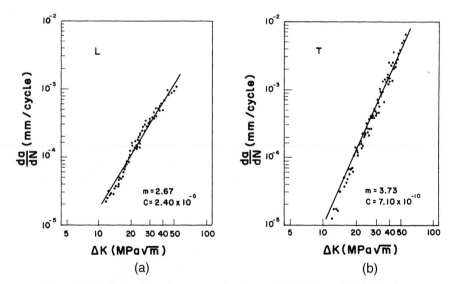

Figure 14.23 Fatigue crack propagation in an AISI 4140 steel. (a) Longitudinal direction (parallel to rolling direction). (b) Transverse direction (perpendicular to rolling direction). (Reprinted with permission from E. G. T. De Simone, K. K. Chawla, and J. C. Miguez Suárez, *Proc. 4th CBECIMAT* (Florianópolis, Brazil, 1980), p. 345)

AISI 4140 steel. The exponent m has a higher value in the transverse direction than in the longitudinal (rolling) direction, due to the presence of elongated inclusions.

EXAMPLE 14.3

Consider long crack propagation under fatigue. Develop an expression for the number of cycles, ΔN, required for a crack to grow from an initial length a_i to a final length a_f, Given that $\Delta K = Y \Delta\sigma\sqrt{\pi a}$ and $da/dN = C \Delta K^m$, where the symbols have their usual significance. Discuss the implications of the expression.

Solution:

$$da/dN = C \Delta K^m,$$

$$\Delta N = \int_{a_i}^{a_f} \frac{da}{C \Delta K^m}$$

$$= \int_{a_i}^{a_f} \frac{da}{C(Y \Delta\sigma\sqrt{\pi})^m a^{m/2}}$$

$$= \frac{a_f^{(1-m/2)} - a_i^{(1-m/2)}}{C(Y \Delta\sigma\sqrt{\pi})^m (1 - m/2)}$$

$$= \frac{1 - (a_i/a_f)^{m/2-1}}{C(Y \Delta\sigma\sqrt{\pi})^m (m/2 - 1)} \left(\frac{1}{a_i^{m/2} - 1} \right).$$

The implications of the expression are that it is not valid for $m = 2$ and that ΔN is more sensitive to the initial crack length a_i than the final crack length a_f.

EXAMPLE 14.4

The fatigue crack markings shown in Figure E14.4 were found in a fractured part. Determine the time to rupture of this part if the loading frequency

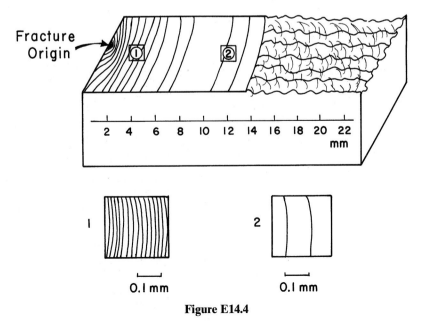

Figure E14.4

is 10 Hz, the maximum stress applied to the part is 300 MPa, and the minimum stress is zero. The initiation stage of the flaw is 50 percent of the life of the part.

Solution: We have

$$\left(\frac{da}{dN}\right)_1 = 0.02 \text{ mm}, \qquad \left(\frac{da}{dN}\right)_2 = 0.1 \text{ mm},$$

$$a_1 = 2 \text{ mm}, \qquad a_2 = 10 \text{ mm}.$$

Fracture occurs when $a_f = 14$ mm. Assuming that we have plane strain,

$$K_{Ic} = 1.12\sigma\sqrt{\pi a}$$
$$= 1.12 \times 300\sqrt{\pi \times 0.014}$$
$$= 70 \text{ MPa m}^{1/2}.$$

We now find the parameters for the Paris equation:

$$\frac{da}{dN} = C(\Delta K)^4,$$

$$\Delta K_1 = 1.12\sigma\sqrt{\pi a_1} = 1.12 \times 300\sqrt{\pi \times 2 \times 10^{-3}} = 26.6 \text{ MPa m}^{1/2},$$

$$\Delta K_2 = 1.12\sigma\sqrt{\pi a_2} = 1.12 \times 300\sqrt{\pi \times 10 \times 10^{-3}} = 59.55 \text{ MPa m}^{1/2},$$

$$0.02 \times 10^{-3} = C(26.6)^m,$$

$$0.1 \times 10^{-3} = C(59.55)^m,$$

$$m = 1.997,$$

$$C = 2.8 \times 10^{-8} \text{ MPa}^{-1},$$

$$\frac{da}{dN} = 2.8 \times 10^{-8}(\Delta K)^2 = 2.8 \times 10^{-8} \times 1.12 \times 300\sqrt{\pi a},$$

$$\frac{da}{a^{1/2}} = 1.67 \times 10^{-5} \, dN.$$

Integrating, we obtain

$$\int_0^{a_f} \frac{da}{a^{1/2}} = 1.67 \times 10^{-5} \int dN = 1.67 \times 10^{-5} \, N,$$

$$2a_f^{1/2} = 1.67 \times 10^{-5} \, N.$$

Each cycle corresponds to 0.1s, so

$$t = 0.1 \, N$$

$$= \frac{2 \times (14 \times 10^{-3})^{1/2}}{1.67 \times 10^{-5} \times 0.1} = 1.4 \times 10^5 \text{ s}.$$

The total time is equal to the initiation time plus the propagation time:

$$t = 2t = 2.8 \times 10^5 \text{ s}$$

$$= 77.7 \text{ hours.}$$

EXAMPLE 14.5

An aluminum alloy has a plane-strain fracture toughness K_{Ic} of 50 MPa $\sqrt{\text{m}}$. A service engineer has detected a 1-mm-long crack in an automotive component made of this alloy. The component will be subjected to cyclic fatigue with $\Delta\sigma = 100$ MPa with $R = 0$. How many more cycles can this component endure? Take $K = 1.05\,\sigma\sqrt{\pi a}$ and $da/dN = 1.5 \times 10^{-24}\,\Delta K$.

Solution: The final crack length a_f can be obtained from K_{Ic} with 100 MPa of applied stress:

$$K_{Ic} = 50 = 1.05 \times 100 \times \sqrt{\pi a_f},$$

$$a_f = \left[\frac{50}{1.05 \times 100\sqrt{\pi}}\right]^2 \text{ m}$$

$$= 0.072 \text{ m}$$

$$= 72 \text{ mm.}$$

We thus have

$$N_f = \frac{1}{(1.5 \times 10^{-35})(1.05^4)\pi^2(100 \times 10^6)^4}\left[\frac{1}{a_i} - \frac{1}{a_f}\right]$$

$$= \frac{100}{(1.5)(1.22)(9.87)}\left[1 - \frac{1}{72}\right]$$

$$= 10^4\,[0.055][0.986]$$

$$= 546 \text{ cycles.}$$

EXAMPLE 14.6

When subjected to fatigue under a $\Delta\sigma = 140$ MPa, an alloy showed the following Paris-type fatigue crack propagation relationship:

$$\frac{da}{dN}\text{ (m/cycle)} = 0.66 \times 10^{-8}(\Delta K)^{2.25}$$

where ΔK is in MPa$\sqrt{\text{m}}$. Estimate the number of cycles required for the crack to grow from 2 mm to 8.8 mm.

Solution:

$$\frac{da}{dN} = 0.66 \times 10^{-8}(1.12\,\Delta\sigma\sqrt{\pi a})^{2.25},$$

$$a^{-1.125}da = 0.66 \times 10^{-8} \times (1.12)^{2.25} \times (140)^{2.25}(\pi)^{1.25}dN.$$

Integrating, we get

$$\int_{a_0}^{a_c} a^{-1.25}\,da = 2.0815 \times 10^{-3} \int_0^{N_f} dN,$$

$$N_f \times 2.0815 \times 10^{-3} = -\left[\frac{a^{-0.125}}{0.125}\right]_{0.002}^{0.088},$$

$$N_f = -\frac{\dfrac{1.3548 - 2.1745}{0.125}}{2.0815 \times 10^{-3}} = 3.15042 \times 10^3 \text{ cycles,}$$

or

$$N_f = 3{,}150 \text{ cycles.}$$

A Paris–Erdogan type of relationship can be used to describe the fatigue crack propagation rate da/dN in polymers also. Figure 14.24 shows the fatigue crack rates as a function of ΔK for a number of thermoplastic polymers. Note that semicrystalline nylon 66 is superior in resistance to fatigue crack growth than amorphous, but ductile, polymers such as polyvinyl chloride (PVC) and polycarbonate (PC), which in turn are superior to brittle, amorphous polymers such as PS and PMMA. The latter both show deformation by crazing. Metals such as aluminum alloy and steel (not shown in the figure) would have curves to the right of that of nylon. That is, polymers show a lower resistance to fatigue crack propagation than metals do: Unlike metals, the range of the exponent m for polymeric materials can be quite large, from 4 to 20.

The molecular weight of a polymer is a very important variable for a number of properties, including fatigue crack propagation. In general, as the molecular weight increases, the fatigue strength increases and the fatigue crack propagation rate decreases. Figure 14.25 shows the variation in the fatigue crack propagation rate in PMMA and PVC, at a

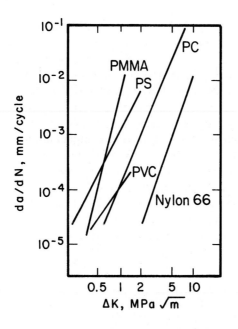

Figure 14.24 Fatigue crack propagation rates for a number of polymers. (After R. W. Hertzberg, J. A. Manson, and M. Skibo, *Polymer Eng. Sci.*, 15 (1975) 252)

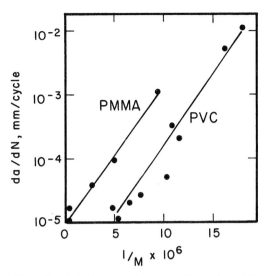

Figure 14.25 Variation in fatigue crack propagation rates, at fixed values of ΔK (= 0.6 MPa$\sqrt{m}$) and test frequency ν_c (= 10 Hz), as a function of reciprocal of molecular weight for PMMA and PVC. (After S. L. Kim, M. Skibo, J. A. Manson, and R. W. Hertzberg, *Polymer Eng. Sci.,* 17 (1977) 194)

constant value of ΔK (0.6 MPa$\sqrt{m}$) and at 10 Hz, as a function of $1/M$, where M is the molecular weight.

Earlier, it was thought that cyclic fatigue in ceramics did not occur, at least not in an inert atmosphere. This was based on the fact that, in a ceramic, no dislocation-based plastic deformation occurred at the crack tip. Although dislocation plasticity is generally absent in ceramics, many ceramics show subcritical crack growth, under cyclic loading, at room temperature and at elevated temperatures. The deformation mechanisms under cyclic loading are generally different from those under static loading. While dislocation-based cyclic slip is responsible for fatigue in metals, phenomena such as microcracking, phase transformations, interfacial sliding, and creep can promote an inelastic constitutive response in brittle solids, leading to cycle fatigue. Work done by Suresh and coworkers, as well as others, on fatigue crack growth in a variety of brittle solids in compression, tension, and tension–compression fatigue shows that mechanical fatigue effects—that is, stable crack propagation—under cyclic fatigue conditions—can occur in ceramics at room temperature and in brittle solids as well.[7] Ewart and Suresh were the first ones to show such a cyclic fatigue effect in ceramics by subjecting them to cyclic compression. Researchers have used a variety of loading techniques to obtain fatigue crack growth data in ceramics under cyclic loading, such as four-point flexure, compact tension, and wedge-opening load specimens. Figure 14.26 shows the fatigue crack growth in an alumina sample (grain size $\approx 10\mu m$) subjected to tension–compression fatigue ($R = -1$) at a frequency of 5 Hz, in terms of da/dN vs. K_{max}, the maximum stress intensity factor. If we take $\Delta K = K_{max}$, it is easy to see that the data correspond to a Paris–Erdogan type of power law, $da/dN = C\Delta K^m$. The figure also shows the data obtained from static loading in terms of crack growth per cycle. We use the relationship $da/dN = (da/dt)/\nu$, where ν is the frequency, and plot this against the maximum stress intensity factor K_{max} at which the static test was

[7]L. Ewart and S. Suresh, *J. Mater. Sci.* 22 (1987) 1173; S. Suresh and J. R. Brockenbrough, *Acta Met.,* 36 (1988) 1455; M. J. Reece, F. Guiu, and M. F. R. Sammur, *J. Amer. Ceram. Soc.,* 72 (1989) 348; R. H. Dauskardt, D. B. Marshall, and R. O. Ritchie, *J. Amer. Ceram. Soc.* 73 (1990) 893.

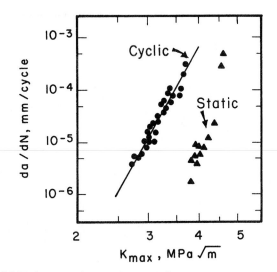

Figure 14.26 Fatigue crack growth rate da/dN in alumina as a function of the maximum stress intensity factor K_{max} under fully reversed cyclic loads ($\nu = 5$ Hz). Also indicated are the rates of crack growth per cycle derivedc from static-load fracture data. (After M. J. Reece, F. Guiu, and M. F. R. Sammur, *J. Amer. Ceram. Soc.,* 72 (1989) 348)

performed. The idea of putting the two curves together is to show that the crack growth rate under cyclic loading is much faster than that under static loading. According to a compilation by Suresh (see same in Suggested Readings, p. 450), the value of the exponent m for ceramics varies from 8 to 42. Dauskardt et al. studied the fatigue crack growth behavior of zirconia partially stabilized with magnesia after four different heat treatments.[8] Their results indicated that the crack growth rate followed the Paris–Erdogan type of power law, with values of m ranging from 21 to 42, much higher than for metals (2–4).

14.7 HYSTERETIC HEATING IN FATIGUE

An important aspect of fatigue behavior has to do with hysteretic heating. During each loading cycle, we get a hysteresis loop. The area of the loop represents the energy spent in the cycle. Most of the hysteretic energy is dissipated as heat, and if the material is not a good thermal conductor, as is the case with most polymers, then a temperature rise can occur. Such hysteretic heating and the resultant thermal softening can be important at high strain rates in most insulators. In extreme cases, the temperature rise can be large enough that the material fails by viscous flow or melting. Even room temperature can be quite high for polymers. It is convenient to examine the temperature effects in terms of the homologous temperature T_H. Thermally activated phenomena become operative in most materials at $T_H > 0.4$–0.5. Consider a thermoplastic polymer with a melting point of 300°C. Taking room temperature to be 300 K, we find that T_H for this polymer is $300/573 = 0.52$. Thus, even a small increase in temperature above the ambient temperature will put such a polymer in the regime where significant thermal softening can occur. Thermoplastics commonly have low thermal diffusivity and show nonlinear viscosity. The degree of thermal softening during fatigue will depend on the magnitude and frequency of the applied stress and on the viscoelastic characteristics of the polymer. It is easy to see that an appreciable

[8]R. H. Dauskardt, D. B. Marshall, and R. O. Ritchie, *J. Amer. Ceram. Soc.,* 73 (1990) 893.

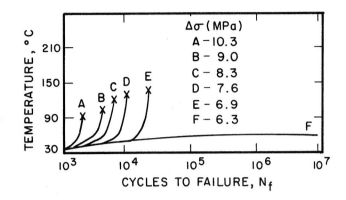

Figure 14.27 Effect of the applied stress range $\Delta\sigma$ on temperature rise in PTFE subjected to stress-controlled fatigue. The symbol x denotes failure of the specimen. (After M. N. Riddell, G. P. Koo, and J. L. O'Toole, *Polymer Eng. Sci.* 6 (1966))

rise in temperature can result at high stress amplitudes or frequencies. This can lead to a reduced fatigue life at frequencies greater than 10 Hz at room temperature. In fact, the ASTM standard specification D-671-71 calls for the measurement of the temperature at fatigue failure. If the temperature goes above the glass transition temperature of the polymer, thermal softening, fracture, and resolidification of the fractured material can occur.

The hysteretic temperature rise is a function of the dimensions of the specimen also. In a thinner specimen, a greater proportion of the heat generated will be lost to the environment. A thicker specimen, on the other hand, will retain a larger fraction of heat and thus show a lower fatigue endurance limit than a thinner specimen of the same material. Figure 14.27 shows the effect of applied stress on the temperature rise in polytetrafluoroethylene (PTFE) subjected to cycling at 30 Hz, at room temperature under stress control. The endurance limit $\Delta\sigma_L$ for this material under these conditions is 6.5 MPa. For $\Delta\sigma > \Delta\sigma_L$, indicated by the curves marked *A, B, C, D,* and *E* in the figure, a rapid increase in temperature occurred with an increasing number of cycles. For $\Delta\sigma < \Delta\sigma_L$, represented by curve *F* in the figure, the temperature rise was not high enough to cause thermal softening.

Hysteretic heating effects during high-frequency fatigue have also been observed in continuous fiber-reinforced ceramic matrix composites.[9] Unlike the heating effect observed in polymers or polymer matrix composites, the origin of heating in ceramic matrix composites is the frictional sliding between two mating surfaces, such as a fiber/matrix interface or an interlaminar shear.

14.8 FATIGUE CRACK CLOSURE

Under certain circumstances, surfaces of a fatigue crack can contact each other, and the crack will close even when the far-field stress field is still tensile. The crack does not reopen until a sufficiently high tensile stress is reached in the next loading cycle. This phenomenon, called *crack closure,* was said by a number of researchers to occur as result of crack-tip plasticity. As the applied stress on a material is increased, a plastic zone develops at the crack tip. (See chapter 7.) As the crack grows, a plastically deformed zone is produced in its wake, while the material surrounding this zone is still elastic. The explanation of this

[9]See, for example, J. W. Holmes and C. Cho, *J. Amer. Ceram. Soc.,* 75 (1992) 929; N. Chawla, Y. K. Tur, J. W. Holmes, J. R. Barber, and A. Szweda, *J. Amer. Ceram. Soc.,* 81 (1998) 1221.

phenomenon was that the plastically deformed zone caused the crack surfaces to close before zero stress was reached. However, for fatigue crack growth to occur, the crack must be fully open. Thus, premature contact between the crack surfaces—i.e., crack closure—results in a lowering of the crack driving force. It follows that one should use an *effective* stress intensity factor range, ΔK_{eff}, rather than ΔK in fatigue crack growth analysis. If the stress at which the crack is just open is σ_{op}, and the corresponding stress intensity factor is K_{op}, then we can define the effective cyclic stress intensity factor as

$$\Delta K_{eff} = K_{max} - K_{op}.$$

Recall that the applied cyclic stress intensity factor is given by $\Delta K = K_{max} - K_{min}$ and that $K_{op} > K_{min}$. Therefore, we will have

$$\Delta K > \Delta K_{eff}.$$

Elber proposed that ΔK_{eff} explains the R effect on the fatigue crack growth rate.[10] At high values of R, the crack closure effect is small because K_{op} approaches K_{min}, and ΔK_{eff} becomes closer to ΔK. Later, other explanations besides crack-tip plasticity were proposed for the crack closure effect. Among the various phenomena held to be responsible for crack closure are crack surface roughness, asperities in the crack wake from oxides or corrosion products, viscous fluid, and phase transformation ahead of the crack tip. The oxide-induced crack closure is possible in a material that forms an oxide film on the surface easily. When such a material is subjected to cyclic stress near the threshold regime at low load ratios R and in a moist environment, corrosion products (i.e., oxides) of thickness comparable to the crack-tip opening displacements can build up at and near the tip. The oxide film continually breaks and forms behind the tip due to the crack surfaces coming together as a result of plasticity-induced closure and mode-I displacements characteristic of near-threshold crack growth. The crack closes at stress intensities above K_{min}. This mechanism is less likely to operate in a dry, oxygen-free environment and high load ratios. (Plasticity-induced closure is small.) The formation of an oxide film is time dependent and not likely to occur at high frequencies. Roughness-induced crack closure is thought to occur when the fracture surface roughness is comparable in size to the crack-tip opening displacement (CTOD) and significant mode-II deformation occurs. In such a case, cracks can become wedge–closed at contact points above the crack faces. Crack closure causes an increase in stiffness and a decrease in compliance. High values of R result in less crack closure; that is, ΔK_{eff} is closer to ΔK for higher R.

14.9   THE TWO-PARAMETER APPROACH

In a series of papers, Vasudevan et al. proposed a new, two-parametric approach to fatigue crack propagation.[11] Among the features of their approach are the following:

- It is not necessary to invoke crack closure to explain fatigue crack propagation.
- Plasticity at the crack tip cannot contribute to crack closure.
- Crack closure induced by oxide, corrosion, or roughness is very local and small.

There are five local parameters: the cyclic stress intensity factor ΔK; the maximum stress intensities K_{max}; the minimum stress intensity K_{min}; the mean cyclic stress intensity factor

---

[10]W. Elber, *Eng. Fract. Mech.*, 2 (1970) 37.

[11]See, for example, A. K. Vasudevan, K. Sadananda, and N. Louat, *Mater. Sci. & Eng.*, A188 (1994) 1.

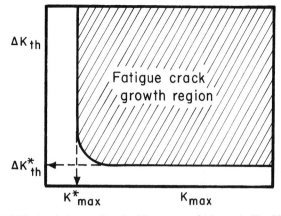

Figure 14.28 A fatigue threshold curve. (After A. K. Vasudevan, K. Sadananda, and N. Louat, *Mater. Sci. & Eng.,* A188 (1994) 1)

K_{mean}; and the R ratio. Out of these five, Vasudevan et al. used the applied driving force ΔK and the peak stress intensity K_{max} as the two parameters that are *sufficient and necessary* to analyze fatigue crack propagation data. There are thus two threshold quantities, one in each of the two parameters: the alternating stress intensity factor ΔK^* and K^*_{max}. These two must be satisfied *simultaneously* for crack propagation to occur. Based on data available in the literature on a wide range of alloys, Vasudevan and colleagues constructed a fundamental fatigue threshold curve or a fatigue map, as shown in Figure 14.28. The fundamental curve is independent of testing and geometric parameters; it depends only the material and environmental parameters. The shape and magnitude of such fundamental curves can vary, depending on the curve of ΔK_{th} vs. the load ratio R. On the basis of these new concepts, Vasudevan et al. classified the fatigue crack growth data into five different classes, using the experimental data on ΔK_{th} vs. R. Such a classification could provide a new basis for understanding the synergistic effects of various driving forces (mechanical, chemical, and microstructural) upon fatigue crack growth.

14.10 THE SHORT-CRACK PROBLEM IN FATIGUE

For long cracks, under conditions of applicability of linear elastic fracture mechanics (LEFM), there exists a threshold stress intensity range ΔK_{th}, below which no fatigue crack growth occurs. The value of the cyclic threshold stress intensity depends on a variety of factors: the microstructure of the material; the test environment; the load ratio R ($= \sigma_{min}/\sigma_{max} = K_{min}/K_{max}$); various crack-tip factors such as the amount of overload, cold work, etc.; experimental techniques; and the geometry of the specimen. It has been observed that *short* fatigue cracks, in metals and polymers, can propagate at rates different from those of the corresponding long fatigue cracks under the influence of the same driving force. Generally, for a given ΔK, the growth rates of small cracks are higher than those of long cracks.[12] A *short* crack is a crack that is smaller than the microstructural unit of the materials; for instance, a crack of length equal to grain or precipitate size is a short crack. In practice, one finds that long cracks can be between 1 and 20 mm, while short cracks are smaller than 0.1 mm. The anomalous growth of short cracks is explained

---

[12]See, for example, R. A. Smith and K. J. Miller, *Int. J. Mech. Eng.,* 20 (1978) 201; and S. Suresh and R. O. Ritchie, *Intl. Met. Rev.,* 29 (1984) 445.

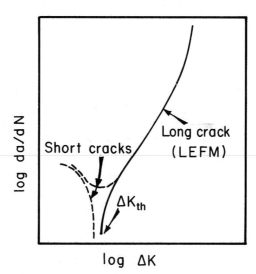

Figure 14.29 Fatigue crack growth rates for long and short cracks.

in Figure 14.29, a plot of log da/dN vs. log ΔK. Long cracks do not grow below a constant threshold ΔK_{th}. For long cracks, we have

$$K_I = Y\sigma\sqrt{\pi a},$$

and the threshold ΔK_{th} is a constant, indicated in Fig. 14.29. This is in accord with LEFM, as K_I alone determines the stress state at the crack tip. However, in the short-crack regime, cracks grow below this threshold value, as indicated by the deviation from the straight line in the figure. Short cracks propagate below the long-crack threshold (ΔK_{th}). The fatigue crack growth rate of short cracks decreases progressively, until a minimum in crack velocity occurs at a crack length on the order of the grain size; that is, $a \approx d$, where d is the grain size. This so-called short-crack anomaly arises when the crack size approaches the dimension of the microstructural feature (e.g., grain size, inclusions, etc.). Under such circumstances, homogeneity is lost, which is implicit in the LEFM treatment of the long crack.

Sadananda and Vasudevan have extended their two-parametric framework (see Section 14.9) to explain short-crack behavior.[13] According to these authors, a crack grows when both thresholds, ΔK^* and K_{max}^*, are met simultaneously. A short crack is no exception, and for it to propagate, it must meet these requirements, too. The short crack is different from the long crack in terms of the internal stresses that it encounters: All short cracks grow in internal stress fields that accentuate the applied stress to the level at which the crack propagates. An important conclusion of Sadananda and Vasudevan's work is that a similitude between the long and short crack is maintained!

14.11 FATIGUE TESTING

Among the reasons for carrying out fatigue testing on a material, we may include the need to develop a better understanding (fundamental or empirical) of the fatigue behavior of the material and the need to obtain more practical information on the fatigue response of a component or structure of the material. The fatigue test samples may thus range from

[13]K. Sadananda and A. K. Vasudevan, in *Fatigue '96, Berlin, May 6–10, 1996,* (Oxford: Pergamon Press, 1996), p. 375; K. Sadananda and A. K. Vasudevan, in *Twenty-seventh National Symp. Fracture Mech,* ASTM STP-1296, (Philadelphia: ASTM, 1996).

tiny samples tested within, say, the specimen chamber of a scanning electron microscope to complete aircraft wings weighing many tons. It would be futile to try to include everything known about fatigue testing here; instead, we present some of the common techniques and point out some of their salient aspects.

14.11.1 Conventional Fatigue Tests

Conventionally, fatigue testing has been done by cycling a given material through ranges of stress amplitude and recording the number of cycles of failure. The results are reported in the form of S–N curves (Figure 14.3). There are two main types of loading: rotating bending tests and direct stress tests (Figure 14.30). In direct stress machines, the stress distribution over any cross section of the specimen is uniform, and we can easily apply a static mean tensile or compressive load (i.e., the R ratio can be varied). However, the more common and popular type of loading has been the rotating bending beam test, described next. Direct loading machines are discussed in Section 14.11.4.

14.11.2 Rotating Bending Machine

Rotating bending tests are perhaps one of the most simple and oldest types of fatigue test. They provide a simple method of determining fatigue properties at zero mean load by

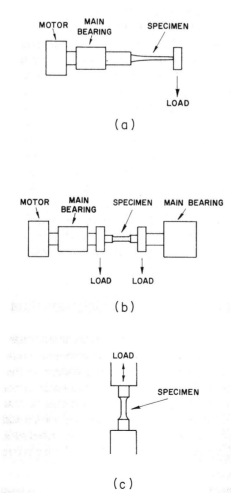

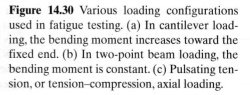

Figure 14.30 Various loading configurations used in fatigue testing. (a) In cantilever loading, the bending moment increases toward the fixed end. (b) In two-point beam loading, the bending moment is constant. (c) Pulsating tension, or tension–compression, axial loading.

applying known bending moments to rotating round specimens. Commercially, many versions are available, the main difference being in the application of the load: at a single point, as in a cantilever loading machine, or by some kind of two- or four-point loading (Figure 14.30). In the latter case, the bending moment is constant over the entire test section of the specimen, and thus, we use a specimen of constant diameter. In the cantilever type of loading machine, the specimen either has a narrow waist, so that the maximum bending stress occurs at the smallest diameter, or has a tapered cross section, such that the maximum bending stresses are constant at all cross sections. The stress at a point on the surface of a rotating bending specimen varies sinusoidally between numerically equal maximum tensile and compressive values in every cycle. Assuming the specimen to be elastic, we have

$$\pm S = \frac{32M}{\pi d^3}, \tag{14.7}$$

where S is the maximum surface stress, M is the bending moment at the cross section under consideration, and d is the diameter of the specimen. In such a test we obtain the number of cycles to failure at a given stress level. The stress level S is continually reduced, and the number of cycles to failure, N_f, increases. A logarithmic scale is used for N, and we obtain an S–N curve. In the case of ferrous materials, we generally attain a fatigue limit or endurance limit S_L (Figure 14.3). Cycling below S_L can be done indefinitely, without resulting in failure of the material. Such an endurance limit is not encountered in nonferrous metals or polymers. In these cases, one sets an arbitrary number of cycles, say, 10^7, and takes the corresponding stress to be the fatigue life of the material.

14.11.3 Statistical Analysis of S–N Curves

It has been observed that, if a sufficiently large number of identical specimens is fatigue tested at the same stress amplitude, a Gaussian or normal distribution describes the logarithm of the fatigue life distribution. Figure 14.31 shows a schematic S–N diagram with a log-normal distribution of lives at various stress levels. There is more of a spread in the lives of a group of specimens tested at a stress level greater than their fatigue limit than in the stress levels necessary to cause failure at a given life. The data from cyclic loading tests (whether rotating bending beam, pulsating tension, or axial tension–compression) must be analyzed statistically. The mean value $\bar{x}$ and the standard deviation σ for a given set of data are given by

$$\bar{x} = \frac{\sum x}{n} \tag{14.8}$$

and

$$\sigma = \left[\frac{\sum (x - \bar{x})^2}{n - 1} \right]^{1/2}, \tag{14.9}$$

where x is the cyclic life of the material at a given stress (the test value) and n is the number of test values (i.e., the number of samples tested to failure at a given stress). With these

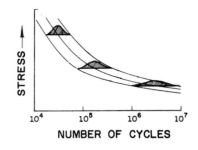

Figure 14.31 *S–N* curve showing log-normal distribution of lives at various stress levels.

statistical parameters, one can obtain confidence limits for the probability of survival of the material. The anticipated fatigue life, with a desired level of confidence (*C*%) that at least *P*% of the samples will not fail, may be written as

$$\text{anticipated life } (C, P) = \bar{x} - q\sigma, \tag{14.10}$$

where *q* is a function of *C*%, *P*%, and the number of test samples used to determine $\bar{x}$ and σ. The selection of a particular confidence limit depends on the importance of the component to the structural integrity of the material. The more important the component, the higher should be the confidence limit and the lower the stress. The *q* values for a given distribution are available in tabulated form in the literature. Table 14.1 presents the *q*-values, assuming a normal distribution. With anticipated life (Equation 14.10) and the *q* tables, we can develop a family of curves showing the probability of survival or failure of a component (Figure 14.32).

TABLE 14.1 *q*-values for *S–N* Data, Assuming a Normal Distribution*

P(%)	75	90	95	99	99.9	75	90	95	99	99.9
n		C = 0.50					C = 0.75			
4	0.739	1.419	1.830	2.601	3.464	1.256	2.134	2.680	3.726	4.910
6	0.712	1.360	1.750	2.483	3.304	1.087	1.860	2.336	3.243	4.273
8	0.701	1.337	1.719	2.436	3.239	1.010	1.740	2.190	3.042	4.008
10	0.694	1.324	1.702	2.411	3.205	0.964	1.671	2.103	2.927	3.858
12	0.691	1.316	1.691	2.395	3.183	0.933	1.624	2.048	2.851	3.760
15	0.688	1.308	1.680	2.379	3.163	0.899	1.577	1.991	2.776	3.661
18	0.685	1.303	1.674	2.370	3.150	0.846	1.544	1.951	2.723	3.595
20	0.684	1.301	1.671	2.366	3.143	0.865	1.528	1.933	2.697	3.561
25	0.682	1.297	1.666	2.357	3.132	0.842	1.496	1.895	2.647	3.497
		C = 0.90					C = 0.95			
4	1.972	3.187	3.957	5.437	7.128	2.619	4.163	5.145	7.042	9.215
6	1.540	2.494	3.091	4.242	5.556	1.895	3.006	3.707	5.062	6.612
8	1.360	2.219	2.755	3.783	4.955	1.617	2.582	3.188	4.353	5.686
10	1.257	2.065	2.568	3.532	4.629	1.465	2.355	2.911	3.981	5.203
12	1.188	1.966	2.448	3.371	4.420	1.366	2.210	2.736	3.747	4.900
15	1.119	1.866	2.329	3.212	4.215	1.268	2.068	2.566	3.520	4.607
18	1.071	1.800	2.249	3.106	4.078	1.200	1.974	2.453	3.370	4.415
20	1.046	1.765	2.208	3.052	4.009	1.167	1.926	2.396	3.295	4.319
25	0.999	1.702	2.132	2.952	3.882	1.103	1.838	2.292	3.158	4.143

*Reprinted with permission from ASTM STP No. 91, 1963, p. 67.

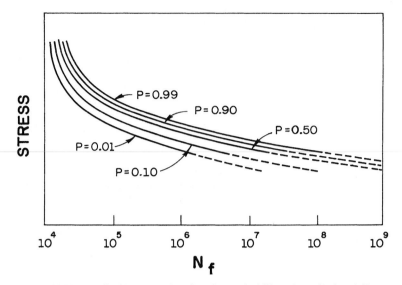

Figure 14.32 Family of curves showing the probability of survival or failure of a component.

14.11.4 Nonconventional Fatigue Testing

In the category of nonconventional fatigue testing, we include practically all modern fatigue testing other than that involving the determination of S–N curves. The machines used are direct loading machines. The drive system of the load train receives a time-dependent signal from the controls, converts it into a force, or displacement–time excitation, and transfers this excitation to the fatigue specimen. The three common control parameters are force, deflection or displacement, and strain. For most constant-amplitude fatigue tests, a simple harmonic motion is programmed into the drive system. Electronic function generators are commonly used; they generate an electrical signal that varies with time in the way the fatigue control parameter is desired to vary with time. A variety of signals can be programmed—for example, constant amplitude, constant frequency, and zero mean stress; constant amplitude, constant frequency, with a non-zero-mean stress level; random loading; and so on.

14.11.5 Servohydraulic Machines

Servohydraulically operated fatigue machines have become increasingly popular over the years. Figure 14.33 shows a line diagram of a servohydraulically operated closed-loop system. The load, applied through a hydraulic actuator, is measured by a load cell in series with the specimen. The amplified signal from the load cell is compared in a differential amplifier with the desired signal obtained from, say, a function generator. Thus, this system forms a closed-loop load control system. We can also have a displacement or strain control from a transducer or a strain gage on the specimen instead of the load cell. The actual value measured by a load cell, displacement transducer, or strain gage is continuously compared with the desired value and continuously corrected by the high-response electromagnetic servovalve. The energy is provided by a hydraulic power supply. The main advantage of such machines is a higher degree of flexibility. Larger specimen deflections are possible than are possible in electromechanical machines. Thus, we can test components involving large deflections, as well as conventional, stiff specimens. Another

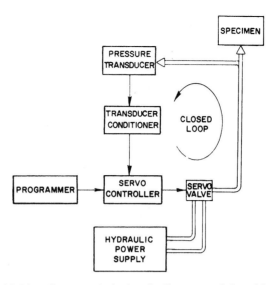

Figure 14.33 Line diagram of a hydraulically operated closed-loop system.

major advantage has to do with the versatility of the system in regard to the input signal that can be used. Virtually any analog signal from a function generator, magnetic tape, or a random noise generator is acceptable. This enables us to use not only constant-amplitude waveforms or block-program spectrum loading, but also random waveforms, such as those obtained from actual service conditions. The upshot is that the materials or components can be subjected to more realistic fatigue testing. The main disadvantage of servohydraulic machines is, of course, that they require much higher power consumption than conventional devices.

14.11.6 Low-cycle Fatigue Tests

Under conditions of high nominal stresses (i.e., short lifetimes, less than 10^4 cycles), the constant-stress amplitude test gives only limited information. This is because rather large plastic strain components are involved in such cases. Under such conditions, the cyclic stress–strain curves obtained under strain control become more useful. Servohydraulic machines are generally used in a closed-loop mode. Figure 14.34 shows schematically a cyclic straining facility. Axial tension–compression is generally employed. We measure stress as a function of the number of strain reversals. Usually, stress and strain signals are fed to an X–Y recorder, and a complete hysteresis loop is obtained. The area of the loop is the plastic strain energy per cycle.

Cyclic Stress–Strain Curves. We can obtain cyclic stress–strain curves by linking the tips of a series of hysteresis loops obtained from equivalent specimens tested at different plastic strain amplitudes ($\Delta\varepsilon_p$). There are also methods of obtaining cyclic stress–strain curves from a single specimen. The hysteresis loop adjusts rather quickly following a sudden change in $\Delta\varepsilon_p$. Thus, we can obtain a cyclic stress–strain curve from one specimen tested at several strain amplitudes. This is called a *multiple-step test*. Another method is the *incremental step test* with one specimen. This method consists of gradually increasing the cyclic strain range until a cyclic strain of about $\pm1\%$ is attained. The strain range is then slowly reduced, and the procedure is repeated until the material is stabilized.

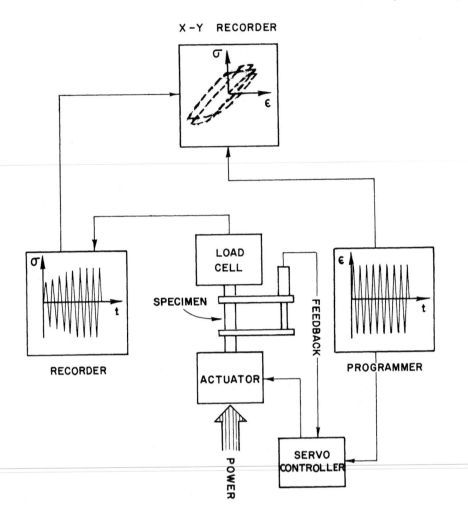

Figure 14.34 Block diagram of a low-cycle fatigue-testing system.

14.11.7 Fatigue Crack Propagation Testing

As pointed out earlier, the process of fatigue failure consists of the following two stages:

1. A certain number of cycles N_i in which a small crack is initiated. Some people include in this stage early growth of the microcrack to a somewhat larger crack.
2. Propagation of a major crack. Generally, this occurs in such a way that we are able to describe the propagation behavior by some kind of standard relationship, say, the Paris–Erdogan relation. There is a substage of this propagation stage wherein the final rupture occurs, namely, when the crack has reached a certain critical length for the material, the applied stress, and the test piece or structural component.

Much attention has been paid to the crack propagation behavior of materials in fatigue. Fatigue crack growth rates under service conditions can be of great importance, especially in determining inspection intervals. For example, wheels on large aircraft may have an ample safe lifetime after the appearance of detectable cracks. What we want to be sure of is that these cracks will not grow to a size that is critical for the part during the time available before the next periodic inspection.

Flat-sheet specimens are commonly chosen for crack propagation studies. The starter notch can be a side edge notch, a central through-the-thickness hole, or some other shape appropriate to the form of defects observed in service. These notches can be cut by a mechanical saw, electrical discharge machining, and so on. Usually, crack growth measurements are made after a small initial propagation in which there is an atomically sharp fatigue crack. The crack length is measured as a function of the number of cycles, and subsequent analysis is carried out in terms of fracture mechanics concepts. Synchronized strobe lighting can be used to illuminate the surface of the sample in order to provide a stable, vibration-free crack-length reading capability, or, in more sophisticated cases, a movie record can be obtained of the increase in the length of the crack. Traveling or stereo zoom microscopes are used in manual monitoring crack length. Such devices typically can read up to 0.01 mm. We may have scale markings photographically prepared on the sample or have a scale inserted in the ocular piece of the microscope. We can also use crack propagation gages, consisting of a series of 20 or 25 parallel, equally spaced resistance wires in the form of a grid. Crack length is measured by monitoring the overall change in resistance. In the electric potential drop method, a constant direct current is passed through the specimen containing a crack. The resistance of the specimen changes as the crack grows and is detected by measuring the potential drop across the mouth of the starter notch. Figure 14.35 shows the setup for a bend and a compact-tension specimen. As a crack is observed propagating, the number of cycles required for each increment is recorded, and a crack growth rate da/dN is computed from the curve of the crack length a versus the number of cycles N. The cyclic stress intensity factor at the crack tip (ΔK) can be computed from the crack length and the load. By plotting da/dN versus ΔK, we can obtain the fatigue crack growth characteristics of the material.

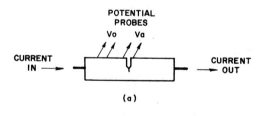

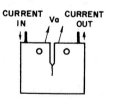

Figure 14.35 Electric potential drop method for crack growth measurements. (a) Bend specimen. (b) Compact-tension specimen.

SUGGESTED READINGS

D. L. Davidson and J. Lankford, *Intl. Mater. Rev.,* 37 (1992) 65.

A Guide for Fatigue Testing and the Statistical Analysis of Fatigue Data. ASTM STP 91. Philadelphia: ASTM, 1963.

R. W. Hertzberg, *Deformation and Fracture Mechanics of Engineering Materials,* 4th ed. New York: Wiley, 1996.

R. W. HERTZBERG, and J. A. MASON. *Fatigue of Engineering Plastics.* New York: Academic Press, 1980.

J. G. JOHNSON. *The Statistical Treatment of Fatigue Experiments.* New York: Elsevier, 1964.

R. O. RITCHIE, *Mater. Sci. & Eng.,* 103 (1988) 15.

S. T. ROLFE and J. M. BARSOM. *Fracture and Fatigue Control in Structures,* 2d ed. Englewood Cliffs, NJ: Prentice-Hall, 1987.

S. SURESH. *Fatigue of Materials.* Cambridge, U.K.: Cambridge University Press, 1991.

EXERCISES

14.1   Many operations, such as machining, grinding, electroplating, and case-hardening, may induce residual stresses in a material. Discuss, in general terms, the effect of such residual stresses on the fatigue life of the material.

14.2   A steel has the following properties:

$$\text{Young's modulus } E = 210 \text{ GPa},$$

$$\text{Monotonic fracture stress } \sigma_f = 2.0 \text{ GPa},$$

$$\text{Monotonic strain at fracture } \varepsilon_f = 0.6,$$

$$\text{Exponent } n' \text{ (cyclic)} = 0.15.$$

Compute the total strain that a bar of this steel will be subject to under cyclic straining before failing at 1,500 cycles.

14.3   The low-cycle fatigue behavior of a material can be represented by

$$\sigma_L = \sigma_{UTS} N_f^{-0.1},$$

where σ_L is the endurance limit, σ_{UTS} is the ultimate tensile strength, and N_f is the number of cycles to failure. If σ_{UTS} for this material is 500 MPa, find its endurance limit. A sample of the material is subjected to block loading consisting of 40, 30, 20, and 10% of fatigue life at σ_L, $1.10\sigma_L$, $1.2\sigma_L$, and $1.3\sigma_L$, respectively. Use the Palmgren–Miner relationship to estimate the fatigue life of the sample under this block loading.

14.4   A microalloyed steel was subjected to two fatigue tests at ± 400 MPa and ± 250 MPa. Failure occurred after 2×10^4 and 1.2×10^6 cycles, respectively, at these two stress levels. Making appropriate assumptions, estimate the fatigue life at ± 300 MPa of a part made from this steel that has already suffered 2.5×10^4 cycles at ± 350 MPa.

14.5   (For the discriminating student) Calculate the life of a 15-cm-thick panel between 0 and $0.2\sigma_y$, where $\sigma_y = 345$ MPa. The photomicrograph of Figure E14.5 shows the fatigue striations at 4 cm of the initial flaw (which had a diameter of 200 μm) that gave rise to failure.

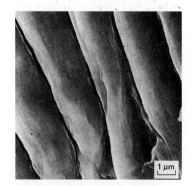

Figure E14.5 Striations on fatigue fracture surface in AISI 304 stainless steel. (Reprinted with permission from K. K. Chawla and P. K. Liaw, *J. Mater. Sci.,* 14 (1979) 2143.)

The frequency of cycling is 10 Hz, and the plane-strain fracture toughness is 44 MPa $\sqrt{m}$. The separation between fatigue striations at 4 mm from the flaw is 100 times smaller than the ones shown in the figure.

14.6   The curve of crack growth rate da/dN vs. cyclic stress intensity ΔK for a material in the Paris regime is shown in Figure E14.6. Determine the parameters C and m for this material (*Hint:* Take any two points on the curve, and determine the slope m of the line.)

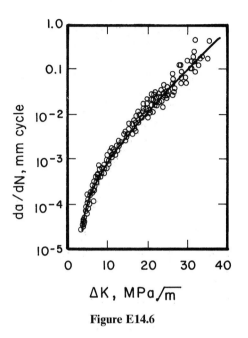

Figure E14.6

14.7   A steel has the following properties:

$$\text{Yield stress } \sigma_y = 700 \text{ MPa,}$$

$$\text{Fracture toughness } K_{Ic} = 165 \text{ MPa } \sqrt{m}.$$

A plate of this steel containing a single edge crack was tested in fatigue under $\Delta\sigma = 140$ MPa, $R = 0.5$, and $a_0 = 2$ mm. It was observed experimentally that fatigue crack propagation in the steel could be described by the Paris-type relationship

$$\frac{da}{dN} \text{ (m/cycle)} = 0.66 \times 10^{-8}(\Delta K)^{2.25},$$

where ΔK is measured in MPa $\sqrt{m}$.

(a)   What is the critical crack size a_c at σ_{max}?

(b)   Compute the fatigue life of the steel.

14.8   Obtain the parameters for the Paris-type relationship for the data shown in Figure E14.8 for the aluminum alloy 7075-T6.

14.9   For the 7075-T6 alloy in Exercise 14-8, determine the length of a crack after 10^5 cycles if the initial crack size was equal to 0.2 mm and the cyclic loading was such that, at the onset of fatigue, $\Delta K = 10$ MPa $\sqrt{m}$.

14.10   Suppose that the fatigue life of a precracked specimen is totally occupied by crack propagation. If, in a certain case, the initial crack growth rate is given by

$$\frac{da}{dN} = C\Delta K^2$$

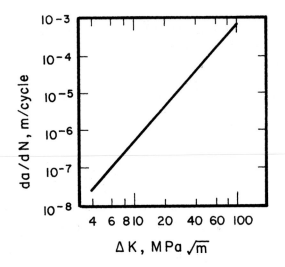

Figure E14.8 Adapted from *ASM Metals Handbook,* 9th ed. (Metals Park, Ohio: ASM, 1986) p. 103.

and $\Delta K = 2\alpha\,\Delta\sigma\sqrt{\pi a}$, show that

$$C = \frac{1}{4N\Delta\sigma^2\alpha^2\pi}\ln\frac{a_f}{a_0},$$

where N is the fatigue life, a_0 is the initial crack length, and a_f is the final crack length, of the specimen.

14.11 Fatigue crack propagation in a polymer can be described by the relationship

$$\frac{da}{dN} = 0.5\times10^6\,\Delta K^{3.5},$$

where da/dN is in m/cycle and ΔK is in MPa $\sqrt{m}$. A sample with the following dimensions and a central through-the-thickness crack was subjected to fatigue under a maximum load of 200 N and a minimum load of zero:

thickness $B = 10$ mm,

width $W = 50$ mm,

crack length $2a = 10$ mm.

Using an appropriate expression for ΔK (see Chapter 7), calculate da/dN for this sample.

14.12 Estimate the life of a hip implant made of 304L stainless steel if it contains initial flaws with length $2c = 200\ \mu m$ and a height $2a = 100\ \mu m$. Assume that the force applied on the artificial hip is

walking: $3W$,

running: $7W$,

where W is the weight of the person. The fatigue response of 304L can be represented by

$$\frac{da}{dN} = 5.5\times10^{-9}\Delta K^3,$$

where da/dN is in mm/cycle and K is in MPa$\sqrt{m}$. The person is assumed to

(a) walk 3 hours per day

(b) walk 3 hours and jog 20 minutes per day.

Make all necessary assumptions.

14.13 Assuming that fatigue failures are initiated at the "weakest link," we may use the Weibull frequency distribution function to represent the fatigue lives of a group of specimens tested under identical conditions. We have

$$f(N) = \frac{b}{N_a - N_0} \left(\frac{N - N_0}{N_a - N_0}\right)^{b-1} \exp\left[-\left(\frac{N - N_0}{N_a - N_0}\right)^b\right]$$

where N is the specimen's fatigue life, N_0 is the minimum life ≥ 0, N_a is the characteristic life at 36.8% survival of the population (36.8% = $1/e$, where e = 2.718), and b is the shape parameter of the Weibull distribution curve. Letting $x = (N - N_0)/(N_a - N_0)$, plot frequency curves $f(N)$ versus x for b = 1, 2, and 3.

14.14 Fatigue data are, generally, analyzed cumulatively to determine the survival percentage. The Weibull cumulative function for the fraction of population failing at N is an integration of the expression for $f(N)$ in the preceding exercise. Show that this function is

$$F(N) = 1 - \exp\left[-\left(\frac{N - N_0}{N_a - N_0}\right)^b\right].$$

Transform F into a straight-line relationship by taking the logarithm of the logarithm of the equation. Show how this relationship can be used on log–log paper for graphically fitting the Weibull cumulative distribution and for graphically estimating the parameters b, N_0, and N_a.[14]

14.15 One of the worst single aircraft accidents in history resulted in the loss of 520 lives. It was produced by the growth of a fatigue crack in the back of the bulkhead of a Boeing 747 plane. (See Figure E14.15.) The fatigue fracture was caused by a repair that replaced a double row of rivets by a single row in certain places. The accident occurred after the plane reached an altitude of 7,200 m. The atmospheric pressure decreases by 12 Pa for every meter increase in altitude.

(a) Calculate the stress cycle to which the pressurized cabin and bulkhead were subjected in each takeoff–landing sequence of the plane.

(b) Establish the critical crack length for which catastrophic growth would occur.

(c) Assuming that fatigue failure started at one of the rivet holes (which had a diameter of 12 mm) and that it propagated through subsequent holes, calculate the number of cycles necessary to bring down the "big bird," given the following data:

Paris relationship constants, $C = 5 \times 10^{-8}$ and $m = 3.6$,

$$\sigma_y = 310 \text{ MPa},$$

$$\sigma_{UTS} = 345 \text{ MPa}.$$

14.16 An alloy steel plate is subjected to constant-amplitude uniaxial tension–compression fatigue. The stress amplitude is 100 MPa. The plate has a yield strength of 1,500 MPa, a fracture toughness of 50 MPa$\sqrt{m}$, and an edge crack of 0.5 mm. Estimate the number of fatigue cycles to cause fracture if da/dN (m/cycle) = $1.5 \times 10^{-24}\ \Delta K^4$, where the units of K are MPa$\sqrt{m}$. Use $Y = 1$ in the fracture toughness equation.

---

[14]See, for example, C. S. Yen, in *Metal Fatigue: Theory and Design*, A. F. Madayag (ed.) (New York: Wiley, 1969), p. 140.

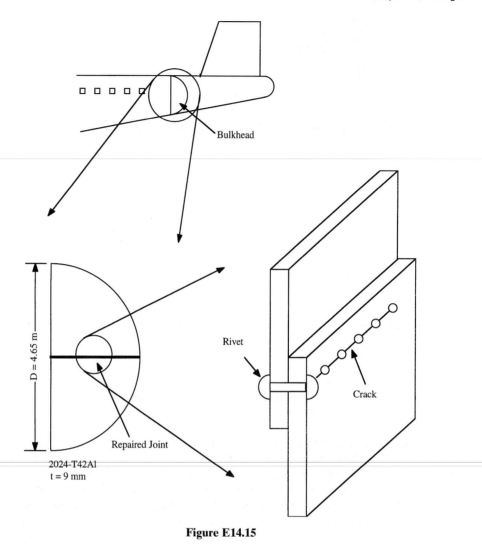

Figure E14.15

14.17 A part is subjected to cyclic loading at 50 Hz. The stress intensity at the tip of a flaw is just above ΔK_{th}, and da/dN is equal to 10^{-8} mm/s. What is the length of crack after one week? Take $m = 3$.

15

Composite Materials

15.1 INTRODUCTION

We can define a *composite* material as a material consisting of two or more physically and/or chemically distinct phases, suitably arranged or distributed. A composite material usually has characteristics that are not depicted by any of its components in isolation. Generally, the continuous phase is referred to as the *matrix,* while the distributed phase is called the *reinforcement.* Three things determine the characteristics of a composite: the reinforcement, the matrix, and the interface between them. In this chapter, we provide a brief survey of different types of composite materials, highlight some of their important features, and indicate their various applications.

15.2 TYPES OF COMPOSITES

We may classify composites on the basis of the type of matrix employed in them—for example, polymer matrix composites (PMCs), metal matrix composites (MMCs), and ceramic matrix composites (CMCs). We may also classify composites on the basis of the type of reinforcement they employ:

1. Particle reinforced composites.
2. Short fiber, or whisker reinforced, composites.
3. Continuous fiber, or sheet reinforced, MMCs.

Figure 15.1 shows typical microstructures of some composites: B/Al (Figure 15.1a), C/polyester (Figure 15.1b), NbC/Ni–Cr, an in situ (eutectic) composite (Figure 15.1c), and

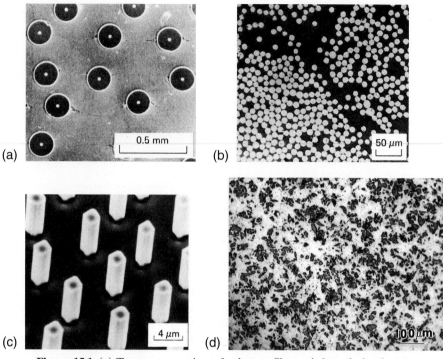

Figure 15.1 (a) Transverse section of a boron fiber reinforced aluminum composite. $V_f = 10\%$. (b) Transverse section of a carbon fiber reinforced polyester resin. $V_f = 50\%$. (Optical.) (c) Deeply etched transverse section of a eutectic composite showing NbC fibers in an Ni–Cr matrix. (Courtesy of S. P. Cooper and J. P. Billingham, GEC Turbine Gnerators Ltd, U.K.) (d) SiC particles in an Al alloy matrix (SEM). $V_p = 17\%$.

SiC_p/Al, a particle-reinforced composite (Figure 15.1d). Volume fractions of fiber (V_f) or particle (V_p) are indicated in the figure caption.

15.3 IMPORTANT REINFORCEMENTS AND MATRIX MATERIALS

Many reinforcement materials are available in a variety of forms: continuous fibers, short fibers, whiskers, particles, etc. Some of the important ones are listed in Table 15.1, along with a summary of their salient characteristics. Reinforcements include organic fibers such as polyethylene and aramid, metallic fibers, and ceramic fibers and particles.

A variety of materials—polymers (thermoset and thermoplastic), metals and their alloys, intermetallics, glasses, glass–ceramics, and crystalline ceramics—can be used as matrices. Most polymer matrix composites consist of cross-linked thermoset polymers such as epoxy, phenolic, and polyester resins. Cross-linked thermoset polymers have an amorphous structure. (See Chapter 1.) Phenolics have the advantage of being cheaper than epoxy and polyester resins. Their main disadvantage is that toxic by-products are liberated during the curing process. Cross-linking of polymer chains during curing in polyester and epoxy occurs by an addition mechanism, without any by-product produced. Glass fiber-reinforced phenolic, polyester, or epoxy have been in use in a variety of automotive components. Epoxy resins have the added attractive feature that they can be partially cured to make fiber/epoxy prepregs, which are subsequently consolidated into a component, usually in an autoclave. A *prepreg* is a thin lamina of unidirectional (or sometimes woven)

TABLE 15.1 Properties of Some Important Reinforcements

Materials (Fibers)	Tensile Modulus (GPa)	Tensile Strength (GPa)	Compressive strength (GPa)	Density (g/cm³)
Alumina	350-380	1.7	6.9	3.9
Boron	415	3.5	5.9	2.5–2.6
SiC	200	2.8	3.1	2.8
E-Glass	71	1.8–3.0	0.7–1.4	2.5
Carbon P100 (pitch based)	725	2.2	0.48	2.15
Carbon M60J (PAN based)	585	3.8	1.67	1.94
Kevlar 49*	125	3.5	0.39–0.48	1.45
Spectra 1000**	172	3	0.17	1.0

*Kevlar is DuPont's trade name for an aramid fiber.
**Spectra 1000 is AlliedSignal's trade name for a polyethylene fiber.

fiber/polymer composite protected on both sides with easily removable separators. A typical unidirectional prepreg comes in the form of a roll that is 300–1,500 mm wide, 0.125 mm thick, and 50–250 m long. Typically, the polymer content is approximately 35% by volume. It is not uncommon to use 50 or even more such plies in a component.

To a lesser extent than thermosets, thermoplastic resins such as poly(ether ether ketone) or PEEK and poly(phenylene sulfide), or polysulfone, are used as matrix materials. PEEK is a high-performance semicrystalline thermoplastic that has been used as a matrix for carbon fibers. It is attractive as a polymeric matrix material because of its superior toughness and impact properties, compared to those of epoxies. Such properties are a function of the crystalline content and morphology of the thermoplastic.

There are many other important matrix materials. Among metallic matrix composite, we have aluminum and its alloys, mainly because of their low density and excellent strength, toughness, and resistance to corrosion, as well as titanium alloys, magnesium alloys, copper, etc. Among intermetallic and ceramic matrix composites are a variety of intermetallic compounds such as molybdenum disilicide and aluminides of nickel and titanium, silica-based glasses, glass–ceramics, and crystalline ceramics such as alumina and silicon carbide.

EXAMPLE 15.1

Carbon black is frequently used as a particulate filler in polymers, both thermoplastic and thermoset. Describe some of the important effects of the addition of carbon black to polymers.

Solution: Carbon black is stronger than the polymer matrix; thus, we get a stronger and harder composite. Carbon black is also thermally more stable than the polymer matrix; therefore, its addition results in a thermally stable composite—that is, improved creep resistance. In addition, carbon black leads to an enhanced dimensional stability. (It has a higher modulus and lower expansion coefficient than the polymer.)

15.3.1 Microstructural Aspects and Importance of the Matrix

As we have said, the differential thermal expansion between the reinforcement and the metal matrix can introduce a high dislocation density in a metallic matrix, especially in the near-interface region of the matrix. This high matrix dislocation density, as well as the reinforcement/matrix interfaces, can provide high diffusivity paths in a composite. A semicrystalline thermoplastic matrix can have its crystallization kinetics altered by the presence of a reinforcement such as a carbon fiber. In the case of a ceramic matrix composite, the brittle matrix can undergo cracking in response to such thermal stresses. Thus, the characteristics of a matrix material are changed by the very process of making a composite. Such is not commonly the case with the reinforcement, however; only in rare instances of very high temperature processing, as, for example, in the case of a CMC, can the reinforcement also undergo a change in its microstructure. Hence, the matrix is much more than a mere medium or glue to hold the reinforcement, be that fibers, whiskers, or particles. Accordingly, it should be chosen after due consideration of its chemical compatibility and thermal mismatch with the reinforcement. Processing-induced chemical reactions and thermal stresses can cause changes in the microstructure of the matrix. These microstructural changes in the matrix, in turn, can affect the mechanical and physical behavior of the composite. The matrix strength in the composite (the in situ strength) will not be the same as that determined from a test of an unreinforced matrix sample in isolation, because the matrix is likely to suffer several microstructural alterations during processing and, consequently, changes in its mechanical properties.

The final matrix microstructure is a function of the type, diameter, and distribution of the fiber, as well as conventional solidification parameters. For example, Mortensen et al. observed normal dendritic structure in the unreinforced region of the matrix of a silicon carbide fiber/Al–4.5% Cu matrix, while in the reinforced region, the dendritic morphology was controlled by the fiber distribution.[1] Second phase appeared preferentially at the fiber/matrix interface or in the narrow interfiber spaces. In short, the microstructure of the matrix in the fiber composite is likely to differ significantly from that of the unreinforced matrix material processed in an identical manner.

Porosity is a critical defect that is likely to be present in the matrix. Porosity can be highly deleterious to the overall performance of a composite. The main sources of porosity are any gas evolution, shrinkage occurring upon solidification, and, in the case of CMCs, incomplete elimination of any binder material. In a composite made by liquid infiltration of a preform, a high volume fraction of reinforcement may impede the flow of the liquid and inhibit any "bulk movement" of the semisolid matrix material. The desirability of having a low porosity in a PMC can be appreciated by the fact that the final stage in any PMC fabrication is called *debulking,* which serves to reduce the number of voids. A low number of voids is necessary for improved interlaminar shear strength. In the case of a CMC, made by sintering of glass or glass–ceramic powder and fibrous reinforcements, the reinforcements can form a network that impedes the transfer of mass required for sintering. Depending on the thermal expansion coefficients of the components, there is also the possibility of developing hydrostatic tensile stresses in the matrix that will counter the driving force for sintering. Following are some of the common structural defects in composites:

- Matrix-rich (fiber-poor) regions.
- Voids.
- Microcracks (which may form due to thermal mismatch between the components, curing stresses, or the absorption of moisture during processing).
- Debonded regions.
- Delaminated regions.
- Variations in fiber alignment.

[1]A. Mortensen, M. N. Gungor, J. A. Cornic, and M. C. Flemings, *J. Met.,* 38 (Mar 1986) 30.

15.4 INTERFACES IN COMPOSITES

The interface region in a particular composite has a great deal of importance in determining the ultimate properties of the composite, essentially for two reasons: The interface occupies a very large area per unit volume in a composite, and, in general, the reinforcement and the matrix form a system that is not in thermodynamic equilibrium. We can define an interface as a boundary surface between two phases in which a discontinuity in one or more material parameters occurs. According to this definition, an interface is a bi-dimensional region across which a discontinuity occurs in one or more material parameters. In practice, there is always some volume associated with the interface region, and a gradual transition in material parameters occurs over the thickness of this interfacial zone. Some of the important parameters that can show a discontinuity at the interface are the elastic moduli, strength, chemical potential, coefficient of thermal expansion of the composite, and others. A discontinuity in chemical potential is likely to cause a chemical interaction, leading to an interdiffusion zone or the formation of a chemical compound at the interface. A discontinuity in the thermal expansion coefficient means that the interface will be in equilibrium only at the temperature at which the reinforcement and the matrix were brought into contact. At any other temperature, biaxial or triaxial stress fields will be present, because of the thermal mismatch between the components of a composite. Thermal stresses due to a thermal mismatch will generally have an expression of the form

$$\sigma = f(E, a, b, r)\, \Delta\alpha\, \Delta T, \tag{15.1}$$

where $f(E, a, b, r)$ is a function of the elastic constants E and the geometric parameters a, b, and r; $\Delta\alpha$ is the difference in the expansion coefficients of the components, and ΔT is the change in temperature of the material. The term $(\Delta\alpha\Delta T)$ is, of course, the thermal strain. (Detailed expressions for thermal stresses in composites can be found in textbooks on composites.[2])

It can be shown that, for a given diameter and volume fraction of reinforcement, a fibrous composite will have a larger interfacial area than a particulate composite will. The important point, however, is that the interfacial area in a composite increases with a decreasing reinforcement diameter. It is easy to visualize the interfacial area becoming very large for reinforcements less than 10–20 μm in diameter. Since chemical and/or mechanical interactions between the reinforcement and the matrix occur at interfaces, an extremely large area of interface has an enormous importance in determining the final properties and performance of a composite.

15.4.1 Crystallographic Nature of the Fiber/Matrix Interface

In crystallographic terms, ceramic/metal interfaces in composites are, generally, incoherent and high-energy interfaces. Accordingly, they can act as very efficient vacancy sinks and provide rapid diffusion paths, segregation sites, and sites of heterogeneous precipitation, as well as sites for precipitate-free zones. Among the possible exceptions are some eutectic composites and the XD™ type of particulate composites. The in situ or eutectic composites do show semicoherent interfaces, that is, the lattice mismatch between the matrix phase and the reinforcement phase is accommodated by creating a network of dislocations. The XD™ process gives particulate composites with stable and clean interfaces. By contrast, interfaces in ceramic matrix composites are generally incoherent, as are interfaces in polymer matrix composites.

---

[2]See, for example, K. K. Chawla, *Composite Materials: Science and Engineering,* second edition (New York: Springer-Verlag, 1998).

15.4.2 Interfacial Bonding in Composites

Some bonding must exist between the reinforcement and the matrix for load transfer from matrix to fiber to occur. Neglecting any direct loading of the reinforcement, the applied load is transferred from the matrix to the reinforcement via a well-bonded interface. However, the degree of bonding desired in different types of composites is not the same. In general, one would like to have a strong interfacial bonding in the case of PMCs and MMCs, with which one aims at exploiting the high stiffness and load-bearing capacity of a fibrous reinforcement. In CMCs, on the other hand, one would like to have a weak interfacial bonding, such that an advancing crack gets deflected there rather than pass through unimpeded. This is because the main objective in CMCs is to enhance their toughness instead of their strength. Crack deflection, crack bridging by fiber, and fiber pullout lead to an increased toughness and a noncatastrophic failure.

An important parameter in regard to the interface is the wettability of reinforcement by the matrix. *Wettability* refers to the ability of a liquid to spread on a solid substrate. Frequently, the contact angle between a liquid drop and a solid substrate is taken as a measure of wettability, a contact angle of 0° indicating perfect wettability and a contact angle of 180° indicating no wettability. *Wettability is only a measure of the possibility of attaining an intimate contact between a liquid and a solid.* Good wetting is a necessary, but not sufficient, condition for strong bonding. One needs a good wetting even for purely mechanical bonding or weak van der Waals bonding; otherwise voids may form at the interface. Besides wettability, other important factors, such as chemical, mechanical, thermal, and structural factors, affect the nature of the bonding between reinforcement and matrix. As it happens, these factors frequently overlap, and it may not always be possible to isolate their effects.

In PMCs, the surfaces of fibers are generally treated to promote chemical or mechanical adhesion with the matrix. For example, glass fiber, a common reinforcement for a variety of polymeric resins, invariably has a treated surface. The treatment is called *sizing*. The *size* is applied to protect glass fiber from the environment, for ease of handling, and to avoid introducing surface defects into the material. Common sizes are starch gum, hydrogenated vegetable oil, gelatin, and polyvinyl alcohol. These sizes are removed before putting in resin matrix by heat cleaning at approximately 340°C and washing. After cleaning, organometallic or organosilane coupling agents are applied. Organosilane compounds have the chemical formula, $R–SiX_3$, with X typically is Cl and R is a resin-compatible group capable of interacting with hydroxylated silanes on the glass surface.

A mechanical keying effect between two surfaces can also contribute to bonding. Chawla and Metzger observed such mechanical bonding effects in alumina/aluminum, a metal-matrix composite system.[3] Bonding due to mechanical interlocking at a rough interface can be equally important in PMCs and CMCs. Carbon fibers are given an oxidation treatment to provide, among other things, a rough surface that aids in bonding with the polymer matrix.

15.4.3 Interfacial Interactions

As mentioned earlier, most composite systems are nonequilibrium systems in the thermodynamic sense; that is, there exists a chemical potential gradient across the fiber/matrix interface. This means that, given favorable kinetic conditions (which, in practice, means a high enough temperature or long enough time), diffusion and/or chemical reactions will occur between the components. The interface layer(s) formed because of such a reaction will generally have characteristics different from those of either one of the components.

[3]K. K. Chawla and M. Metzger, in *Advances in Research on Strength and Fracture of Materials 3* (New York: Pergamon Press, 1978), p. 1039.

At times, however, some controlled amount of reaction at the interface may be desirable for obtaining strong bonding between the fiber and the matrix, but too thick an interaction zone will adversely affect the properties of the composite. Metal and ceramic matrix composites are generally fabricated at high temperatures, because diffusion and chemical reaction kinetics are faster at elevated temperatures than at low temperatures.

A very important factor in regard to reinforcement–matrix compatibility has to do with the mismatch between the coefficient of thermal expansion of the reinforcement and that of the matrix. This thermal mismatch can lead to thermal stresses large enough to cause plastic deformation in a soft metallic matrix and cracking in a brittle ceramic or polymeric matrix. Plastic deformation in the metallic matrix leads to the introduction of defects such as dislocations, vacancies, etc., in the matrix, especially in the region near the interface. The introduction of such defects can and does affect the phenomena responsible for chemical reactions at the interface, as well as characteristics of the matrix such as the precipitation kinetics. Chawla and Metzger showed, in a definitive matter, the importance of thermal stresses in composites.[4] They used a large-diameter tungsten fiber, (225 μm)/copper single-crystal matrix and low-fiber volume fractions. An etch-pitting technique was employed to observe dislocations in the matrix. The researchers observed that the dislocation density near the fiber was much higher than the dislocation density far away from the fiber. The enhanced dislocation density in the copper matrix near the fiber arose because of the plastic deformation of the matrix, in response to the thermal stresses generated by the thermal mismatch between the fiber and the matrix. The existence of a plastically deformed zone containing a high dislocation density in the metallic matrix in the vicinity of the reinforcement was confirmed by transmission electron microscopy by a number of other researchers, both in fibrous and particulate metal matrix composites. Figure 15.2, a TEM micrograph, shows dislocations in an aluminum matrix near a particle of silicon carbide. Such high densities of defects in the metal matrix will, of course, lead to a different set of properties of the matrix. In non-precipitation-hardening metals, this will simply cause a strengthening due to a higher dislocation density. In precipitation-hardenable matrix alloys, such as aluminum-copper, one would expect faster aging kinetics. Preferential precipitation at the reinforcement/matrix interface in an age-hardenable matrix has been observed by many researchers.

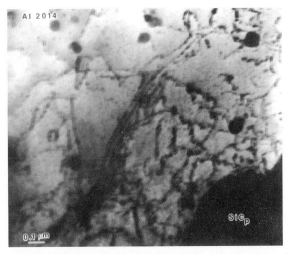

Figure 15.2 TEM micrograph showing dislocations in aluminum in the region near a silicon carbide particle (SiC$_p$).

[4]K. K. Chawla and M. Metzger, *J. Mater. Sci.,* 7 (1972) 34.

EXAMPLE 15.2

Ultrahigh-molecular-weight polyethylene (UHMWPE) fibers such as the Spectra fiber are very hard to bond with most matrix materials. Why?

Solution: High-modulus polyethylene fibers have a very highly oriented and extended chain structure. This high degree of crystallinity and a lack of polar surface do not allow good bonding of the fibers to most matrix materials. Indeed, the surface of such fibers must be treated with plasma to provide roughness or else a copolymer should be used to provide a polar surface, both of which approaches allow for bonding with the matrix. A similar problem exists, to varying degrees, with aramid and carbon fibers as well.

15.5 PROPERTIES OF COMPOSITES

We next describe some of the important properties of composites. In particular, we present expressions that allow us to predict the properties of composites in terms of the properties of their components, their amounts, and their geometric distribution in the composite. We also discuss the limitations of such expressions.

15.5.1 Density and Heat Capacity

Density and heat capacity are two properties that can be predicted rather accurately by a rule-of-mixtures type of relationship, irrespective of the arrangement of one phase in another. The simple relationships predicting these properties of a composite are as follows:

Density. The density of a composite is given by the rule-of-mixture equation

$$\rho_c = \rho_m V_m + \rho_r V_r,$$

where ρ designates the density and V represents volume fraction, with the subscripts c, m, and r denoting the composite, matrix, and reinforcement, respectively.

Heat capacity. The heat capacity of a composite is given by the expression

$$C_c = (C_m \rho_m V_m + C_r \rho_r V_r)/\rho_c,$$

where C denotes heat capacity and the other symbols have the significance given in the equation for density.

15.5.2 Elastic Moduli

The simplest model for predicting the elastic properties of a fiber/reinforced composite is shown in Figure 15.3. In the longitudinal direction, the composite is represented by a system of "action in parallel" (Figure 15.3a). For a load applied in the direction of the fibers, assuming equal deformation in the components, the two (or more) phases are viewed as being deformed in parallel. This is the classic case of Voigt's average, in which one has

$$P_c = \sum_{i=1}^{n} P_i V_i, \tag{15.2}$$

where P is a property, V denotes volume fraction, and the subscripts c and i indicate, respectively, the composite and the ith component of the total of n components. For

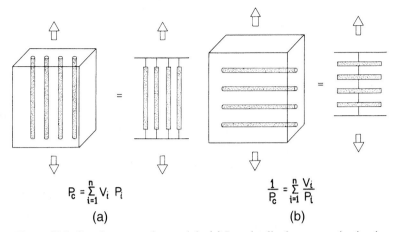

$$P_c = \sum_{i=1}^{n} V_i\, P_i \qquad\qquad \frac{1}{P_c} = \sum_{i=1}^{n} \frac{V_i}{P_i}$$

(a) (b)

Figure 15.3 Simple composite models. (a) Longitudinal response (action in parallel). (b) Transverse response (action in series).

the case under study, $n = 2$, and the property P is Young's modulus. We can write, in extended form,

$$E_c = E_f V_f + E_m V_m, \tag{15.3}$$

where the subscripts f and m indicate the fiber and matrix, respectively.

The elastic properties of such unidirectional composites in the transverse direction can be represented by a system of "action in series" [Figure 15.3b]. Upon loading in a direction transverse to the fibers, then, we have equal stress in the components. This model is equivalent to Reuss' classic treatment. We may write

$$\frac{1}{P_c} = \sum_{i=1}^{n} \frac{V_i}{P_i}. \tag{15.4}$$

Once again, for the case of $n = 2$, and taking the property P to be Young's modulus, we obtain, for the composite,

$$\frac{1}{E_c} = \frac{V_f}{E_f} + \frac{V_m}{E_m}. \tag{15.5}$$

The simple relations expressed Equations 15.4 and 15.5 are commonly referred to as the "rule of mixtures." The reader is warned that this rule is nothing more than a first approximation; more elaborate models have been proposed. Following is a summary of various methods of obtaining composite properties:

1. *The mechanics-of-materials method.* This deals with the specific geometric configuration of fibers in a matrix—for example, hexagonal, square, and rectangular—and we introduce large approximations in the resulting fields.

2. *The self-consistent field method.* This method introduces approximations in the geometry of the phases. We represent the phase geometry by a single fiber embedded in a material whose properties are equivalent to those of a matrix or an average of a composite. The resulting stress field is thus simplified.

3. *The variational calculus method.* This method focuses on the upper and lower limits of the properties of the composite and does not predict those properties

directly. Only when the upper and the lower bounds coincide is a particular property determined. Frequently, however, the upper and lower bounds are well separated.

4. *The numerical techniques method.* Here we use series expansion, numerical analysis, and finite-element techniques.

The variational calculus method do not give exact results. But these results can be used only as indicators of the behavior of the material when the upper and lower bounds are close enough. Fortunately, this is the case for longitudinal properties. Hill has put rigorous limits on the value of E in terms of the bulk modulus in plane strain, k_p, Poisson's ratio v, and the shear modulus G of the two phases.[5] One notes that k_p is the modulus for lateral dilatation with zero longitudinal strain (k_p is not equal to K) and is given by

$$k_p = \frac{E}{2(1 - 2v)(1 + v)}.$$

According to Hill, the bounds on E_c are

$$\frac{4V_f V_m (v_m - v_f)^2}{V_f/k_{pm} + V_m/k_{pf} + 1/G_m} \leq E_c - (E_f V_f + E_m V_m)$$

$$\leq \frac{4V_f V_m (v_m - v_f)^2}{V_f/k_{pm} + V_m/k_{pf} + 1/G_f}. \tag{15.6}$$

It is worth noting that this treatment of Hill does not have restrictions on the form of the fiber, the packing geometry, and so on. We can see, by substituting in Equation 15.6, that the deviations from the rule of mixtures (Equation 15.3) are rather small, for all practical purposes. For example, take $E_f/E_m = 100$, $v_f = 0.25$, $v_m = 0.4$. Then the deviation of the Young's modulus of the composite from that predicted by the rule of mixtures is, at most, 2%. For a metallic fiber (e.g., tungsten in a copper matrix), the deviation is less than 1%. Of course, the rule of mixture becomes exact when $v_f = v_m$.

The transverse properties and the shear moduli are not amenable to such simple reductions. Indeed, they do not obey the rule of mixtures, even to the first approximation. The bounds on them are well spaced. Numerical analysis results show that the behavior of the composite depends on the form and packing of the fiber and on the spacing between fibers.

Unidirectionally reinforced, continuous fiber composites show a linear increase in their longitudinal Young's modulus as a function of the volume fraction of fiber. For materials with a low modulus, such as polymers and metals, reinforcement by high-modulus and high-strength ceramic fibers can result in a significant increase in the composite's elastic modulus and strength. Figure 15.4 shows an example of a linear increase in the longitudinal flexural modulus as a function of the volume fraction of fiber for a glass fiber reinforced epoxy. In the case of CMCs, an increase in the elastic modulus or strength is rarely the objective, because most monolithic ceramics already have very high modulus and strength. However, an increase in the elastic modulus and strength can be a welcome attribute for low-modulus matrix materials—for example, glasses, glass–ceramics, and some crystalline ceramics, such as MgO. Still, the main objective of continuous fiber reinforcement of ceramic matrix materials is to toughen them. In a manner similar to that in PMCs and MMCs, various glass matrix compositions reinforced with carbon fibers have been shown to increase in strength and modulus with the volume fraction of fiber, in accordance with the rule of mixtures. Young's modulus increases linearly with V_f, but at a higher V_f it may deviate from linearity, owing to porosity in the matrix and possible misalignment of the fibers.

[5]R. Hill, *J. Mech. Phys. Solids,* 12 (1964) 199.

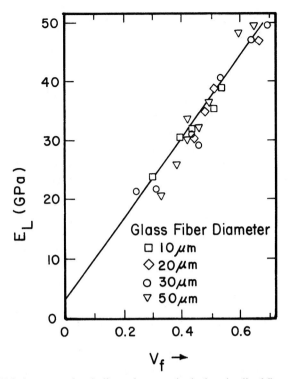

Figure 15.4 An example of a linear increase in the longitudinal flexural modulus as a function of the volume fraction of fiber for a glass fiber-reinforced epoxy. (After R. D. Adams and D. G. C. Bacon, *J. Comp. Mater.,* 7 (1973) 53)

Particle reinforcement also results in an increase in the modulus of the composite—much less, however, than that predicted by the rule of mixtures. This is understandable, inasmuch as the rule of mixtures is valid only for continuous fiber reinforcement. A schematic of the increase in modulus in a composite with volume fraction for the same reinforcement, but a different form of reinforcement—continuous fiber, whisker, or particle—is shown in Figure 15.5. This schematic shows the loss of reinforcement efficiency as one goes from continuous fiber to particle.

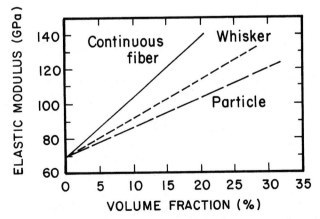

Figure 15.5 Schematic of increase in modulus in a composite with reinforcement volume fraction for a different form of reinforcement—continuous fiber, whisker, or particle. Note the loss of reinforcement efficiency as one goes from continuous fiber to particle.

In sum, we can say that the increase in the longitudinal elastic modulus of a fibrous composite as a function of the reinforcement volume fraction is fairly straightforward. The modulus of a composite is reasonably independent of the reinforcement packing arrangement, as long as all the fibers are parallel. For discontinuous reinforcement, the modulus is also quite independent of the particle clustering, etc. The modulus of a composite does show a dependence on temperature, which enters the picture essentially through the dependence of the matrix modulus on temperature.

EXAMPLE 15.3

Consider a glass fiber reinforced nylon composite. Let the volume fraction of the glass fiber be 65%. The density of glass is 2.1 g cm^{-3}, while that of nylon is 1.15 g cm^{-3}. Compute the density of the composite. Does it matter whether the glass fiber is continuous?

Solution:

The density of the composite is given by

$$\rho_c = \rho_f V_f + \rho_m V_m = 0.65 \times 2.1 + 0.35 \times 1.15 = 1.76 \text{ g cm}^{-3}.$$

It does not matter what the exact form of the glass fiber is in the composite. In fact, it could be in the form of equiaxial particles. A rule-of-mixture type of expression is valid for all composites, irrespective of the precise geometrical distribution of phases.

EXAMPLE 15.4

A carbon fiber reinforced epoxy composite consists of unidirectionally aligned fibers and has $V_f = 65\%$. Calculate the longitudinal and transverse Young's modulus of this composite. $E_f = 200$ GPa, $E_m = 5$ GPa.

Solution:

$$E_{cl} = V_f E_f + (1 - V_f)E_m$$
$$= 0.65 \times 200 + 0.35 \times 5 \text{ GPa}$$
$$= 131.75 \text{ GPa}.$$

In the transverse direction, we have the expression

$$\frac{I}{E_{ct}} = \frac{V_f}{E_f} + \frac{1 - V_f}{E_m}.$$

Rearranging yields

$$E_{ct} = \frac{E_f E_m}{E_f(1 - V_f) + E_m V_f}$$
$$= \frac{200 \times 5}{200 \times 0.35 + 5 \times 0.65}$$
$$= 13.65 \text{ GPa}.$$

Note the high degree of anisotropy.

EXAMPLE 15.5

Alumina particle (15 volume %) reinforced aluminum composite is used for making some special mountain bicycles. The density of alumina is 3.97 g cm^{-3}, while that of aluminum is 2.7 g cm^{-3}. Why is this composite used to make the mountain bicycle?

Solution: The alumina–aluminum composite will, of course, be slightly heavier than the unreinforced aluminum. The driving force for using the composite in this case is the enhanced stiffness: $E_{Al_2O_3} = 380$ GPa, while $E_{Al} = 70$ GPa. We can estimate the stiffness of the composite as

$$E_{composite} = E_{Al_2O_3} V_{Al_2O_3} + E_{Al} V_{Al}$$
$$= 380 \times 0.15 + 70 \times 0.85$$
$$= 57 + 59.50 = 116.50 \text{ GPa}.$$

This estimate is somewhat higher than that realized in practice, because the expression is valid for an unidirectional fiber reinforced composite, whereas the composite under consideration is a particulate composite. Even so, there is an almost 50% gain in stiffness by adding 15 volume % of alumina particles to aluminum.

15.5.3 Strength

Unlike elastic moduli, it is difficult to predict the strength of a composite by a simple rule-of-mixture type of relationship, because strength is a very structure-sensitive property. Specifically, for a composite containing continuous fibers and that is unidirectionally aligned and loaded in the fiber direction, the stress in the composite is written as

$$\sigma_c = \sigma_f V_f + \sigma_m V_m, \tag{15.7}$$

where σ is the axial stress, V is the volume fraction, and the subscripts c, f, and m refer to the composite, fiber, and matrix, respectively. The reason that the rule of mixture does not work for properties such as strength, compared to its reasonable application in predicting properties such as Young's modulus in the longitudinal direction, is the following: The elastic modulus is a relatively structure-insensitive property, so, the response to an applied stress in the composite state is nothing but the volume-weighted average of the individual responses of the isolated components. Strength, on the contrary, is an extremely structure-sensitive property. Thus, synergism can occur in the composite state. Let us now consider the factors that may influence, in one way or the other, composite properties. First, the matrix or fiber structure may be altered during fabrication; and second, composite materials generally consist of two components whose thermomechanical properties are quite different. Hence, these materials suffer residual stresses and/or alterations in structure due to the internal stresses. The differential contraction that occurs when the material is cooled from the fabrication temperature to ambient temperature can lead to rather large thermal stresses, which, in turn, lead a soft metal matrix to undergo extensive plastic deformation. The deformation mode may also be influenced by rheological interaction between the components. The plastic constraint on the matrix due to the large difference in the Poisson's ratio of the matrix compared with that of the fiber, especially in the stage wherein the fiber deforms elastically while the matrix deforms plastically, can alter the stress state in the composite. Thus, the alteration in the microstructure of one or both of the components or the interaction between the components during straining can give rise to synergism in the strength properties of the composite. In view of this, the rule of mixture would be, in the best of the circumstances, a lower bound on the maximum stress of a composite.

Having made these observations about the applicability of the rule of mixture to the strength properties, we will still find it instructive to consider this lower bound on the mechanical behavior of the composite. We ignore any negative deviations from the rule of mixtures due to any misalignment of the fibers or due to the formation of a reaction product between fiber and matrix. Also, we assume that the components do not interact during straining and that these properties in the composite state are the same as those in the isolated state. Then, for a series of composites with different fiber volume fractions, σ_c would be linearly dependent on V_f. Since $V_f + V_m = 1$, we can rewrite Equations 15.7 as

$$\sigma_c = \sigma_f V_f + \sigma_m (1 - V_f). \tag{15.8}$$

We can put certain restrictions on V_f in order to have real reinforcement. For this, a composite must have a certain minimum-fiber (continuous) volume fraction, V_{min}. Assuming that the fibers are identical and uniform (that is, all of them have the same ultimate tensile strength), the ultimate strength of the composite will be attained, ideally, at a strain equal to the strain corresponding to the ultimate stress of the fiber. Then, we have

$$\sigma_{cu} = \sigma_{fu} V_f + \sigma'_m (1 - V_f), \qquad V_f \geq V_{min}, \tag{15.9}$$

where σ_{fu} is the ultimate tensile of stress of the fiber in the composite and σ'_m is the matrix stress at the strain corresponding to the fiber's ultimate tensile stress. Note that σ'_m is to be determined from the stress–strain curve of the matrix alone; that is, it is the matrix flow stress at a strain in the matrix equal to the breaking strain of the fiber. As already indicated, we are assuming that matrix stress–strain behavior in the composite is the same as in isolation. At low volume fractions, if a work-hardened matrix can counterbalance the loss of load-carrying capacity as a result of fiber breakage, the matrix will control the strength of the composite. Assuming that all the fibers break at the same time, in order to have a real reinforcement effect, one must satisfy the relation

$$\sigma_{cu} = \sigma_{fu} V_f + \sigma'_m (1 - V_f) \geq \sigma_{mu} (1 - V_f), \tag{15.10}$$

where σ_{mu} is the ultimate tensile stress of the matrix. The equality in this expression serves to define the minimum fiber volume fraction, V_{min}, that must be surpassed in order to have real reinforcement. In that case,

$$V_{min} = \frac{\sigma_{mu} - \sigma'_m}{\sigma_{fu} + \sigma_{mu} - \sigma'_m}. \tag{15.11a}$$

The value of V_{min} increases with decreasing fiber strength or increasing matrix strength.

In case we require that the composite strength should surpass the matrix ultimate stress, we can define a critical fiber volume fraction, V_{crit}, that must be exceeded. V_{crit} is given by the equation

$$\sigma_{cu} = \sigma_{fu} V_f + \sigma'_m (1 - V_f) > \sigma_{mu}$$

In this case,

$$V_{crit} = \frac{\sigma_{mu} - \sigma'_m}{\sigma_{fu} - \sigma'_m} \tag{15.11b}$$

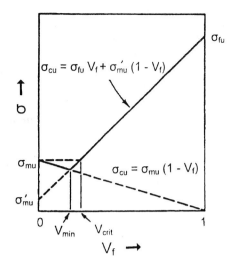

Figure 15.6 Determination of V_{min} and V_{crit}.

V_{crit} increases with increasing degree of matrix work-hardening ($\sigma_{mu} - \sigma'_m$). Figure 15.6 shows graphically the determination of V_{min} and V_{crit}. One notes that V_{crit} will always be greater than V_{min}.

In general, by incorporating fibers, we can increase the strength of the composite in the longitudinal direction. The strengthening effect in the transverse direction is not significant. Particle reinforcement can result in a more isotropic strengthening, provided that we have a uniform distribution of particles. Carbon, aramid, and glass fibers are used in epoxies to obtain high-strength composites. Such PMCs, however, have a maximum use temperature of about 150°C. Metal matrix composites, such as silicon carbide fiber in titanium, can take us to moderately high application temperatures. For applications requiring very high temperatures, we must resort to ceramic matrix composites. Silicon carbide whisker reinforced alumina composites show a good combination of mechanical and thermal properties: substantially improved strength, fracture toughness, thermal shock resistance, and high-temperature creep resistance over that of monolithic alumina. Figure 15.7 gives an example of the improvement in strength in the silicon carbide whisker/alumina composites as a function of the whisker volume fraction and test temperature. Similar results have been obtained with silicon carbide whisker/reinforced silicon nitride composites.

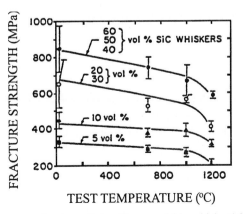

Figure 15.7 Increase in strength in silicon carbide whisker/alumina composites as a function of the whisker volume fraction and test temperature. (After G. C. Wei and P. F. Becher, *Am. Ceram. Soc. Bull.*, 64 (1985) 333)

Finally, we should mention the strength of in situ composites. In Figure 15.1c, we showed the microstructure of an in situ composite. Such a composite is generally made by the unidirectional withdrawal of heat during the solidification of a eutectic alloy. This controlled solidification allows for one phase to appear in an aligned fibrous form in a matrix of the other phase. The strength σ of such an in situ metal matrix composites made by directional solidification of eutectic alloys is given by a relationship similar to the Hall–Petch relationship used for grain-boundary strengthening.[6]

$$\sigma = \sigma_0 + k\lambda^{-1/2}$$

Here σ_0 is a friction stress term, k is a material constant, and λ is the interfiber spacing between rods, or lamellae. It turns out that one can vary λ rather easily by controlling the solidification rate R, because $\lambda^2 R$ equals a constant. Thus, one can easily control the strength of these in situ composites.

EXAMPLE 15.6

Consider a uniaxial fiber reinforced composite of aramid fibers in an epoxy matrix. The volume fraction of fibers is 60%. The composite is subjected to an axial strain of 0.1%. Compute the modulus and strength along the axial direction of the composite.

Solution: Both fiber and matrix deform elastically to a strain of 0.1%. Thus, we have

$$E_{cl} = E_f V_f + E_m(1 - V_f)$$
$$= 140 \times 0.6 + 5 \times 0.4$$
$$= 84 + 2 = 86 \text{ GPa};$$

$$\sigma_{cl} = \sigma_f V_f + \sigma_m(1 - V_f)$$
$$= eE_f V_f + eE_m(1 - V_f)$$
$$= e(E_{cl}) = 0.001 \times 86 \text{ GPa}$$
$$= 86 \text{ MPa}.$$

15.5.4   Anisotropic Nature of Fiber Reinforced Composites

Fiber reinforced composites are highly anisotropic. We derive an expression for the variation in the Young's modulus with the orientation of the fiber for a unidirectionally aligned composite.

The generalized Hooke's law can be written as (see Section 2.8)

$$\varepsilon_i = S_{ij}\sigma_j,$$

where ε_i is the strain, σ_j is the stress, S_{ij} is the compliance matrix, and i and j take values from 1 to 6, with summation indicated by a repeated suffix.

The compliances S_{11}, S_{22}, and S_{33} are reciprocals of the generalized stiffness moduli, and it can be shown (see Chapter 2) that they transform with rotation about a principal axis, say, the x_3-axis, according to relations of the type

$$S'_{11} = m^4 S_{11} + n^4 S_{22} + m^2 n^2 (2S_{12} + S_{66}) + 2mn (m^2 S_{16} + n^2 S_{26}), \qquad (15.13a)$$

---

[6]H. E. Cline, E. F. Walter, E. F. Koch, and L. M. Osika, *Acta Met.,* 19 (1971), 405.

where $m = \cos\theta$, and $n = \sin\theta$, in which θ is the angle of rotation.

For the discriminating reader, we should point out that this equation follows from the transformation relationship for the fourth rank elasticity tensor:

$$S'_{ijkl} = \ell_{im}\,\ell_{jn}\,\ell_{ko}\,\ell_{lp}\,S'_{mnop}$$

where ℓ_{im}, ℓ_{jn}, ℓ_{ko}, and ℓ_{lp} are the transformation coefficients. For S'_{1111}, we have

$$S'_{1111} = \ell_{1m}\,\ell_{1n}\,\ell_{1o}\,\ell_{1p}\,S'_{mnop}$$

Care should be exercised when changing from the tensorial to matrix notation (see section 2.8). For more details on such mathematical operations, the students should consult a text (for example, J. F. Nye. *Physical Properties of Crystals.* London: Oxford University Press, 1975). After converting to matrix notation we arrive at

$$S_{1111} = S_{11}; \quad S_{2222} = S_{22}; \quad 4S_{1212} = S_{66}; \quad 2S_{2212} = S_{26}.$$

For an orthotropic sheet material, for which the x_3-axis is normal to the plane of the sheet, we have $S_{16} = S_{26} = 0$; then, assuming that the properties in the directions 1 and 2 are the same, equation (15.13a) becomes

$$S'_{11} = (m^4 + n^4)\,S_{11} + m^2n^2\,(2S_{12} + S_{66})$$

$$= \frac{1}{2}S_{11} + \left(\frac{1}{2}S_{12} + \frac{1}{4}S_{66}\right) + \left[\frac{1}{2}S_{11} - \left(\frac{1}{2}S_{12} + \frac{1}{4}S_{66}\right)\right]\cos^2 2\theta.$$

Now let E_0 and E_{45} be Young's modulus for $\theta = 0°$ and $\theta = 45°$, respectively. Then $S_{11} = 1/E_0$ and $\frac{1}{2}S_{11} + \frac{1}{2}S_{12} + \frac{1}{4}S_{66} = 1/E_{45}$. Using these relationships, we get

$$S' = \frac{1}{E_\theta} = \frac{1}{E_{45}} - \left(\frac{1}{E_{45}} - \frac{1}{E_0}\right)\cos^2 2\theta, \tag{15.13b}$$

where E_θ is the modulus of the composite when the loading direction makes an angle θ with the fiber direction.

We can also write the compliances S_{12} and S_{66} in terms of the shear modulus G and Poisson's ratio ν for stresses applied in the plane of the sheet in the directions 1 and 2. From this, we obtain the relationship

$$\frac{1}{2G} = \frac{2}{E_{45}} - \frac{1}{E_0}(1 - \nu).$$

15.5.5 Aging Response of Matrix in MMCs

We have pointed out that the microstructure of a metallic matrix is modified by the presence of a ceramic reinforcement (particle, whisker, or fiber). In particular, a higher dislocation density in the matrix metal or alloy than that in the unreinforced metal or alloy has been observed. The higher dislocation density in the matrix has its origin in the thermal mismatch ($\Delta\alpha$) between the reinforcement and the metal matrix. For example, the thermal mismatch in the case of SiC/Al has a high value of $21 \times 10^{-6}\mathrm{K}^{-1}$, which will lead to thermal stress high enough to deform the matrix plastically and thus leave the matrix work-hardened. One expects that the quenching from the solutionizing temperature to room temperature, a change of about 450°C, will result in a large zone of matrix plastically deformed around each ceramic particle in which the dislocation density will be very high. This high dislocation density will tend to accelerate the aging kinetics of the matrix. Age-hardening treatment can contribute a considerable increment in strength to a precipitation-hardenable aluminum alloy composite. It should be borne in mind that the particle and whisker types of reinforcement, such as SiC, B_4C, Al_2O_3, etc., are unaffected by the aging process. These particles, however, can affect the precipitation behavior of the matrix quite significantly. The dislocations generated by thermal mismatch form heterogeneous nucleation sites for the precipitates in the matrix during subsequent aging treatments. This in turn alters the precipitation kinetics in the matrix of the composite, compared to the precipitation kinetics in the unreinforced material. Most metal matrix composite work has involved *off-the-shelf* metallic alloys, especially in the case of particle reinforced metal matrix composites. It is important to bear in mind that in such cases using the standard heat treatment practices given in the manuals and handbooks for unreinforced alloys can lead to drastically different results.

15.5.6 Toughness

The toughness of a given composite depends on the following factors:

- Composition and microstructure of the matrix.
- Type, size, and orientation of the reinforcement.
- Any processing done on the composite, insofar as it affects microstructural variables (e.g., the distribution of the reinforcement, porosity, segregation, etc.).

Continuous fiber reinforced composites show anisotropy in toughness just as in other properties. The 0° and 90° arrangements of fibers result in two extremes of toughness, while the 0°/90° arrangement (i.e., alternating laminae of 0° and 90°) gives a sort of pseudo random arrangement with a reduced degree of anisotropy. Using fibers in the form of a braid can make the crack propagation toughness increase greatly due to extensive matrix deformation, crack branching, fiber bundle debonding, and pullout. The composition of the matrix can also have a significant effect on the toughness of a composite: The tougher the matrix, the tougher will be the composite. Thus, a thermoplastic matrix would be expected to provide a higher toughness than a thermoset matrix. In view of the importance of toughness enhancement in CMCs, we offer a summary of the rather extensive effort that has been expended in making tougher ceramics. Some of the approaches to enhancing the toughness of ceramics include the following:

1. *Microcracking.* If microcracks form ahead of the main crack, they can cause crack branching, which in turn will distribute the strain energy over a large area. Such microcracking can thus decrease the stress intensity factor at the principal crack tip. Crack branching can also lead to enhanced toughness, because the stress required to drive a number of cracks is more than that required to drive a single crack.

2. *Particle Toughening.* The interaction between particles that do not undergo a phase transformation and a crack front can result in toughening due to crack bowing between particles, crack deflection at the particle, and crack bridging by ductile particles. Incremental increases in toughness can also result from an appropriate thermal mismatch between particles and the matrix. Taya et al. examined the effect of thermal residual stress in a TiB_2 particle reinforced silicon

carbide matrix composite.[7] They attributed the increased crack growth resistance in the composite vis-à-vis the unreinforced SiC to the existence of compressive residual stress in the SiC matrix in the presence of TiB_2 particles.

3. *Transformation toughening.* This involves a phase transformation of the second-phase particles at the crack tip with a shear and a dilational component, thus reducing the tensile stress concentration at the tip. In particulate composites, such as alumina containing partially stabilized zirconia, the change in volume associated with the phase transformation in zirconia particles is exploited to obtain enhanced toughness. In a partially stabilized zirconia (e.g., $ZrO_2 + Y_2O_3$), the stress field at the crack tip can cause a stress-induced martensitic transformation in ZrO_2 from a tetragonal phase (t) to a monoclinic one (m); that is,

$$ZrO_2\,(t) \rightarrow ZrO_2\,(m).$$

This transformation causes an expansion in volume (by approximately 4%) and a shear (0.16). The transformation in a particle at the crack tip results in stresses that tend to close the crack, and thus a portion of the energy that would go to fracture is spent in the stress-induced transformation. Also, the dilation in the transformed zone around a crack is opposed by the surrounding untransformed material, leading to compressive stresses that tend to close the crack. This results in increased toughness. The phenomenon of transformation toughening was discussed in Chapter 11. Transformation in the wake of a crack can result in a closure force that tends to resist the crack opening displacement. Crack deflection at zirconia particles can also contribute to toughness.

4. *Fiber or whisker reinforcement.* Toughening by long fibers or whiskers can bring into play a series of energy-absorbing mechanisms in the fracture process of CMCs and thus allow these materials to tolerate damage.

It appears that the effectiveness of various toughening mechanisms for structural ceramics decreases in the following order: continuous fiber reinforcement; transformation toughening; whiskers, platelets, and particles; microcracking. Many researchers have shown that if we add continuous C or SiC fibers to a glass or ceramic matrix, we can obtain a stress–strain curve of the type shown in Figure 15.8. This curve has the following salient features:

- Damage-tolerant behavior in a composite consisting of two brittle components.
- Initial elastic behavior.
- At a stress σ_0, the brittle matrix cracks.
- The crack bypasses the fibers and leaves them bridging the crack.
- Under continued loading, we have regularly spaced cracks in the matrix, bridged by the fibers.
- Noncatastrophic failure occurs. Fiber pullout occurs after the peak load, followed by failure of the composite when the fibers fail.

The final failure of the composite is not the result of the passage of a single crack; that is, *self-similar* crack propagation does not occur. Thus, it is difficult to define an unambiguous fracture toughness value, such as a value for K_{Ic}.

---

[7]M. Taya, S. Hayashi, A. S. Kobayashi, and H. S. Yoon, *J. Am. Ceram. Soc.,* 73 (1990) 1382.

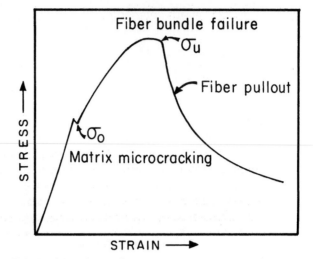

Figure 15.8 A schematic tensile stress–strain curve of a unidirectionally re-inforced ceramic matrix composite.

Under some circumstances, σ_0 is independent of the *preexisting flaw size* and is a material property. In general, σ_0 is a function of the matrix fracture toughness, interfacial shear strength, fiber volume fraction, fiber radius, and elastic constants of the fiber and matrix. For fully bridged cracks, the value of σ_0 can be independent of the length of the crack.

Although the appearance of the first matrix crack does *not* signify a complete failure of the composite, it does result in a reduced slope of the stress–strain curve, i.e., a decreased modulus. It also implies an easy access path for any aggressive environmental species that might be present. Matrix microcracking will cause a reduction in strength and modulus and might lead to internal oxidation and spalling as well. Thus, it would appear that, in practical terms, any microcrack toughening is not going to be very high, and the reliability of such composites containing microcracks also will not be very high.

It has been amply demonstrated that reinforcement with continuous fibers such as carbon, alumina, silicon carbide, and mullite fibers in brittle matrix materials (e.g., cement, glass, and glass–ceramic matrix) can result in toughening.[8] Not all of these failure mechanisms need operate simultaneously in a given fiber–matrix system, and often, in many composite systems, only one or two of the mechanisms will dominate the total fracture toughness. We discuss this topic further in Section 15.8.

15.6 *LOAD TRANSFER FROM MATRIX TO FIBER*

The matrix has the important function of transmitting the applied load to the fiber. Recall that we emphasized the idea that in fiber reinforced composites, the fibers are the principal load-carrying members. No direct loading of fibers from the ends is admitted. One imagines each fiber to be embedded inside a matrix continuum; the state of stress (and, consequently, that of strain) of the matrix is perturbed by the presence of the fiber (Figure 15.9). When the composite is loaded axially, the axial displacements in the fiber and in the matrix are locally different due to the different elastic moduli of the components. Macroscopically, the composite is deformed homogeneously.

[8]See, for example, K. K. Chawla, *Ceramic Matrix Composites* (London: Chapman & Hall, 1993).

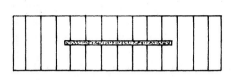
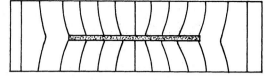

BEFORE DEFORMATION   **AFTER DEFORMATION**

Figure 15.9 Perturbation of the matrix stress state due to the presence of fiber.

EXAMPLE 15.7

The presence of voids in a composite is a serious, but commonly encountered, flaw. Suggest a simple method of determining the void content in a composite.

Solution: A simple method involves determining the density of the composite and getting an accurate estimate of values of the density of the reinforcement and matrix, most likely from the literature. We can write, for the volume of the voids in a composite,

$$V_v = V_c - (V_r + V_m)$$

where V is the volume and the subscripts, v, c, r, and m denote the void, composite, reinforcement, and matrix, respectively. Then, knowing the mass and density values, we can write

$$V_v = (M_c/\rho_c) - (V_m/\rho_r + M_m/\rho_m),$$

where M is the mass and ρ is the density, and the subscripts have the significance as before. The density of the composite can then be determined experimentally by Archimedes' method. The amount of reinforcement can be obtained by simply dissolving the matrix in a suitable chemical or by using a thermal method and weighing the residue.

Another simple method of determining the void content is by quantitative microstructural analysis.

The difference in the axial displacements in the fiber and the matrix implies that shear deformations are produced on planes parallel to the fiber axis and in the direction of this axis. These shear deformations are the means by which the applied load is distributed between the two components.

Let us consider the distribution of the longitudinal stress along the fiber–matrix interface. There are two distinct cases: (1) The matrix is elastic and the fiber is elastic, and (2) the matrix is plastic and the fiber is elastic.

15.6.1 Fiber and Matrix Elastic

We follow the treatment due to Cox.[9] Consider a fiber of length l embedded in a matrix subjected to a strain. Consider a point a distance x from one end of the fiber. It is assumed that (1) there exists a perfect contact between fiber and matrix (i.e., there is no sliding

[9]H. L. Cox, *Brit. J. App. Phys.*, 3 (1952) 72.

between them) and (2) Poisson's ratios of fiber and matrix are equal. Then the displacement of the point a distance x from one extremity of the fiber can be defined in the following manner; u is the displacement of point x in the presence of the fiber, and v is the displacement of the same point in the absence of the fiber.

The transfer of load from the matrix to the fiber may be written as

$$\frac{dP}{dx} = H(u - v), \tag{15.14a}$$

where P is the load on the fiber and H is a constant to be defined later. (H depends on the geometric arrangement of fibers, the matrix, and their moduli.)

Differentiating Equation 15.14a, we obtain

$$\frac{d^2P}{dx^2} = H\left(\frac{du}{dx} - \frac{dv}{dx}\right). \tag{15.14b}$$

Now, it follows from the definition that

$$\frac{dv}{dx} = \text{strain in matrix} = e,$$

$$\frac{du}{dx} = \text{strain in fiber} = \frac{P}{A_f E_f}, \tag{15.15}$$

where A_f is the transverse-sectional area of the fiber. From Equations 15.14 and 15.15, we obtain

$$\frac{d^2P}{dx^2} = H\left(\frac{P}{A_f E_f} - e\right). \tag{15.16}$$

A solution of this differential equation is

$$P = E_f A_f e + S \sinh \beta x + T \cosh \beta x, \tag{15.17}$$

where

$$\beta = \left(\frac{H}{A_f E_f}\right)^{1/2}. \tag{15.18}$$

The boundary conditions we need to evaluate the constants S and T are

$$P = 0 \text{ at } x = 0 \text{ and } x = 1.$$

Putting in these values and using the "half-angle" trigonometric formulas, we get the equation

$$P = E_f A_f e\left\{1 - \frac{\cosh \beta[(l/2) - x]}{\cosh \beta(l/2)}\right\} \qquad \text{for } 0 < x < \frac{l}{2}, \tag{15.19}$$

or

$$\sigma_f = \frac{P}{A_f} = E_f e\left\{1 - \frac{\cosh \beta[(l/2) - x]}{\cosh \beta(l/2)}\right\} \qquad \text{for } 0 < x < \frac{l}{2}. \tag{15.20}$$

The maximum possible value of strain in the fiber is the imposed strain e, and thus, the maximum stress is eE_f. Hence, as long as we have a sufficiently long fiber, the stress in the fiber will increase from the two ends to a maximum value, $\sigma_f^{max} = E_f e$. It can readily be shown that the average stress in the fiber will be

$$\overline{\sigma}_f = E_f e\left[1 - \frac{\tanh(\beta l/2)}{\beta l/2}\right]. \tag{15.21}$$

The variation in the shear stress τ along the fiber–matrix interface is obtained by considering the equilibrium of forces acting over an element of fiber (with radius r_f). Thus,

$$\frac{dP}{dx} dx = 2\pi r_f \, dx \, \tau. \tag{15.22}$$

P is the tensile load on the fiber and is equal to $\pi r_f^2 \sigma_f$, so

$$\tau = \frac{1}{2\pi r_f} \frac{dP}{dx} = \frac{r_f}{2} \frac{d\sigma_f}{dx}, \tag{15.23}$$

or

$$\tau = \frac{E_f r_f e \beta}{2} \frac{\sinh \beta[(l/2) - x]}{\cosh \beta(l/2)}. \tag{15.24}$$

The variation in τ and σ_f with x is shown in Figure 15.10.

The shear stress τ in Equation 15.24 will be the smaller of the following two shear stresses:

1. Strength of fiber/matrix interface in shear.
2. Shear yield stress of matrix.

Of these two shear stresses, the one that has a smaller value will control the load transfer phenomenon and should be used in Equation 15.24.

The constant H remains to be determined. An approximate value of H is derived next for a particular geometry. Let the fiber length l be much greater than the fiber radius

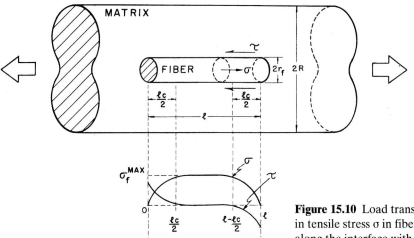

Figure 15.10 Load transfer to fiber. Variation in tensile stress σ in fiber and shear stress τ along the interface with the fiber length l.

r_f, and let $2R$ be the average fiber spacing (center to center). Let $\tau(r)$ be the shear stress in the direction of the fiber axis at a distance r from the axis. Then, at the fiber surface $(r = r_f)$,

$$\frac{dP}{dx} = -2\pi r_f \tau(r_f) = H(u - v).$$

Thus,

$$H = -\frac{2\pi r_f \tau(r_f)}{u - v}. \tag{15.25}$$

Now, let w be the real displacement in the matrix. Then at the fiber–matrix interface, without sliding, $w = u$. At a distance R from the center of a fiber, $w = v$. Considering equilibrium of the matrix between r_f and R, we get

$$2\pi r \tau(r) = \text{constant} = 2\pi r_f \tau(r_f),$$

or

$$\tau(r) = \frac{\tau(r_f)r_f}{r}. \tag{15.26}$$

The shear strain γ in the matrix is given by $\tau(r) = G_m\gamma$, where G_m is the matrix shear modulus. Then

$$\gamma = \frac{dw}{dr} = \frac{\tau(r)}{G_m} = \frac{\tau(r_f)r_f}{G_m r}. \tag{15.27}$$

Integrating from r_f to R, we get

$$\Delta w = \frac{\tau(r_f)r_f}{G_m} \ln\left(\frac{R}{r_f}\right). \tag{15.28}$$

But, by definition,

$$\Delta w = v - u = -(u - v). \tag{15.29}$$

Then

$$\frac{\tau(r_f)r_f}{u - v} = -\frac{G_m}{\ln(R/r_f)}. \tag{15.30}$$

From Equations 15.25 and 15.30, we get

$$H = \frac{2\pi G_m}{\ln(R/r_f)}, \tag{15.31}$$

and from Equation 15.18, we obtain an expression for the load transfer parameter:

$$\beta = \left(\frac{H}{E_f A_f}\right)^{1/2} = \left[\frac{2\pi G_m}{E_f A_f \ln(R/r_f)}\right]^{1/2}.$$ (15.32)

Note that the greater the value of G_m/E_f, the more rapid is the increase in fiber stress from the two ends.

The foregoing analysis is an approximate one—particularly with regard to the evaluation of the load transfer parameter β. More exact analysis give similar results and differ only in the value of β. In all the analyses, however, β is proportional to $\sqrt{G_m/E_f}$, and the differences occur only in the term involving the fiber volume fraction, $\ln(R/r_f)$.

15.6.2   Fiber Elastic and Matrix Plastic

It should be clear from the preceding discussion that, in order to load high-strength fibers to their maximum strength in the matrix, the shear strength must correspondingly be large. A metallic matrix will flow plastically in response to the high shear stress developed. Should the fiber/matrix interface be weaker, it will fail first. Plastic deformation of a matrix implies that the shear stress at the fiber surface, $\tau(r_f)$, will never go above τ_y, the matrix shear yield strength (ignoring any work-hardening effects). In such a case, we get, from an equilibrium of forces, the equation

$$\sigma_f \pi \frac{d^2}{4} = \tau_y \pi d \frac{l}{2},$$

or

$$\frac{l}{d} = \frac{\sigma_f}{2\tau_y}.$$

We consider $l/2$, and not l, because the fiber is being loaded from both ends. If the fiber is sufficiently long, it should be possible to load it to its breaking stress, σ_{fb}, by means of load transfer through the matrix flowing plastically around it. Let $(l/d)_c$ be the minimum fiber length-to-diameter ratio necessary to accomplish this. We call this ratio l/d the aspect ratio of the fiber and $(l/d)_c$ the critical aspect ratio necessary to attain the breaking stress of the fiber, σ_{fb}. Then we can write

$$\left(\frac{l}{d}\right)_c = \frac{\sigma_{fb}}{2\tau_y}.$$ (15.33)

Or we can think of a critical fiber length l_c for a given fiber diameter d:

$$\frac{l_c}{d} = \frac{\sigma_{fb}}{2\tau_y}$$ (15.34)

Thus, the fiber length l must be equal or greater than l_c for the fiber to be loaded to its maximum stress. If $l < l_c$, the matrix will flow plastically around the fiber and will load it to a stress in its central portion given by

$$\sigma_f = 2\tau_y \frac{l}{d} < \sigma_{fb}.$$ (15.35)

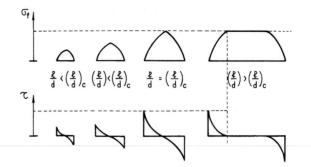

Figure 15.11 Variation in the fiber load transfer length as a function of the aspect ratio l/d.

This is shown in Figure 15.11. An examination of the figure shows that, even for $l/d > (l/d)_c$, the average stress in the fiber will be less than the maximum stress to which it is loaded in its central region. In fact, we can write, for the average fiber stress,

$$\bar{\sigma}_f = \frac{1}{l} \int_0^l \sigma_f \, dx$$

$$= \frac{1}{l} [\sigma_f(l - l_c) + \phi\sigma_f l_c]$$

$$= \frac{1}{l} [\sigma_f l - l_c(\sigma_f - \phi\sigma_f)],$$

or

$$\bar{\sigma}_f = \sigma_f \left(1 - \frac{1 - \phi}{l/l_c}\right), \tag{15.36}$$

where $\phi\sigma_f$ is the average stress in the fiber over a portion $l_c/2$ of its length at both the ends. We can thus regard ϕ as a load transfer function where value will be precisely 0.5 for an ideally plastic matrix (i.e., the increase in stress in the fiber over the portion $l_c/2$ will be linear).

EXAMPLE 15.8

(a) Consider an alumina fiber-reinforced polymer matrix composite. If the strength of the fiber is 1 GPa and the fiber/matrix interface has a shear strength of 10 MPa, compute the critical fiber length l_c. Take the diameter of the alumina fiber to be 10 μm.

(b) The composite in Part a is made of short (1-cm-long), but aligned, alumina fibers. Assuming that each fiber is loaded from both ends in a linear manner, compute the average stress in the fiber in this composite.

Solution: (a) Critical length:

$$l_c/d = \sigma_{fb}/2\tau_i = 1{,}000/2 \times 10 = 50,$$

$$l_c = 50 \times 10 \ \mu\text{m} = 0.5 \ \text{mm}.$$

(b) Average fiber stress:
From the solution to Part a, we have

$$l/l_c = 10/0.5 = 20,$$
$$\overline{\sigma}_f = \sigma_f[1 - (1 - \phi)/(l/l_c)]$$
$$= 1,000[1 - (1 - 0.5)/20] = 1,000\,(1 - 0.025) = 975 \text{ MPa}.$$

15.7 FRACTURE IN COMPOSITES

Fracture is a complex subject, even in monolithic materials. (See Chapters 7–9.) Undoubtedly, it is even more complex in composite materials. A great variety of deformation modes can lead to failure in a composite. The operative failure mode will depend, among other things, on loading conditions and the particular composite system. The microstructure has a very important role in the mechanics of rupture of a composite. For example, the fiber diameter, its volume fraction and alignment, damage due to thermal stresses that may develop during fabrication or service—all these factors can contribute to, and directly influence, crack initiation and propagation. A multiplicity of failure modes can exist in a composite under different loading conditions.

15.7.1 Single and Multiple Fracture

In general, the two components of a composite will have different values of strain to fracture. When the component that has the smaller breaking strain fractures, the load carried by this component is thrown onto the other one. If the latter component, which has a higher strain to fracture, can bear the additional load, the composite will show multiple fracture of the brittle component (the one with smaller fracture strain); eventually, a particular transverse section of composite becomes so weak, that the composite is unable to carry the load any further, and it fails.

Let us consider the case of a fiber reinforced composite in which the fiber fracture strain is less than that of the matrix. Then the composite will show a single fracture when

$$\sigma_{fu}V_f > \sigma_{mu}V_m - \sigma'_m V_m, \tag{15.37}$$

where σ'_m is the matrix stress corresponding to the fiber fracture strain and σ_{fu} and σ_{mu} are the ultimate tensile stresses of the fiber and matrix, respectively. This equation says that when the fibers break, the matrix will not be in a condition to support the additional load, a condition that is commonly encountered in composites of high V_f, brittle fibers, and a ductile matrix. All the fibers break in more or less one plane, and the composite fails in that plane.

If, on the other hand, we have a system that satisfies the condition

$$\sigma_{fu}V_f < \sigma_{mu}V_m - \sigma'_m V_m, \tag{15.38}$$

the fibers will be broken into small segments until the matrix fracture strain is reached. An example of this type of breakage is shown in Fig. 15.12, an optical micrograph of an Fe–Cu matrix containing a small volume fraction of W fibers.

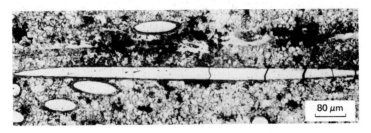

Figure 15.12 Optical micrograph of multiple fracture of tungsten fibers in an Fe–Cu matrix.

In case the fibers have a fracture strain greater than that of the matrix (an epoxy resin reinforced with metallic wires), we would have a multiplicity of fractures in the matrix, and the condition for this may be written as

$$\sigma_{fu} V_f > \sigma_{mu} V_m + \sigma_f' V_f, \tag{15.39}$$

where σ_f' is now the fiber stress corresponding to the matrix fracture strain.

15.7.2 Failure Modes in Composites

Two failure modes are commonly encountered in composites:

1. The fibers break in one plane, and, the soft matrix being unable to carry the load, the composite failure will occur in the plane of fiber fracture. This mode is more likely to be observed in composites that contain relatively high fiber volume fractions and fibers that are strong and brittle. The latter condition implies that the fibers do not show a distribution of strength with a large variance, but show a strength behavior that can be characterized by the Dirac delta function.
2. When the adhesion between fibers and matrix is not sufficiently strong, the fibers may be pulled out of the matrix before failure of the composite. This fiber pullout results in the fiber failure surface being nonplanar.

More commonly, a mixture of these two modes is found: fiber fracture together with fiber pullout. Fibers invariably have defects distributed along their lengths and thus can break in regions above or below the crack tip. This leads to separation between the fiber and the matrix and, consequently, to fiber pullout with the crack opening up. Examples of this mixed fracture mode are shown in Figure 15.13.

One of the attractive characteristics of composites is the possibility of obtaining an improved fracture toughness behavior together with high strength. Fracture toughness can

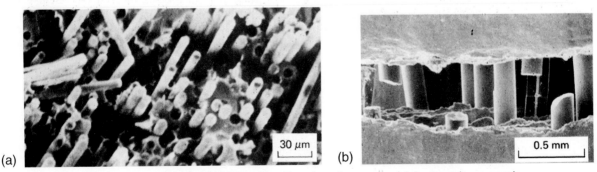

(a)   (b)

Figure 15.13 Scanning electron micrographs of fracture in composites, showing the fiber pullout phenomenon. (a) Carbon polyester. (b) Boron aluminum 6061.

be defined loosely as resistance to crack propagation. In a fibrous composite containing a crack transverse to the fibers, the crack propagation resistance can be increased by doing additional work by means of any or all of the following:

1. Plastic deformation of the matrix.
2. The presence of weak interfaces, fiber/matrix separation, and deflection of the crack.
3. Fiber pullout.

It would appear that debonding of the fiber/matrix interface is a prerequisite for phenomena such as crack deflection, crack bridging by fibers, and fiber pullout. It is of interest to develop some criteria for interfacial debonding and crack deflection. Crack deflection at an interface between materials of identical elastic constants (i.e., the same material joined at an interface) can be analyzed on the basis of the strength of the interface. The deflection of the crack along an interface or the separation of the fiber–matrix interface is an interesting mechanism of augmenting the resistance to crack propagation in composites. Cook and Gordon analyzed the stress distribution in front of a crack tip and concluded that the maximum transverse tensile stress σ_{11} is about one-fifth of the maximum longitudinal tensile stress σ_{22}. They suggested, therefore, that when the ratio σ_{22}/σ_{11} is greater than 5, the fiber/matrix interface in front of the crack tip will fail under the influence of the transverse tensile stress, and the crack would be deflected 90° from its original direction. That way, the fiber/matrix interface would act as a crack arrester. This is shown schematically in Figure 15.14. The improvement in fracture toughness due to the presence of weak interfaces has been confirmed qualitatively.

Another treatment of this subject is based on a consideration of the fracture energy of the constituents.[10] Two materials that meet at an interface are more than likely to have *different* elastic constants. This mismatch in moduli causes shearing of the crack surfaces, which leads to a mixed-mode stress state in the vicinity of an interface crack tip involving both the tensile and shear components. This, in turn, results in a mixed-mode fracture, which can occur at the crack tip or in the wake of the crack. Figure 15.15 shows crack front and crack wake debonding in a fiber-reinforced composite. Because of the mixed-mode fracture, a single-parameter description by the critical stress intensity factor K_{Ic} will not do; instead, one needs a more complex formalism of fracture mechanics to describe the situation. In this case, the parameter K becomes scale sensitive, but the critical strain energy release rate G_{Ic} is not a scale-sensitive parameter. G is a function of the phase angle Ψ, which, in turn, is a function of the normal and shear loading. For the opening mode, or mode I, $\Psi = 0°$, while for mode II, $\Psi = 90°$. One needs to specify both G and Ψ to analyze the debonding at the interface. Without going into the details, we present here the final results of such an analysis, in the form of a plot of G_i/G_f vs. α, where G_i is the mixed-mode interfacial fracture energy of the interface, G_f is the mode-I fracture energy of the fiber, and α is a measure of the elastic mismatch between the matrix and the reinforcement, defined as

$$\alpha = \left(\frac{\overline{E}_1 - \overline{E}_2}{\overline{E}_1 + \overline{E}_2}\right), \tag{15.40}$$

where

$$\overline{E} = \frac{E}{1 - \nu^2}. \tag{15.41}$$

[10]See M. Y. He and J. W. Hutchinson, *J. App. Mech.*, 56 (1989) 270; and A. G. Evans and D. B. Marshall, *Acta Met.*, 37 (1989) 2567.

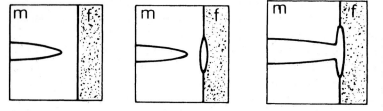

Figure 15.14 Fracture of weak interface in front of crack tip due to transverse tensile stress; m and f indicate the matrix and fiber, respectively. (After J. Cook and J. E. Gordon, *Proc. Roy. Soc. (London),* A 228 (1964) 508)

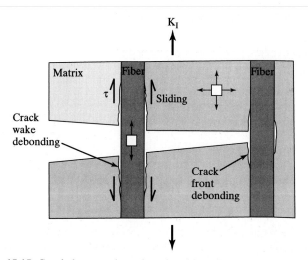

Figure 15.15 Crack front and crack wake debonding in a fiber-reinforced composite.

The plot in Figure 15.16 shows the conditions under which the crack will deflect along the interface or propagate through the interface into the fiber. For all values of G_i/G_f below the cross-hatched boundary, interface debonding is predicted. For the special case of zero elastic mismatch (i.e., for $\alpha = 0$), the fiber–matrix interface will debond for G_i/G_f less than about 0.25. Conversely, for G_i/G_f greater than 0.25, the crack will propagate across the fiber. In general, for elastic mismatch, with α greater than zero, the minimum interfa-

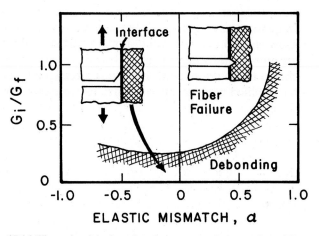

Figure 15.16 The ratio of the interface fracture toughness to that of fiber, G_i/G_f, vs. the elastic mismatch α. Interfacial debonding occurs under the curve, while for conditions above the curve, the crack propagates through the interface.

cial toughness required for interface debonding increases (i.e., high-modulus fibers tend to favor debonding). One shortcoming of this analysis is that it treats the fiber and matrix as isotropic materials; this is not always true, especially for carbon fiber.

Gupta et al.[11] derived strength and energy criteria for crack deflection at a fiber/matrix interface for several composite systems, taking due account of the anisotropic nature of the fiber. They used an experimental technique—spallation by means of a laser Doppler displacement interferometer—to measure the tensile strength of a planar interface. Through this technique, these researchers have tabulated the required values of the interface strength and fracture toughness for delamination in a number of ceramic, metal, intermetallic, and polymer matrix composites.

15.8 SOME FUNDAMENTAL CHARACTERISTICS OF COMPOSITES

Composite materials are not like any other common type of material. They are inherently different from monolithic materials, and consequently, these basic differences must be taken into account when one designs or fabricates any article from composite materials. In what follows, we give a brief description of some of the fundamental characteristics of composites.

15.8.1 Heterogeneity

Composite materials are inherently heterogeneous, consisting as they do of two components of different elastic moduli, different mechanical behavior, different expansion coefficients, and so on. For this reason, the analysis of, and the design procedures for, composite materials are quite intricate and complex, compared to those for ordinary materials. The structural properties of composites are functions of:

1. The properties of their components.
2. The geometric arrangement of their components.
3. The interface between the components.

Given two components, we can obtain a great variety of properties by manipulating items 2 and 3.

15.8.2 Anisotropy

In general, monolithic materials are reasonably isotropic; that is, their properties do not show any marked preference for any particular direction. The unidirectional composites are anisotropic due to their very nature. Once again, the analysis and design of composites should take into account this strong directionality of properties—properties that cannot be specified without any reference to some direction. Figure 15.17 shows, schematically, the elastic moduli of a monolithic material and a composite as a function of fiber orientation θ. A monolithic material (e.g., Al) is an isotropic material; therefore, its moduli do not vary with the angle of testing, and the graphs are horizontal.

For an ordinary material (say, aluminum), the designer only needs to open a manual and find one unique value of strength or one unique value of the modulus of the material. But for composite materials, the designer has to consult performance charts representing the specific strength and the specific rigidity of the various composite systems. (See Figure 15.18.)

Ordinary materials, such as aluminum or steel, can be represented by a fixed point. For a composite material, however, there does not exist a unique combination of these properties.

---

[11]V. Gupta, J. Yuan, and D. Martinez, *J. Am. Ceram. Soc.*, 76 (1993) 305.

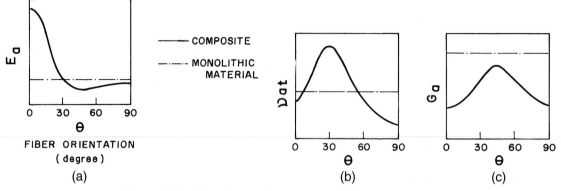

Figure 15.17 Schematic of variation in elastic moduli of a fiber composite and a monolithic material with the angle of reinforcement. E_a is the axial Young's modulus, ν_{at} is the principal Poisson's ratio, and G_a is the axial shear modulus.

Instead, the composite contains a system of properties and must be represented by an area instead of a point. We call these graphs "carpet plots." The highest point on the graph represents the longitudinal properties of the composite, while the lowest point represents quasi-isotropic properties. The important point to make is that, depending on the construction of a composite and the appropriate quantity of fiber, the characteristics of the composite can be varied. In other words, composites can be tailormade, in accord with the final objective.

15.8.3 Shear Coupling

The properties of a composite are very sensitive functions of the fiber orientation. They display what is called *shear coupling:* shear strains produced by axial stress and axial strains produced by shear stress. (See Figure 15.19.) In response to a uniaxially applied load, an isotropic material produces only axial and transverse strains. In fiber reinforced composites, however, a shear strain γ is also produced in response to an axial load, because the fibers tend to align themselves in the direction of the applied load. This shear distortion can be eliminated if one makes a *cross-ply* composite—a composite containing an equal number of parallel fibers, alternately aligned at a given angle and at a complementary angle with respect to the loading axis (Figure 15.20). That is, we have the various layers in a composite arranged at $\pm\,\theta$ degrees to the loading axis, and thus, the shear distortion due to one layer is compensated for by an equal and opposite shear distortion due to the other. However, this balance occurs only in two dimensions, whereas the real-life composites are three-dimensional materials. This leads to an "edge effect" in which the individual layers deform differently under tension and in the neighborhood of the free edges, giving rise to out-of-plane shear and bending. The stacking

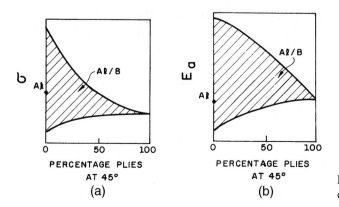

Figure 15.18 Schematic of a performance chart of a composite.

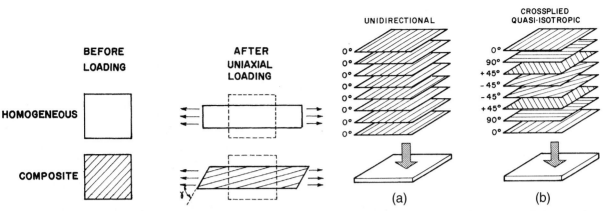

Figure 15.19 Shear coupling in a fiber composite

Figure 15.20 Unidirectional and cross-plied composites.

sequence of the various layers in the composite is important. For example, in a laminate composite consisting of fibers at $+90°$, $+45°$, $-45°$, $-45°$, $+45°$, and $+90°$, subjected to an in-plane tensile stress, there occur compressive stresses in the direction of thickness, in the vicinity of the edges. Should the same composite have the sequence $+45°$, $-45°$, $+90°$, $+90°$, $-45°$, $+45°$, however, these stresses in the direction of thickness are of a tensile nature and thus tend to delaminate the composite, clearly an undesirable effect.

15.9 FUNCTIONALLY GRADED MATERIALS

There is a good deal of interest in making materials that are graded in some respect. The gradient may be of the chemical composition, density, or coefficient of thermal expansion of the material, or it may involve microstructural features—for example, a particular arrangement of second-phase particles or fibers in a matrix. Such materials are called *functionally graded materials*, and the acronym FGM is commonly applied to them in the literature. Strictly speaking, though, the term "graded material" ought to be enough to convey the meaning; that is, the word "functionally" is redundant. The idea, however, is a very general one, viz., instead of having a step function, say, in composition at an interface, we should have a gradually varying composition from component A to component B. Figure 15.21 shows schematically the microstructure of a functionally graded material. Such a graded interface can be very useful in ameliorating high mechanical and thermal stresses. The concept of a functionally graded material is applicable to any material, polymer, metal, or ceramic.[12]

15.10 APPLICATIONS

It is convenient to divide the applications of all composites into aerospace and non-aerospace categories. In the category of aerospace applications, low density coupled with other desirable features, such as a tailored thermal expansion and conductivity, and high stiffness and strength, are the main drivers. Performance, rather than cost, is an important item as well. We next give a brief description of various applications of composites.

15.10.1 Aerospace Applications

Reduction in the weight of a component is as major driving force for any application in the aerospace field. The Boeing 757 and 767 jets were the first large commercial aircraft to

---

[12]See B. Ilschner, *J. Mech. Phys. Solids,* 44 (1996) 647; S. Suresh and A. Mortensen, *Intl. Mater. Rev.,* 42 (1997) 85.

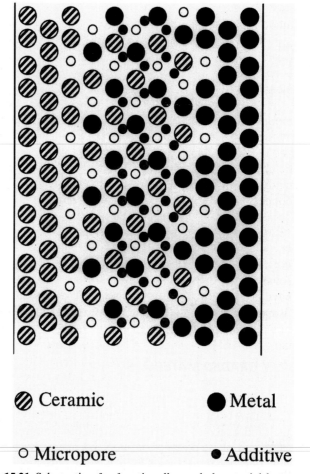

Figure 15.21 Schematic of a functionally graded material between a ceramic on the left-hand side and a metal on the right-hand side. Also shown are micropores and additives.

make widespread use of structural components made of PMCs. About 95% of the visible interior parts in Boeing 757 and 767 cabins are made from nonconventional materials. One of the main reasons for the decision to use such materials was the steadily dropping price of carbon fibers. Similarly, there has been an increasing use of composites in aircraft, including helicopters, used by defense services. Weight and cost savings are the driving forces for these applications. Consider, for example, the Sikorsky H-69 helicopter. For this helicopter, manufacturing the conventional fuselage, of metal construction, is very labor intensive. In comparison, the composite fuselage, of carbon, aramid, and glass fiber/epoxy, has much fewer parts, assemblies, and fasteners. PMCs are also lighter and cheaper to use than metals in the manufacture of fuselages. The use of lighter composites in aircraft results in energy savings: For a given aerodynamic configuration of an aircraft, there is a direct correlation between the weight of the airplane and fuel consumption. Weight savings resulting from the use of new, lighter materials lead to great increases in fuel economy.

In examining the applications of composites in space, it should be recognized that environment of space is not benign. Among the hazardous items that may be encountered in space are orbital debris, meteorites, and atomic oxygen. It appears that metal matrix composites can withstand the space environment better than polymer matrix composites. In the Hubble telescope, pitch-based continuous carbon fiber reinforced aluminum was used for waveguide booms because this composite is very light and has a high elastic modulus and a low coefficient of thermal expansion.

Other aerospace applications of MMCs involve the replacement of light, but toxic, beryllium by various composites. For example, in the U.S. Trident missile, beryllium has been replaced by an SiC_p/Al composite, which is also used in aircraft electronic equipment racks.

CMCs can lead to potential improvements in aircraft, helicopters, missiles, reentry modules of spacecraft, and other aerospace vehicles. Projected skin temperatures in future hypersonic aircraft are over 1600°C. Other parts, such as radomes, nose tips, leading edges, and control surfaces, will have only slightly lower temperatures. Currently, one uses sacrificial, non-load-bearing thermal protection CMC materials on load-bearing components made of conventional materials. With the use of CMCs, one can have load-bearing components that are reusable at operating temperatures.

15.10.2 Nonaerospace Applications

Polymer composites based on aramid, carbon, and glass fibers are routinely used in civil construction and in marine and sporting goods. Applications in the sporting goods industry have burgeoned in the last quarter of the 20th century, all the way from tennis rackets to fishing poles to a whole variety of equipment used in downhill as well as cross-country skis, boots, poles, gloves, etc. The main advantages that the use of composites brings to the sporting goods industry are safety, less weight, and higher strength than conventional materials. Ski poles made of polymer composites are lighter and stiffer than aluminum poles. Frequently, hybrid composites are used, such as carbon fibers laid over a small sleeve of aramid.

Composites are also used in rifle stocks for biathlons because both weight and strength are important in the rifles, which may have to be carried over distances of up to 20 km. The automobile industry is a major user of PMCs, mainly because of the cost advantage over other types of composites.

One of the important applications of MMCs in the automotive area is in the diesel piston crown. This application involves the incorporation of short fibers of alumina or alumina plus silica into the crown. The conventional diesel engine piston has an Al–Si casting alloy with a crown made of nickel cast iron. The replacement of the nickel cast iron by an aluminum matrix composite resulted in a lighter, more abrasion-resistant, and cheaper product. Yet another application of MMCs is in the automobile engine of the Honda Prelude. In the conventional automobile, the major part of the engine, and also the heaviest part, is the cast iron engine block. In the general quest for high performance combined with a light vehicle, the cast iron engine block has been replaced by light aluminum alloy in some automobiles, resulting in a weight reduction of 15–35 kg. But even in these aluminum engines, the liners are generally made of cast iron. This is because cast iron has superior sliding characteristics (pistons sliding in the cylindrical bores) than aluminum alloys do. The Honda Motor Company has developed an aluminum engine (used in the Prelude), with cylinder liners made of alumina– and carbon-fiber reinforced aluminum. The most important characteristic for this application is resistance against sliding. Seizure occurs when the coefficient of friction increases very rapidly. According to the researchers at Honda, a hybrid composite consisting of alumina and carbon fibers gave the best results. This was attributed to the self-lubricating properties of carbon fiber and the sliding resistance of alumina fiber. In composites containing only alumina fibers, when a scratch appeared, it easily worsened. In the case of hybrid alumina and carbon fibers in Al, the scratch did not grow. Particulate metal matrix composites—especially light ones such as aluminum and magnesium—also find applications in automotive and sporting goods. In this regard, it is important to remember that the price per kilogram becomes the driving force for the application.

Copper-based composites having Nb, Ta, or Cr as the second phase in a discontinuous form are of interest for certain applications requiring high thermal conductivity and high strength. Sometimes we refer to these composites as Cu–X composites, where X, which is insoluble in copper at room temperature, forms the second phase. One specific example is a high heat-flux application in the thrust chambers of rocket engines. Cu-X systems are very useful for processing such composites. At room temperature, the second phase appears in a dendritic form, which can be converted into a filamentary or ribbon form by mechanical working. Note that the ribbon morphology is thermodynamically unstable at high temperatures, because the ribbons tend to form spheroids with time, as a function of temperature.

Conventional commercial superconductors are referred to as *niobium-based superconductors* because Nb–Ti and Nb$_3$Sn are superconducting materials. These conventional superconductors are nothing but copper matrix composites.

An area in which CMCs have found application is that of cutting tools. Silicon carbide whisker reinforced alumina (SiC$_w$/Al$_2$O$_3$) is used as a cutting-tool insert for high-speed cutting of superalloys. For example, in the cutting of Inconel 718, SiC$_w$/Al$_2$O$_3$ composite tools perform three times better than conventional ceramic tools and eight times better than cemented carbides.

Carbon/carbon composites are used as implants, as well as for internal fixation of bone fractures, because of their excellent biocompatibility. They are also used for making molds for hot pressing. Carbon/carbon molds can withstand higher pressures and offer a longer service life than does polycrystalline graphite. However, their high cost limits them to aerospace and other specialty applications. The low oxidation resistance of carbon/carbon composites is a serious limitation, but is not a problem for short term applications such as shields, rocket nozzles, and reentry vehicles.

SUGGESTED READINGS

R. J. ARSENAULT AND R. EVERETT, EDS. *Metal Matrix Composites,* vols. 1 and 2. San Diego: Academic Press, 1991.

K. K. CHAWLA. *Composite Materials: Science and Engineering.* Second edition. New York: Springer-Verlag, 1998.

K. K. CHAWLA. *Ceramic Matrix Composites.* London: Chapman & Hall, 1993.

K. K. CHAWLA. *Fibrous Materials.* Cambridge, U.K.: Cambridge University Press, 1998.

T. W. CLYNE AND P. WITHERS. *Metal Matrix Composites.* Cambridge, U.K.: Cambridge University Press, 1994.

M. R. PIGGOTT. *Load Bearing Fibre Reinforced Composites.* Oxford: Pergamon Press, 1980.

L. N. PHILLIPS, ED. *Design with Advanced Composite Materials.* London: The Design Council, 1989.

S. SURESH, A. NEEDLEMAN, and A. M. MORTENSEN, EDS. *Metal Matrix Composites.* Stoneham, MA: Butterworth–Heinemann, 1993.

M. TAYA AND R. J. ARSENAULT. *Metal Matrix Composites,* Oxford: Pergamon Press, 1989.

EXERCISES

15.1   Describe some composite materials that occur in nature. Describe their structure and properties.

15.2   To promote wettability and avoid interfacial reactions, protective coatings are sometimes applied to fibers. Any improvement in the behavior of a composite will depend on the stability of the layer of coating. The maximum time t for the dissolution of this layer can be estimated by the diffusion distance

$$x \approx \sqrt{Dt},$$

where D is the diffusivity of the matrix in the protective layer. Making an approximation that the matrix diffusion in the protective layer can be represented by self-diffusion, compute the time required for a 0.1-μm-thick protective layer on the fiber to be dissolved at T_m and $0.75 T_m$, where T_m is the matrix melting point in kelvin. Assume a reasonable value of D for self-diffusion in metals, taking into account the variation in D with temperature.

15.3 A fibrous form represents a higher energy form vis-à-vis a spherical form. Hence, a fibrous phase produced by unidirectional solidification of a euectic will tend to form spheroids because such a change of shape results in a decrease in the surface energy of the material. Compute the energy released when a 10-cm-long, 20-μm-diameter fiber becomes spheroidal. The specific surface energy of the fibrous phase is 500 m Jm^{-2}.

15.4 One can obtain two-dimensional isotropy in a fiber composite plate by having randomly oriented fibers in the plane of the plate. Show that the average in-plane modulus is

$$\overline{E}_\theta = \frac{\displaystyle\int_0^{\pi/2} E_\theta \, d\theta}{\displaystyle\int_0^{\pi/2} d\theta}.$$

Plot E_u / E_{11} versus V_f for fiber reinforced composites with $E_f / E_m = 1, 10$, and 100.

15.5 Consider a carbon fiber reinforced epoxy composite. The fibers are continuous, unidirectionally aligned and 60% by volume. The tensile strength of carbon fibers is 3 GPa, and the Young's modulus is 250 GPa. The tensile strength of the epoxy matrix is 50 MPa, and its Young's modulus is 3 GPa. Compute the Young's modulus and the tensile strength of the composite in the longitudinal direction.

15.6 A steel wire of diameter 1.25 mm has an aluminum coating such that the composite wire has a diameter of 2.50 mm. Some other pertinent data are as follows:

Property	Steel	Aluminum
Elastic modulus E	210 GPa	70 GPa
Yield stress σ_y	200 MPa	70 MPa
Poisson ratio ν	0.3	0.3
Coefficient of thermal expansion (linear)	$11 \times 10^{-6} \mathrm{K}^{-1}$	$23 \times 10^{-6} \mathrm{K}^{-1}$

(a) If the composite wire is loaded in tension, which of the two components will yield first? Why?

(b) What tensile load can the composite wire support without undergoing plastic strain?

(c) What is the elastic modulus of the composite wire?

(d) What is the coefficient of thermal expansion of the composite wire?

15.7 A boron/aluminum composite has the following characteristics:

Unidirectional reinforcement
Fiber volume fraction $V_f = 50\%$
Fiber length $l = 0.1$ m
Fiber diameter $d = 100\ \mu$m
Fiber ultimate stress $\sigma_{fu} = 3$ GPa
Fiber strain corresponding to σ_{fu}, $e_{fu} = 0.75\%$ (uniform elongation)
Fiber Young's modulus $E_f = 415$ GPa
Matrix shear yield stress $\tau_{ym} = 75$ MPa
Matrix stress at $e = e_{fu}$, $\sigma'_m \big|_{e/fu} = 93$ MPa
Matrix ultimate stress $\sigma_{mu} = 200$ MPa

Compute:

(a) The critical fiber length l_c for the load transfer.

(b) The ultimate tensile stress of the composite.

(c) V_{min} and V_{crit} for this composite system.

15.8 Determine Young's modulus for a steel fiber/aluminum matrix composite material, parallel and perpendicular to the fiber direction. The reinforcement steel has $E = 210$ GPa and $V_f = 0.3$, and the aluminum matrix has $E = 70$ GPa and $V_m = 0.7$.

15.9 An injection molded composite has short, aligned fibers. The fiber volume fraction is 40%. The length and diameter of the fibers are 500 and 10 μm, respectively. Assume a square distribution of fibers in the cross section of the composite. The Young's modulus of the fiber and matrix are 230 and 3 GPa, respectively. The shear modulus of the matrix is 1 GPa. Compute the mean strength of this composite in the fiber directions.

15.10 A glass fiber reinforced polymer matrix composite has the following characteristics:

fiber maximum strength = 2 GPa
Interfacial shear strength = 50 MPa
Fiber radius = 10 μm

Compute the critical fiber length for this system.

15.11 A unidirectionally reinforced fiber reinforced composite has the following characteristics:

$$E_f = 380 \text{ GPa}, \qquad \sigma_{fu} = 3 \text{ GPa}, \quad \nu_f = 0.4,$$

$$\tau_i = 50 \text{ MPa}, \qquad \text{fiber length } l = 10 \text{ cm},$$

$$\text{fiber diameter } d = 15 \ \mu\text{m}.$$

The matrix strength at fiber failure is $\sigma'_m = 200$ MPa. Assuming that the fibers are aligned, compute the critical length for load transfer in this composite. If the load transfer coefficient $\beta = 0.5$, what is the strength of the composite along the fiber direction?

15.12 Consider a fiber of radius r embedded up to a length l in a matrix. (See Figure E15.12.) When the fiber is pulled, the adhesion between the fiber and the matrix produces a shear stress τ at the interface. In a composite system containing a fiber of fracture stress σ_f equal to eight times the maximum shear stress τ_{max} that the interface can bear, what fiber aspect ratio is required to break the fiber rather than pull it out?

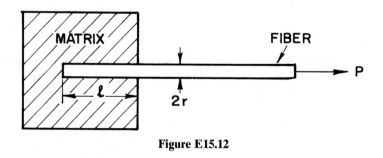

Figure E15.12

15.13 List some nonstructural applications of composite materials.[13]

[13]See M. B. Bever, P. E. Duwez, and W. A. Tiller, *Mater. Sci. Eng.,* 6 (1970), 149.

Index